SELECTED PAPERS ON PHYSICAL PROCESSES IN IONIZED PLASMAS

Edited by

DONALD H. MENZEL

Director, Harvard College Observatory

DOVER PUBLICATIONS, INC.

NEW YORK

Published simultaneously in Canada by McClelland and Stewart, Ltd.

Published in the United Kingdom by Constable and Company Limited, 10 Orange Street, London W. C. 2.

This new Dover edition, first published in 1962, is a selection of papers published for the first time in collected form. The editor and publisher are indebted to the original authors and to the journals which originally published the papers for assistance and permission to reproduce the articles.

Manufactured in the United States of America

Dover Publications, Inc.
180 Varick Street
New York 14, N. Y.

CONTENTS

PAGE

INTRODUCTION vi

PART 1. THEORY OF HYDROGENIC SPECTRA

ABSORPTION COEFFICIENTS AND HYDROGEN LINE INTENSITIES by Donald H. Menzel and Chaim L. Pekeris.
Monthly Notices of the Royal Astronomical Society, Vol 96, No. 1, November, 1935. 3

PART 2. PHYSICAL PROCESSES IN GASEOUS NEBULAE

1 ABSORPTION AND EMISSION OF RADIATION by Donald H. Menzel.
The Astrophysical Journal, Vol. 85, No. 4, May, 1937. 40

2 THEORY OF THE BALMER DECREMENT by Donald H. Menzel and James G. Baker.
The Astrophysical Journal, Vol. 86, No. 1, July, 1937. 50

3 THE BALMER DECREMENT by James G. Baker and Donald H. Menzel.
The Astrophysical Journal, Vol. 88, No. 1, July, 1938. 58

4 THE MECHANISTIC AND EQUILIBRIUM TREATMENT OF NEBULAR STATISTICS by Donald H. Menzel, Lawrence H. Aller, and James G. Baker.
The Astrophysical Journal, Vol. 88, No. 3, October, 1938. 71

5 ELECTRON TEMPERATURES by James G. Baker, Donald H. Menzel, and Lawrence H. Aller.
The Astrophysical Journal, Vol. 88, No. 4, November, 1938. 77

6 THE EQUATIONS OF RADIATIVE TRANSFER by Lawrence H. Aller, James G. Baker, and Donald H. Menzel.
The Astrophysical Journal, Vol. 89, No. 5, June, 1939. 84

7 THE TRANSFER OF RADIATION IN THE LYMAN CONTINUUM by James G. Baker, Lawrence H. Aller, and Donald H. Menzel.
The Astrophysical Journal, Vol. 90, No. 2, September, 1939. 91

8 THE ULTRAVIOLET RADIATION FIELD AND ELECTRON TEMPERATURE OF AN OPTICALLY THICK NEBULA by Lawrence H. Aller, James G. Baker, and Donald H. Menzel.
The Astrophysical Journal, Vol. 90, No. 4, November, 1939. 101

9 ON THE EXCITATION OF FRACTIONAL MULTIPLETS BY ELECTRON CAPTURE by George H. Shortley and Donald H. Menzel.
The Astrophysical Journal, Vol. 91, No. 3, April, 1940. 111

10 COLLISIONAL EXCITATION OF NEBULIUM by Malcolm H. Hebb and Donald H. Menzel.
The Astrophysical Journal, Vol. 92, No. 3, November, 1940. 124

11 STRENGTHS OF FORBIDDEN LINES IN p^2, p^3, AND p^4 AS A FUNCTION OF COUPLING by G. H. Shortley, L. H. Aller, J. G. Baker and D. H. Menzel.
The Astrophysical Journal, Vol. 93, No. 1, January, 1941. 140

12 THE ELECTRON DENSITIES OF SOME BRIGHT PLANETARY NEBULAE by Donald H. Menzel and Lawrence H. Aller.
The Astrophysical Journal, Vol. 93, No. 1, January, 1941. 157

13 THE ELECTRON TEMPERATURES OF SOME TYPICAL PLANETARY NEBULAE by Donald H. Menzel, Lawrence H. Aller, and Malcolm H. Hebb.
The Astrophysical Journal, Vol. 93, No. 2, March, 1941. 164

14 SPECTROPHOTOMETRY OF SOME TYPICAL PLANETARY NEBULAE by Lawrence H. Aller.
The Astrophysical Journal, Vol. 93, No. 2, March, 1941. 170

15 THE STATISTICAL EQUILIBRIUM OF NEUTRAL HELIUM by Leo Goldberg.
The Astrophysical Journal, Vol. 93, No. 2, March, 1941. 178

16 THE ABUNDANCE OF *O* III by Donald H. Menzel and Lawrence H. Aller.
The Astrophysical Journal, Vol. 94, No. 1, July, 1941. 184

17 FLUORESCENCE IN HIGH-EXCITATION PLANETARIES by Donald H. Menzel and Lawrence H. Aller.
The Astrophysical Journal, Vol. 94, No. 3, November, 1941. 191

18 THE CHEMICAL COMPOSITION OF THE PLANETARY NEBULAE by Lawrence H. Aller and Donald H. Menzel.
The Astrophysical Journal, Vol. 102, No. 2, September, 1945. 205

PART 3. INTERPRETATION OF NEBULAR SPECTRA

1 THE ELECTRON VELOCITY DISTRIBUTION IN GASEOUS NEBULAE AND STELLAR ENVELOPES by David Bohm and Lawrence H. Aller.
The Astrophysical Journal, Vol. 105, No. 1, January, 1947. 233

2 MULTIPLET INTENSITIES FOR THE NEBULAR LINES ^4S-^2D OF *O* II by L. H. Aller, C. W. Ufford, and J. H. Van Vleck.
The Astrophysical Journal, Vol. 109, No. 1, January, 1949. 253

3 SPECTROPHOTOMETRY OF REPRESENTATIVE PLANETARY NEBULAE by Lawrence H. Aller.
The Astrophysical Journal, Vol. 113, No. 1, January, 1951. 265

4 THERMAL EFFECTS OF COLLISIONAL LINE EXCITATION IN GASEOUS NEBULAE by Lawrence H. Aller.
The Astrophysical Journal, Vol. 118, No. 3, November, 1953. 281

5 PHOTOELECTRIC SPECTROPHOTOMETRY OF PLANETARY NEBULAE by William Liller and Lawrence H. Aller.
The Astrophysical Journal, Vol. 120, No. 1, July, 1954. 289

6 THE COMPOSITION OF THE PLANETARY NEBULA NGC 7027 by Lawrence H. Aller.
The Astrophysical Journal, Vol. 120, No. 3, November, 1954. 299

7 THE SPECTRUM OF NGC 7027 by L. H. Aller, I. S. Bowen, and R. Minkowski.
The Astrophysical Journal, Vol. 122, No. 1, July, 1955. 311

8 SPECTROPHOTOMETRY OF PLANETARY NEBULAE by R. Minkowski and L. H. Aller.
The Astrophysical Journal, Vol. 124, No. 1, July, 1956. 321

9 THE INTERPRETATION OF THE SPECTRUM OF NGC 7027 by L. H. Aller and R. Minkowski.
The Astrophysical Journal, Vol. 124, No. 1, July, 1956. 338

10 CHEMICAL COMPOSITIONS OF SELECTED PLANETARY NEBULAE by Lawrence H. Aller.
The Astrophysical Journal, Vol. 125, No. 1, January, 1957. 344

PART 4. THE INTENSITIES OF "FORBIDDEN" SPECTRAL LINES

THE COMPUTATION OF QUADRUPOLE AND MAGNETIC-DIPOLE TRANSITION PROBABILITIES by George H. Shortley.
The Physical Review, Vol. 57, No. 3, February 1, 1940. 365

INTRODUCTION

FROM THE viewpoint of the astronomer, hydrogen is the most important atom in the universe. Spectroscopic analysis reveals that about 90 per cent of all the atoms are hydrogen. Some 9 per cent are helium. The remaining 1 per cent comprises all the remaining chemical elements. These proportions may vary slightly from star to star and in different regions of the universe as the result of nuclear reactions occurring deep in stellar interiors. But the fact remains that, for the astronomer, hydrogen assumes special importance. Even when hydrogen is relatively unexcited, as in the atmospheres of cooler stars, its very abundance causes it to influence the physical state of other gases. For example, many molecular hydrides tend to form in the hot gases.

We often tend to forget that the principal series of hydrogen lies in the far ultraviolet with Lyman alpha occurring at 1215A and the long wave limit of the continuum falling at 912A. Thus, in this short-wave region, for all classes of stars—hot or cold—hydrogen atoms exert a controlling influence upon the opacity of the gas. And, as the temperature rises, the influence on opacity moves down into the longer wave lengths. In gases whose temperatures exceed approximately 8,000° K, hydrogen exerts a dominant role. When hydrogen is highly ionized, its very abundance insures that most of the electrons will derive from this source. Through the action of these electrons, which collide with and excite other atoms, many interesting and significant atomic processes can occur, which influence the spectra of heavenly objects and give information concerning the physical conditions in the volumes of gas where the spectra arise.

The hydrogen atom, with its single electron, is the simplest of all the atomic systems and, therefore, easiest to study theoretically. Recognizing the importance of hydrogen in astrophysics, C. L. Pekeris and I, back in 1935, checked and extended existing formulas for the intensities of lines and continua in the spectrum of hydrogen. This study led me to initiate, in 1937, a series of eighteen papers, all published in *The Astrophysical Journal* under the heading: "Physical Processes in Gaseous Nebulae." During the next eight years a number of physicists and astrophysicists collaborated with me in studying many types of physical processes occurring in highly ionized gases, including both radiative and collisional interactions with atoms other than hydrogen. And when, in 1945, I finally terminated what was clearly a non-convergent series, other scientists continued the analysis. Lawrence H. Aller, of the University of Michigan, who had played a major role in the original series, took the lead in the newer series.

Although the reprints from the original series were quickly exhausted, the demand for them has risen with the passage of time. The studies are basic to spectro-chemical analysis in many cases of physics and geophysics. Since modern terminology refers to such ionized media as plasmas, I have altered the title to indicate the general applicability of the developed theory to problems other than those of gaseous nebulae.

With the assistance of Dr. Aller, I have selected additional representative papers that contribute to the broad understanding of physical processes in highly ionized plasmas, especially those containing large amounts of hydrogen. To these and the original series of papers, I have added the basic study of the hydrogen spectrum that made the investigation possible in the first place. I have taken advantage of this new edition to correct a few misprints that occurred in the original papers. To the many authors who have generously given me permission to reprint the papers, and for their advice in the selection, I express my deep appreciation. The basic principles of spectroscopic analysis as developed within these papers should be useful to many students in many areas of science.

Cambridge, Massachusetts
April, 1961

Donald H. Menzel
Director
Harvard College Observatory

PART 1. THEORY OF HYDROGENIC SPECTRA

Reprinted from the MONTHLY NOTICES OF THE ROYAL ASTRONOMICAL SOCIETY, *Vol.* 96, *No.* 1

ABSORPTION COEFFICIENTS AND HYDROGEN LINE INTENSITIES

DONALD H. MENZEL AND CHAIM L. PEKERIS

(Communicated by Harlow Shapley)

1. *General Formulæ for Transition Probabilities in a Coulomb Field between States of Positive or Negative Energy*

The problem of the absorption of light by electrons in the neighbourhood of an ion has been discussed by many authors in connection with the theory of the stellar absorption coefficient. The general expressions for the absorption coefficient obtained from the wave-mechanical treatment of the problem are in the form of double series the summation of which has been carried out only in a number of limiting cases. Closed expressions for the intensities of lines in the Lyman and Balmer series were first obtained by

Pauli * and later checked by Sugiura,† who extended the formulæ to the Paschen series and included the case of the bound-free transitions. The original work of Kramers,‡ which was based on the classical theory of light and dealt with intensities arising from bound-free and free-free transitions, was shown by Gaunt § to be valid in the region of large quantum numbers. Oppenheimer || has given a general solution of the problem, though his formulæ are not reduced to the simplest form and his asymptotic approximations are in error. Maue ¶ has further extended the results of Gaunt.

Recently, L. McLean ** has derived a closed expression for the intensities in the case of transitions between discrete states. We give, without formal proof, a somewhat more general formula than that of McLean, which applies to the bound-free and free-free transitions as well, and have derived convenient asymptotic approximations for these formulæ.

The frequency of a line in a hydrogenic atom is given by the usual formula:

$$\nu_{nn'} = R\left(\frac{1}{n'^2} - \frac{1}{n^2}\right), \tag{1.1}$$

where n' and n are the respective quantum numbers of the lower and upper states. This expression may be extended to include the bound-free and free-free transitions if we let $n \to i\kappa$ and $n' \to il$, where κ and l are real positive numbers lying in the range 0 to ∞. They are not necessarily integers. When we adopt this convention, the distinction between an ionized and a neutral atom is only formal. We may regard an ionized atom as a neutral atom with its electron in a hyperbolic instead of in an elliptic orbit.

The total energy emitted per cm.³ per second by an assembly containing N_n atoms per cm.³ in state n is, by definition,

$$E_\nu = N_n A_{nn'} h\nu_{nn'}, \tag{1.2}$$

where $A_{nn'}$ is the Einstein probability coefficient of spontaneous transition. By Ladenburg's relation

$$A_{nn'} = \frac{\varpi_{n'}}{\varpi_n}\frac{8\pi^2\epsilon^2\nu^2{}_{nn'}}{mc^3}f_{n,\,n'}, \tag{1.3}$$

where $\varpi_{n'}$ and ϖ_n are the statistical weights of the respective states and $f_{n,\,n'}$ is the Ladenburg f, sometimes called the oscillator strength or number of dispersion electrons per atom. It does not appear to have been realized that equation (1.3) is applicable to transitions from continuous as well as from discrete states, when the weight factor is properly defined.

* *Cf.* Schroedinger, *Ann. der Phys.*, **80**, 437, 1926.

† Sugiura, *Jour. de Phys.*, **8**, 113, 1927; *Sci. Pap. Inst. Phys. and Chem. Research (Tokyo)*, **11**, 1, 1929.

‡ Kramers, *Phil. Mag.*, **44**, 836, 1923.

§ Gaunt, *Phil. Trans.*, A, **229**, 163, 1930.

|| Oppenheimer, *Zs. für Phys.*, **41**, 768, 1927; *ibid.*, **55**, 725, 1929.

¶ Maue, *Ann. der Phys.*, **13**, 161, 1932.

** McLean, *Nature*, **124**, 25, 1932. Also, *Phil. Mag.* (7), **18**, 845, 1934.

For a discrete state of hydrogen

$$\varpi_n = 2n^2. \tag{1.4}$$

The factor 2 arises from electron spin, for which we shall sometimes write ϖ_e. For a continuous state the choice of weight factor depends upon the definition of N_n, the number of atoms in a continuous state. Several possible choices are available. We may, for example, define N_n as the number of atoms per cm.^{3} whose electrons in hyperbolic orbits have velocities at infinity ranging from v to $v+dv$. As Fowler * has shown, the proper weighting in this region of phase space, where the distribution is essentially classical, is obtained by assigning a weight ϖ' to the cell characterized by component momenta lying between p_1 and p_1+dp_1, p_2 and p_2+dp_2, p_3 and p_3+dp_3, and located in the volume element $dq_1dq_2dq_3$, where

$$\varpi' = \varpi_e \frac{dp_1dp_2dp_3dq_1dq_2dq_3}{h^3}. \tag{1.5}$$

Expressing the component momenta in terms of the actual velocity v of the electron, we find, on introducing polar co-ordinates,

$$\varpi' = \varpi_e \frac{mdv \,.\, mvd\theta \,.\, mv \sin\theta d\phi dq_1dq_2dq_3}{h^3}. \tag{1.6}$$

To obtain the total weight for the electronic state corresponding to the velocity range v to $v+dv$, we integrate this expression over the purely geometrical variables ϕ and θ, respectively from 0 to 2π and from 0 to π, and likewise over the average volume occupied by an electron. If there are N_e electrons per cm.^{3}, then this volume is clearly $1/N_e$. Thus, we obtain finally

$$\varpi_n = \varpi_e \frac{4\pi m^3v^2dv}{N_eh^3}, \tag{1.7}$$

which is probably the most convenient form. In order to compare certain of our results with those of Gaunt and Maue, however, we may define N_n as the number of electrons per unit of frequency range, and also refer the results to unit electron density. By the equations of the photoelectric effect,

$$h(\nu-\nu_0) = \tfrac{1}{2}mv^2 \tag{1.8}$$

and

$$hd\nu = mvdv, \tag{1.9}$$

which, combined with (1.7), give

$$\varpi_n'' = \varpi_e \frac{4\pi m^2 v}{h^2}, \tag{1.10}$$

on setting N_e and $d\nu$ equal to unity.

In thermodynamic equilibrium the absorption and emission must be equal. We have

* Fowler, *Statistical Mechanics*, p. 14 (Cambridge University Press, 1929).

*

$$4\pi N_{n'}\int_0^\infty a_{nn'}(\nu)I_\nu(1-e^{-h\nu/kT})d\nu = 4\pi N_{n'}I_\nu(1-e^{-h\nu/kT})\int_0^\infty a_{nn'}(\nu)d\nu = E_\nu, \quad (1.11)$$

where $a_{nn'}(\nu)$ is the absorption coefficient per atom in state n', and I_ν is given by Planck's formula

$$I_\nu = \frac{2h\nu^3}{c^2}\frac{1}{(e^{h\nu/kT}-1)}. \quad (1.12)$$

The factors I_ν and $(1-e^{-h\nu/kT})$ have been removed from the parentheses in the customary manner, on the assumption that the emission line is narrow so that ν does not change appreciably over the effective range of integration. The factor in parentheses takes care of the "stimulated emissions," which should always be treated as if they were negative absorptions, since they act to decrease the opacity. With weights defined as above, Boltzmann's formula is generally applicable to discrete and continuous states :

$$\frac{N_n}{N_{n'}} = \frac{\varpi_n}{\varpi_{n'}}e^{-h\nu_{nn'}/kT}. \quad (1.13)$$

When (1.7) is substituted in (1.13) and the resulted expression integrated over the velocity range from 0 to ∞, the ordinary dissociation formula is immediately obtained.* These equations are easily seen to be consistent with the definition of the integrated absorption coefficient

$$\int_0^\infty a_{nn'}(\nu)d\nu = \frac{\pi\epsilon^2}{mc}f_{nn'}. \quad (1.14)$$

Our general formula for the oscillator strength is as follows :—

$$f_{nn'} = \frac{2^6}{3}\frac{D}{\varpi_{n'}}\left|\frac{[(n-n')/(n+n')]^{2n+2n'}}{n^2n'^2\left(\frac{1}{n'^2}-\frac{1}{n^2}\right)^3}\frac{\Delta(n, n')}{(n-n')}\right|. \quad (1.15)$$

$$\Delta \equiv \left[F\left(-n+1, -n', 1, -\frac{4nn'}{(n-n')^2}\right)\right]^2 - \left[F\left(-n'+1, -n, 1, -\frac{4nn'}{(n-n')^2}\right)\right]^2, \quad (1.16)$$

where F is the ordinary hypergeometric function :

$$F(\alpha, \beta, \gamma, \chi) = 1 + \frac{\alpha\beta}{1!\gamma}\chi + \frac{\alpha(\alpha+1)\beta(\beta+1)}{2!\gamma(\gamma+1)}\chi^2 + \ldots \quad (1.17)$$

D is the normalizing factor. For transitions between discrete states,

$$D_{\text{I}} = 1. \quad (1.18)$$

For bound-free transitions,

$$D_{\text{II}} = (1-e^{-2\pi\kappa})^{-1}; \quad (1.19)$$

* *Cf.* Menzel, *Proc. N.A.S.*, **19**, 40, 1933.

while for free-free transitions,*

$$D_{\text{III}} = (1 - e^{-2\pi\kappa})^{-1}(1 - e^{-2\pi l})^{-1}e^{2\pi l}. \tag{1.20}$$

The vertical bars signify, as usual, that the absolute value is to be taken of the complex expression that results when $i\kappa$ or il is substituted for n or n'. When n and n' are both real no difficulty arises, and McLean's formula is immediately obtained. When $n \equiv i\kappa$, (1.15) becomes

$$f_{\kappa n'} = \frac{2^5}{3}\frac{1}{n'^2}\frac{e^{-4\kappa \tan^{-1}(n'/\kappa)}}{\kappa^3 n'^3\left(\frac{1}{n'^2} + \frac{1}{\kappa^2}\right)^{7/2}(1 - e^{-2\pi\kappa})} \,|\, \Delta(i\kappa, n') \,|, \tag{1.21}$$

for bound-free transitions. For the free-free case,

$$f_{\kappa l} = \frac{2^6}{3}\frac{1}{\varpi_l}\frac{e^{-2\pi\kappa} \,|\, \Delta(i\kappa, il) \,|}{\kappa^2 l^2\left(\frac{1}{\kappa^2} - \frac{1}{l^2}\right)^3 (l - \kappa)(1 - e^{-2\pi\kappa})(1 - e^{-2\pi l})}, \tag{1.22}$$

where ϖ_l is given by (1.7) or (1.10).

The problem of calculating the various Δ's remains. As long as n' is real, the various hypergeometric functions degenerate to polynomials, that become increasingly complicated as n' increases. We have computed the Δ's for a number of cases and have found that they may be factorized as follows :—

$$\Delta(n, 1) = 8n(n+1)/(n-1)^2, \tag{1.23}$$

$$\Delta(n, 2) = [16n(n+2)/(n-2)^6](3n^2 - 4)(5n^2 - 4), \tag{1.24}$$

$$\Delta(n, 3) = [24n(n+3)/(n-3)^{10}](13n^4 - 78n^2 + 81)(29n^4 - 126n^2 + 81), \tag{1.25}$$

$$\begin{aligned}\Delta(n, 4) = {} & [32n(n+4)/9(n-4)^{14}](197n^6 - 3152n^4 + 13056n^2 - 12288) \cdot \\ & \cdot (539n^6 - 6800n^4 + 20736n^2 - 12288),\end{aligned} \tag{1.26}$$

$$\begin{aligned}\Delta(n, 5) = {} & [40n(n+5)/9(n-5)^{18}](1083n^8 - 36100n^6 + 372250n^4 - 1312500n^2 + \\ & + 1171875) \cdot \\ & \cdot (3467n^8 - 95700n^6 + 786250n^4 - 2062500n^2 + 1171875),\end{aligned} \tag{1.27}$$

$$\begin{aligned}\Delta(n, 7) = {} & [56n(n+7)/2025(n-7)^{26}](567409n^{12} - 5606082n^{10} + \\ & + 199343265 1n^8 - 32628840888n^6 + 242899890135n^4 - \\ & - 737260399890n^2 + 622857924045) \cdot \\ & (2297425n^{12} - 196890722n^{10} + 6067218955n^8 - \\ & - 82901603148n^6 + 500240606775n^4 - 1144024758450n^2 + \\ & + 622857924045).\end{aligned} \tag{1.28}$$

---

* This is the appropriate factor if we write

$$\left(\frac{n - n'}{n + n'}\right) = e^{i\pi}\left(\frac{l - \kappa}{l + \kappa}\right),$$

so that

$$\left|\left[\frac{(n - n')}{(n + n')}\right]^{2n+2n'}\right| = e^{-2\pi\kappa - 2\pi l}.$$

These expressions are valid also for $n = i\kappa$, and hence may be used for obtaining $|\Delta(i\kappa, n')|$. In this case the bracketed factors in the foregoing equations, aside from numerical constants, reduce to the form

$$\left|\frac{i\kappa(i\kappa+n')}{(i\kappa-n')^{4n'-2}}\right| = \frac{\kappa}{(\kappa^2+n'^2)^{2n'-3/2}}. \tag{1.29}$$

The factors in parentheses contain only even powers of $i\kappa$ and are, therefore, real. The signs, however, all become positive. Thus

$$\left|\Delta(i\kappa, 2)\right| = \frac{16\kappa}{(\kappa^2+4)^{5/2}}(3n^2+4)(5n^2+4), \tag{1.30}$$

and similarly for other values of n'. The resulting expressions agree with those of Sugiura * for the discrete-discrete and bound-free cases, where $n' = 1$, 2 and 3, as given in his equations (79), (84), (94) and (80), (88), (100) respectively. Values of $f_{n, n'}$, computed from the exact equation (1.15) for all possible discrete transitions up to and including $n = 14$, are given in Tables IA and IB. The values for the Lyman, Balmer and Paschen series have been extended to $n = 35$. The corresponding Einstein A's, from (1.3), are given in Tables IIA and IIB. These values agree closely with those given by Slack † up to and including $n = 7$.

TABLE IA

Values of $f_{n, n'}$ computed from equation (1.15)

n \ n'	1	2	3	n \ n'	1	2	3
2	4.162×10^{-1}			19	2.295×10^{-4}	5.167×10^{-4}	8.364×10^{-4}
3	7.910×10^{-2}	6.408×10^{-1}		20	1.966×10^{-4}	4.418×10^{-4}	7.117×10^{-4}
4	2.899×10^{-2}	1.193×10^{-1}	8.420×10^{-1}	21	1.698×10^{-4}	3.803×10^{-4}	6.111×10^{-4}
5	1.394×10^{-2}	4.467×10^{-2}	1.506×10^{-1}	22	1.476×10^{-4}	3.302×10^{-4}	5.286×10^{-4}
6	7.800×10^{-3}	2.209×10^{-2}	5.585×10^{-2}	23	1.276×10^{-4}	2.885×10^{-4}	4.608×10^{-4}
7	4.814×10^{-3}	1.271×10^{-2}	2.768×10^{-2}	24	1.137×10^{-4}	2.534×10^{-4}	4.040×10^{-4}
8	3.184×10^{-3}	8.037×10^{-3}	1.604×10^{-2}	25	1.005×10^{-4}	2.240×10^{-4}	3.558×10^{-4}
9	2.216×10^{-3}	5.429×10^{-3}	1.023×10^{-2}	26	8.931×10^{-5}	1.987×10^{-4}	3.155×10^{-4}
10	1.605×10^{-3}	3.851×10^{-3}	6.981×10^{-3}	27	7.963×10^{-5}	1.772×10^{-4}	2.809×10^{-4}
11	1.201×10^{-3}	2.836×10^{-3}	4.996×10^{-3}	28	7.138×10^{-5}	1.587×10^{-4}	2.513×10^{-4}
12	9.215×10^{-4}	2.150×10^{-3}	3.711×10^{-3}	29	6.431×10^{-5}	1.427×10^{-4}	2.243×10^{-4}
13	7.226×10^{-4}	1.672×10^{-3}	2.839×10^{-3}	30	5.809×10^{-5}	1.288×10^{-4}	2.034×10^{-4}
14	5.774×10^{-4}	1.326×10^{-3}	2.223×10^{-3}	31	5.260×10^{-5}	1.167×10^{-4}	1.840×10^{-4}
15	4.687×10^{-4}	1.070×10^{-3}	1.776×10^{-3}	32	4.784×10^{-5}	1.060×10^{-4}	1.670×10^{-4}
16	3.855×10^{-4}	8.770×10^{-4}	1.443×10^{-3}	33	4.359×10^{-5}	9.654×10^{-5}	1.521×10^{-4}
17	3.211×10^{-4}	7.273×10^{-4}	1.189×10^{-3}	34	3.982×10^{-5}	8.829×10^{-5}	1.389×10^{-4}
18	2.703×10^{-4}	6.098×10^{-4}	9.914×10^{-4}	35	3.656×10^{-5}	8.084×10^{-5}	1.272×10^{-4}

* Sugiura, *op. cit.*

† Slack, *Phys. Rev.*, **31**, 527, 1928.

TABLE IB

Values of $f_{nn'}$ computed from equation (1.15)

n' \ n	5	6	7	8	9
4	1.038×10^{0}	1.794×10^{-1}	6.551×10^{-2}	3.229×10^{-2}	1.872×10^{-2}
5		1.231×10^{0}	2.070×10^{-1}	7.455×10^{-2}	3.644×10^{-2}
6			1.425×10^{0}	2.340×10^{-1}	8.318×10^{-2}
7				1.615×10^{0}	2.595×10^{-1}
8					1.810×10^{0}
9					
10					
11					
12					
13					

n' \ n	10	11	12	13	14
4	1.195×10^{-2}	8.185×10^{-3}	5.888×10^{-3}	4.394×10^{-3}	3.375×10^{-3}
5	2.102×10^{-2}	1.345×10^{-2}	9.208×10^{-3}	6.630×10^{-3}	4.959×10^{-3}
6	4.035×10^{-2}	2.320×10^{-2}	1.479×10^{-2}	1.013×10^{-2}	7.299×10^{-3}
7	9.162×10^{-2}	4.415×10^{-2}	2.525×10^{-2}	1.605×10^{-2}	1.096×10^{-2}
8	2.876×10^{-1}	1.001×10^{-1}	4.792×10^{-2}	2.725×10^{-2}	1.733×10^{-2}
9	2.001×10^{0}	3.140×10^{-1}	1.084×10^{-1}	5.147×10^{-2}	2.915×10^{-2}
10		2.191×10^{0}	3.410×10^{-1}	1.180×10^{-1}	5.514×10^{-2}
11			2.380×10^{0}	3.658×10^{-1}	1.245×10^{-1}
12				2.572×10^{0}	3.928×10^{-1}
13					2.764×10^{0}

TABLE IIA

Values of $A_{nn'}$ computed from equation (1.3)

n \ n'	1	2	3	n \ n'	1	2	3
2	4.67×10^{8}			19	5.05×10^{3}	2.79×10^{3}	1.95×10^{3}
3	5.54×10^{7}	4.39×10^{7}		20	3.91×10^{3}	2.16×10^{3}	1.51×10^{3}
4	1.27×10^{7}	8.37×10^{6}	8.94×10^{6}	21	3.06×10^{3}	1.69×10^{3}	1.18×10^{3}
5	4.10×10^{6}	2.52×10^{6}	2.19×10^{6}	22	2.42×10^{3}	1.34×10^{3}	9.33×10^{2}
6	1.64×10^{6}	9.68×10^{5}	7.74×10^{5}	23	1.92×10^{3}	1.07×10^{3}	7.47×10^{2}
7	7.53×10^{5}	4.37×10^{5}	3.34×10^{5}	24	1.57×10^{3}	8.66×10^{2}	6.03×10^{2}
8	3.85×10^{5}	2.20×10^{5}	1.64×10^{5}	25	1.28×10^{3}	7.06×10^{2}	4.91×10^{2}
9	2.13×10^{5}	1.21×10^{5}	8.85×10^{4}	26	1.05×10^{3}	5.80×10^{2}	4.07×10^{2}
10	1.26×10^{5}	7.08×10^{4}	5.13×10^{4}	27	8.70×10^{2}	4.80×10^{2}	3.33×10^{2}
11	7.79×10^{4}	4.37×10^{4}	3.14×10^{4}	28	7.25×10^{2}	4.00×10^{2}	2.78×10^{2}
12	5.04×10^{4}	2.82×10^{4}	2.01×10^{4}	29	6.09×10^{2}	3.35×10^{2}	2.32×10^{2}
13	3.37×10^{4}	1.88×10^{4}	1.34×10^{4}	30	5.14×10^{2}	2.83×10^{2}	1.96×10^{2}
14	2.33×10^{4}	1.30×10^{4}	9.16×10^{3}	31	4.36×10^{2}	2.40×10^{2}	1.67×10^{2}
15	1.65×10^{4}	9.16×10^{3}	6.45×10^{3}	32	3.72×10^{2}	2.05×10^{2}	1.42×10^{2}
16	1.19×10^{4}	6.63×10^{3}	4.65×10^{3}	33	3.19×10^{2}	1.76×10^{2}	1.22×10^{2}
17	8.81×10^{3}	4.89×10^{3}	3.43×10^{3}	34	2.74×10^{2}	1.51×10^{2}	1.05×10^{2}
18	6.62×10^{3}	3.67×10^{3}	2.57×10^{3}	35	2.38×10^{2}	1.31×10^{2}	9.08×10^{1}

TABLE IIB

Values of $A_{nn'}$ computed from equation (1.3)

n' \ n	5	6	7	8	9
4	2.68×10^6	7.67×10^5	3.03×10^5	1.42×10^5	7.43×10^4
5		1.02×10^6	3.24×10^5	1.38×10^5	6.87×10^4
6			4.50×10^5	1.55×10^5	7.03×10^4
7				2.26×10^5	8.15×10^4
8					1.23×10^5
9					
10					
11					
12					
13					

n' \ n	10	11	12	13	14
4	4.21×10^4	2.54×10^4	1.61×10^4	1.06×10^4	7.25×10^3
5	3.78×10^4	2.23×10^4	1.39×10^4	9.10×10^3	6.15×10^3
6	3.67×10^4	2.10×10^4	1.28×10^4	8.23×10^3	5.50×10^3
7	3.88×10^4	2.11×10^4	1.24×10^4	7.80×10^3	5.12×10^3
8	4.65×10^4	2.29×10^4	1.28×10^4	7.77×10^3	5.00×10^3
9	7.12×10^4	2.80×10^4	1.42×10^4	8.14×10^3	5.05×10^3
10		4.35×10^4	1.77×10^4	9.29×10^3	5.39×10^3
11			2.78×10^4	1.15×10^4	6.14×10^3
12				1.85×10^4	7.82×10^3
13					1.26×10^4

It is easy to show that Kramers' formula for the absorption coefficient is equivalent to taking

$$f'_{nn'} = \frac{2^6}{3\sqrt{3}\pi}\frac{1}{\varpi_{n'}}\frac{1}{\left(\frac{1}{n'^2}-\frac{1}{n^2}\right)^3}\left|\frac{1}{n^3}\frac{1}{n'^3}\right|. \tag{1.31}$$

When n is imaginary,

$$\left|\frac{1}{(i\kappa)^3}\right| = \frac{1}{\kappa^3}, \tag{1.32}$$

and similarly for n'. Equation (1.31) is a useful approximation in the discrete-discrete case, as well as for transitions involving continuous states. So close is the agreement between Kramers' expression and the exact expression derived from wave-mechanics, that we find it convenient to write

$$f_{nn'} = gf'_{nn'}, \tag{1.33}$$

where g is a correction factor depending upon n and n'. g is usually close to unity. From (1.15) and (1.33),

$$g = \pi\sqrt{3}D\left|\frac{[(n-n')/(n+n')]^{2n+2n'}nn'\Delta(n, n')}{(n-n')}\right|. \tag{1.34}$$

We may now distinguish three cases, g_{I}, g_{II} and g_{III}, which refer respectively to the discrete-discrete, bound-free and free-free transitions. In these cases we have, accurately,

$$g_{\mathrm{I}} = \pi\sqrt{3}\frac{[(n-n')/(n+n')]^{2n+2n'}nn'\Delta(n, n')}{(n-n')}, \tag{1.35}$$

$$g_{\mathrm{II}} = \frac{\pi\sqrt{3}\kappa n' e^{-4\kappa \tan^{-1}(n'/\kappa)}}{(\kappa^2+n'^2)^{1/2}(1-e^{-2\pi\kappa})}\,|\,\Delta(i\kappa, n')\,| \tag{1.36}$$

and

$$g_{\mathrm{III}} = \frac{\pi\sqrt{3}\kappa l e^{-2\pi\kappa}}{(l-\kappa)(1-e^{-2\pi\kappa})(1-e^{-2\pi l})}\,|\,\Delta(i\kappa, il)\,|. \tag{1.37}$$

When n' is large, the exact equations become somewhat cumbersome to handle. We have, accordingly, applied contour integration and the method of steepest descents to obtain the values of $|\,\Delta\,|$ in certain limiting cases. The details of the derivations are given in the Appendix. The results follow, with the conditions under which the formulæ hold indicated in parentheses.

$$g_{\mathrm{I}} \simeq \left[1 - \frac{0.1728\left(1+\frac{n'^2}{n^2}\right)}{(1-n'^2/n^2)^{2/3}(n')^{2/3}} - \frac{0.0496\left(1+\frac{484}{15}\frac{n'^2}{n^2}+\frac{n'^4}{n^4}\right)}{(1-n'^2/n^2)^{4/3}n'^{4/3}} + \right], \quad \left(\frac{n'}{n} << 1\right). \tag{1.38}$$

$$g_{\mathrm{II}} \simeq \left[1 - \frac{0.1728\left(1-\frac{n'^2}{\kappa^2}\right)}{(1+n'^2/\kappa^2)^{2/3}(n')^{2/3}} - \frac{0.0496\left(1-\frac{484}{15}\frac{n'^2}{\kappa^2}+\frac{n'^4}{\kappa^4}\right)}{(1+n'^2/\kappa^2)^{4/3}(n')^{4/3}} + \right], \quad \left(\frac{\kappa}{n'} >> 1\right). \tag{1.39}$$

$$g_{\mathrm{II}} \simeq \left[1 + \frac{0.1728\left(1-\frac{\kappa^2}{n'^2}\right)}{(1+\kappa^2/n'^2)^{2/3}\kappa^{2/3}} - \frac{0.0496\left(1-\frac{484}{15}\frac{\kappa^2}{n'^2}+\frac{\kappa^4}{n'^4}\right)}{(1+\kappa^2/n'^2)^{4/3}\kappa^{4/3}} + \right], \quad \left(\frac{\kappa}{n'} << 1\right). \tag{1.40}$$

$$g_{\mathrm{III}} \simeq \left[1 + \frac{0.1728\left(1+\frac{\kappa^2}{l^2}\right)}{(1-\kappa^2/l^2)^{2/3}\kappa^{2/3}} - \frac{0.0496\left(1+\frac{484}{15}\frac{\kappa^2}{l^2}+\frac{\kappa^4}{l^4}\right)}{(1-\kappa^2/l^2)^{4/3}\kappa^{4/3}} + \right], \quad \left(\frac{l}{\kappa} >> 1\right). \tag{1.41}$$

The following alternative expressions for the g factors, which take into account only the first two terms of the approximation and are valid only over the ranges indicated above, are easily obtained.

$$g_{\mathrm{I}} \simeq 1 - 0.1728\left(\frac{\nu}{RZ^2}\right)^{1/3}\left[\frac{2}{n'^2}\left(\frac{RZ^2}{\nu}\right)-1\right]$$

$$= 1 - 0.1728\left(\frac{\nu}{RZ^2}\right)^{1/3}\left[\frac{2}{n^2}\left(\frac{RZ^2}{\nu}\right)+1\right]. \tag{1.42}$$

The formulæ for g_{II} and g_{III} may be deduced in a similar manner, or they may be obtained directly from (1.42) by allowing $n \rightarrow i\kappa$ and $n' \rightarrow il$, as before.

$$g_{\mathrm{II}} \simeq 1 - 0{\cdot}1728\left(\frac{\nu}{RZ^2}\right)^{1/3}\left[\frac{2}{n'^2}\left(\frac{RZ^2}{\nu}\right) - 1\right]$$

$$= 1 + 0{\cdot}1728\left(\frac{\nu}{RZ^2}\right)^{1/3}\left[\frac{2}{\kappa^2}\left(\frac{RZ^2}{\nu}\right) - 1\right] \qquad (1.43)$$

and

$$g_{\mathrm{III}} \simeq 1 + 0{\cdot}1728\left(\frac{\nu}{RZ^2}\right)^{1/3}\left[\frac{2}{l^2}\left(\frac{RZ^2}{\nu}\right) + 1\right]$$

$$= 1 + 0{\cdot}1728\left(\frac{\nu}{RZ^2}\right)^{1/3}\left[\frac{2}{\kappa^2}\left(\frac{RZ^2}{\nu}\right) - 1\right]. \qquad (1.44)$$

When $n' \rightarrow \infty$ in (1.40), which corresponds to the short-wave limit of the cont.-cont. spectrum, we have

$$g_{\mathrm{II}}(i\kappa, \infty) = 1 + \frac{0{\cdot}1728}{\kappa^{2/3}} - \frac{0{\cdot}0496}{\kappa^{4/3}}. \qquad (1.45)$$

A comparison between the values of $g_{\mathrm{II}}(i\kappa, \infty)$, as computed from (1.45), and those given by Gaunt in (6.45) is shown in Table III, where it is to

TABLE III

Values of $g_{\mathrm{II}}(i\kappa, \infty)$

κ	Gaunt (6.45)	(1.17)
$\frac{1}{2}$	1·1	1·1
1	1·2	1·12
2	1·1	1·09

be noted that the last figures in Gaunt's values are stated to be uncertain. When $\kappa \rightarrow \infty$, we have from (1.39)

$$g_{\mathrm{II}}(i\,\infty, n') = 1 - \frac{0{\cdot}1728}{(n')^{2/3}} - \frac{0{\cdot}0496}{(n')^{4/3}}. \qquad (1.46)$$

Values of $g_{\mathrm{II}}(i\,\infty, n')$ computed from (1.46) are shown in Table IV, together with the values computed by Gaunt (6.50), Maue (Table 2), and the exact values obtained with the aid of equations (1.23) to (1.28). The exact values of $g_{\mathrm{II}}(i\kappa, n')$, obtained from the latter equations, are plotted in fig. 1. Of these the curves for $n' = 1$, 2, 3 were given by Maue in his fig. 6 and were also tabulated by Strömgren.* The lines at the right of the diagram indicate the respective values of $g_{\mathrm{II}}(i\,\infty, n')$.

* B. Strömgren, *Zs. für Astr.*, 4, 118, 1932.

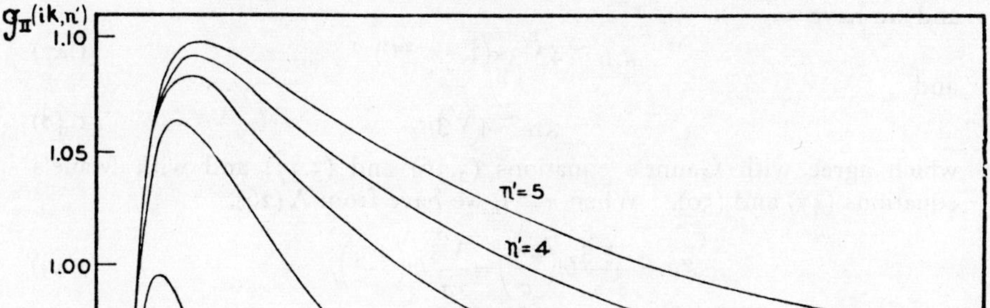

FIGURE 1

Values of g_{II} (iκ, n′) computed from (1.36) and (1.23) to (1.27). The lines at the right show the respective values of g_{II} (i ∞, n′).

TABLE IV

Values of $g_{II}(i\infty, n')$

n'	Exact	(1.18)	Maue	Gaunt
1	0·7973	0·7776	0·797	0·8
2	0·8762	0·8714	0·876	0·9
3	0·9075	0·9055	0·91	0·9
4	0·9247	0·9236	$0·92_5$	1·0
5	0·9358	0·9351	$0·93_5$	
6		0·9431	$0·94_5$	
7	0·9488	0·9491		

Further checks on our formulæ may be made in the limiting cases of $\kappa \rightarrow 0$ and when $\kappa \simeq l$. When $\kappa << 1$, $|\Delta(i\kappa, n')| \simeq |\Delta(i\kappa, il)| \simeq 8\kappa$,

**

and we have

$$g_{\text{III}} \simeq 4\sqrt{3}\kappa(1 - e^{-2\pi l})^{-1} \tag{1.47}$$

and

$$g_{\text{II}} \simeq 4\sqrt{3}\kappa, \tag{1.48}$$

which agree with Gaunt's equations (5.36) and (5.37) and with Maue's equations (47) and (50). When $\kappa \simeq l$, we have from A (26),

$$g_{\text{III}} \simeq \frac{\sqrt{3}}{\pi} ln\left(\frac{2\kappa}{c}\right) = \frac{\sqrt{3}}{\pi} ln\left(\frac{4\nu_\kappa}{\nu}\right), \tag{1.49}$$

in agreement with Gaunt (6.44) and Maue (61).

When $n - n'$ is small, the value of g_{I} may be computed most simply by equations A (27) and A (28). The approximation A (29) is also useful when n' is large.

At this point we should mention that the factor, $3\sqrt{3}$, by which Sugiura * multiplies g_{III}, is apparently erroneous. His derivation is supposed to take account of retardation, but elementary considerations show that this effect should lead to a correction term of the order v^2/c^2. Sugiura's work contains several errors, one of which is in the normalizing factors.

The quantities $A_{\kappa n'}$ for the bound-free transitions are related to the target areas for capture $\sigma_{\kappa n'}$ and to Milne's † capture probability coefficients $F(v)$. If $f(v)$ is the velocity-distribution function, the number of captures per cm.³ per second per velocity range dv is

$$N_i N_e f(v) v \sigma_{\kappa n'} dv, \tag{1.50}$$

where N_i is the number of ionized atoms per cm.³.

In the present notation, the number of transitions from continuous states in the range $d\kappa$ corresponding to the velocity range dv is

$$N_\kappa A_{\kappa n'} d\kappa = N_i f(v) A_{\kappa n'} dv d\kappa, \tag{1.51}$$

whence

$$A_{\kappa n'} = N_e v \sigma_{\kappa n'} = 2\pi N_e v F(v) \tag{1.52}$$

and

$$\sigma_{\kappa n'} = \frac{2^4}{3\sqrt{3}} \frac{\epsilon^2 h}{m^2 c^3} \frac{1}{\left(\frac{1}{n'^2} + \frac{1}{\kappa^2}\right)} \frac{\kappa^2}{n'^3} g_{\text{II}}, \tag{1.53}$$

by (1.7), (1.31), (1.33) and the definition of κ:

$$\tfrac{1}{2} m v^2 = \frac{hRZ^2}{\kappa^2} = h\nu, \tag{1.54}$$

$$m v dv = -\frac{2hRZ^2}{\kappa^3} d\kappa = h d\nu. \tag{1.55}$$

In these equations, $h\nu$ obviously refers to the energy of the electrons and not to the frequency of the emitted radiation.

* Sugiura, *Sci. Pap. Inst. Phys. and Chem. Research* (*Tokyo*), **17**, 89, 1931.

† Milne, *Phil. Mag.*, **47**, 209, 1924.

2. *The Atomic Coefficients of General Absorption*

We turn now to discussion of the absorption coefficients themselves. The continuous absorption coefficient per atom in state n' may be defined as follows :— *

$$a_{\kappa n'}(\nu)=\frac{\pi\epsilon^2}{mc}\frac{df_{\kappa n'}}{d\nu}, \tag{2.1}$$

where

$$df_{\kappa n'}=f_{\kappa n'}d\kappa. \tag{2.2}$$

Employing (1.31), (1.33), (1.54) and (1.55) we obtain

$$a_{n'}(\nu)=\frac{\pi\epsilon^2}{mc}\frac{1}{2RZ^2}\frac{2^5}{3\sqrt{3}\pi}\frac{1}{n'^5}\frac{1}{\left(\frac{1}{n'^2}+\frac{1}{\kappa^2}\right)^3}g_{II}=\frac{2^4\pi^2}{3\sqrt{3}}\frac{\epsilon^6Z^2}{ch^4\nu^3}\frac{1}{n'^2}\frac{2hRZ^2}{n'^3}g_{II}, \tag{2.3}$$

since

$$R=\frac{2\pi^2\epsilon^4m}{h^3}. \tag{2.4}$$

If for the moment we omit the factor $(1-e^{-h\nu/kT})$, which allows for the effect of negative absorptions, the total contribution at frequency ν to the optical depth $d\tau_\nu$ by atoms in the layer of thickness dH is

$$d\tau_\nu=\sum_{n_0'}^{\infty}N_{n'}a_{n'}(\nu)dH=N_0dH\frac{\varpi_\epsilon\varpi'}{b_0(T)}e^{-\chi_0/kT}\frac{2^4\pi^2}{3\sqrt{3}}\frac{\epsilon^6Z^2}{ch^4\nu^3}\cdot 2hRZ^2\times$$
$$\times\sum_{n_0'}^{\infty}\frac{e^{hRZ^2/n'^2kT}}{n'^3}g_{II}, \tag{2.5}$$

where N_0 is the total number of neutral atoms, irrespective of quantum state. $b_0(T)$ is the partition function of the neutral atom and χ_0 is the ionization potential,

$$b_0(T)=\sum_1^{\infty}\varpi_{n'}e^{-E_{n'}/kT}. \tag{2.6}$$

For hydrogen, ionized helium, etc., $\varpi_{n'}$ is given by (1.4), and $E_{n'}$, the excitation potential, is

$$E_{n'}=hRZ^2\left(1-\frac{1}{n'^2}\right). \tag{2.7}$$

For complex atoms it is unlikely that equation (2.3) holds exactly for the lower atomic states. The upper atomic states of any atom are, however, essentially hydrogenic. If ϖ' is the weight of the ion, the total weight of all sub-levels produced by coupling an electron in all possible orbits of given total quantum number n' is

$$\varpi_{n'}=\varpi_e\varpi'n'^2. \tag{2.8}$$

If neutral atomic terms built on ionic states other than the lowest come into account, ϖ' in equation (2.8) should be replaced by $b'(T)$, the

* *Cf.* Menzel, *Publ. Lick Obs.*, **17**, 222, 1930.

partition function of the ion. The excitation potentials of the higher states will be

$$E_{n'} = \chi_0 - \frac{hRZ^2}{n'^2}. \tag{2.9}$$

This assumes zero quantum defect. n_0', the lower limit of the summation, is defined so that

$$\nu \geqslant \frac{RZ^2}{n_0'^2}. \tag{2.10}$$

Thus, in the case of hydrogen, if we are dealing with ν to the violet of $R/4$, the Balmer limit, $n_0' = 2$; if the ν under consideration lies between $R/4$ and $R/9$, $n_0' = 3$, etc. This equation may be further simplified if N_0 is eliminated by means of the equation of dissociative equilibrium,

$$\frac{N_i}{N_0} P_e = \frac{(2\pi m)^{3/2}(kT)^{5/2}}{h^3} \frac{\varpi_\epsilon b_1(T)}{b_0(T)} e^{-\chi_0/kT}. \tag{2.11}$$

Or

$$d\tau_\nu = N_i dH \frac{2^4\pi^2}{3\sqrt{3}} \frac{\epsilon^6 Z^2}{ch\nu^3} \frac{P_\epsilon}{(2\pi mkT)^{3/2}} \frac{2hRZ^2}{kT} \sum_{n_0'}^{\infty} \frac{e^{hRZ^2/n'^2kT}}{n'^3} g_{\text{II}}$$

$$= N_i dHC_0 \frac{Z^2 P_\epsilon T^{-3/2}}{\nu^3} \left[\frac{2hRZ^2}{\kappa T} \sum_{n_0'}^{\infty} \frac{e^{hRZ^2/n'^2kT}}{n'^3} g_{\text{II}} \right], \tag{2.12}$$

where

$$C_0 = \frac{2^4\pi^2}{3\sqrt{3}} \frac{\epsilon^6}{ch} \frac{1}{(2\pi mk)^{3/2}} = 2{\cdot}67 \times 10^{24}. \tag{2.13}$$

In the case of complex atoms it would appear to be permissible to evaluate the sum by means of integration. The energy levels formed by coupling an electron of total quantum number n' will not be coincident. The series levels of various atoms in a composite stellar atmosphere will overlap. Integration is equivalent to assuming a continuous set of quantum defects between successive integral quantum numbers:

$$\frac{2hRZ^2}{kT} \sum_{n'_0}^{\infty} \frac{e^{hRZ^2/n'^2kT}}{n'^3} g_{\text{II}} \simeq \frac{2hRZ^2}{kT} \int_{\sqrt{\frac{RZ^2}{\nu}}}^{\infty} \frac{e^{hRZ^2/n'^2kT}}{n'^3} g_{\text{II}} dn', \tag{2.14}$$

$$d\tau_\nu = N_i dHC_0 \frac{Z^2 P_\epsilon T^{-3/2}}{\nu^3} \bar{g}_{\text{II}} e^{h\nu/kT}, \tag{2.15}$$

$$\bar{g}_{\text{II}} \simeq 1 - 0{\cdot}1728 \left(\frac{\nu}{RZ^2}\right)^{1/3} \left(1 - \frac{2kT}{h\nu}\right) - e^{-h\nu/kT} \left[1 + 0{\cdot}1728 \left(\frac{\nu}{RZ^2}\right)^{1/3} \left(1 + \frac{2kT}{h\nu}\right)\right], \tag{2.16}$$

when use is made of (1.43). For the free-free transitions, the absorption coefficient per ionized atom for electronic transitions from one interval dv' of electronic energy to another interval dv becomes

$$a_{\kappa,\,l}(\nu)=\frac{\pi\epsilon^2}{mc}\frac{d^2f_{\kappa l}}{d\nu d\nu'},\tag{2.17}$$

where

$$d^2f=f_{\kappa l}d\kappa dl.\tag{2.18}$$

Employing (1.31)–(1.33), (1.54), (1.55) and

$$\tfrac{1}{2}mv^2=\frac{hRZ^2}{l^2}=h\nu',\tag{2.19}$$

we obtain

$$a_{\kappa l}(\nu)=\frac{\pi\epsilon^2}{mc}\frac{1}{4R^2Z^4}\frac{2^6}{3\sqrt{3}\pi}\frac{1}{\varpi_e}\frac{1}{\left(\frac{1}{\kappa^2}-\frac{1}{l^2}\right)^3}g_{\text{III}}.\tag{2.20}$$

In this case the optical depth at ν is obtained by integrating over all values of l (*i.e.* over ν'), with g_{III} as given in equation (1.44).

$$d\tau_\nu=N_idH\frac{2^5\pi^2\epsilon^6Z^2}{3\sqrt{3}ch^3\nu^3}\int_0^\infty\frac{f(v)dvg_{\text{III}}}{\varpi_e}d\nu'=N_idHC_0\frac{Z^2P_eT^{-3/2}}{\nu^3}\int_0 e^{-h\nu'/kT}g_{\text{III}}d\left(\frac{h\nu'}{kT}\right)\tag{2.21}$$

in thermodynamic equilibrium. $f(v)$ is given by Maxwell's law:

$$f(v)dv=4\pi\left(\frac{m}{2\pi kT}\right)^{3/2}e^{-\frac{1}{2}mv^2/kT}v^2dv.\tag{2.22}$$

If we express the electron pressure in dynes per cm.^{2},

$$P_e=N_ekT.\tag{2.23}$$

Introducing equation (1.44), we find the following expression:—

$$d\tau_\nu=N_idHC_0\frac{Z^2P_eT^{-3/2}}{\nu^3}\bar{g}_{\text{III}},\tag{2.24}$$

where

$$\bar{g}_{\text{III}}=1+0{\cdot}1728\left(\frac{\nu}{RZ^2}\right)^{1/3}\left[1+\frac{2kT}{h\nu}\right].\tag{2.25}$$

The effective absorption coefficient $\alpha'(\nu)$ *per ionized atom* is obtained by summing the contributions of the bound-free and free-free transitions and multiplying by $(1-e^{-h\nu/kT})$ to allow for the negative absorptions. Thus equations (2.15) and (2.24) give

$$\alpha'(\nu)=\frac{C_0Z^2P_e}{\nu^3T^{3/2}}(e^{h\nu/kT}-1)g',\tag{2.26}$$

where

$$g'\simeq 1-0{\cdot}1728\left(\frac{\nu}{RZ^2}\right)^{1/3}\left(1-\frac{2kT}{h\nu}\right).\tag{2.27}$$

For hydrogen-like atoms we may write

$$\alpha'(\nu)=\frac{C_0Z^2P_e}{\nu^3T^{3/2}}\left[\frac{2hRZ^2}{kT}\sum_{n_0'}^{\infty}\frac{e^{hRZ^2/n'^2kT}}{n'^3}g_{\text{II}}+\bar{g}_{\text{III}}\right](1-e^{-h\nu/kT})\tag{2.28}$$

from (2.12) and (2.24). Equations (2.26) and (2.28) should be employed in place of those given by Pannekoek,* whose formulæ are in error through omission of a weight factor n'^2 in (2.14) *et seq.* He also sets $g=1$ throughout.

It is interesting to apply equation (2.26) to the calculation of opacity in the solar atmosphere. We shall assume that hydrogen is the predominant constituent, but that most of the opacity comes from metallic atoms, for which $Z=1$. The validity of this last assumption will be discussed further in Section 3. Under these physical conditions the total optical depth down to a layer where the electron pressure is P_0 is given by †

$$\tau_1 = \frac{1}{\bar{m}g}\frac{C_0 P_0{}^2}{\nu^3 T^{3/2}}(e^{h\nu/kT}-1)g'\left[1-\frac{K}{P_0}+\frac{K^2}{P_0{}^2}\log\left(1+\frac{P_0}{K}\right)\right], \tag{2.29}$$

where $\bar{m}$ is the mean atomic mass, here taken to be that of hydrogen, and g is the surface gravity, $2{\cdot}74\times 10^4$ cm./sec.$^{-2}$. K is the dissociation constant of the opacity-producing ion; it is equivalent to the right-hand side of equation (2.11). For *Mg* and *Fe*, the most abundant of the metals, $K>P$, for solar conditions, and the bracketed factor is practically $\frac{1}{2}$. Milne ‡ has shown that the effective photosphere lies approximately at $\tau_1=\frac{1}{3}$. Taking $\nu=6\times 10^{14}$, which corresponds approximately to λ 5000 A., and $T=6000°$, we find $P_0=32$ dynes or $3{\cdot}2\times 10^{-5}$ atmospheres. From an analysis of the relative intensities of lines from ionized and neutral lines of the same element, Russell § derives the value $P_0=46\pm 9$ dynes. The close agreement between observed and calculated values is encouraging indeed. The present result is in good agreement with an earlier result obtained by one of us,‖ wherein a value of $P_0=40$ dynes was deduced. The same general wave-mechanical formulæ, however, formed the basis of both computations.

3. *Mean Values of the Absorption Coefficient*

Equation (2.26) may also be used to calculate a_R', the reciprocal or Rosseland mean absorption coefficient, defined as follows :—

$$a_R' = \frac{\displaystyle\int_0^\infty \frac{\partial}{\partial T}\frac{\nu^3}{e^{h\nu/kT}-1}d\nu}{\displaystyle\int_0^\infty \frac{1}{a'(\nu)}\frac{\partial}{\partial T}\frac{\nu^3}{(e^{h\nu/kT}-1)}d\nu} = \frac{C_0 h^3 Z^2 P_e}{k^3 T^{9/2}}\,\frac{24\theta(4)}{7\,!\,\frac{1}{2}[\theta(6)-\theta(7)]}g_R', \tag{3.1}$$

where

$$g_R' = 1 - \frac{0{\cdot}1728\left(\dfrac{kT}{hRZ^2}\right)^{1/3}\{\Gamma(25/3)[\theta(19/3)-\theta(22/3)]-2\Gamma(22/3)[\theta(16/3)-\theta(19/13)]\}}{7\,!\,[\theta(6)-\theta(7)]}$$

$$= 1 - 0{\cdot}1098\left(\frac{kT}{hRZ^2}\right)^{1/3}, \tag{3.2}$$

* Pannekoek, *M.N.*, **91**, 162, 1931.
† Russell, *Ap. J.*, **78**, 239, 1933; *Contr. Mt. Wilson Obs.*, No. 477.
‡ Milne, *Phil. Trans.*, A, **228**, 421, 1929.
§ Russell, *op. cit.*; *cf.* also *Ap. J.*, **70**, 11, 1929; *Contr. Mt. Wilson Obs.*, No. 383.
‖ Menzel, *Publ. Lick Obs.*, **17**, 245, 1930.

and

$$\left.\begin{array}{l} \theta(a) = 1 + \frac{1}{2^a} + \frac{1}{3^a} + \ldots, \\ \theta(4) = \frac{\pi^4}{90}, \quad \theta(6) = 1\cdot017343, \quad \theta(7) = 1\cdot008349, \\ \theta(\tfrac{16}{3}) = 1\cdot028597, \quad \theta(\tfrac{19}{3}) = 1\cdot013564, \quad \theta(\tfrac{22}{3}) = 1\cdot006567. \end{array}\right\} \quad (3\cdot3)$$

Thus, we have numerically

$$a_R' = C_R \frac{Z^2 P_e}{T^{9/2}} g_R' \quad (3\cdot4)$$

$$g_R' = 1 - 0\cdot1098 \left(\frac{T}{1\cdot57 \times 10^5 Z^2} \right)^{1/3} \quad (3\cdot5)$$

$$C_R = \frac{16\pi^2 \theta(4)}{315\sqrt{3}[\theta(6) - \theta(7)]} \frac{\epsilon^6 h^2}{c\kappa^3} \frac{1}{(2\pi mk)^{3/2}} = 3\cdot32 \times 10^{-7}. \quad (3\cdot6)$$

The final result should be valid as long as the second term in g_R' is small.

With the exception of the small correction factor g_R', equation (3.4) is identical with that given by Rosseland,* derived by quite different logic and stated by him to be an upper limit to the value of the opacity coefficient. Chandrasekhar † criticizes this value on several counts, among which are the following :—

(1) Rosseland's method amounts to an extrapolation of the Kramers-Gaunt formula for free-free transitions to the case of the bound-free transitions.

(2) The assumption of a continuous series of negative energy states is fictitious and, since the Rosseland mean introduces a weighting factor $\nu^4 e^{-h\nu/kT}$, the resultant value of the absorption coefficient will be too large.

(3) The extension of the Maxwellian velocity distribution to states of negative energy leads to a gross exaggeration of the absorption arising from bound states of large negative energy.

(4) The total absorption at frequency ν yields an infinite mean ‡ absorption coefficient.

Chandrasekhar's criticisms do not appear to be well founded. First, the extrapolation of the free-free to the bound-free case is seen to be justified. Second, the convention of adopting a continuous series of negative states is equivalent to our substitution of an integration for the more accurate summation. Further, the weighting factor proves to be exceptionally low. From equations (2.26) and (3.1) we find that the value of $\bar{\nu}$, corresponding to the average frequency used in obtaining the Rosseland mean, is given by

$$\frac{\bar{\nu}^3}{(e^{h\bar{\nu}/kT} - 1)} = \left(\frac{kT}{h}\right)^3 \frac{7!\,\frac{1}{2}[\theta(6) - \theta(7)]}{24\theta(4)}, \quad (3\cdot7)$$

* Rosseland, *Astrophysik*; *cf.* also *Handb. der Astrophysik*, Band III, Part 1, pp. 469 *et seq.*

† Chandrasekhar, *Proc. Roy. Soc.*, A, **133**, 241, 1931 ; *ibid.*, **135**, 472, 1932.

‡ The "straight," not the Rosseland mean.

whence

$$\bar{\nu} \simeq 1{\cdot}36\frac{kT}{h}, \quad \text{or} \quad \lambda \simeq 1{\cdot}05T. \tag{3.8}$$

This value of $\bar{\nu}$ is well within the limit where (2.26) would be expected to fail seriously, and the formula should be quite accurate. Third, the extension of the Maxwellian distribution to negative energy states is quite proper, since it is nothing more than an expression of the Boltzmann law. Finally, the infinite value of the straight mean absorption coefficient arises from the extension of the integration to values of $h\nu \geqslant \chi_0$, where χ_0 is the ionization potential of the atom in question. An approximate value of the absorption coefficient for frequencies in this range is obtained by setting $h\nu = \chi_0$ in the exponential term of (2.26). Our failure to do this in the case of the Rosseland mean, however, introduces no appreciable error in the result, since the exponential then falls in the denominator.

The chief objection that may be urged against the validity of the formulæ (2.26) and (2.28) is the use we have made of the Boltzmann and dissociation formulæ, since these are strictly applicable only in the case of thermodynamic equilibrium. The absorption coefficient is usually employed to calculate the *flux* of radiation, which in turn violates the condition of thermodynamic equilibrium, for which state the flux is zero. Nevertheless, the first appreciable deviation from thermodynamic equilibrium occurs in the lowest atomic states, which do not enter into the formulæ (2.26) and (2.28). To the first approximation the partition of atoms among the higher atomic states may be represented as a Boltzmann distribution characterized by a parameter T_e, the electron temperature, instead of T_0, the temperature of the radiation.

The absorption coefficient depends upon the value of Z. In stellar atmospheres, where most of the opacity comes from the ionization of neutral atoms, $Z = 1$. Russell * sets $Z = (\chi_0/R)^{\frac{1}{2}}$, where χ_0 is the ionization potential of the opacity-producing atom. The opacity we are considering, however, comes chiefly from atomic states of high total and high azimuthal quantum number, for which the energy levels are essentially hydrogenic.

It is interesting to apply our results to an assembly consisting of ionized complex atoms. Consider an atom of atomic number Z and therefore capable of existing in $Z+1$ different stages of ionization. Denote the number of electrons that have been removed from a given atom by q and the percentage of such atoms by x_q. Then, at a given temperature and density, the various equilibria become

$$\frac{x_1}{x_0} = \frac{K_1}{P}, \qquad \frac{x_2}{x_1} = \frac{K_2}{P}, \qquad \ldots \quad \frac{x_q}{x_{q-1}} = \frac{K_2}{P}, \tag{3.9}$$

where P is the electron-pressure and

$$K_q = \frac{(2\pi m)^{3/2}(kT)^{5/2}}{h^3} \frac{\varpi_\epsilon b_q(T)}{b_{q-1}(T)} e^{-\chi_{q-1}/kT}. \tag{3.10}$$

* Russell, *Ap. J.*, 78, 239, 1927.

The notation is analogous to that used in equation (2.11). Thus

$$x_q = x_0 \frac{K_1 K_2 \ . \ . \ . \ K_q}{P^q} \tag{3.11}$$

and

$$\sum_{q=0}^{Z} x_q = 1. \tag{3.12}$$

We easily find that

$$x_0 = \frac{1}{\Phi}, \tag{3.13}$$

where

$$\Phi = 1 + \frac{K_1}{P} + \frac{K_1 K_2}{P^2} + \ . \ . \ . \ \frac{K_1 K_2 \ . \ . \ . \ K_Z}{P^Z}. \tag{3.14}$$

Also

$$P = N(x_1 + 2x_2 + \ . \ . \ . \ Zx_Z)kT. \tag{3.15}$$

N is the total number of atoms per cm.^{3} in all stages of ionization. The optical depth arising from a gas of this type is, by equation (2.5) *et seq.*,

$$d\tau = NdH \sum_{0}^{Z-1} x_q \varpi_e \frac{b_{q+1}(T)}{b_q(T)} e^{-\chi_q/kT} \frac{2^4 \pi^2}{3\sqrt{3}} \frac{\epsilon^6 (q+1)^2}{ch^4 \nu^3} kTe^{h\nu/kT}, \tag{3.16}$$

where we have set $g' = 1$. This equation includes the free-free transitions but not the stimulated emissions. For accurate solution of the problem it is necessary to know all the ionization potentials, χ_q. Approximate values have been assigned in certain cases, but the mathematical relationships between successive ionization potentials is quite complex. We may, however, consider an idealized atom, where

$$\chi_q = hR(q+1)^2. \tag{3.17}$$

This gives the correct value for

$$\chi_{Z-1} = hRZ^2, \tag{3.18}$$

and also the correct order of magnitude for χ_0. Writing

$$K_q = ae^{-bq^2}, \tag{3.19}$$

and setting the factor $\omega_e b_q(T)/b_{q-1}(T)$ equal to unity, we find

$$\Phi = 1 + \frac{a}{P}e^{-b} + \frac{a^2}{P^2}e^{-b(1+4)} + \ . \ . \ . \ + \frac{a^q}{P^q}e^{-b(1+4+ \ . \ . \ . \ +q^2)} + \ . \ . \ . \ = \sum_{0}^{Z} e^{q\ln\frac{a}{P} - \frac{b}{3}\left(q^3 + \frac{3}{2}q^2 + \frac{q}{2}\right)} \tag{3.20}$$

$$\simeq \int_0^\infty e^{q\ln\frac{a}{P} - \frac{b}{3}\left(q^3 + \frac{3}{2}q^2 + \frac{q}{2}\right)} dq, \tag{3.21}$$

as long as the atoms of the assembly are not completely "stripped." To evaluate this integral we first determine q_0, the value of q for which the exponent has a maximum:

$$q_0 = -\frac{1}{2} + \left(\frac{1}{12} + \frac{\ln a/P}{b}\right)^{\frac{1}{2}}. \tag{3.22}$$

By means of the substitution

$$y \equiv (q-q_0)\left(\frac{b^2}{12}+b\ln\frac{a}{P}\right)^{\frac{1}{2}},$$

the integral of (3.21) may be expressed in the form

$$\Phi = \frac{\left(\frac{a}{P}\right)^{q_0} e^{-\frac{b}{3}\left(-\frac{1}{6}+\frac{\ln\frac{a}{P}}{b}\right)\left(\frac{1}{12}+\frac{\ln\frac{a}{P}}{b}\right)^{\frac{1}{2}}}}{\left(\frac{b^2}{12}+b\ln\frac{a}{P}\right)^{\frac{1}{2}}}\int_{-y_0}^{\infty} e^{-y^2-\frac{y^3}{\alpha}}dy, \tag{3.23}$$

where

$$\alpha \equiv 3b^{\frac{1}{2}}\left(\frac{1}{12}+\frac{\ln\frac{a}{P}}{b}\right)^{\frac{3}{2}} \tag{3.24}$$

and

$$y_0 = q_0[b(q_0+\tfrac{1}{2})]^{\frac{1}{2}}. \tag{3.25}$$

If y_0 and α are large,

$$\int_{-y_0}^{\infty} e^{-y^2-y^3/\alpha}dy \simeq \int_{-\infty}^{\infty} e^{-y^2}\left(1-\frac{y^3}{\alpha}+\frac{y^6}{2\alpha^2}-\frac{y^9}{6\alpha^3}+\frac{y^{12}}{24\alpha^4}\ldots\right)dy$$

$$=\sqrt{\pi}\left[1+\frac{15}{16\alpha^2}+\frac{15}{16}\frac{(7.9.11)}{8.24\alpha^4}+\ldots\right], \tag{3.26}$$

whence

$$\Phi = \frac{\left(\frac{a}{P}\right)^{q_0} e^{-\frac{b}{3}\left(-\frac{1}{6}+\frac{\ln\frac{a}{P}}{b}\right)\left(\frac{1}{12}+\frac{\ln\frac{a}{P}}{b}\right)^{\frac{1}{2}}}\sqrt{\pi}}{\left(\frac{b^2}{12}+b\ln\frac{a}{P}\right)^{\frac{1}{2}}}\left[1+\frac{15}{16\alpha^2}+\ldots\right]. \tag{3.27}$$

Similarly

$$P = \frac{NkT}{\Phi}\sum_0^Z q\left(\frac{a}{P}\right)^q e^{-\frac{b}{3}\left(q^3+\frac{3}{2}q^2+\frac{q}{2}\right)} = \frac{NkT}{\Phi}\frac{\partial\Phi}{\partial\ln\left(\frac{a}{P}\right)} = NkT\left(q_0-\frac{1}{\frac{b}{3}+4\ln\frac{a}{P}}\right). \tag{3.28}$$

$$a_\nu = \frac{24\pi^2}{3\sqrt{3}}\frac{\epsilon^6}{ch^4\nu^3}kTe^{h\nu/kT}\frac{1}{\Phi}\sum^{Z-1}\left(\frac{a}{P}\right)^q(q+1)^2 e^{-\frac{b}{3}\left(q^3+\frac{3}{2}q^2+\frac{q}{2}\right)-b(q+1)^2}. \tag{3.29}$$

Using the same method of asymptotic integration as in (3.26), we obtain

$$\sum_0^{Z-1}\left(\frac{a}{P}\right)^q(q+1)^2 e^{-\frac{b}{3}\left(q^3+\frac{3}{2}q^2+\frac{q}{2}\right)-b(q+1)^2} \cong \left(1+\frac{2\partial}{\partial\ln\left(\frac{a}{P}\right)}+\frac{\partial^2}{\partial\ln\left(\frac{a}{P}\right)^2}\right)\times$$

$$\times\int_0^{\infty} e^{q\ln(a/P)-\frac{b}{3}\left(q^3+\frac{3}{2}q^2+\frac{q}{2}\right)-b(q+1)^2}dq \cong \frac{\Phi}{\left(\frac{a}{P}\right)}\left[\frac{\ln\left(\frac{a}{P}\right)}{b}-\sqrt{\frac{1}{12}+\frac{\ln(a/P)}{b}}+\frac{1}{3}\right]. \tag{3.30}$$

If we now multiply (3.29) by $(1 - e^{-h\nu/kT})$ to take account of the stimulated emission, and by (3.7) use $0\cdot873\left(\frac{kT}{h}\right)^3$ for $\frac{\bar{\nu}^3}{(e^{h\bar{\nu}/kT}-1)}$, we get

$$\bar{a}_R = \frac{2^4\pi^2}{3\sqrt{3}} \frac{\epsilon^6 kT\left[\frac{ln(a/P)}{b} - \sqrt{\frac{1}{12} + \frac{lna/P}{b} + \frac{1}{3}}\right]}{ch^4 0\cdot873\left(\frac{kT}{h}\right)^3\left(\frac{a}{P}\right)}. \tag{3.31}$$

We shall now apply (3.28) and (3.31) to the centre of Capella. Using Eddington's data,* $T = 7\cdot2 \times 10^6$, $\rho = 0\cdot0547$ and $N = 5\cdot88 \times 10^{20}$ for the number of (iron) atoms per cm.^{3}, we get from (3.28) $P = 1\cdot11 \times 10^{13}$ dynes/cm.^{2}. q_0, the most abundant stage of ionization, is 19. From (3.31) we obtain $a_R = 1\cdot86 \times 10^{-22}$ for the mean absorption coefficient per atom. The absorption coefficient per gram $\kappa = a_R N/\rho = 2\cdot00$, as compared with Eddington's value of 4·95 (*loc. cit.*, p. 229). The largest term in the sum of (3.29) is the 19th ($q = 18$), so that the effective value of Z is about 19. Eddington uses a value of 26 for Z, which introduces a factor of two in κ, since a_ν is proportional to Z^2. Agreement with the astronomical value of 50 is no longer expected. The method of extending these formulæ to the cases discussed by B. Strömgren, who assumes an excess of H, is obvious.

The effect upon the mean absorption coefficient that arises from abrupt changes in opacity at each series limit has been widely discussed. The absorption beyond the series limits runs continuously into the overlapping lines to the redward of the limit. Furthermore, at the extremely high temperatures and densities of stellar interiors it appears probable that even the lowest atomic levels will be appreciably broadened. Probably there is no line absorption at all. All of the series members overlap and form a continuum, the effect of which should be added to that of the ordinary continuous absorption. Though the magnitude of the correction cannot be estimated with precision, elementary considerations based on the value of the ratio $\int df : \sum f$ for the predominant series indicate that the factor may be as much as 2 or 3.

Here there is no difficulty with the so-called "guillotine effect," since the lower limit, $n = 1$, is specified and the pseudo-classical quantum range $0 \leqslant n < 1/2$ does not come into account.

4. *Absorption Coefficient for Neutral Hydrogen Atoms in Stellar Atmospheres*

Pannekoek † has suggested that part, if not all, of the opacity of stellar atmospheres, especially for the later-type stars, may come from the interaction of neutral hydrogen atoms with free electrons. The electrons are supposed to come from the more easily ionized metals, and tend to interact

* Eddington, *Internal Constitution of the Stars*, p. 229.

† Pannekoek, *op. cit*.

with the neutral hydrogen rather than with their own positive ions because of the relatively high abundance of the former. Russell,* who found Pannekoek's formula, the analogy of our equation (2.26), inadequate to account for the observed opacity, has also discussed the possibility of a contribution by neutral atoms to the atmospheric opacity. We have shown that our formula removes the old discrepancy between theory and observation. Nevertheless, it would appear worth while to investigate the absorption effects of neutral atoms.

The problem of the continuous emission during the free-free transitions of electrons colliding with neutral atoms has been treated by L. Nedelsky.† He assumes that the potential field V of the atoms is given by

$$\begin{aligned} V &= Ze^2/a - Ze^2/r \qquad &\text{for } r \leqslant a \\ &= 0 &\text{for } r > a, \end{aligned} \tag{4.1}$$

which is equivalent to the field due to a point charge Ze surrounded by a spherical shell of radius a carrying a charge $-Ze$. His formulæ can be applied directly to our case by making $Z=1$ and taking the radius of the first Bohr orbit for a. Now in the spectroscopically accessible range of frequencies the effective range of electron velocities for temperatures between 3000° A. and 8000° A. is about from 0.7×10^8 to 1.5×10^8 cm./sec. In this velocity range Nedelsky's approximation for slow electrons is inapplicable, since his quantity y, which is supposed to be much less than one, varies here from about 0.3 to 0.7. In using his exact formula (24) it was found, however, that the first two terms in the summation were sufficient:

$$\mathcal{J}_\nu = \frac{16e^2h^2n'^2}{3c^3\pi^3m^2}\frac{y'}{y}[\,|\delta_0'|^2\,|\delta_1|^2\,|I_{1,0}|^2 + |\delta_1'|^2\,|\delta_0|^2\,|I_{0,1}|^2\,], \tag{4.2}$$

where $\mathcal{J}_\nu$ denotes the radiation emitted by an electron during a single encounter with an atom. This is in accordance with his observation on p. 653 that the contribution from the terms with $1 > y$ is negligible. Further, it was found semi-empirically that the individual factors appearing in the brackets of (4.2) could be represented for the effective velocity range by linear or quadratic expressions in v^2 and v'^2, v and v' being the initial and final velocities respectively. Thus the quantity in brackets was conveniently expressed by a polynomial of the sixth order in v'^2 with coefficients depending on v^2. Since $n'^2 \simeq 0.506(1 + 0.105v'^2 + 0.0111v'^4)$, and $(y'/y) = (v'/v)$, we have

$$\mathcal{J}_\nu = \frac{4e^2h^2}{3c^3\pi m^2}(v'/v)P(v, v'^2), \tag{4.3}$$

P standing for $(\pi^2/4)n'^2$ times the brackets in (4.2).

We wish to compute the absorption coefficient per hydrogen atom in a gas containing N_e electrons per unit volume. We do this, in the usual way,‡ by first computing the emission. Since an electron of velocity v

* Russell, *Ap. J.*, 78, 239, 1927.
† Nedelsky, *Phys. Rev.*, 42, 641, 1932.
‡ Milne, *M.N.*, 85, 750, 1925.

emits light of all frequencies up to $\nu = mv^2/2h$, we have for the intensity of radiation, E_ν,

$$E_\nu d\nu = 4\pi N_e \int_{v_0}^{\infty} (m/2\pi kT)^{3/2} e^{-mv^2/2kT} v^3 J_\nu(v^2, v'^2) dv d\nu =$$

$$= 4\pi N_e (4e^2h^2/3c^3\pi m^2)(m/2\pi kT)^{3/2} \int_{v_0}^{\infty} e^{-mv^2/2kT} v^2 v' P(v^2, v'^2) dv d\nu, \tag{4.4}$$

where $v_0{}^2 \equiv \dfrac{2h\nu}{m} = v^2 - v'^2$. In thermodynamic equilibrium,

$$E_\nu d\nu = 4\pi I_\nu a_\nu d\nu = 4\pi(2h\nu^3/c^2)e^{-h\nu/kT} a_\nu d\nu, \qquad P_e = N_e kT.$$

Hence

$$a_\nu = \frac{2e^2 h P_e (m/2\pi kT)^{3/2}}{3c\pi m^2 \nu^3 kT} e^{h\nu/kT} \int_{v_0}^{\infty} e^{-mv^2/2kT} v^2 v' P(v^2, v'^2) dv d\nu.$$

Let

$$y \equiv v'^2/v_0{}^2, \qquad z \equiv (mv_0{}^2/2kT),$$

then

$$e^{h\nu/kT} \int_{v_0}^{\infty} e^{-mv^2/2kT} v^2 v' P(v^2, v'^2) dv = (v_0{}^4/2) \int_0^{\infty} e^{-zy}(y+y^2)^{\frac{1}{2}} P(v_0{}^2, v_0{}^2 y) dy \equiv$$

$$\equiv (v_0{}^4/2) A(z, v_0{}^2), \tag{4.5}$$

where we have first transformed $P(v^2, v'^2)$ into $P(v_0{}^2, v'^2)$ through the relation $v^2 = v_0{}^2 + v'^2$, and then written $v_0{}^2 y$ for v'^2. Integrals of the form $\int_0^{\infty} e^{-zy} y^n (y+y^2)^{\frac{1}{2}} dy \equiv \psi_n(z)$ can be expressed in terms of Hankel functions of the first kind with imaginary arguments, for which extensive tables exist.* We have

$$\psi_0(x) = -(\pi/4x) e^{x/2} H_1{}^{(1)}\left(\frac{ix}{2}\right),$$

$$\psi_1(x) = -\frac{d\psi_0(x)}{dx} = \psi_0(x)[(2/x) - (\tfrac{1}{2})] + (i\pi/8x) e^{x/2} H_0{}^{(1)}(ix/2). \tag{4.6}$$

The higher-order functions may be obtained from the identity

$$\psi_n(x) = (-)^n \frac{d^n \psi_0(x)}{dx^n},$$

and the relations

$$\frac{d^n \psi_0(x)}{dx^n} = [1 - (n+1)/x] \frac{d^{n-1}\psi_0(x)}{dx^{n-1}} + [(n - \tfrac{1}{2})/x] \frac{d^{n-2}\psi_0(x)}{dx^{n-2}}.$$

Thus we computed the ψ's for a set of values of z corresponding to the temperature and velocity ranges in Table V. As a check on the calculations, we developed asymptotic series for the ψ's for large values of z and found that the results agree. Further, the contribution from the high-order terms

* Jahnke and Emde, *Funktionen Tafeln* ; G. N. Watson, *Bessel Functions*.

in P becomes increasingly small, which substantiates the validity of the original approximation for $\mathcal{J}_\nu$. Values of a_ν computed from

$$a_\nu = \frac{2e^2hP_e(m/2kT)^{1/2}z^2A(z, v_0^2)}{3c\pi^{5/2}m^3\nu^3} = aP_e \tag{4.7}$$

are given in Table V. This should be multiplied, also, by the factor $(1 - e^{-h\nu/kT})$ to allow for the negative absorptions.

TABLE V

Values of a_ν in units of $10^{-28}P_e$. [Correct to within 5 per cent.]

T \ λ	8900 A.	6810 A.	5390 A.	4360 A.	3030 A.
3000°	6·82	4·12	2·70	1·88	1·04
4000	5·81	3·41	2·20	1·50	0·81
6000	4·83	2·71	1·69	1·12	0·55
8000	4·33	2·36	1·44	0·93	0·47

Russell has derived the following formula for τ_2, the optical depth arising from the interaction of electrons with neutral atoms:—

$$\tau_2 = \frac{aP_0^2}{2\bar{m}g\epsilon}(1 - e^{-h\nu/kT})\left(\frac{4}{3}\frac{P_0}{K} + 1\right), \tag{4.8}$$

where the neutral atoms are assumed to be in great abundance. ϵ is the relative abundance, by volume, of the element giving rise to the electrons. From equation (2.29) we have

$$\frac{\tau_2}{\tau_1} \simeq \frac{a\nu^3T^{3/2}}{C_0\epsilon g'}e^{-h\nu/kT}\left(\frac{4}{3}\frac{P_0}{K} + 1\right). \tag{4.9}$$

For metallic atoms in the sun $\epsilon \simeq 10^{-3}$ and $P_0/K << 1$. Hence, for $\lambda = 5000°$A.,

$$\frac{\tau_2}{\tau_1} = 6{\cdot}5 \times 10^{-2}. \tag{4.10}$$

For a star where $T \sim 3000°$ a much greater percentage of atoms will be neutral. *Na* and *K* are the only atoms of any abundance that approach complete ionization, and for these atoms $\epsilon \sim 10^{-5}$. The change in the other factors, however, appears to offset the effect, for we find, with $P_0/K \sim 10$,

$$\frac{\tau_2}{\tau_1} \sim 0{\cdot}25. \tag{4.11}$$

The exact values are very sensitive to the factors assumed for the abundances. The marked spectral differences between dwarf and giant M stars may possibly be due to the opacity of neutral atoms operating in the former. If we may rely on the values of ϵ, then the effect of neutral hydrogen on the opacity is not important for stars of spectral types G5 and earlier.

We should, however, take such opacity into consideration in the later types, especially for members of the dwarf sequence.

5. *Discussion of Woolley's Work on Absorption Coefficients*

As this paper nears completion there comes to hand a discussion, by Woolley,* of the general problem of continuous absorption coefficients. He introduces an oscillator strength, f_c, for the continuum :

$$f_c = \int df = \int_{RZ^2/n^2}^{\infty} \frac{df}{d\nu} d\nu = \frac{8}{3\sqrt{3}\pi n} \bar{g} \tag{5.1}$$

by equation (2.3). A mean value of g is employed. Woolley substitutes f_c in place of part of the constant of $\dfrac{df}{d\nu}$.

The resulting absorption coefficient, for hydrogen, becomes

$$a_\nu = \frac{2\pi\epsilon^2}{mc\nu^3}\left(\frac{R}{n^2}\right)^2 f_c. \tag{5.2}$$

In applying this formula to complex spectra Woolley replaces $(R/n^2)^2$ by $\nu_0{}^2$, the particular series limit under consideration. From the equation,

$$\sum_n f_{nn'} + f_c = 1, \tag{5.3}$$

and from Korff's † experiments, which indicate a value of f very close to unity for the D lines, Woolley concludes that f_c must be very small. The Kramers formula applied to continuous absorption beyond the limit of the principal series of sodium will give, therefore, too large a result, unless, as Woolley suggests, a large g value is adopted. We note that equations (5.1) and (5.3) would rule out even this hope of escape. We cannot agree, however, with the sweeping generalization that Woolley draws from this single example : that Kramers' formula gives too high an opacity for optical spectra of all complex atoms.

The absorption beyond the principal series is of no particular significance in the astronomical opacity problem ; it comes into account only where rates of ionization are important. The transfer of radiation through the atmospheric layers of the star is mainly governed by continuous absorption by atoms in atomic states only two or three volts below the ionization level. These states are practically hydrogenic, and the quantum defects will be practically zero for the states of greatest weight (highest value of azimuthal quantum number). Korff's result cannot be used as an argument against the validity of the formulæ we give in the preceding sections.

Woolley attempts to evaluate f_c for the metals on the basis of a most surprising assumption. Milne ‡ has found that the spectral distribution of

* Woolley, *M.N.*, **95**, 101, 1934.
† Korff, *Ap. J.*, **76**, 124, 1932.
‡ Milne, *Phil. Trans.*, A, **223**, 213, 1922.

solar radiation may be predicted approximately on the assumption that the absorption coefficient is independent of ν. To attain this approximate constancy Woolley equates the absorption coefficients at wave-lengths 4000 A. and 6000 A. He further assumes that the opacity at the respective wave-lengths arises from absorption by metallic atoms and by hydrogen atoms in state 3. He finds, from this condition, that $f_c = 2{\cdot}37 \times 10^{-6}$. It is obvious that this low value, whether it be correct or not, arises directly from the high excitation potential of the 3-quantum state of hydrogen. So few hydrogen atoms exist in this state that the absorbing power of the metallic atoms, despite their relative scarcity, must be cut far below the value to be expected from Kramers' theory. The value is made even lower through omission of series limits from higher quantum states.

We have already presented arguments against the small values of f_c It should be noted, however, that Woolley's assumptions lead to extremely low opacities and would require electron pressures at the base of the reversing layer at least 10^3 times greater than those observed. While we do not wish to place any special emphasis on the fact, it is interesting to note that our formula, (2.26), approximately fulfills Woolley's requirement of equal opacities at wave-lengths 4000 A. and 6000 A.

In his paper Woolley also discusses the question of continuity of the absorption coefficient. Near the series limit the overlapping lines give an absorption coefficient that goes over, without discontinuity, into the spectral region to the violet of the limit. Some of the values given in Woolley's Table III do not exhibit this continuity. Woolley does not question the existence of the continuity, but he lays the disagreement to possible defects in the wave-mechanical calculations. We find that Sugiura's formulæ for the individual sub-series of hydrogen, as well as our general formulæ for the integrated transitions, require *exact* continuity in all cases. Any disagreement must be attributed to numerical errors. Woolley's formula for testing continuity holds only when the ν^{-3} law is strictly obeyed. In the general case, the condition for continuity requires

$$\lim_{\kappa \to \infty} \left(\frac{df}{d\nu}\right) = \lim_{n \to \infty} \left(\frac{n^3}{2RZ^2} f_{nn'}\right). \tag{5.4}$$

Woolley's f_c should not be confused with our f values for the continuum. The reader should be warned that it will not generally be legitimate to treat the continuum as if it were a single discrete state with an f value equal to f_c. The appropriate weighting factors, ϖ_e, must be introduced to take account of the velocity distribution of the electrons and the effect of electron pressure.

Appendix

Asymptotic Expansions for

$$\Delta \equiv F(-n+1, -n', 1, x)^2 - F(-n'+1, -n, 1, x)^2, \; x = -4nn'/(n-n')^2$$

Asymptotic series for Δ for large values of n and n' can be obtained by the method of steepest descent. For this purpose it is necessary to represent the hypergeometric functions by integrals in the complex plane. Starting with the second function appearing in Δ, we first write $F(-n'+1, -n, 1, x) = (1-x)^{n+n'}F(n', n+1, 1, x) \equiv (1-x)^{n+n'}P$, where P satisfies the differential equation

$$x(1-x)\ddot{P}+[1-x(2+n+n')]\dot{P}-n'(n+1)P=0. \qquad \text{A (1)}$$

If we now assume that

$$P=(1/2\pi i)\int[(1-z)^n/z^{n'}(1-x-z)^{n+1}]dz, \qquad \text{A (2)}$$

and substitute in (1), we shall obtain the expression

$$\int\frac{d}{dz}\left[\frac{(1-z)^{1+n}}{z^{n'-1}(1-x-z)^{2+n}}\right]dz=0.$$

Hence the above form for P will be correct provided we choose the path of integration such that the difference in the values of the quantity

$$[(1-z)^{1+n}/z^{n'-1}(1-x-z)^{2+n}]$$

at the beginning and the end of the path is zero. Similarly one finds that

$$F(-n+1, -n', 1, x)=(1-x)^{n+n'}(1/2\pi i)\int[(1-z)^{n-1}/(1-x-z)^n z^{n'+1}]dz \equiv$$
$$\equiv (1-x)^{n+n'}Q, \qquad \text{A (3)}$$

provided that the increment in $[(1-z)^n/z^{n'}(1-x-z)^{n+1}]$ over the path is zero.

We shall take up first the case when n and n' are positive integers. The F's in Δ then reduce to Jacobi polynomials,* for which we have

$$F(-n', -n+1, 1, x)=\frac{(1-x)^{n+n'}}{(n-1)!}\frac{d^{n-1}}{dx^{n-1}}[x^{n-1}(1-x)^{-n'-1}]=$$
$$=\frac{(1-x)^{n+n'}}{n'!}\frac{d^{n'}}{dx^{n'}}[x^{n'}(1-x)^{-n}]=\frac{(1-x)^{n+n'}}{2\pi i}\oint_{t=x}\frac{t^{n'}(1-t)^{-n}dt}{(t-x)^{n'+1}}=$$
$$=\frac{(1-x)^{n+n'}}{2\pi i}\oint_{z=1-x}\frac{(1-z)^{n'}dz}{z^n(1-x-z)^{n'+1}}, \qquad \text{A (4)}$$

where the path in the z plane is a negative (clockwise) loop enclosing $z=(1-x)$ but not z and 1. Since, with n and $n' \geqslant 1$, the integral in A (4) vanishes on a circle of infinite radius, the above loop can be deformed into a positive loop enclosing 1 and 0, but not $(1-x)$. This path, which evidently satisfies the aforementioned conditions, we shall adopt for P and Q.

Following the usual procedure of the method of steepest descent, we write:

$$Q\equiv\left(\frac{1}{2\pi i}\right)\oint\frac{(1-z)^{n-1}dz}{z^{n'+1}(1-x-z)^n}=\left(\frac{1}{2\pi i}\right)\oint\frac{e^{-n'\left[lnz-\frac{1}{\alpha}ln\left(\frac{1-z}{1-x-z}\right)\right]}dz}{z(1-z)},$$

$$P\equiv\left(\frac{1}{2\pi i}\right)\oint\frac{(1-z)^{n}dz}{z^{n'}(1-x-z)^{n+1}}=\left(\frac{1}{2\pi i}\right)\oint\frac{e^{-n'\left[lnz-\frac{1}{\alpha}ln\left(\frac{1-z}{1-x-z}\right)\right]}dz}{(1-x-z)}, \qquad \text{A (5)}$$

* Courant and Hilbert, *Meth. d. Math. Physik*, p. 74, 1924.

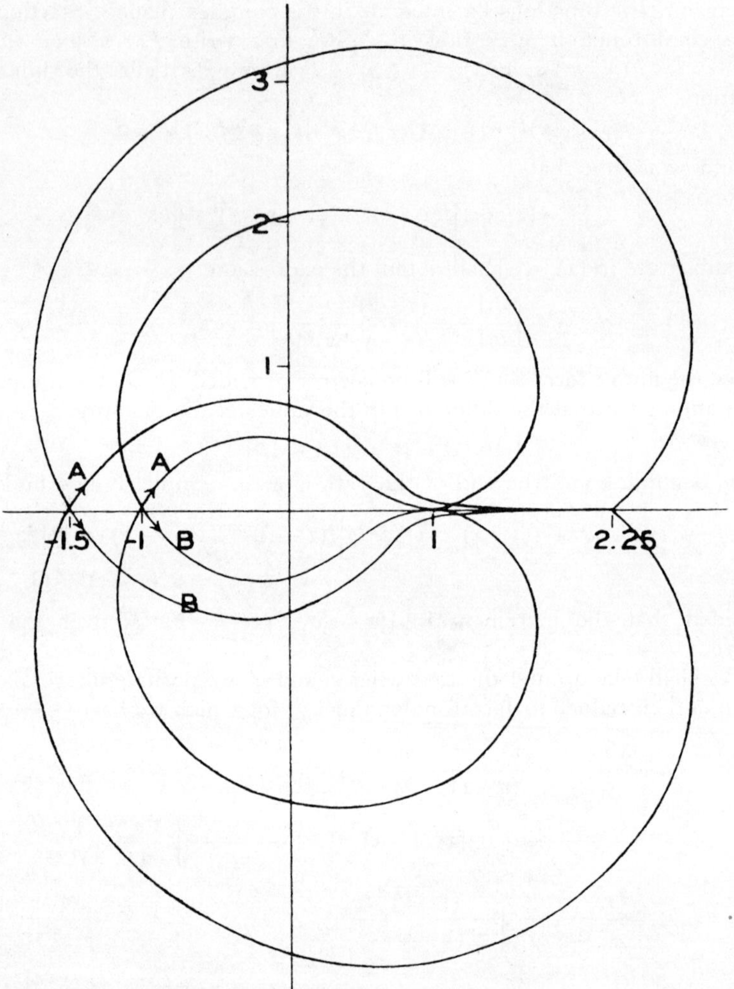

FIGURE 2

Paths of steepest descent and ascent in case n, n′ are real and $\alpha = 0$, 1/5.

with $\alpha \equiv (n'/n)$. The derivative with respect to z of the exponents in A (5) vanishes at $z = -(1+\alpha)/(1-\alpha) \equiv -\beta$. At this point

$$f(z) \equiv \ln z - (1/\alpha)[\ln(1-z) - \ln(1-x-z)] = i\pi + (1+1/\alpha)\ln\beta.$$

The path of steepest descent, which starts at $z = -\beta$, is such that along it the imaginary part of $f(z)$ equals $i\pi$. If we write $z = \rho e^{i\theta}$, this requires that

$$\theta - (1/\alpha)\tan^{-1}\{x\rho\sin\theta/[(1-x)(1-2\rho\cos\theta) - x\rho\cos\theta + \rho^2]\} = \pi, \qquad \text{A (6)}$$

or

$$\rho = \tfrac{1}{2}[(2-x)\cos\theta - x\sin\theta/\tan(\alpha\pi - \alpha\theta)] \pm$$
$$\pm \{(\tfrac{1}{4})[(2-x)\cos\theta - x\sin\theta/\tan(\alpha\pi-\alpha\theta)]^2 - 1 + x\}^{\frac{1}{2}}. \qquad \text{A (7)}$$

A (7) determines two loops, symmetrical around the real axis, both starting at $-\beta$ and ending at $(1-x)$. These are shown in fig. 2 for the cases $\alpha = 0$, $\alpha = (1/5)$. To find the branches along which $f(z)$ increases we define $\tau(z)$, real along the path, by

$$\tau(z) \equiv lnz - (1/\alpha)[ln(1-z) - ln(1-x-z)] - i\pi - (1+1/\alpha)ln\beta, \qquad \text{A (8)}$$

and develop τ in a power series of $u \equiv (z+\beta)$. One finds that

$$-\tau/(1-\alpha)^2 \equiv y^3 = (1/12\beta^3)u^3 + (1/8\beta^4)u^4 + (1/80\beta^5)(11+\alpha^2)u^5 +$$
$$+ (1/96\beta^6)(13+3\alpha^2)u^6 + \dots \qquad \text{A (9)}$$

When τ is positive,

$$y = |\,\tau/(1-\alpha^2)\,|^{1/3}e^{i\pi},\ e^{i\pi/3},\ e^{i5\pi/3}.$$

Now when $|\,u\,|$ is small $y \simeq (1/12\beta^3)^{1/3}u$, hence τ is positive along the direction indicated by arrows in fig. 2, where, in fact, it increases in a monotone fashion from zero to ∞. Clearly, a path which encloses 1 and 0 is represented by $B-A$. To avoid the singularity at $(1-x)$ one may deform the path in its vicinity to a vertical line, such a modification being immaterial to the result on account of the exponential factor in A (5). With τ as an independent variable, we now have

$$Q = (1/2\pi i)e^{-i\pi n'}\beta^{-(n'+n)}\int_0^\infty e^{-n'\tau}\Big[(1-x-z)/(z+\beta)^2\Big]_A^B d\tau, \qquad \text{A (10)}$$

$$P = (1/2\pi i)e^{-i\pi n'}\beta^{-(n'+n)}\int_0^\infty e^{-n'\tau}\Big[(z-z^2)/(z+\beta)^2\Big]_A^B d\tau. \qquad \text{A (11)}$$

The quantities in brackets in A (10) and A (11) can be obtained as functions of τ, or y, by first expanding in a power series in u and then expressing u in terms of y by inverting A (9). One thus obtains

$$\Big[(1-x-z)/(z+\beta)^2\Big]_A^B = \Big\{\Big(1+\frac{1}{\beta}\Big)12^{-2/3}y^{-2} + 12^{-1/3}y^{-1} + \Big[\Big(1+\frac{1}{\beta}\Big)\Big(\frac{7}{20}+\frac{\alpha^2}{10}\Big) - \frac{1}{2\beta}\Big] +$$

$$+\frac{12^{1/3}(1+\alpha^2)}{20}y + \Big(1+\frac{1}{\beta}\Big)12^{2/3}\Big(\frac{3}{560}+\frac{121}{700}\alpha^2+\frac{3\alpha^4}{560}\Big)y^2 + \dots\ \Bigg|_{y=\frac{e^{i\pi/3}\tau^{1/3}}{(1-\alpha^2)^{1/3}}}^{y=\frac{e^{i5\pi/3}\tau^{1/3}}{(1-\alpha)^2}} =$$

$$= i\sqrt{3}\Big\{\Big(1+\frac{1}{\beta}\Big)12^{-2/3}(1-\alpha^2)^{2/3}\tau^{-2/3} + 12^{-1/3}(1-\alpha^2)^{1/3}\tau^{-1/3} - \frac{12^{1/3}(1+\alpha^2)\tau^{1/3}}{20(1-\alpha^2)^{1/3}} -$$

$$-\Big(1+\frac{1}{\beta}\Big)12^{2/3}\Big(\frac{3}{560}+\frac{121}{700}\alpha^2+\frac{3\alpha^4}{560}\Big)\frac{\tau^{2/3}}{(1-\alpha)^{2/3}} + c\tau^{4/3}\Big\}. \qquad \text{A (12)}$$

Similarly one finds that

$$\left[\frac{z(1-z)}{(z+\beta)^2}\right]_A^B = i\sqrt{3}\Bigg\{-\left(1+\frac{1}{\beta}\right)12^{-2/3}(1-\alpha^2)^{2/3}\tau^{-2/3}+12^{-1/3}(1-\alpha^2)^{1/3}\tau^{-1/3}-\frac{12^{1/3}(1+\alpha^2)\tau^{1/3}}{20(1-\alpha^2)^{1/3}}+$$

$$+\left(1+\frac{1}{\beta}\right)12^{2/3}\left(\frac{3}{560}+\frac{121}{700}\alpha^2+\frac{3\alpha^4}{560}\right)\frac{\tau^{2/3}}{(1-\alpha^2)^{3/2}}+c\tau^{4/3}+\dots\Bigg\}. \quad \text{A (13)}$$

Upon carrying out the integration in A (10) and A (11) we now obtain

$$F(-n', -n+1, 1, x) = \frac{e^{-i\pi n'}\beta^{(n+n')}\sqrt{3}}{2\pi \, . \, 12^{2/3}}\Bigg\{\frac{\left(1+\frac{1}{\beta}\right)(1-\alpha^2)^{2/3}\Gamma(1/3)}{(n')^{1/3}}+\frac{12^{1/3}(1-\alpha^2)^{1/3}\Gamma(2/3)}{(n')^{2/3}}-$$

$$-\frac{3(1+\alpha^2)\Gamma(4/3)}{5(1-\alpha^2)^{1/3}(n')^{4/3}}-\frac{(1+1/\beta)12^{1/3}\Gamma(5/3)9}{(1-\alpha^2)^{2/3}(n')^{5/3}140}\left(1+\frac{484\alpha^2}{15}+\alpha^4\right)+\frac{c}{(n')^{7/3}}\Bigg\}, \quad \text{A (14)}$$

$$F(-n'+1, -n, 1, x) = -\frac{e^{-i\pi n'}\beta^{(n+n')}\sqrt{3}}{2\pi \, . \, 12^{2/3}}\Bigg\{\frac{(1+1/\beta)(1-\alpha^2)^{2/3}\Gamma(1/3)}{(n')^{1/3}}-\frac{12^{1/3}(1-\alpha^2)^{1/3}\Gamma(2/3)}{(n')^{2/3}}+$$

$$+\frac{3(1+\alpha^2)\Gamma(4/3)}{5(1-\alpha^2)^{1/3}(n')^{4/3}}-\frac{(1+1/\beta)12^{1/3}\Gamma(5/3)9}{(1-\alpha^2)^{2/3}(n')^{5/3}140}\left(1+\frac{484\alpha^2}{15}+\alpha^4\right)-\frac{c}{(n')^{7/3}}\Bigg\}. \quad \text{A (15)}$$

Finally we have

$$\Delta = F^2(-n', -n+1, 1, x)^2 - F^2(-n'+1, -n, 1, x) = \frac{\left(\frac{n+n'}{n-n'}\right)^{2n+2n'}(n-n')}{\sqrt{3}\pi n n'} \, .$$

$$\cdot\left[1-\frac{0{\cdot}1728(1+\alpha^2)}{(1-\alpha^2)^{2/3}(n')^{2/3}}-\frac{0{\cdot}0496\left(1+\frac{484\alpha^2}{15}+\alpha^4\right)}{(1-\alpha^2)^{4/3}(n')^{4/3}}+\dots\right]. \quad \text{A (16)}$$

The degree of approximation of A (16) is shown in Table VI, where the last column gives the quantity in brackets. In the first case, the third term is of the order of magnitude of the second, so that a better approximation is obtained by stopping with the second term. Clearly the approximation is good when n' is large and α small.

TABLE VI

n'	n	α	Δ exact	Δ appr.	
10	20	$\frac{1}{2}$	$3{\cdot}6672 \times 10^{26}$	$3{\cdot}676 \times 10^{26}$	$1 - 0{\cdot}0564 - 0{\cdot}0286$
20	100	$\frac{1}{5}$	$1{\cdot}3087 \times 10^{40}$	$1{\cdot}3069 \times 10^{40}$	$1 - 0{\cdot}0251 - 0{\cdot}0022$
10	∞	0	$4{\cdot}1557 \times 10^{17}$	$4{\cdot}1548 \times 10^{17}$	$1 - 0{\cdot}03723 - 0{\cdot}00230$
20	∞	0	$4{\cdot}9654 \times 10^{32}$	$4{\cdot}96710 \times 10^{32}$	$1 - 0{\cdot}02346 - 0{\cdot}00091$

There now remains the discussion of the case of free-free transitions, where $n = ik$, $n' = il$, and $l > k$. Here too x is a negative real number. The integrands in A (5), which now have branch points at 0, 1 and $1-x$, can be made single-valued by cutting the z plane along the real axis from 1 to $(1-x)$ and from 0 to $-\infty$ and assigning values for $arg(z)$ as shown in fig. 3. In this plane the expressions $[(1-z)^{1+n}/z^{n'-1}(1-x-z)^{2+n}]$ and $[(1-z)^{n}/z^{n'}(1-x-z)^{n+1}]$ return to their original values after passing a loop around the cut from 1 to $(1-x)$. The integrals A (2) and A (3), when taken along this path, are then solutions of the differential equations satisfied by P and Q respectively, and since, when we pass the loop in the negative direction, they assume the value $+1$ for $x=0$, they actually represent P and Q. Having chosen the appropriate path, we find that P and Q are still given by A (5), where now $n'=ik$ and $\alpha=(k/l)$. If we write

$$\psi(z)\equiv ilnz-i(1/\alpha)[ln(1-z)-ln(1-x-z)],$$

we find again that $\psi(z)$ vanishes at $z_0=-\beta$ and that $\psi(z_0)=-\pi+i(1+1/\alpha)ln\beta$. The path of steepest descent is then chosen such that along it $Im\,.\,\psi(z)=(1+1/\alpha)ln\beta$. With $z=\rho e^{i\theta}$ this implies that

$$ln\rho+\left(\frac{1}{2\alpha}\right)ln\{[(1-x)^2-2\rho(1-x)\cos\theta+\rho^2]/(1-2\rho\cos\theta+\rho^2)\}=(1+1/\alpha)ln\beta,$$

or that

$$\cos\theta=\frac{(1-x)^2+\rho^2-u(1+\rho^2)}{2\rho(1-x-u)},\qquad u\equiv\beta^{2+2\alpha}/\rho^{2\alpha}. \qquad \text{A (17)}$$

In the limit of $\alpha\to 0$ the path is given by

$$\cos\theta=[(1+\rho^2)(ln\rho-2)+4]/(2\rho ln\rho). \qquad \text{A (18)}$$

These consist of two loops, symmetrical about the axis of reals, starting at $z_0=-\beta$ (either on the upper or on the lower bank) and ending on the positive axis of reals at distances less than 1 and greater than $(1-x)$ respectively, as is shown for the case $\alpha=0$ in fig. 3. We now define $\tau(z)$, real along the path, by

$$\tau(z)=ilnz-i(1/\alpha)[ln(1-z)-ln(l-x-z)]-i(1+1/\alpha)ln\beta+\pi, \qquad \text{A (19)}$$

and develop $\tau(z)$ into a power series in $u=(z+\beta)$. Clearly, we have

$$y^3=i\tau/(1-\alpha^2),\quad y=\left|\frac{\tau}{1-\alpha^2}\right|^{1/3}e^{i\pi/6},\ e^{i5\pi/6},\ e^{i3\pi/2},\ \text{with } y^3 \text{ given by A (9).}$$

Hence, starting with $z=\beta e^{i\pi}$, τ will be positive along A and B, along each of which it in fact increases from 0 to 2π, its final value at $z=\beta e^{-i\pi}$. The appropriate path for our integrals is $B-A$, since this forms a closed negative loop around the cut $1\to(1-x)$, and does not cross the cut to the left of the origin. We now have instead of A (10) and A (11),

$$Q=-\left(\frac{1}{2\pi}\right)e^{\pi\kappa}\beta^{-i(\kappa+l)}\int_0^{2\pi}e^{-\kappa\tau}\Big[(1-x-z)/(z+\beta)^2\Big]_A^B d\tau, \qquad \text{A (20)}$$

$$P=-\left(\frac{1}{2\pi}\right)e^{\pi\kappa}\beta^{-i(\kappa+l)}\int_0^{2\pi}e^{-\kappa\tau}\Big[(z-z^2)/(z+\beta)^2\Big]_A^B d\tau, \qquad \text{A (21)}$$

where at B $y=e^{i5\pi/6}\left|\frac{\tau}{(1-\alpha^2)}\right|^{1/3}$ and at A $y=e^{i\pi/6}\left|\frac{\tau}{(1-\alpha^2)}\right|^{1/3}$, but the

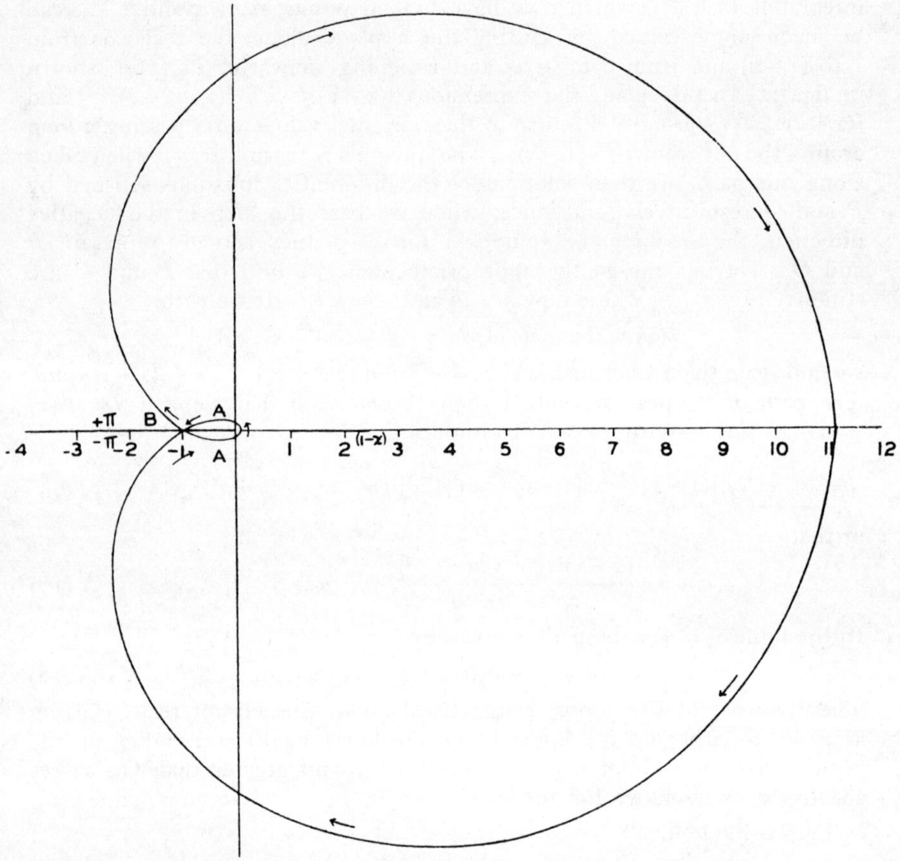

FIGURE 3

Path of steepest descent in case $n = i\kappa$, $n' = il$, and $a = 0$. The positive sense of loop A is clockwise. When $a \neq 0$, $(1 - x)$ will be to the right of 1, as indicated in the figure.

expressions in brackets in A (20) and A (21) are the *same* functions of y as before. In the subsequent integrations we neglect terms of the order of $e^{-2\pi k}$ by making the upper limits in the integral ∞, since for large values of κ they are of lower order of magnitude than any power of κ. We finally get

$$\Delta = \frac{ie^{2\pi\kappa}(l-\kappa)\left(\frac{l+\kappa}{l-\kappa}\right)^{2i(\kappa+l)}}{lk\pi\sqrt{3}} R,$$

$$R = \left[1 + \frac{0{\cdot}1728(1+\kappa^2/l^2)}{(1-\kappa^2/l^2)^{2/3}\kappa^{2/3}} - \frac{0{\cdot}0496\left(1 + \frac{484\kappa^2}{15l^2} + \kappa^4/l^4\right)}{(1-\kappa^2/l^2)^{4/3}\kappa^{4/3}}\right], \qquad \text{A (22)}$$

$$|\Delta| = \frac{e^{2\pi\kappa}(l-\kappa)R}{\sqrt{3}\pi l\kappa}. \qquad \text{A (23)}$$

From A (22) we now get an expression for Δ for the case of a free-discrete transition, by substituting n' for il. The result is

$$|\Delta| = \frac{e^{4\kappa \tan^{-1}(n'/\kappa)}(\kappa^2+n'^2)^{1/2}}{\sqrt{3}\pi\kappa n'}\left[1 - \frac{0\cdot1728(1-n'^2/\kappa^2)}{(1+n'^2/\kappa^2)^{2/3}(n')^{2/3}} - \frac{0\cdot0496\left(1-\frac{484}{15}\frac{n'^2}{\kappa^2}+\frac{n'^4}{\kappa^4}\right)}{(1+n'^2/\kappa^2)^{4/3}(n')^{4/3}}\right], \qquad \text{A (24)}$$

a formula which is applicable for $\kappa \gtreqless n'$, provided n' is large. When $\kappa < n'$, A (24) can be written in the convenient form

$$|\Delta| = \frac{e^{4\kappa \tan^{-1}(n'/\kappa)}(\kappa^2+n'^2)^{1/2}}{\sqrt{3}\pi\kappa n'}\left[1 + \frac{0\cdot1728(1-\kappa^2/n'^2)}{(1+\kappa^2/n'^2)^{2/3}\kappa^{2/3}} - \frac{0\cdot0496\left(1-\frac{484\kappa^2}{n'^2}+\frac{\kappa^4}{n'^4}\right)}{(1+\kappa^2/n'^2)^{4/3}\kappa^{4/3}}\right]. \qquad \text{A (25)}$$

It should be noticed that A (24) and A (25) can also be derived from A (16) by putting $n = i\kappa$; * also that A (16) and A (23) are inapplicable when $n \simeq n'$ and $\kappa \simeq l$, while this is not true for A (24) and A (25) when $\kappa \simeq n'$. This difference is to be expected on physical grounds, since the former two cases represent transitions between states of nearly equal energies, while the other does not.

The case $\kappa \simeq l$ (κ and l finite) deserves special consideration since it corresponds to the long-wave limit in continuous-continuous transitions. The usual procedure used in the method of steepest descent for this case fails because of the infinity in x: with $c \equiv l - \kappa$, $x = -4\kappa l/c^2$. On account of this very fact, however, we can use here the asymptotic expansions of the hypergeometric functions in powers of (l/x). We have †

$$\Delta = F(-n+1, -n', 1, x)^2 - F(-n'+1, -n, 1, x)^2 =$$

$$= \left\{\frac{\Gamma(-1+n-n')(4\kappa l/c^2)^{-1+n}}{\Gamma(n)\Gamma(-n')} F(1-n, 1-n, 2-n+n', 1/x) + \right.$$

$$\left. + \frac{\Gamma(1-n+n')(4\kappa l/c^2)^{n'}}{\Gamma(1-n)\Gamma(1+n')} F(-n', -n', -n'+n, 1/x)\right\}^2 -$$

* One can, in fact, obtain A (22) from A (16) by retaining in the latter the factor $e^{-i2\pi n'}$ and writing il for n and $i\kappa$ for n'. This purely formal substitution is merely suggested by the fact that $|n| > |n'|$ and $|l| > |\kappa|$, and has no physical significance. Further, one has to write $(n')^{2/3} = -(\kappa^{2/3})$ in the expression in the brackets. With these substitutions one gets the form A (22) for Δ, where we should keep in mind that with the present significance of n and n' the Δ in A (16) is the negative of the Δ in A (22).

† Whittaker and Watson, *Modern Analysis*, p. 289, 1920. This formula should read:

$$\frac{\Gamma(a)\Gamma(b)}{\Gamma(c)}F(a, b, c, z) = \frac{\Gamma(a)\Gamma(b-a)}{\Gamma(c-a)}(-z)^{-a} . F(a, 1-c+a, 1-b+a, 1/z) +$$

$$+ \frac{\Gamma(b)\Gamma(a-b)}{\Gamma(c-b)}(-z)^{-b}F(b, 1-c+b, 1-a+b, 1/z).$$

$$-\left\{\frac{\Gamma(-1+n'-n)(4\kappa l/c^2)^{-1+n'}}{\Gamma(n')\Gamma(-n)}F(1-n',1-n',2-n'+n,1/x)+\right.$$
$$\left.+\frac{\Gamma(1-n'+n)(4\kappa l/c^2)^{n}}{\Gamma(1-n')\Gamma(1+n)}F(-n,-n,-n+n',1/x)\right\}^2\simeq$$
$$\simeq\left[\frac{\Gamma(1+ic)(4\kappa l/c^2)^{n'}}{\Gamma(1-n)\Gamma(1+n')}\right]^2-\left[\frac{\Gamma(1-ic)(4\kappa l/c^2)^{n}}{\Gamma(1-n')\Gamma(1+n)}\right]^2.$$

Further, since

$$\Gamma(1+ic)\simeq 1,\qquad \Gamma(1-n)\Gamma(1+n')\simeq\kappa l\Gamma(-i\kappa)\Gamma(i\kappa)\simeq\frac{2\pi\kappa}{e^{\pi\kappa}(1-e^{-2\pi\kappa})},$$

we get

$$\Delta=\frac{(4\kappa l/c^2)^{2n}}{\kappa^4\Gamma^2(-i\kappa)\Gamma^2(i\kappa)}[e^{i2cln(4\kappa l/c^2)}-1]=icln(2\kappa/c)(4\kappa l/c^2)^{2i\kappa}e^{2\pi\kappa}(1-e^{-2\pi\kappa})^2/(\pi^2\kappa),$$

$$|\Delta|=\frac{cln(2\kappa/c)e^{2\pi\kappa}(1-e^{-2\pi\kappa})^2}{\pi^2\kappa^2}.\qquad \text{A (26)}$$

Finally we shall discuss the case where n and n' are integers and $c(\equiv n-n'\geqslant 1)$ is relatively small. The expansion A (16) is then inapplicable since α is close to unity; in fact, the expression in brackets in A (16) for (g_I) deviates here considerably from unity. We use again the hypergeometric functions of $(1/x)$, which in this case can be obtained by writing the terminating series in the reverse order and factoring out the term with the highest power of x. Thus

$$F(-n',-n+1,1,x)=\frac{(n-1)!\,x^{n'}}{n'!\,(c-1)!}F(-n',-n',c,1/x),\qquad \text{A (27)}$$

$$F(-n'+1,-n,1,x)=\frac{n!\,x^{n'-1}}{(n'-1)!\,(c+1)!}F(-n'+1,-n'+1,c+2,1/x).\qquad \text{A (28)}$$

When c is small, the F's on the right-hand sides of A (27) and A (28) converge rapidly, and one can quickly compute the individual terms appearing in Δ. In this connection we may note that

$$(1-x)^{2n'}=x^{2n'}(1-1/x)^{2n'}=x^{2n'}(1+c^2/2n\,.\,2n')^{2n'}\simeq x^{2n'}e^{\frac{c^2}{2n}}.\qquad \text{A (29)}$$

Abstract

1. A general formula is given for $f_{nn'}$, the oscillator strength of a transition between atomic states n and n'. When the upper state lies in the continuum, n is set equal to $i\kappa$, where κ is real, and the absolute value of the resulting complex expression is taken. Asymptotic expressions for various limiting cases are worked out by the method of steepest descents (see Appendix). Gaunt's limiting formula is verified and extended. The expression

$$f_{nn'}=\frac{2^6}{3\sqrt{3}\pi}\frac{1}{\varpi_{n'}}\frac{1}{\left(\frac{1}{n'^2}-\frac{1}{n^2}\right)}\left|\frac{1}{n^3}\cdot\frac{1}{n'^3}\right|g,$$

where $\varpi_{n'}$ is the weight of the lower state and g a correction factor of the order unity, holds for any atomic transition. Tables and formulæ for g are given for all important cases. Tables of $f_{nn'}$ and $A_{nn'}$ (Einstein transition probability) are given for the first fourteen series of H. Those for the Lyman, Balmer and Paschen series are particularly complete. A new formulation is given of the general theory of statistical equilibrium and the photoelectric effect.

2. Various expressions for the continuous absorption coefficient arising from free-free and bound-free transitions are derived. These are summed over all quantum states, extended to cover the case of complex atoms, and the resulting formulæ are applied to calculation of the opacity of the solar atmosphere. The electron pressure so derived is in good agreement with observation.

3. The Rosseland mean opacity coefficient is calculated and extended to include the case of a gas that is capable of being multiply ionized.

4. The opacity that arises from interaction of neutral hydrogen atoms and free electrons is evaluated and is shown to be negligible except, perhaps, in the atmospheres of very cool stars.

5. Woolley's conclusion, that the opacity arising from ionization of metallic atoms is 10^{-6} times less than that to be expected on Kramers' theory, is shown to be false.

Harvard College Observatory and
Massachusetts Institute of Technology :
April 1935.

PART 2. PHYSICAL PROCESSES IN GASEOUS NEBULAE

Reprinted from *The Astrophysical Journal,* Vol. 85, No. 4, May, 1937.

PHYSICAL PROCESSES IN GASEOUS NEBULAE

I. ABSORPTION AND EMISSION OF RADIATION

DONALD H. MENZEL

ABSTRACT

In this paper, the first of a series dealing with the physical state of gaseous nebulae, various fundamental formulae are derived. The total emission and absorption of radiation by atomic hydrogen are evaluated, together with the number of transitions to and from any quantum level, discrete or continuous. The equations are thrown into simple homogeneous form. The general equations that determine the statistical equilibrium of the assembly and the partition of atoms into various atomic states are developed. Solution of these equations is deferred until a later paper.

This paper is the first of a series in which the physical conditions of gaseous nebulae will be discussed. While a gaseous nebula is undoubtedly a fairly complicated assembly, the physical state is probably simpler than that obtaining in a stellar atmosphere. Not only are the nebular problems more tractable theoretically, but the entire body of the nebula, because of its transparency, is open to observation.

These papers will be concerned chiefly with the emission, both continuous and monochromatic, that arises from atomic processes. The directly scattered or absorbed stellar light will be discussed where it appears to have any direct bearing on the atomic emission. The source of the luminosity is taken to be from the star (or stars) imbedded in the nebula. This assumption is probably justified, despite our failure to locate the exciting star in a few cases, as for NGC 6960, the well-known Network Nebula, where perhaps the star is variable or obscured by cosmic dust.

Fundamental to any discussion of physical processes is the evaluation of the rates of emission and absorption of radiation by atoms in various quantum states. The relation of the radiation field to the transitions that may occur was laid down by Einstein[1] for discrete states and was generalized by Milne[2] to include continuous atomic states. Numerical calculations, however, require a knowledge

[1] *Phys. Zs.*, **18**, 121, 1917.

[2] *Phil. Mag.*, **47**, 209, 1924; *ibid.*, 547, 1925.

of various atomic constants, which are known explicitly only for hydrogenic atoms.[3] The general formulae are easily derived and may be reduced to a very simple form.

Let N_n be the number of atoms per cubic centimeter in quantum state n, and $A_{nn'}$ the Einstein probability coefficient for spontaneous transition from state n to n'. We have the formula

$$A_{nn'} = \frac{\tilde{\omega}_{n'}}{\tilde{\omega}_n} \frac{8\pi^2\epsilon^2\nu^2}{mc^3} f_{nn'} , \tag{1}$$

where $\tilde{\omega}_n$ and $\tilde{\omega}_{n'}$ are the statistical weights of the respective quantum states and $f_{nn'}$ is the so-called "oscillator strength," which may be written

$$f_{nn'} = \frac{2^6}{3\sqrt{3\pi}} \frac{1}{\tilde{\omega}_{n'}} \frac{1}{\left(\frac{1}{n'^2} - \frac{1}{n^2}\right)^3} \left| \frac{1}{n^3} \frac{1}{n'^3} \right| g . \tag{2}$$

The quantity g is the Kramer-Gaunt factor.

The intensity of the incident radiation may be represented by the expression

$$I_\nu = W \frac{2h\nu^3}{c^2} \frac{1}{(e^{h\nu/kT_1} - 1)} , \tag{3}$$

where W is the "dilution factor," not necessarily constant with respect to frequency, and T_1 some convenient parameter that can be interpreted as a true temperature only if the radiation is "black" or "gray."

In general, the only part of the assembly likely to approach the condition of thermodynamic equilibrium will be the free electrons, which may possess approximately a Maxwellian velocity distribution, defined by a kinetic temperature, T_ϵ. In such an ionized assembly, electron captures will occur at about the same rate as in thermodynamic equilibrium. Considerable simplification results if we refer all other processes to the electron captures as a standard. When the velocity distribution of the electrons is Maxwellian, the

[3] Menzel and Pekeris, *M.N.*, **96**, 77, 1935.

ionized portion of the assembly may actually be regarded as being in thermodynamic equilibrium at temperature T_ϵ. Accordingly, we may write N_n, the number of atoms in quantum state n, in terms of N_i and N_ϵ, the respective number of ions and electrons per unit volume, by means of the dissociation formula

$$N_n = b_n N_i N_\epsilon \frac{h^3}{(2\pi mkT_\epsilon)^{3/2}} \frac{\bar{\omega}_n}{2} e^{hRZ^2/n^2kT_\epsilon}, \tag{4}$$

where b_n is a factor indicating the degree of departure from thermodynamic equilibrium at temperature T_ϵ. For strict thermodynamic equilibrium $b_n = 1$.

The atomic absorption coefficient per atom in state n' may be integrated over the line in question on the assumption that the intensity of the incident radiation does not change appreciably over the half-width of the absorption coefficient. This width will generally be governed by the kinetic velocities of the atoms.

In the following theoretical discussion the calculated emission or number of excitations is referred to unit volume per unit time. The total radiation, absorbed in the transition $n' \to n$, including the effect of the "stimulated emissions," which must be counted as negative absorptions, is easily found to be

$$\left.\begin{aligned} E_{n'n} &= 4\pi \left(N_{n'} - \frac{\bar{\omega}_{n'}}{\bar{\omega}_n} N_n\right) W \frac{2h\nu^3}{c^2} \frac{1}{(e^{h\nu/kT_1} - 1)} \frac{\pi\epsilon^2}{mc} f_{nn'} \\ &= N_i N_\epsilon \frac{KZ^4}{T_\epsilon^{3/2}} \left[W \frac{b_{n'} e^{h\nu/kT_\epsilon} - b_n}{e^{h\nu/kT_1} - 1}\right] \frac{g}{n'^3} \frac{2hRZ^2}{n^3} e^{hRZ^2/n^2kT_\epsilon}, \end{aligned}\right\} \tag{5}$$

where

$$K = \frac{h^3}{(2\pi mk)^{3/2}} \frac{8\pi^2\epsilon^2R^2}{mc^3} \frac{2^4}{3\sqrt{3}\pi}$$

$$= 4.1413 \times 10^{-16} \times 8.0236 \times 10^9 \times 0.98014$$

$$= 3.257 \times 10^{-6} \tag{6}$$

The second factor in the foregoing expression for K is easily seen to be three times the classical damping factor for frequency R. The third factor is very close to unity. Dividing by $h\nu$, we find that the effective number of transitions per second, from state n' to state n, is equal to

$$F_{n'n} = N_i N_\epsilon \frac{KZ^4}{T_\epsilon^{3/2}} \left[W \frac{b_{n'} e^{h\nu/kT_\epsilon} - b_n}{e^{h\nu/kT_1} - 1} \right] \frac{g}{n'} \frac{2e^{hRZ^2/n^2kT_\epsilon}}{n(n^2 - n'^2)} . \tag{7}$$

The stimulated emissions from the given state are counted as *negative arrivals* in that state.

In these equations

$$\nu = RZ^2 \left(\frac{1}{n'^2} - \frac{1}{n^2} \right) , \tag{8}$$

where Z is the atomic number. In thermodynamic equilibrium the factor in brackets reduces to unity. The total radiation emitted in spontaneous transitions from state n to n' is

$$E_{nn'} = N_i N_\epsilon \frac{KZ^4}{T_\epsilon^{3/2}} b_n \frac{g}{n'^3} \frac{2hRZ^2}{n^3} e^{hRZ^2/n^2kT_\epsilon} , \tag{9}$$

and the corresponding number of downward transitions is

$$F_{nn'} = N_i N_\epsilon \frac{KZ^4}{T_\epsilon^{3/2}} b_n \frac{g}{n'} \frac{2e^{hRZ^2/n^2kT_\epsilon}}{n(n^2 - n'^2)} . \tag{10}$$

The Kramer-Gaunt factor, g, which is a function of n and n', may be computed exactly from equations given by Menzel and Pekeris,[4] or may be represented to a high degree of accuracy by the formula

$$g = 1 - \frac{0.1728 \left(1 + \frac{n'^2}{n^2} \right)}{\left(1 - \frac{n'^2}{n^2} \right)^{2/3} n'^{2/3}} = 1 - 0.1728 \left(\frac{\nu}{RZ^2} \right)^{1/3} \left[\frac{2}{n'^2} \left(\frac{RZ^2}{\nu} \right) - 1 \right] . \tag{11}$$

Menzel and Pekeris have shown that the foregoing equations may also be applied to transitions involving continuous states, by means of the simple substitution

$$n = i\kappa , \tag{12}$$

[4] *Ibid.*

where κ is a real, but not necessarily integral, quantum number. The absolute value of the resulting expression is taken, whenever the result happens to be complex. κ is defined by

$$\left.\begin{aligned} &\frac{hRZ^2}{\kappa^2} = \tfrac{1}{2}\,mv^2\,, \qquad -\frac{2hRZ^2}{\kappa^3}\,d\kappa = mvdv = hd\nu\,, \\ &\text{and} \\ &\qquad\qquad \nu = RZ^2\left(\frac{1}{n'^2}+\frac{1}{\kappa^2}\right). \end{aligned}\right\} \tag{13}$$

v is the velocity of the photoelectron. To obtain the emission per frequency interval $d\nu$ in any transition from a discrete state, n', to a state, κ, in the continuum, multiply the right-hand side of equation (5) by $d\kappa$ and substitute from (12) and (13). The resulting expression[5] is

$$E_{n'\kappa}\,d\nu = N_i N_\epsilon \frac{KZ^4}{T_\epsilon^{3/2}}\left[W\frac{b_{n'}e^{h\nu/kT_\epsilon}-b_\kappa}{(e^{h\nu/kT_1}-1)}\right]\frac{g}{n'^3}\,e^{-h(\nu-\nu_{n'})/kT_\epsilon}h\,d\nu\,, \tag{14}$$

where

$$\nu_{n'} = \frac{RZ^2}{n'^2}\,. \tag{15}$$

For a Maxwellian velocity distribution at temperature T_ϵ, $b_\kappa = 1$. When the distribution is not Maxwellian, b_κ may be defined conveniently in terms of the parameter T_ϵ, as follows, where the true distribution is $f(v)$:

$$f(v)dv = b_\kappa\, 4\pi\left(\frac{m}{2\pi kT_\epsilon}\right)^{3/2} v^2 e^{-mv^2/2kT_\epsilon}dv\,, \tag{16}$$

[5] For absolute homogeneity of the E's and F's the variable κ should be retained. ν is, however, a more convenient variable. For formal consistency we should have written $E_{n'\kappa}\,d\kappa = E_{n'\nu_{\kappa'}}\,d\nu_{\kappa'}$. The latter notation has been discarded in favor of the less cumbersome form: $E_{n'\kappa}\,d\nu$. The physical dimensions of $E_{n'\kappa}$ are thus not identical with those of $E_{n'n}$. The change should cause no confusion.

subject to the normalizing relation,

$$\int_0^{\infty} f(v)dv = 1 \; . \tag{17}$$

The total energy absorbed in the continuum beyond the n'-th quantum limit will be

$$\int_{\nu_{n'}}^{\infty} E_{n'\kappa} d\nu = N_i N_\epsilon \frac{KZ^4}{T_\epsilon^{3/2} n'^3} kT_1 \bar{W} \bar{g} \left[e^{h\nu_{n'}/kT_\epsilon} b_{n'} \sum_{a=1}^{\infty} \frac{e^{-ah\nu_{n'}/kT_1}}{a} - b_\kappa \sum_{a=1}^{\infty} \frac{1}{\frac{T_1}{T_\epsilon} + a} e^{-ah\nu_{n'}/kT_1} \right] . \tag{18}$$

For thermodynamic equilibrium, the bracketed factor reduces to unity, and

$$\int_{\nu_{n'}}^{\infty} E_{n'\kappa} d\nu \sim N_i N_\epsilon \frac{kKZ^4}{T_1^{1/2}} \frac{\bar{g}}{n'^3} \; . \tag{19}$$

The first summation may be written in closed form:

$$\sum_{a=1}^{\infty} \frac{e^{-ah\nu_{n'}/kT_1}}{a} = - ln(1 - e^{-h\nu_n/kT_1}) \; . \tag{20}$$

Outside of thermodynamic equilibrium, the condition may conceivably arise when the value of the integral turns out to be negative. The physical significance of such a result is that energy is emitted rather than absorbed. This energy must be distinguished, however, from that arising in random emissions. The process merely puts energy back into the original beam, as if the atmosphere had a negative opacity. This extreme will probably never occur in practice.

No astrophysical masers in 1937!

The total number of excitations, per second, from atoms in state n' to the continuum is

$$\left.\begin{aligned}\int_{\nu_{n'}}^{\infty} F_{n'\kappa} d\nu = \int_{\nu_{n'}}^{\infty} \frac{E_{n'\kappa}}{h\nu} d\nu = N_i N_\epsilon \frac{KZ^4}{T_\epsilon^{3/2} n'^3} \bar{W} \bar{g}\, e^{h\nu_{n'}/kT_\epsilon} \\ \times\left[b_n \sum_{a=1}^{\infty}\left[-E_i\left(-\frac{a h\nu_{n'}}{kT_1}\right)\right] - \bar{b}_\kappa \sum_{a=1}^{\infty}\left\{ -E_i\left[-\frac{h\nu_{n'}}{kT_1}\left(\frac{T_1}{T_\epsilon}+a\right)\right]\right\}\right],\end{aligned}\right\} \tag{21}$$

where

$$-E_i(-x_n) = \int_{x_n}^{\infty} \frac{e^{-x}}{x} dx , \tag{22}$$

the well-known exponential integral.

Calculating the random emissions in the same way, we find

$$E_{\kappa n'} d\nu = N_i N_\epsilon \frac{KZ^4}{T_\epsilon^{3/2}} b_\kappa \frac{g}{n'^3} e^{-h(\nu-\nu_{n'})/kT_\epsilon} h\, d\nu , \tag{23}$$

$$\int_{\nu_{n'}}^{\infty} E_{\kappa n'} d\nu = N_i N_\epsilon \frac{kKZ^4}{T_\epsilon^{1/2}} b_\kappa \frac{\bar{g}}{n'^3} , \tag{24}$$

and

$$\int_{\nu_{n'}}^{\infty} F_{\kappa n'} d\nu = N_i N_\epsilon \frac{KZ^4}{T_\epsilon^{3/2}} \bar{b}_\kappa \frac{\bar{g}}{n'^3} e^{h\nu_{n'}/kT_\epsilon}[-Ei(-h\nu_{n'}/kT_\epsilon)] . \tag{25}$$

The results for the free-free transitions may be derived from equation (14) in the same way that (14) was derived from (5). For continuous states, $n' \rightarrow i\kappa'$. Multiply the right-hand side of (14) by $d\nu_{\kappa'}$ and substitute from the "primed" analogues of (13). The resulting expression is

$$E_{\kappa'\kappa} d\nu\, d\nu_{\kappa'} = N_i N_\epsilon \frac{KZ^4}{T_\epsilon^{3/2}} \frac{g}{2RZ^2}\left[W \frac{b_{\kappa'} e^{h\nu/kT_\epsilon} - b_\kappa}{e^{h\nu/kT_1} - 1}\right] e^{-h(\nu+\nu_{\kappa'})/kT_\epsilon} h\, d\nu\, d\nu_{\kappa'} . \tag{26}$$

This formula gives the absorption from continuous states in the frequency range $d\nu_{\kappa'}$ to upper continuous states in the range $d\nu$. Since both κ' and κ vary continuously, there is an infinite number of possible transitions that will result in the absorption of quanta of the same frequency. To calculate the total absorption of the given frequency, we shall first hold ν constant and integrate with respect to $\nu_{\kappa'}$. The result is

$$\int_0^\infty E_{\kappa'\kappa} d\nu\, d\nu_{\kappa'} = N_i N_\epsilon \frac{kKZ^2}{2RT_\epsilon^{1/2}} \bar{g} \left[W \frac{\bar{b}_{\kappa'} e^{h\nu/kT_\epsilon} - \bar{b}_\kappa}{e^{h\nu/kT_1} - 1} \right] e^{-h\nu/kT_\epsilon} d\nu \;. \quad (27)$$

The total absorption in all frequencies is given by a second integration over all ν:

$$\left.\begin{aligned} \int_0^\infty \int_0^\infty E_{\kappa'\kappa} d\nu\, d\nu_{\kappa'} = N_i N_\epsilon \frac{k^2 KZ^2 T_1}{2hRT_\epsilon^{1/2}} \bar{\bar{g}} \bar{W} \Bigg[\bar{\bar{b}}_{\kappa'} + \left(\frac{\bar{\bar{b}}_{\kappa'}}{2} - \frac{\bar{\bar{b}}_\kappa}{1 + \dfrac{T_1}{T_\epsilon}} \right) \\ + \left(\frac{\bar{\bar{b}}_{\kappa'}}{3} - \frac{\bar{\bar{b}}_\kappa}{2 + \dfrac{T_1}{T_\epsilon}} \right) + \ldots \Bigg] , \end{aligned} \right\} \quad (28)$$

where the double bars signify that an average has been taken twice, first with respect to $\nu_{\kappa'}$ and secondly with respect to ν. As before, the bracketed factor reduces to unity for thermodynamic equilibrium. When we divide (27) by $h\nu$ and integrate to evaluate the total number of free-free absorptions, we find that the number is infinite. The significance of this result is not entirely clear. It does not appear to be a consequence of any approximation, such as the adoption of a mean value for g. The correct interpretation of this result is probably akin to that given for the well-known nonconvergence of the so-called "partition function," when excluded volumes are not taken into account. In assuming $d\nu$ and $d\nu_\kappa$ to be continuous, we have in effect taken the number of allowed quantum levels as infinite, whereas a precise quantum-mechanical enumeration of them, in a restricted volume, will lead to a finite number of states, only one of which can be occupied at any one time by an electron. The difficulty arises in the spectral region of very long wave-lengths, where the expression "quantum jump" may have no physical interpretation. Fortunately, evaluation of the integral is of no immediate importance.

The spontaneous free-free emission in the range $d\nu$ is given by

$$\int_0^\infty E_{\kappa\kappa'} d\nu \, d\nu_{\kappa'} = N_i N_\epsilon \frac{kKZ^2}{2RT_\epsilon^{1/2}} \bar{g}b_\kappa e^{-h\nu/kT_\epsilon} d\nu \, , \tag{29}$$

and the total emission, obtained by integrating (29), is

$$\int_0^\infty \int_0^\infty E_{\kappa\kappa'} d\nu \, d\nu_{\kappa'} = N_i N_\epsilon \frac{k^2 T_\epsilon^{1/2} Z^2}{2R} \bar{g}b_\kappa \, . \tag{30}$$

The foregoing equations are mathematically sufficient for determining the statistical distribution of atoms in the various atomic states in a pure hydrogen atmosphere, where collisional processes are inoperative, except perhaps to preserve a Maxwellian velocity distribution among the free electrons. It is necessary to equate the numbers of atoms entering to those leaving each atomic level. Thus, for each quantum state, n, we have an equation of the form

$$\left.\begin{aligned} \sum_{=n+1}^{\infty} F_{n''n} + \int_{\nu_n}^{\infty} F_{\kappa n} d\nu + \sum_{n'=1}^{n-1} F_{n'n} = \sum_{n''=n+1}^{\infty} F_{nn''} + \int_{\nu_n}^{\infty} F_{n\kappa} d\nu \\ + \sum_{n'=1}^{n-1} F_{nn'} \, . \end{aligned}\right\} \tag{31}$$

The significance of the notation is shown by the inequality

$$n'' > n > n' \, . \tag{32}$$

We also have the condition that the energy absorbed must be equal to that emitted, in order to preserve equilibrium:

$$\left.\begin{aligned} \sum_{n=1}^{\infty} \sum_{n''=n+1}^{\infty} E_{nn''} + \sum_{n=1}^{\infty} \int_{\nu_n}^{\infty} E_{n\kappa} d\nu + \int_0^\infty \int_0^\infty E_{\kappa'\kappa} d\nu \, d\nu_{\kappa'} \\ = \sum_{n=1}^{\infty} \sum_{n''=n+1}^{\infty} E_{n''n} + \sum_{n=1}^{\infty} \int_{\nu_n}^{\infty} E_{\kappa n} d\nu + \int_0^\infty \int_0^\infty E_{\kappa\kappa'} d\nu \, d\nu_{\kappa'} \, . \end{aligned}\right\} \tag{33}$$

The problem is as follows: Given a gaseous assembly, with a certain number of hydrogen atoms per cubic centimeter, and incident radiation of a specified spectral intensity and dilution. Calculate the quantities N_i,N_ϵ, T_ϵ, and the respective values of b_n. We have exactly the requisite number of conditions for solving the problem. The only difficulty that arises, aside from the mathematical complexity that results from the summations, is the fact that we have an infinite number of equations to solve. Each equation is linear, and each involves the b_n for all states, higher and lower. In a subsequent paper Mr. James G. Baker and I shall show how the equations may be solved in certain simple cases. The form into which the equations have been thrown, with the introduction of the factors b_n, is surprisingly simple. The equations may, with simple changes, be adapted to the problem of transfer of radiation in a stellar atmosphere or in a nebula.

HARVARD OBSERVATORY
February 1937

Reprinted from *The Astrophysical Journal,* Vol. 86, No. 1, July, 1937.

PHYSICAL PROCESSES IN GASEOUS NEBULAE

II. THEORY OF THE BALMER DECREMENT

DONALD H. MENZEL AND JAMES G. BAKER

ABSTRACT

An exact algebraic solution of the equations given in the first paper of this series is carried out for the statistical equilibrium of an assembly of hydrogenic atoms. The exciting radiation is assumed to come only from beyond the limit of the Lyman series. Hence discrete states are filled only by capture from the continuum and by cascade from higher discrete states. The electrons are assumed to have a Maxwellian velocity distribution. The formula is given in terms of a quantity b_n, the ratio between the actual population of state n and the population under conditions of thermodynamic equilibrium. An asymptotic expression is given for b_n in the case of large quantum number, from which it is shown that b_n approaches unity as n becomes infinite. Numerical solutions are deferred until the next paper.

The problem of the theoretical decrement of the Balmer series has been attacked by several investigators, notably Plaskett,[1] Carroll,[2] and Cillié.[3] General equations for the statistical equilibrium of an assembly not in thermodynamic equilibrium were first given by Rosseland.[4] For each atomic state one writes down the condition that the number of transitions from that state, by all possible processes, is equal to the number of transitions to that state, per unit time. Equilibrium is thus insured. Atoms possess, however, an infinite number of states, and an accurate solution of the problem requires the treatment of an infinite number of simultaneous linear equations.

To simplify the problem, Plaskett and Carroll have assumed that the atom possesses but seven discrete levels, the eighth denoting ionization. The effect of the remaining levels was entirely neglected. For the atomic emission probabilities they used the values of the Einstein A's as given by Slack[5] and of the capture probabilities as given by Gaunt.[6] Cillié[7] has recently employed the more extensive data on the emission probabilities, as given by Menzel and Pekeris,[8] to extend his original calculations to fourteen states.

[1] *Harvard Circ.*, No. 335, 1928.

[2] *M.N.*, **90**, 588, 1930.

[3] *Ibid.*, **92**, 820, 1932.

[4] *Ap. J.*, **63**, 218, 1926.

[5] *Phys. Rev.*, **31**, 527, 1928.

[6] *Phil. Trans.*, A, **229**, 163, 1930.

[7] *M.N.*, **96**, 771, 1936.

[8] *M.N.*, **96**, 77, 1935.

One further simplifying assumption, justified to some extent by physical reasoning, is made. In the problem of the planetary nebula the argument is that the Lyman line radiation from the nuclear star is so depleted by the star's atmosphere and the lower nebular layers that the only radiation available for atomic excitation lies beyond the limit of the Lyman series.

For the present we shall accept the conditions of the preceding paragraph. It is instructive to carry out the accurate solution of the equations in this particular case. The general procedure followed by the earlier investigators is, however, open to question. The numerical solutions are none too satisfactory, because of the approximations involved. One has the feeling that more attention should be paid to the neglected transitions. Continuity of the energy levels through the higher terms and on into the continuum is an essential feature of every atom. Furthermore, the use of the exact values of the Einstein A's is incompatible with the use of the Kramer-Gaunt approximation to the emission in the continuum, since such procedure involves a discontinuity at the series limit. A further and serious objection is that this method gives no information about the intensities of high series members.

The exact equations for the statistical equilibrium have been given in the first paper[9] of this series, hereinafter referred to as "I." An atom in quantum state n may make spontaneous transitions to lower discrete states n', or, under the influence of radiation, to either higher discrete states n'' or higher continuous states κ. Let $F_{nn'}$ denote the number of transitions n to n' per unit volume per second. Then the condition of equilibrium (I, equation 31) requires that

$$\left.\begin{aligned}\sum_{n''=n+1}^{\infty} F_{n''n} + \int_{\nu_n}^{\infty} F_{\kappa n}d\nu + \sum_{n'=1}^{n-1} F_{n'n} = \sum_{n''=n+1}^{\infty} F_{nn''}& \\ + \int_{\nu_n}^{\infty} F_{n\kappa}d\nu + \sum_{n'=1}^{n-1} F_{nn'}\,.&\end{aligned}\right\} \quad (1)$$

Under the assumed conditions, there is no radiational excitation from any quantum state but the first; and, since the effect of Lyman

[9] Menzel, *Ap. J.*, 85, 330, 1937.

line radiation is neglected, the only excitation of the 1-quantum atoms is to the continuum. Hence, for each state other than the first we have an equation of the form

$$\sum_{n''=n+1}^{\infty} F_{n''n} + \int_{\nu_n}^{\infty} F_{\kappa n} d\nu = \sum_{n'=1}^{n-1} F_{nn'} . \tag{2}$$

Substituting from I, equations (10) and (25), we find that

$$N_i N_\epsilon \frac{KZ^4}{T_\epsilon^{3/2}} \left\{ \sum_{n''=n+1}^{\infty} b_{n''} e^{X_{n''}} g_{n''n} u_{n''n} + \bar{g} S_n - b_n e^{X_n} t_n \right\} = 0 , \tag{3}$$

where

$$u_{n''n} = \frac{2n^2}{n''(n''^2 - n^2)} ,$$

$$S_n = e^{X_n}[-E_i(-X_n)] ,$$

$$t_n = \sum_{1}^{n-1} - u_{n'n} ,$$

$$X_n = \frac{hRZ^2}{n^2 kT_\epsilon} = \frac{1.5703 \times 10^5}{n^2 T_\epsilon} ,$$

for hydrogen.

In the foregoing equations N_i and N_ϵ are, respectively, the number of ions and electrons per unit volume; T_ϵ is the electron temperature; g, the Kramer-Gaunt factor for the transition in question; and b_n, a factor that denotes the departure of the assembly from thermodynamic equilibrium at temperature T_ϵ. It is defined by the relation (I, 4)

$$N_n = b_n N_i N_\epsilon \frac{h^3}{(2\pi mkT_\epsilon)^{3/2}} n^2 e^{X_n} . \tag{4}$$

For thermodynamic equilibrium, $b_n = 1$.

$$K = \frac{h^3}{(2\pi mk)^{3/2}} \frac{8\pi^2 \epsilon^2 R^2}{mc^3} \frac{2^4}{3\sqrt{3\pi}} = 3.257 \times 10^{-6} \tag{5}$$

To duplicate Cillié's solution we have merely to substitute the accurate g values for the first fourteen equations, cut off all the summations at $n''=14$, set $\bar{g}$ (the averaged g for the bound-free transitions) equal to unity, and, neglecting all higher equations, solve simultaneously for the various b_n's. We prefer to start from the opposite end and obtain the b_n's for the higher levels first.

A consistent procedure is to set, as a first approximation, g equal to unity throughout. The second of the two summations is easily evaluated for large values of n:

$$t_n = \sum_{n'=1}^{n-1} \frac{2n^2}{n'(n^2 - n'^2)} = \sum_{n'=1}^{n-1} \left(\frac{2}{n'} + \frac{1}{n - n'} - \frac{1}{n + n'}\right) = 3\sum_{n'=1}^{n-1} \frac{1}{n'} - \sum_{n'=1}^{n-1} \frac{1}{n + n'} \quad (6)$$

$$= \left[3 \ln n + 3\gamma - \ln 2 - \frac{3}{4n} - \frac{5}{16n^2} - \dots \right] \quad (7)$$

$$\sim 3 \ln n + 1.04 \, . \quad (8)$$

For n as small as 10 the bracket in (7) is correct to five significant figures. In deriving these formulae we have noted the identity

$$\sum_{1}^{n-1} \frac{1}{n'} = \sum_{1}^{n-1} \frac{1}{n - n'} \quad (9)$$

and also the well-known expression for the sum to a terms of the harmonic series

$$\sum_{1}^{a} \frac{1}{a} = \ln a + \gamma + \frac{1}{2a} - \frac{1}{12a^2} + \dots, \quad (10)$$

where γ is Euler's constant, 0.577 . . . . Various functional expansions of an obvious nature have also been used.

For the middle term of equation (3), certain relationships may be derived.

For all values of X_n different from zero,

$$S_n = e^{X_n}[-E_i(-X_n)] = e^{X_n}\left[-\gamma - lnX_n + \frac{X_n}{1\cdot 1!} - \frac{X_n^2}{2\cdot 2!} + \frac{X_n^3}{3\cdot 3!} - \ldots\right]. \quad (11)$$

For $X_n >> 1$, it is convenient to employ the following asymptotic expansion:

$$S_n = e^{X_n}[-E_i(-X_n)] = \frac{1}{X_n}\left[1 - \frac{1!}{X_n} + \frac{2!}{X_n^2} - \ldots\right]. \quad (12)$$

Equation (12) is needed only for the lowest quantum states.

Thus, the only troublesome term in equation (3) is the first. The difficulty arises from the occurrence of the factor $b_{n''}$ under the summation sign. Before proceeding to the exact solution of the equation, we shall employ an approximation, which indicates clearly the physical nature of b_n. From the character of the summation we note that the principal contribution to its value comes from n'' terms bordering on n. If, therefore, $b_{n''}$ is a slowly changing function of n'', we may remove it from the summation sign in order to obtain the first approximation.

The first term may be written as follows:

$$b_n \sum_{n''=n+1}^{\infty} \frac{2e^{X_{n''}}}{\left(\frac{1}{n^2} - \frac{1}{n''^2}\right)n''^3} \sim b_n \int_{n+1}^{\infty} \frac{e^{X_{n''}}}{\left(\frac{1}{n^2} - \frac{1}{n''^2}\right)} \frac{2dn''}{n''^3}, \quad (13)$$

since, when n'' is large, we may approximate the sum by an integral. If we change variables by

$$\left.\begin{aligned} \frac{hRZ^2}{kT}\left(\frac{1}{n^2} - \frac{1}{n''^2}\right) &= X_n - X_{n''} = y\,, \\ \frac{2hRZ^2}{kT}\frac{dn''}{n''^3} &= dy\,, \end{aligned}\right\} \quad (14)$$

the summation becomes, by equation (11),

$$\left.\begin{aligned} &= b_n e^{X_n} \int_{X_n - X_{n+1}}^{X_n} \frac{e^{-y}}{y}\, dy = b_n e^{X_n}[-E_i(-X_n + X_{n+1}) + E_i(-X_n)] \\ &\qquad \sim b_n e^{X_n}\left[\ln n - \ln 2 - X_n + \frac{3}{2n} - \frac{7}{8n^2} \dots \right]. \end{aligned}\right\} \tag{15}$$

When the limiting formulae (8), (11), and (15) are substituted back into equation (3) and the common factors are canceled, we may solve for b_n and find

$$b_n \sim \frac{2 \ln n - \gamma - \ln X_1 + X_n + \dots}{2 \ln n + 3\gamma - \frac{9}{4n} + \frac{9}{16n^2} + X_n - \dots}. \tag{16}$$

As $n \to \infty$, $b_n \to 1$. This limiting value of unity is in accord with our physical intuition, which leads us to expect a continuity between states of negative and positive energy near the ionization limit of the atom.[10] Since our calculations are based on the assumption that the electrons possess a Maxwellian velocity distribution, the value of b, appropriate to the continuum where thermodynamic equilibrium obtains, is unity. The approximate value of (16) can be employed, however, only for the uppermost states.

Equation (3) may be written as an infinite set of simultaneous equations involving an infinite number of unknowns. One may write down symbolically the infinite determinants for the general case and expand by minors. Since equation (3) does not contain n' directly, the solution for any b_n is of precisely the same form. Accordingly, the equations as written for the solution of b_n begin with the nth equation and not with $n = 2$, although the answer is the same. This fact simplifies the solution considerably. The determinant of the coefficients of the unknowns has the value $\prod_n^{\infty} t_i$. Since $t_i > 1$, this expression is, of course, infinite in value. The determinants in the numerator, however, contain the same minors from some point on, and hence, the ratios reduce to infinite series.

[10] The limiting form of b_n was given by Menzel and Cillié, *Ap. J.*, 85, 88, 1937.

An expansion of the terms and a recombination of the cyclical subscripts reduce to the exact solution

$$b_n = \frac{S_n + V_n}{e^{X_n t_n}}, \tag{17}$$

where

$$\left. \begin{aligned} V_n = \sum_{n+1}^{\infty} \frac{S_i u_{in}}{t_i} + \sum_{n+1}^{\infty}\sum_{j+1}^{\infty} \frac{S_i u_{ij} u_{jn}}{t_i t_j} \\ + \sum_{n+1}^{\infty}\sum_{\kappa+1}^{\infty}\sum_{j+1}^{\infty} \frac{S_i u_{ij} u_{j\kappa} u_{\kappa n}}{t_i t_j t_\kappa} + \ldots . \end{aligned} \right\} \tag{18}$$

It is evident that these summations converge individually, for

$$\frac{S_i}{t_i} \to \tfrac{2}{3} \quad \text{as} \quad i \to \infty$$

and

$$\sum_{n+1}^{\infty} \frac{S_i}{t_i} u_{in} < \tfrac{2}{3} \sum_{n+1}^{\infty} u_{in} < \tfrac{2}{3}\, ln\, n\, .$$

Indeed, since

$$\lim_{n\to\infty} b_n \to 1\,, \qquad \lim_{n\to\infty} V_n \to ln\, n + \text{const.}$$

An examination of equation (3) shows that formula (17) may be derived directly by the method of successive approximations. This procedure is by far the shortest. The fact that these two entirely different methods yield precisely the same formula for b_n is interesting in itself. What is so obvious here, however, will not be so at all for the more complicated solutions to come. Accordingly, we have employed the determinant method to indicate how the future refinements will go.

We are now enabled to determine precisely the numerical values of b_n. The solution (17) is of the final form for the nebular problem, since, upon introducing the Gaunt factor and collision and radiation terms from I, we need only to redefine the functions given above.

The next paper will contain useful tables of the important functions of the preceding equations and, in particular, will contain a tabulation of the values of b_n over a wide range of temperature and of quantum number, on the basis of the straight capture spectrum of hydrogenic atoms. A comparison with observations and results of other investigators will also be included.

HARVARD COLLEGE OBSERVATORY
May 14, 1937

Reprinted from *The Astrophysical Journal*, Vol. 88, No. 1, July, 1938.

PHYSICAL PROCESSES IN GASEOUS NEBULAE

III. THE BALMER DECREMENT

JAMES G. BAKER[1] AND DONALD H. MENZEL

ABSTRACT

This paper contains a numerical solution of the equations derived and formally solved in two earlier papers of the series. Various tables of general interest, including those of functions $X_n = hR/n^2kT_\epsilon$ and $-[Ei(-X_n)]$, for useful astrophysical ranges of n and T_ϵ are given. The Balmer decrement, computed under two alternative hypotheses—**A** for a nebula transparent to Lyman line radiation, and **B** for an opaque nebula—is tabulated. The latter hypothesis agrees better with the observed data. The conclusion is reached that the electron temperature, T_ϵ, of the nebular gas cannot be effectively determined from observed Balmer decrement data, because the decrement is insensitive to temperature. In view of the extreme physical conditions that exist in nebulae, the partition of atoms among the various excited levels approaches surprisingly close to the thermodynamic value.

In papers I and II[2] of this series the fundamental equations for calculating the relative intensities of lines of the Balmer series were set up and formally solved. The present paper gives the numerical results and the comparison of theory with observation.

For the benefit of those whose interest is chiefly in the final results, we shall briefly recapitulate the theory and fundamental hypotheses on which the calculations were based. The energy emitted in a given Balmer line, say in the transition from level n to level 2, is given by

$$E_{n2} = N_n A_{n2} h\nu \, , \tag{1}$$

where N_n is the number of atoms per cubic centimeter in level n, A_{n2} is the Einstein spontaneous probability coefficient, and ν is the frequency. Outside of thermodynamic equilibrium, N_n must be calculated by balancing the number of atoms leaving a given quantum level against those entering, by all possible routes. We thus have an infinite set of simultaneous equations to solve, one equation for each level. The method of solving these equations, for certain specified physical conditions, was given in II. The results are most con-

[1] Society of Fellows, Harvard University.

[2] Menzel, *Ap. J.*, **85**, 330; **86**, 70, 1937.

veniently expressed in terms of a dimensionless parameter, b_n, defined by

$$N_n = b_n N_{nT} = b_n N_i N_\epsilon \frac{h^3}{(2\pi mkT_\epsilon)^{3/2}} n^2 e^{X_n} , \tag{2}$$

(II, Eq. [4]), where

$$X_n = \frac{hRZ^2}{n^2 KT_\epsilon} . \tag{3}$$

b_n gives the departure from thermodynamic equilibrium. For thermodynamic equilibrium, $b_n = 1$.

We have calculated the Balmer decrement on two alternative hypotheses about the nature of nebular excitation. Under **A** we assume that an electron enters quantum level n of an atom either by capture from the free state or by cascade from a higher discrete level. We neglect the radiation field of the star and assume that there is no reabsorption of Lyman line radiation. For **A** we distinguish two cases: $\mathbf{A}_1$, where an idealized hydrogen atom with the Gaunt g factor equal to unity was employed, and $\mathbf{A}_2$, where the exact g values were introduced. Comparison of the results indicates how far calculations based on the simpler assumptions may be trusted and in what directions the deviations will tend. In a nebula that is very opaque to Lyman line radiation, we may take account of the nebular radiation field, but not of the stellar, by assuming that absorptions from level 1 to level n are exactly balanced by the inverse spontaneous transitions. Zanstra has based his theoretical calculations on this assumption, which we shall designate as **B**. In II the radiation field was entirely neglected. We may allow for it, under **B**, by redefining the quantity t_n as

$$t_n = \sum_{2}^{n-1} - u_{n'n} \tag{4}$$

instead of

$$t_n = \sum_{1}^{n-1} - u_{n'n} . \tag{5}$$

The remaining terms of the previous equations are unaltered, and the solution is carried out as before.

The numerical calculations were laborious, and various intermediate results of astrophysical interest have been tabulated with the hope that they may prove useful to others. The tables cover a

TABLE 1

$X_n \times 10^{-p}$

n	5,000°	p	10,000°	p	20,000°	p	40,000°	p	80,000°	p	160,000°	p	320,000°	p
1	3.141	1	1.570	1	7.851	0	3.926	0	1.963	0	9.814	−1	4.907	
2	7.851	0	3.926	0	1.963		9.814	−1	4.907	−1	2.454		1.227	
3	3.490		1.745		8.724	−1	4.362		2.181		1.090		5.450	−2
4	1.963		9.814	−1	4.907		2.454		1.227		6.140	−2	3.070	
5	1.256		6.281		3.141		1.570		7.851	−2	3.930		1.970	
6	8.724	−1	4.362	−1	2.181	−1	1.091	−1	5.453	−2	2.727	−2	1.36	−2
7	6.409		3.205		1.602		8.012	−2	4.006		2.003		1.00	
8	4.907		2.454		1.227		6.134		3.067		1.533		7.67	−3
9	3.877		1.939		9.693	−2	4.847		2.423		1.212		6.06	
10	3.141		1.570		7.851		3.926		1.963		9.815	−3	4.91	
11	2.596	−1	1.298	−1	6.489	−2	3.244	−2	1.622	−2	8.11	−3	4.06	−3
12	2.181		1.091		5.452		2.726		1.363		6.82		3.41	
13	1.858		9.292	−2	4.646		2.323		1.161		5.81		2.90	
14	1.602		8.012		4.006		2.003		1.001		5.01		2.50	
15	1.396		6.979		3.490		1.745		8.725	−3	4.36		2.18	
16	1.227	−1	6.134	−2	3.067	−2	1.534	−2	7.670	−3	3.83	−3	1.92	−3
17	1.087		5.434		2.717		1.358		6.790		3.40		1.70	
18	9.693	−2	4.847		2.423		1.212		6.058		3.03		1.52	
19	8.700		4.350		2.175		1.087		5.437		2.72		1.36	
20	7.851		3.926		1.963		9.814	−3	4.907		2.45		1.23	
21	7.121	−2	3.561	−2	1.780	−2	8.902	−3	4.45	−3	2.23	−3	1.11	−3
22	6.489		3.244		1.622		8.111		4.06		2.03		1.01	
23	5.937		2.968		1.484		7.421		3.71		1.86		9.28	−4
24	5.452		2.726		1.363		6.815		3.41		1.70		8.52	
25	5.025		2.512		1.256		6.281		3.14		1.57		7.85	
26	4.646	−2	2.323	−2	1.161	−2	5.807	−3	2.90	−3	1.45	−3	7.26	−4
27	4.308		2.154		1.077		5.385		2.69		1.35		6.73	
28	4.006		2.003		1.002		5.007		2.50		1.25		6.26	
29	3.734		1.867		9.336	−3	4.668		2.33		1.17		5.83	
30	3.490		1.745		8.724		4.362		2.18		1.09		5.45	

large range of temperatures and quantum numbers. A sufficient number of figures was carried in the calculations to insure the correctness of the values as tabulated.

Table 1 contains values of X_n, equation (3), with Z set equal to unity. When Z is not unity, as for ionized helium, the tabulated temperatures must be multiplied by Z^2, to correspond with the true

TABLE 2

$g_{nn'}$

n'	n = 2	3	4	5	6	7	8	9	10	11	12	13	14	15
1	.717	.765	.780	.786	.790	.792	.793	.794	.795	.795	.795	.796	.796	.796
2	....	.757	.822	.844	.855	.861	.865	.867	.869	.870	.871	.872	.872	.873
3	....	....	.768	.839	.865	.878	.886	.891	.895	.897	.899	.900	.901	.902
4	....	....	....	.772	.847	.874	.890	.898	.903	.908	.911	.914	.915	.916
5	....	....	....	....	.774	.851	.881	.896	.905	.912	.916	.919	.922	.924
6	....	....	....	....	....	.776	.853	.884	.899	.910	.917	.922	.926	.929
7	....	....	....	....	....	....	.777	.852	.887	.904	.915	.922	.926	.930
8	....	....	....	....	....	....	....	.779	.856	.888	.905	.916	.925	.929
9	....	....	....	....	....	....	....	....	.779	.856	.889	.906	.917	.925
10	....	....	....	....	....	....	....	....	....	.780	.858	.888	.907	.919
11	....	....	....	....	....	....	....	....	....	....	.781	.854	.890	.908
12	....	....	....	....	....	....	....	....	....	....	....	.781	.860	.891
13	....	....	....	....	....	....	....	....	....	....	....	....	.781	.861
....	....	....	....	....	....	....	....	....	....	....	....	....	....	.781

n'	n = 16	17	18	19	20	21	22	23	24	25	26	27	28	29	30
1	.796	.796	.796	.796	.796	.797	.797	.797	.797	.797	.797	.797	.797	.797	.797
2	.873	.874	.874	.874	.874	.875	.875	.875	.875	.875	.875	.875	.875	.875	.875
3	.902	.903	.904	.904	.904	.904	.905	.905	.905	.905	.906	.906	.906	.906	.906
4	.917	.918	.919	.920	.920	.921	.921	.921	.922	.922	.922	.922	.922	.923	.923
5	.926	.927	.928	.929	.930	.930	.931	.931	.932	.932	.932	.933	.933	.933	.933
6	.931	.933	.934	.936	.937	.937	.938	.939	.939	.940	.940	.940	.941	.941	.941
7	.933	.935	.937	.939	.940	.941	.942	.943	.943	.944	.945	.945	.945	.946	.946
8	.933	.937	.939	.941	.943	.944	.945	.946	.947	.948	.948	.949	.949	.950	.950
9	.930	.934	.938	.940	.942	.944	.945	.947	.948	.948	.949	.950	.950	.951	.952
10	.927	.932	.936	.940	.942	.944	.946	.948	.949	.950	.951	.952	.953	.953	.954
11	.919	.927	.933	.937	.940	.943	.945	.947	.949	.950	.951	.952	.953	.954	.955
12	.909	.920	.928	.933	.938	.941	.944	.946	.948	.950	.951	.952	.953	.954	.955
13	.892	.910	.921	.929	.935	.939	.943	.946	.948	.950	.952	.953	.954	.955	.956
14	.862	.894	.911	.923	.931	.936	.941	.945	.947	.950	.952	.954	.955	.956	.957
15	.782	.864	.894	.912	.923	.931	.937	.941	.945	.948	.950	.953	.954	.956	.957
16	....	.782	.866	.894	.912	.923	.931	.937	.942	.946	.949	.951	.953	.955	.956
17	....	....	.782	.868	.894	.912	.924	.932	.938	.942	.946	.950	.951	.954	.955
18	....	....	....	.782	.862	.894	.912	.924	.932	.938	.942	.946	.949	.952	.954
19	....	....	....	....	.782	.862	.894	.912	.924	.932	.938	.943	.947	.950	.952
20	....	....	....	....	....	.782	.862	.895	.912	.924	.932	.938	.943	.947	.950
21	....	....	....	....	....	....	.782	.862	.895	.913	.924	.932	.939	.943	.947
22	....	....	....	....	....	....	....	.782	.862	.895	.913	.924	.933	.939	.943
23	....	....	....	....	....	....	....	....	.782	.862	.895	.913	.924	.933	.939
24	....	....	....	....	....	....	....	....	....	.782	.863	.895	.913	.924	.933
25	....	....	....	....	....	....	....	....	....	....	.782	.863	.895	.913	.925
26	....	....	....	....	....	....	....	....	....	....	....	.782	.863	.895	.913
27	....	....	....	....	....	....	....	....	....	....	....	....	.782	.863	.895
28	....	....	....	....	....	....	....	....	....	....	....	....	....	.782	.863
29	....	....	....	....	....	....	....	....	....	....	....	....	....	....	.782

temperatures. For example, the solution for He^+ at $T_\epsilon = 40{,}000^\circ$ is equivalent to that for hydrogen at $T_\epsilon = 10{,}000^\circ$.

The Kramer-Gaunt factors for the discrete-discrete transitions are given in Table 2. The functional form of this factor, $g_{nn'}$, is

TABLE 3*a*

$[-E_i(-X_n)]$

n	T_ϵ						
	5,000°	10,000°	20,000°	40,000°	80,000°	160,000°	320,000°
2	0.0000	0.0041	0.0525	0.2264	0.5712	1.0590	1.6399
3	0.0071	0.0700	0.2727	0.6453	1.1524	1.7449	2.3857
4	0.0525	0.2264	0.5712	1.0590	1.6399	2.2746	2.9377
5	0.1451	0.4296	0.8720	1.4252	2.0443	2.6993	3.3731
6	0.2727	0.6453	1.1524	1.7449	2.3857	3.0521	3.7318
7	0.4188	0.8574	1.4080	2.0256	2.6799	3.3533	4.0365
8	0.5712	1.0590	1.6400	2.2746	2.9377	3.6167	4.3012
9	0.7235	1.2483	1.8512	2.4976	3.1669	3.8481	4.5352
10	0.8720	1.4252	2.0443	2.6993	3.3731	4.0565	4.7448
11	1.0252	1.5905	2.2217	2.8832	3.5703	4.2454	4.9345
12	1.1524	1.7449	2.3857	3.0521	3.7318	4.4182	5.1079
13	1.2832	1.8896	2.5379	3.2082	3.8899	4.5772	5.2675
14	1.4080	2.0256	2.6799	3.3533	4.0365	4.7246	5.4153
15	1.5268	2.1537	2.8128	3.4888	4.1732	4.8620	5.5530
16	1.6400	2.2745	2.9377	3.6157	4.3012	4.9906	5.6818
17	1.7480	2.3890	3.0555	3.7352	4.4216	5.1114	5.8028
18	1.8512	2.4976	3.1669	3.8481	4.5352	5.2253	5.9170
19	1.9498	2.6009	3.2726	3.9550	4.6427	5.3331	6.0249
20	2.0443	2.6993	3.3731	4.0565	4.7448	5.4355	6.1274
21	2.1348	2.7933	3.4689	4.1532	4.8419	5.5328	6.2249
22	2.2217	2.8832	3.5703	4.2454	4.9345	5.6257	6.3178
23	2.3053	2.9694	3.6529	4.3337	5.0231	5.7144	6.4066
24	2.3857	3.0521	3.7318	4.4182	5.1079	5.7994	6.4917
25	2.4632	3.1318	3.8124	4.4993	5.1893	5.8809	6.5732
26	2.5379	3.2082	3.8899	4.5772	5.2675	5.9592	6.6516
27	2.6101	3.2820	3.9645	4.6523	5.3428	6.0346	6.7270
28	2.6799	3.3533	4.0365	4.7247	5.4153	6.1072	6.7997
29	2.7474	3.4221	4.1061	4.7945	5.4853	6.1773	6.8699
30	2.8128	3.4887	4.1732	4.8620	5.5530	6.2450	6.9376

given by Menzel and Pekeris.[3] For n' from 1 to 6, and for n from 2 to 30, the accurate expressions were employed; for all other values, the asymptotic expression

$$g_{nn'} = 1 - \frac{0.1728\left(1 + \dfrac{n'^2}{n^2}\right)}{\left(1 - \dfrac{n'^2}{n^2}\right)^{2/3}(n^2)^{2/3}} \tag{6}$$

[3] *M.N.*, **96**, 77, 1935.

was used. This expression differs from the exact one for n' equal to 6 by a small number in the third decimal place. Consequently, an adjustment was made to the asymptotic values in order to remove the discrepancy. It is evident that this factor converges only very slowly to unity as n becomes infinite.

TABLE 3*b*

$\bar{g}[-E_i(-X_n)]$

n	T_ϵ						
	5,000°	10,000°	20,000°	40,000°	80,000°	160,000°	320,000°
2	0.000	0.004	0.053	0.229	0.578	1.374	1.884
3	0.007	0.071	0.279	0.662	1.185	1.808	2.483
4	0.053	0.231	0.588	1.093	1.699	2.374	3.081
5	0.148	0.440	0.900	1.479	2.130	2.832	3.556
6	0.278	0.663	1.193	1.816	2.496	3.213	3.946
7	0.427	0.882	1.459	2.112	2.811	3.535	4.276
8	0.583	1.091	1.702	2.377	3.088	3.822	4.564
9	0.739	1.287	1.922	2.611	3.331	4.071	4.816
10	0.891	1.470	2.124	2.824	3.550	4.295	5.043
11	1.048	1.641	2.309	3.018	3.760	4.498	5.249
12	1.178	1.800	2.480	3.196	3.933	4.684	5.437
13	1.312	1.950	2.640	3.362	4.102	4.856	5.611
14	1.440	2.091	2.788	3.515	4.259	5.016	5.773
15	1.562	2.224	2.928	3.659	4.406	5.166	5.924
16	1.678	2.349	3.059	3.794	4.544	5.306	6.065
17	1.789	2.468	3.182	3.921	4.674	5.437	6.199
18	1.895	2.581	3.299	4.041	4.797	5.562	6.325
19	1.996	2.688	3.410	4.155	4.913	5.680	6.445
20	2.093	2.790	3.516	4.263	5.023	5.792	6.558
21	2.186	2.888	3.617	4.366	5.128	5.899	6.667
22	2.275	2.981	3.723	4.465	5.229	6.001	6.770
23	2.361	3.071	3.810	4.559	5.325	6.099	6.869
24	2.443	3.157	3.894	4.649	5.417	6.192	6.964
25	2.523	3.240	3.979	4.736	5.505	6.282	7.055
26	2.600	3.319	4.060	4.820	5.590	6.368	7.142
27	2.675	3.396	4.139	4.900	5.672	6.451	7.226
28	2.746	3.470	4.215	4.977	5.751	6.531	7.307
29	2.815	3.542	4.288	5.052	5.827	6.608	7.385
30	2.882	3.611	4.359	5.125	5.900	6.682	7.461

The familiar exponential integral function

$$[-E_i(-X_n)] = \int_{X_n}^{\infty} \frac{e^{-\zeta}}{\zeta}\, d\zeta \tag{7}$$

is given in Table 3*a* as a function of n and T_ϵ. In the computations we made use of formulae (12) and (13) of paper II, and of Jahnke and Emden's tables. The ordinary tables, with uniformly spaced argument, are somewhat difficult to use because of the involved interpolation.

The expression

$$\bar{g}[-Ei(-X_n)] = \int_{X_n}^{\infty} \frac{e^{-\zeta}}{\zeta} g_{\kappa n} d\zeta \tag{8}$$

is given in Table 3*b*. The Kramer-Gaunt factor for the bound-free transitions, $g_{\kappa n}$, was obtained from the formulae of Menzel and Pekeris. The integration was carried out by numerical integration formulae. Tables 3*a* and 3*b*, by multiplication with e^{X_n}, are convertible immediately into the functions S_n and $\bar{g}S_n$, which are given in paper II, equations (11) and (12).

Table 4*a* gives the expression

$$t_n = \sum_{1}^{n-1} \frac{2n^2}{n'(n^2 - n'^2)} \tag{9}$$

and Table 4*b*,

$$t_n = \sum_{1}^{n-1} \frac{2n^2 g_{nn'}}{n'(n^2 - n'^2)} \, . \tag{10}$$

The first four tables of this paper give directly and indirectly all the functions of n and T_ϵ necessary in the evaluation of b_n by equation (17) of paper II, both for **A** and for **B**. Since the function V_n is only very slowly convergent, the calculations proved to be lengthy. An expression of the form

$$\sum_{i=31}^{\infty} \frac{S_i u_{in}}{t_i} = a + bn + cn^2 \tag{11}$$

was developed to complete the first summation accurately. This summation is the kernel of the succeeding double summation, which reduces, therefore, to a single summation. The first three terms of V_n were computed directly. The contribution of the remaining terms was obtained graphically by careful asymptotic extrapolation based on the plot of S_n and of the first three terms of V_n against position in the sequence. The b_n's for the two cases $\mathbf{A}_1$ and $\mathbf{A}_2$, which were smoothed in the last figure to take out slight numerical inaccuracies, are tabulated in Tables 5*a* and 5*b*.

In Table 6, under the column headed b_n, are given the values, computed on hypothesis **B**, for an atom with an infinite number of

TABLE 4*a*

t_n (CASE $\mathbf{A}_1$)

n	t_n	n	t_n	n	t_n
2	2.6667	11	8.1614	21	10.1363
3	4.0500	12	8.4285	22	10.2775
4	4.9905	13	8.6738	23	10.4124
5	5.7044	14	8.9005	24	10.5415
6	6.2801	15	9.1112	25	10.6653
7	6.7627	16	9.3082	26	10.7841
8	7.1782	17	9.4929	27	10.8984
9	7.5430	18	9.6670	28	11.0086
10	7.8682	19	9.8321	29	11.1148
		20	9.9881	30	11.2174

TABLE 4*b*

t_n (CASE $\mathbf{A}_2$)

n	t_n	n	t_n	n	t_n
2	1.911	11	6.943	21	8.886
3	3.084	12	7.205	22	9.026
4	3.929	13	7.442	23	9.160
5	4.589	14	7.668	24	9.287
6	5.132	15	7.876	25	9.410
7	5.590	16	8.070	26	9.528
8	5.989	17	8.253	27	9.641
9	6.339	18	8.426	28	9.750
10	6.657	19	8.589	29	9.859
		20	8.740	30	9.956

levels. For purposes of comparison, we also give values of b_n derived from the work of Cillié.[4] Cillié based his calculations on assumption **B,** but he made two approximations to facilitate the numerical work. A model atom consisting of only fourteen levels was assumed, and the higher discrete states were neglected. For the fourteen levels, accurate values of the Einstein A's were taken; but for the electron captures, the g factor was set equal to unity. For purposes of comparison, we have also given the b_n for a fourteen-state atom at $T_\epsilon =$

[4] *Ibid.*, **96**, 771, 1936.

TABLE 5*a*

b_n (CASE $\mathbf{A}_1$)

n	T_ϵ						
	5,000°	10,000°	20,000°	40,000°	80,000°	160,000°	320,000°
2	0.000034	0.002893	0.03365	0.1377	0.3306	0.5873	0.8809
3	.003273	.02945	.1097	.2447	.4251	.6237	.8303
4	.01824	.07359	.1781	.3182	.4843	.6558	.8268
5	.04280	.1180	.2330	.3705	.5264	.6801	.8293
6	.07199	.1588	.2764	.4115	.5538	.6994	.8335
7	.09926	.1937	.3116	.4431	.5780	.7141	.8472
8	.1259	.2237	.3387	.4663	.5978	.7260	.8532
9	.1499	.2486	.3629	.4888	.6161	.7358	.8555
10	.1710	.2714	.3836	.5062	.6281	.7439	.8604
11	.1912	.2903	.4002	.5190	.6381	.7509	.8632
12	.2074	.3072	.4147	.5302	.6449	.7568	.8657
13	.2233	.3223	.4275	.5399	.6525	.7622	.8697
14	.2376	.3358	.4388	.5485	.6595	.7667	.8699
15	.2507	.3478	.4489	.5561	.6657	.7708	.8718
20	.3016	.3931	.4891	.5899	.6889	.7852	.8780
25	.3359	.4269	.5203	.6121	.7033	.7912	.8801
30	0.3581	0.4487	0.5400	0.6237	0.7085	0.7936	0.8818

TABLE 5*b*

b_n (CASE $\mathbf{A}_2$)

n	T_ϵ						
	5,000°	10,000°	20,000°	40,000°	80,000°	160,000°	320,000°
3	0.00434	0.0393	0.146	0.296	0.576	0.843	1.136
4	.0234	.0961	.233	.420	.639	.861	1.103
5	.0540	.150	.296	.478	.685	.877	1.089
6	.0905	.202	.350	.528	.714	.888	1.079
7	.123	.242	.391	.561	.736	.898	1.068
8	.154	.277	.422	.586	.753	.905	1.061
9	.182	.311	.448	.609	.770	.910	1.058
10	.206	.340	.470	.626	.780	.918	1.056
11	.228	.359	.488	.641	.790	.921	1.054
12	.248	.376	.503	.651	.800	.925	1.052
13	.266	.392	.517	.663	.808	.928	1.050
14	.281	.406	.528	.673	.813	.930	1.048
15	.296	.419	.539	.680	.818	.933	1.046
20	.350	.467	.583	.712	.842	.943	1.041
25	.385	.498	.611	.732	.858	.952	1.037
30	0.408	0.519	0.630	0.750	0.871	0.962	1.031

20,000°, with the capture g factor taken into account. The effect of the neglected higher levels is clearly exhibited by the actual crossing of the true b_n's by Cillié's b_n's at quantum number 6.

It is to be noted that b_n can be greater than unity. Referring to paper II, equation (17), we see that S_n and V_n must increase with T_ϵ. For any T_ϵ, as we have shown in II, $b_n \to 1$ as $n \to \infty$. There must,

TABLE 6

b_n (CASE **B**)

n	T_ϵ: 5,000°		10,000°		20,000°			40,000°	80,000°
	b_n	b_n (Cillié)	b_n	b_n (Cillié)	b_n	b_n (Cillié)	14-State b_n	b_n	b_n
3......	0.0098	0.0120	0.089	0.111	0.330	0.409	0.420	0.670	1.304
4......	.0406	.0448	.166	.188	.404	.453	.466	.729	1.108
5......	.0840	.0878	.233	.251	.460	.494	.510	.743	1.065
6......	.132	.131	.296	.298	.512	.520	.538	.773	1.045
7......	.173	.166	.341	.333	.550	.538	.558	.789	1.036
8......	.211	.195	.379	.357	.577	.547	.568	.802	1.032
9......	.244	.218	.417	.373	.600	.549	.570	.816	1.031
10......	.271	0.236	.448	0.383	.620	0.545	0.565	.825	1.030
15......	.371		.526		.676			.853	1.027
20......	.428		.571		.713			.871	1.025
25......	.464		.600		.736			.882	1.024
30......	0.486		0.618		0.750			0.893	1.024

then, be a temperature for which $b_n \sim 1$, when the conditions of thermodynamic equilibrium are closely approximated by the nebula. For assumption $\mathbf{A}_2$, the temperature at which the pseudo-thermodynamic equilibrium occurs is about 235,000°. For **B** it occurs near 71,000°.

The relative intensities I_n of the Balmer lines, referred, as is customary, to $H\beta$ as unity, are listed in Tables 7*a*, 7*b*, and 8 for the assumptions $\mathbf{A}_1$, $\mathbf{A}_2$, and **B**, respectively. In the last table are given, also, the values of I_n as derived by Cillié. The intensities are calculated from the formula

$$I_n = \left(\frac{b_n}{b_4}\right) e^{X_n - X_4} \frac{(4)^3}{n^3} \frac{g_{n2}}{g_{42}} = \left(\frac{b_n}{b_4}\right) I_{nT} , \qquad (12)$$

TABLE 7*a*

I_n (CASE A₁)

n	T_ϵ						
	5,000°	10,000°	20,000°	40,000°	80,000°	160,000°	320,000°
3	1.953	2.035	2.118	2.205	2.289	2.364	2.438
4	1.000	1.000	1.000	1.000	1.000	1.000	1.000
5	0.592	0.576	0.561	0.546	0.533	0.519	0.508
6	0.389	0.368	0.350	0.334	0.317	0.305	0.294
7	0.271	0.252	0.235	0.220	0.205	0.195	0.187
8	0.198	0.180	0.165	0.152	0.141	0.132	0.126
9	0.149	0.135	0.121	0.111	0.101	0.094	0.089
10	0.115	0.104	0.091	0.083	0.075	0.069	0.066
11	0.092	0.081	0.071	0.063	0.057	0.052	0.049
12	0.074	0.065	0.056	0.050	0.044	0.040	0.038
13	0.060	0.052	0.045	0.040	0.035	0.032	0.030
14	0.050	0.043	0.037	0.032	0.028	0.026	0.024
15	0.042	0.036	0.030	0.026	0.023	0.021	0.019
20	0.020	0.017	0.014	0.012	0.010	0.009	0.008
25	0.011	0.009	0.007	0.006	0.005	0.005	0.004
30	0.007	0.005	0.004	0.004	0.003	0.003	0.002

TABLE 7*b*

I_n (CASE A₂)

n	T_ϵ						
	5,000°	10,000°	20,000°	40,000°	80,000°	160,000°	320,000°
3	1.859	1.915	1.984	2.073	2.165	2.241	2.302
4	1.000	1.000	1.000	1.000	1.000	1.000	1.000
5	0.598	0.576	0.560	0.548	0.539	0.524	0.513
6	0.396	0.374	0.353	0.338	0.321	0.307	0.296
7	0.274	0.255	0.236	0.221	0.207	0.196	0.185
8	0.199	0.182	0.165	0.153	0.141	0.132	0.124
9	0.149	0.136	0.120	0.111	0.101	0.093	0.086
10	0.114	0.105	0.091	0.082	0.075	0.069	0.063
11	0.090	0.081	0.070	0.062	0.056	0.052	0.048
12	0.073	0.065	0.055	0.050	0.045	0.040	0.036
13	0.059	0.052	0.043	0.039	0.035	0.032	0.029
14	0.049	0.042	0.036	0.032	0.029	0.025	0.023
15	0.041	0.035	0.030	0.025	0.023	0.020	0.019
20	0.019	0.016	0.014	0.012	0.010	0.009	0.007
25	0.011	0.009	0.007	0.006	0.005	0.004	0.004
30	0.006	0.005	0.004	0.003	0.003	0.003	0.003

where I_{nT} is the value of the relative intensity for thermodynamic equilibrium. Because of the approximations, the decrements calculated by Cillié for a given temperature actually correspond to those for a temperature much higher.

In Table 8, under the heading I_n (Obs.), are given the means of the observed data for 17 nebulae, as determined by Berman.[5] The

TABLE 8

I_n (CASE B)

n	T_ϵ 5,000° I_n	5,000° I_n (Cillié)	10,000° I_n	10,000° I_n (Cillié)	20,000° I_n	20,000° I_n (Cillié)	40,000° I_n	40,000° I_n (Obs.)	80,000° I_n	160,000° I_n
3	2.43	2.70	2.50	2.78	2.59	2.88	2.71	2.77	2.83	2.93
4	1.00	1.00	1.00	1.00	1.00	1.00	1.00	1.00	1.00	1.00
5	0.53	0.51	0.51	0.50	0.50	0.48	0.49	0.50	0.48	0.47
6	0.33	0.30	0.31	0.29	0.30	0.27	0.29	0.26	0.27	0.26
7	0.223	0.193	0.206	0.180	0.192	0.168	0.179	0.18	0.169	0.159
8	0.157	0.134	0.143	0.120	0.130	0.110	0.120	0.12	0.112	0.104
9	0.115	0.094	0.105	0.084	0.093	0.076	0.085	0.09	0.078	0.072
10	0.087	0.069	0.079	0.061	0.069	0.054	0.062		0.057	0.052
15	0.030		0.025		0.021		0.018		0.017	0.015
20	0.014		0.012		0.010		0.009		0.006	0.006
25	0.007		0.006		0.005		0.004		0.003	0.003
30	0.004		0.003		0.003		0.002		0.002	0.001

means are set between T_ϵ equal to 40,000° and to 80,000°, because the correspondence is closest for these temperatures. We draw attention to the fact that the observations seem to agree best with theory in the same region where thermodynamic equilibrium nearly prevails, although we attach no especial significance to this result. The Balmer decrement shows only a very slow change with temperature. Observations of extreme accuracy will be required if electron temperatures are to be determined from line intensities. Selective interstellar absorption, as discussed by Page[6] and by Berman,[5] only adds to the difficulties. Furthermore, as a comparison of Tables

[5] *Ibid.*, p. 890.

[6] *Ibid.*, p. 604.

7*b* and 8 shows, the decrement is much more sensitive to the mode of excitation than it is to electron temperature.

We conclude, therefore, that measures of the Balmer decrement are unsuited to determinations of the temperatures of the electron gas. The result, disappointing at first sight, proves to be fortunate, after all. The observed Balmer decrement can be used to indicate the physical nature of the excitation. For example, the general agreement of the observational data with the entries of Table 8 probably signifies that assumption **B** lies nearer to the truth than **A**.

Attention is also drawn to the close numerical agreement of Tables 7*a* and 7*b* for quantum numbers greater than 4. For the higher series members, the introduction of the Gaunt factors does little to change the relative intensities, although the absolute intensities change by as much as 20 per cent.

One advantage of the present type of solution, aside from its numerical accuracy, is the fact that the absolute intensities of both low and high series members, and even of the continuum beyond the Balmer limit, are determinate. The emission intensity, depending secondarily upon the temperatures, can be used to evaluate nebular densities to a high degree of astrophysical accuracy. Further discussion of the point is deferred until a later paper of the series. We may conclude the present contribution with the remark that perhaps the most surprising result of our investigation is that the individual b_n's fall as close to unity as they do, under conditions apparently far from those of thermodynamic equilibrium.

HARVARD OBSERVATORY
December 9, 1937

Reprinted from *The Astrophysical Journal,* Vol. 88, No. 3, October, 1938.

PHYSICAL PROCESSES IN GASEOUS NEBULAE

IV. THE MECHANISTIC AND EQUILIBRIUM TREATMENT OF NEBULAR STATISTICS

DONALD H. MENZEL, LAWRENCE H. ALLER, AND JAMES G. BAKER[1]

ABSTRACT

The statistical equilibrium of nebulae has been discussed in the past from two points of view. H. H. Plaskett employed a *mechanistic* argument, where the captures in a given level are subdivided according to the probability of transitions to lower levels and are followed through successive transitions down to the ground level. Carroll criticized Plaskett's treatment and developed equations on an *equilibrium* argument. We have shown that the two points of view, when properly applied, lead to the same result; and we have derived the necessary equations, which will lead to a determination of electron temperature.

There are essentially two methods for approaching the problem of the distribution of atoms among the various atomic states of a hydrogen nebula. The first one, historically, used by H. H. Plaskett,[2] involves counting the number of captures in a given atomic level and then tracing the subsequent history of these captured electrons as they cascade to lower atomic levels, some going to one and some to other levels. This point of view, which may be termed the "mechanistic," was criticized by Carroll,[3] who argued that it did not insure statistical equilibrium. In our opinion the criticism is unjustified.

Carroll took what might be called the "equilibrium" point of view, equating the number of atoms arriving in a given level to the number leaving it, without inquiring into the past or future history of a given transition. We shall show later in this paper that the two points of view are essentially equivalent. Carroll gave a rigorous solution for an atom possessing seven discrete levels and a continuum. Plaskett, using the mechanistic argument in calculating for a seven-level atom, was able to allow in part for the equations neglected by Carroll. His results should be, therefore, in closer agreement with

[1] Society of Fellows, Harvard University.

[2] *Harvard Circ.*, No. 335, No. 1928.

[3] *M.N.*, **90**, 588, 1930.

the exact theory than the later ones of Cillié,[4] derived by the method of equilibria.

In calculating the partition of atoms in higher levels, we have found it more convenient to apply the equilibrium approach. But for certain problems the mechanistic point of view is simpler. In evaluating the temperature T_ϵ of the electron gas, for example, in terms of the temperature T_I of the exciting radiation, one may equate mechanistically the number of ionizations from the ground level to the number of captures on all levels. In thermodynamic equilibrium, detailed balancing would ensue and T_ϵ must equal T_I. One of us has shown[5] that the same equality of electron and radiation temperatures would have to exist in an atom with one discrete state, whatever the dilution of the radiation. But when many or an infinite number of levels are considered, the temperatures will not be equal, and $T_\epsilon < T_\mathrm{I}$. In the present paper we shall neglect collisional excitation of all kinds, although such an effect could be easily included. Our aim is to prove that the mechanistic and equilibrium equations are consistent with one another, and that they may be used interchangeably in various types of calculations.

Four fundamental equations of equilibrium may be written down at once in the functional notation of I,[6] and in the form presented below they are symbolically exact. The first equation is that of statistical equilibrium for the ground level, that is to say, it describes the physical condition that the number of atoms entering the ground level per unit time and volume, by all possible routes, must be precisely equal to the number leaving the ground level:

$$\sum_2^\infty F_{n\mathrm{I}} + \int_{\nu_\mathrm{I}}^\infty F_{\kappa\mathrm{I}} d\nu = \sum_2^\infty F_{\mathrm{I}n} + \int_{\nu_\mathrm{I}}^\infty F_{\mathrm{I}\kappa} d\nu \,. \tag{1}$$

Next in order is the equation of statistical equilibrium for the excited level, n. This equation appeared as I (31), and was em-

[4] *Ibid.*, **92**, 820, 1932; **96**, 771, 1936.

[5] D. H. Menzel, *Pub. A.S.P.*, **43**, 334, 1931.

[6] D. H. Menzel, *Ap. J.*, **85**, 330, 1937; D. H. Menzel and J. G. Baker, *ibid.*, **86**, 70, 1937; J. G. Baker and D. H. Menzel, *ibid.*, **88**, 52, 1938.

ployed in a simplified form in II[6] and III[6] for determining the Balmer decrement. Physically, this equation describes the condition that the number of atoms entering an excited level, n, by all possible routes, must equal the number of atoms leaving that level per unit time and volume:

$$\left.\begin{aligned}\sum_{n+1}^{\infty} F_{n''n} + \int_{\nu_n}^{\infty} F_{\kappa n} d\nu + \sum_{1}^{n-1} F_{n'n} = \sum_{n+1}^{\infty} F_{nn''} \\ + \int_{\nu_n}^{\infty} F_{n\kappa} d\nu + \sum_{1}^{n-1} F_{nn'} .\end{aligned}\right\} \quad (2)$$

The third equation of statistical equilibrium describes the physical condition that the number of atoms entering the continuum must be equal to the number of atoms leaving, per unit time and volume:

$$\sum_{1}^{\infty} \int_{\nu_{n'}}^{\infty} F_{n'\kappa} d\nu = \sum_{1}^{\infty} \int_{\nu_{n'}}^{\infty} F_{\kappa n'} d\nu . \quad (3)$$

The last fundamental equation first appeared as I (33) and is rewritten below in a form in which E is replaced by $Fh\nu$. This equation is a description of the physical condition for radiative equilibrium, namely, that the amount of energy absorbed must equal the amount of energy lost per unit time and volume:

$$\left.\begin{aligned}\sum_{1}^{\infty} \int_{\nu_n}^{\infty} F_{\kappa n} h\nu_{\kappa n} d\nu + \sum_{2}^{\infty} \sum_{1}^{n-1} F_{nn'} h\nu_{nn'} + \int_{0}^{\infty} \int_{0}^{\infty} F_{\kappa''\kappa} h\nu_{\kappa''\kappa} d\nu_{\kappa''} d\nu_{\kappa} \\ = \sum_{1}^{\infty} \int_{\nu_n}^{\infty} F_{n\kappa} h\nu_{n\kappa} d\nu + \sum_{2}^{\infty} \sum_{1}^{n-1} F_{n'n} h\nu_{n'n} + \int_{0}^{\infty} \int_{0}^{\infty} F_{\kappa'\kappa} h\nu_{\kappa'\kappa} d\nu_{\kappa'} d\nu_{\kappa} .\end{aligned}\right\} \quad (4)$$

For the gaseous nebulae exposed to dilute temperature radiation with a dilution factor comparable to that found in planetaries, certain of the terms in the foregoing general equation may be neglected. By far the greater proportion of neutral atoms are in the ground level. The problem of radiative transfer becomes important only for the Lyman lines and Lyman continuum. Even with the second level there are too few atoms for a Balmer transfer problem to exist. In other words, the nebula is relatively transparent to the Balmer radiation. The number of atoms entering level n by absorption of radiation in a subordinate line will be negligibly small in comparison with the number cascading down from higher levels or being excited from the ground level. In view of the small population in the upper levels and of the dilution of the radiation, photoelectric ionization from all levels other than the first can also be safely neglected. Omitting all terms involving upward transitions from $n \geqslant 2$, and in the continuum, we can re-write the four fundamental equations in a simplified form:

$$\sum_{2}^{\infty} F_{n1} + \int_{\nu_1}^{\infty} F_{\kappa 1} d\nu = \sum_{2}^{\infty} F_{1n} + \int_{\nu_1}^{\infty} F_{1\kappa} d\nu \,, \tag{5}$$

$$\sum_{n+1}^{\infty} F_{n''n} + \int_{\nu_n}^{\infty} F_{\kappa n} d\nu + F_{1n} = \sum_{1}^{n-1} F_{nn'} \,, \tag{6}$$

$$\int_{\nu_1}^{\infty} F_{1\kappa}\, d\nu = \sum_{1}^{\infty} \int_{\nu_n}^{\infty} F_{\kappa n} d\nu = \sum_{2}^{\infty} \int_{\nu_n}^{\infty} F_{\kappa n} d\nu + \int_{\nu_1}^{\infty} F_{\kappa 1} d\nu \,, \tag{7}$$

$$\left.\begin{aligned} \sum_{1}^{\infty} \int_{\nu_n}^{\infty} F_{\kappa n} h\nu_{\kappa n} d\nu + \sum_{2}^{\infty} \sum_{1}^{n-1} F_{nn'} h\nu_{nn'} + \int_{0}^{\infty} \int_{0}^{\infty} F_{\kappa\kappa'} h\nu_{\kappa\kappa'} d\nu_{\kappa} d\nu_{\kappa'} \\ = \int_{\nu_1}^{\infty} F_{1\kappa} h\nu_{1\kappa} d\nu + \sum_{2}^{\infty} F_{1n} h\nu_{1n} \,. \end{aligned}\right\} \tag{8}$$

To this set of equations we now add a fifth equation, which may be regarded as mechanistic in nature. The amount of energy gained by the continuum by ionization from the ground level reappears in emission by captures on all the levels of the atom and subsequent cascading to the ground level, and also in free-free emissions. The number of atoms captured in level n per unit time and volume must be equal to the number contributed by level n to the ground level per unit time and volume by downward transitions with a net energy loss of amount $h\nu_{n1}$. Consequently, we have

$$\left.\begin{aligned}\int_{\nu_1}^{\infty} F_{1\kappa} h\nu_{1\kappa} d\nu &= \sum_{1}^{\infty}\int_{\nu_n}^{\infty} F_{\kappa n} h(\nu_{\kappa n}+\nu_{n1}) d\nu + \int_0^{\infty}\int_0^{\infty} F_{\kappa\kappa'} h\nu_{\kappa\kappa'} d\nu_\kappa d\nu_{\kappa'} \\ &= \sum_{1}^{\infty}\int_{\nu_n}^{\infty} F_{\kappa n} h\nu_{\kappa n} d\nu + \sum_{2}^{\infty}\int_{\nu_n}^{\infty} F_{\kappa n} h\nu_{n1} d\nu + \int_0^{\infty}\int_0^{\infty} F_{\kappa\kappa'} h\nu_{\kappa\kappa'} d\nu_\kappa d\nu_{\kappa'} \,.\end{aligned}\right\} \quad (9)$$

We may establish the aforementioned consistency of the mechanistic and statistical points of view very simply by manipulation of the preceding five equations. First, we sum (6) from 2 to infinity in order to render it independent of n, and then we obtain

$$\sum_{2}^{\infty}\sum_{n+1}^{\infty} F_{n''n} + \sum_{2}^{\infty}\int_{\nu_n}^{\infty} F_{\kappa n} d\nu + \sum_{2}^{\infty} F_{1n} = \sum_{2}^{\infty}\sum_{1}^{n-1} F_{nn'} \,. \quad (10)$$

The relationship

$$\sum_{2}^{\infty}\left\{\sum_{n+1}^{\infty} F_{n''n} - \sum_{2}^{n-1} F_{nn'}\right\} = 0 \quad (11)$$

is easily proved by expanding the symbolic summations, which will be found to be identical. Each summation reduces to a series of the form

$$F_{32} + F_{42} + F_{52} + \ldots + F_{43} + F_{53} + F_{63} + \ldots + F_{54} + \ldots + \ldots$$

Hence, (10) becomes

$$\sum_{2}^{\infty}\int_{\nu_n}^{\infty} F_{\kappa n} d\nu + \sum_{2}^{\infty} F_{1n} = \sum_{2}^{\infty} F_{n1} \,. \quad (12)$$

Using (7), we obtain

$$\int_{\nu_1}^{\infty} F_{1\kappa} d\nu + \sum_{2}^{\infty} F_{1n} = \sum_{2}^{\infty} F_{n1} + \int_{\nu_1}^{\infty} F_{\kappa 1} d\nu \,, \quad (13)$$

which is (1). Therefore, equations (5), (6), and (7) are consistent with one another. We now must show that (8) and (9) are consistent with one another and with (5), (6), and (7). Upon substituting (9) into (8), we obtain

$$\left.\begin{aligned}\sum_{2}^{\infty}\sum_{1}^{n-1} F_{nn'}h\nu_{nn'} &= \sum_{2}^{\infty}\int_{\nu_n}^{\infty} F_{\kappa n}h\nu_{n1}d\nu + \sum_{2}^{\infty} F_{1n}h\nu_{n1}\\ &= \sum_{2}^{\infty} h\nu_{n1}\left[\int_{\nu_n}^{\infty} F_{\kappa n}d\nu + F_{1n}\right]\\ &= \sum_{2}^{\infty} h\nu_{n1}\left[\sum_{1}^{n-1} F_{nn'} - \sum_{n+1}^{\infty} F_{n''n}\right]\\ &= \sum_{2}^{\infty}\sum_{1}^{n-1} F_{nn'}h\nu_{n1} - \sum_{2}^{\infty}\sum_{n+1}^{\infty} F_{n''n}h\nu_{n1}\,,\end{aligned}\right\} \tag{14}$$

where (6) has been used.

We next replace $h\nu_{nn'}$ in (14) by $(h\nu_{n1} - h\nu_{n'1})$, and find that

$$\sum_{2}^{\infty}\sum_{2}^{n-1} F_{nn'}h\nu_{n'1} = \sum_{2}^{\infty}\sum_{n+1}^{\infty} F_{n''n}h\nu_{n1}\,. \tag{15}$$

A consideration of the uniform convergence of this expression will readily show that it is really an identity. As was indicated for (11), the double summations on both sides of (15) consist of the same elements. The five equations, (5), (6), (7), (8), and (9), are thus shown to be consistent, and the mechanistic and equilibrium points of view are reconciled. The next paper of this series will employ these equations to find the relationship between the excitation and electron temperatures of a nebula.

HARVARD OBSERVATORY

Reprinted from *The Astrophysical Journal*, Vol. 88, No. 4, November, 1938.

PHYSICAL PROCESSES IN GASEOUS NEBULAE

V. ELECTRON TEMPERATURES

JAMES G. BAKER,[1] DONALD H. MENZEL, AND
LAWRENCE H. ALLER

ABSTRACT

The relation between the electron temperature of a gaseous nebula and the temperature of the exciting star is derived and applied to an optically thin nebula. The b_n's have been recomputed both with the radiation field included and as a function of the temperature of the exciting star. The Balmer decrement proves to be steeper when the influence of Lyman line absorption is included.

The equations derived in papers I, II, and IV of this series may be applied to the problem of calculating the electron temperature of the nebular gas. We make the following assumptions: the electrons are produced by photo-ionization; radiationless encounters between the various components of the assembly are sufficiently frequent that the electrons attain a Maxwellian velocity distribution; the nebula is in a steady state with respect to the distribution of atoms in various levels; the energy emitted per unit time must equal the energy absorbed.

The calculations of paper III gave the values of the quantities $b_2, b_3, \ldots$ etc. The results are based on the equations for statistical equilibrium of quantum levels 2 and higher. These equations are sufficient to determine the absolute nebular emission and the Balmer decrement when N_ϵ and T_ϵ are given; we have still to predict the values of the two latter quantities from the physical parameters of the exciting radiation. In addition, there is the third parameter, b_1—which defines the number of atoms in the ground level—to be determined. We may suppose, for the sake of simplicity, that the radiation from the nuclear star is "black," at temperature T_1, with the dilution factor W specified. There are still two equations from paper I or paper IV that have not been employed, viz., the one defining the statistical equilibrium of level 1 and the one expressing conservation of emitted and absorbed radiation. A third relationship is ob-

[1] Society of Fellows, Harvard University.

tained when the density of the nebula is given, hence the problem is determinate. One may substitute the previously evaluated b_n's in the two remaining equations and calculate b_1, N_ϵ, and T_ϵ.

Fortunately, for the gaseous nebulae, where the radiation is very dilute one may determine T_ϵ directly without first evaluating the intermediate b_n's. T_ϵ proves to be independent either of the density or of the dilution factor, as long as the latter is small. The simplification is obtained by means of the mechanistic approach developed in paper IV. The numerical results are given for an optically thin nebula, by which term we mean a nebula of optical thickness small enough that the spectral energy curve of the star, beyond the Lyman limit, is not seriously depleted by absorption effects.

It now seems advisable to write out the necessary equations in the form in which they are to be used. We shall use the equations and the notation of papers I and II,[2] writing out first the explicit forms of the equations IV (3), IV (4), and IV (9). Adopting as before $b_\kappa = 1$, we find that IV (3) becomes, in view of the definitions of II and equations IV (14), I (25),

$$b_1 \int_{\nu_1}^{\infty} W_{1\kappa}\left(\frac{e^{h\nu_1/kT_\epsilon}}{e^{h\nu/kT_1}-1}\right)\frac{d\nu}{\nu} = \sum_{1}^{\infty}\frac{S_n}{n^3} + \int_{\nu_1}^{\infty} W_{1\kappa}\left(\frac{e^{-h(\nu-\nu_1)/kT_\epsilon}}{e^{h\nu/kT_1}-1}\right)\frac{d\nu}{\nu}. \quad (1)$$

It must be borne in mind that $W_{1\kappa}$ is a very small quantity of the order of 10^{-15}. The second term on the right of the foregoing equation represents the negative absorptions. It follows that

$$e^{+h\nu_1/kT_\epsilon}\int_{\nu_1}^{\infty}\frac{W_{1\kappa}e^{-h\nu/kT_\epsilon}}{e^{h\nu/kT_1}-1}\frac{d\nu}{\nu} < \frac{e^{h\nu_1/kT_\epsilon}}{e^{h\nu_1/kT_1}-1}\int_{\nu_1}^{\infty} W_{1\kappa}e^{-h\nu/kT_\epsilon}\frac{d\nu}{\nu}$$

$$< \frac{\dfrac{kT_\epsilon}{h\nu_1}}{e^{h\nu_1/kT_1}-1}\overline{W}_{1\kappa} < \frac{\dfrac{kT_1}{h\nu_1}}{e^{h\nu_1/kT_1}-1}\overline{W}_{1\kappa} \leq 3W_{1\kappa}\,.$$

Therefore, the effect of induced emissions, i.e., induced captures, is negligibly small. From (1) we see that b_1 is a very large quantity determined principally by the magnitude of $W_{1\kappa}$. Terms that con-

[2] D. H. Menzel, *Ap. J.*, **85**, 330, 1937; D. H. Menzel and J. G. Baker, *Ap. J.*, **86**, 70, 1937; J. G. Baker and D. H. Menzel, *Ap. J.*, **88**, 52, 1938; D. H. Menzel, L. H. Aller, and J. G. Baker, *Ap. J.*, **88**, 313, 1938.

tain both $W_{1\kappa}$ and b_1 as a product cannot be neglected, but the terms that contain $W_{1\kappa}$ without the b_1 factor can readily be neglected. This amounts to a formal proof of our previous statement that the negative absorptions from all levels and the positive absorptions from all levels but the first are negligible. Our equations may now be written

$$b_1 e^{X_1} \int_{y_1}^{\infty} W_{1\kappa} \frac{dy}{y(e^y - 1)} = \sum_{1}^{\infty} \frac{S_n}{n^3} = G_{T_\epsilon} , \tag{2}$$

which defines G_{T_ϵ}. The equation of radiative equilibrium for the assembly is

$$\left.\begin{aligned} &T_\epsilon^2 \left(\frac{k^2}{2hRZ^2}\right) + akT_\epsilon + 2hrZ^2 \sum_{n=2}^{\infty} \sum_{n'=1}^{n-1} b_n \frac{1}{n'^3} \frac{1}{n^3} e^{X_n} \\ &- b_1 e^{X_1} \sum_{2}^{\infty} W_{1n''} \frac{2hRZ^2}{e^{h\nu/kT_1} - 1} \frac{1}{n''^3} - b_1 e^{X_1} kT_1 \int_{y_1}^{\infty} W_{1\kappa} \frac{dy}{e^y - 1} = 0 , \end{aligned}\right\} \tag{3}$$

where

$$a = \sum_{1}^{\infty} \frac{1}{n^3} .$$

The equation of radiative equilibrium for the continuum becomes

$$\left.\begin{aligned} b_1 e^{X_1} \left(\frac{k}{2hRZ^2}\right) T_1 \int_{y_1}^{\infty} W_{1\kappa} \frac{dy}{e^y - 1} &= \left(\frac{k}{2hRZ^2}\right)^2 T_\epsilon^2 \\ &+ \left(\frac{k}{2hRZ^2}\right) aT_\epsilon + \tfrac{1}{2} G_{T_\epsilon} - \tfrac{1}{2} \sum_{1}^{\infty} \frac{S_n}{n^5} . \end{aligned}\right\} \tag{4}$$

Equations IV (1) and IV (2) may be discussed in a similar manner, and we shall treat the problem later in connection with the derivation of the formula for b_n. It is to be pointed out that all these equations contain b_1, T_ϵ, and T_1. One might expect, therefore, that it would be possible to choose any two of them, eliminate b_1, and derive a relation between T_ϵ and T_1. In particular, one might suppose that we could choose any two of equations (2), (3), or (4), solve each for b_1, and equate. Our choice is really somewhat more limited than that. If we suppose, for example, the star to radiate as a black body with no absorption lines in the Lyman series, we do not know the b_n's and therefore cannot use equation (3). Our procedure in this case would be to solve (2) and (4) for b_1 and equate. In the problem where we supposed no radiation to leave the star in the Lyman lines but where the star is supposed to shine as a black body beyond the Lyman limit, we know the b_n's and can therefore apply any two of these three equations. Numerical calculations based on (2) and (3) gave exactly the same relation between T_ϵ and T_1 as those based on (2) and (4). The agreement provides an excellent check on the formulas and computations. Note that the relation between T_ϵ and T_1 does not depend on whether the central star has any line radiation or not. It is to be mentioned that the expressions for b_1 from (2) and (4) are equally good for a static or an expanding nebula because absorption in the continuum is not seriously affected by a Doppler shift. From (2) and (4) we have for the optically thin nebula, where $W_{1\kappa}$ does not depend on ν,

$$\left.\begin{aligned} &\frac{\left(\dfrac{k}{2hRZ^2}\right) T_1 \displaystyle\int_{y_1}^{\infty} \frac{dy}{e^y - 1}}{\displaystyle\int_{y_1}^{\infty} \frac{dy}{y(e^y - 1)}} \\ &\qquad = \frac{\left(\dfrac{k}{2hRZ^2}\right)^2 T_\epsilon^2 + a\left(\dfrac{k}{2hRZ^2}\right) T_\epsilon + \frac{1}{2} G_{T_\epsilon} - \frac{1}{2}\displaystyle\sum_1^{\infty} \frac{S_n}{n^5}}{G_{T_\epsilon}} \end{aligned}\right\} \quad (5)$$

or, symbolically,

$$f_1(T_1) = f_2(T_\epsilon) ,$$

since the left side of the equation depends only on T_1 and the right side only on T_ϵ. For any assigned T_1 the corresponding T_ϵ may be readily found. Each side of the equation is plotted as a function of its appropriate temperature. A specified value of T_1 determines $f_1(T_1)$, which must equal $f_2(T_\epsilon)$. From the graph of $f_2(T_\epsilon)$ the appropriate T_ϵ may then be ascertained. It is to be mentioned that the quadratic $f_2 - f_1$ gives but one real positive root.

The relation between T_ϵ and T_1 is given in the following table for the optically thin nebula. The Gaunt g factor was taken equal to unity in the calculations.

T_1	T_ϵ
5,000	5,000
10,000	9,500
20,000	18,000
40,000	34,000
80,000	57,000
160,000	92,000
320,000	132,000

Note that for the higher temperatures the electron temperature departs farther and farther from the temperature of the source. For very high temperatures the effect of the neglect of the g factor produces an error of at most 5 per cent in the foregoing results. At low temperatures the error is negligible.

From equation (2) or (4) we can determine b_1. The fact that its order of magnitude must be determined by the dilution factor follows at once from elementary considerations. Comparing the gaseous nebula with an assembly in thermodynamic equilibrium at temperature T_ϵ, we see that the captures must go at essentially the same rate but that the photo-ionizations are fewer by a factor of about $W_{1\kappa}$. As a result, there appears an enormous excess of atoms in the ground state as compared with the assembly in thermodynamic equilibrium.

The ionization equation for hydrogen follows at once. We shall defer discussion, however, because ionization is unimportant so far as the present investigation is concerned.

We now write equation IV (2) in terms of the physical quantities involved. In view of I (7), and using the notation of II (3), we find, neglecting the terms multiplied by W and not by $b_1 W$,

$$\left.\begin{aligned} \frac{N_i N_\epsilon K Z^4}{n^3 T_\epsilon^{3/2}} \Bigg\{ & \sum_{n''=n+1}^{\infty} b_{n''} e^{X_{n''}} g_{n''n} u_{n''n} + \bar{g} S_n \\ & + \left[W_{1n} \frac{b_1 e^{h\nu/kT_\epsilon}}{e^{h\nu/kT_1} - 1} \right] \frac{2n^2 e^{X_n}}{n^2 - 1} - e^{X_n} b_n t_n \Bigg\} = 0 \,. \end{aligned} \right\} \quad (6)$$

TABLE 1

STELLAR TEMPERATURE, ELECTRON TEMPERATURE, AND ASSOCIATED VALUES OF b_n

n	T_1 5,000°	10,000°	20,000°	40,000°	80,000°	160,000°
	T_ϵ 5,000°	9,500°	18,000°	34,000°	57,000°	92,000°
	b_n (Neglecting Radiation Field)					
2	0.00003	0.0029	0.027	0.090	0.220	0.388
3	.0033	.025	.089	.195	.324	.471
4	.018	.068	.151	.270	.392	.526
5	.043	.109	.210	.324	.444	.560
6	.072	.150	.251	.366	.482	.585
7	.099	.182	.290	.400	.511	.605
8	.126	.213	.319	.425	.531	.621
9	.150	.237	.341	.449	.549	.639
10	.171	.261	.363	.469	.563	.650
11	.191	.281	.381	.481	.576	.659
12	.209	.299	.396	.492	.588	.666
13	.223	.313	.409	.503	.598	.673
14	.238	.328	.420	.511	.604	.679
15	0.251	0.339	0.430	0.520	0.610	0.685
	b_n (Including Radiation Field)					
2	2.98	2.567	2.369	2.281	2.181	2.393
3	1.72	1.608	1.603	1.561	1.429	1.551
4	1.48	1.414	1.411	1.367	1.258	1.356
5	1.40	1.33	1.33	1.29	1.19	1.27
6	1.37	1.29	1.27	1.25	1.16	1.22
7	1.36	1.25	1.25	1.22	1.14	1.19
8	1.34	1.22	1.22	1.20	1.12	1.17
9	1.32	1.20	1.20	1.19	1.11	1.16
10	1.31	1.18	1.18	1.18	1.10	1.15
11	1.28	1.17	1.17	1.17	1.09	1.14
12	1.26	1.15	1.16	1.16	1.09	1.13
13	1.23	1.14	1.15	1.15	1.08	1.13
14	1.20	1.13	1.14	1.14	1.08	1.12
15	1.19	1.12	1.13	1.14	1.07	1.11

This equation may be solved in a manner similar to the method explained in paper II, once the W_{1n} factors are known. We have

$$b_n = \frac{S_{nr} + V_{nr}}{e^{X_n t_n}}, \tag{7}$$

where

$$\left.\begin{aligned} S_{nr} &= \left[S_n + W_{1n}\left(\frac{b_1 e^{h\nu/kT_\epsilon}}{e^{h\nu/kT_1} - 1}\right)\frac{2n^2 e^{X_n}}{n^2 - 1}\right] \\ &= \left[S_n + \frac{W_{1n}}{W_{1\kappa}} \frac{G_{T_\epsilon}}{\displaystyle\int_{y_1}^{\infty} \frac{dy}{y(e^y - 1)}} \frac{\dfrac{2n^2}{n^2 - 1}}{(e^{h\nu/kT_1} - 1)}\right]. \end{aligned}\right\} \tag{8}$$

For an optically thin nebula surrounding a star radiating as a black body, $W_{1n} = W_{1\kappa}$. When we consider the transfer problem, this relation will no longer hold. V_{nr} is defined in terms of S_{nr} in the same manner that V_n was defined in terms of S_n.

The mathematical grouping of these terms must not be interpreted as due to their physical similarity. S_n arises from captures on the level in question; the W_{1n} term originates from transitions from the ground level to the nth level.

If the star is assumed to radiate as a black body, the b_n's are considerably different from the b_n's discussed in paper III. For small values of n they are larger than unity and gradually approach 1 as $n \rightarrow \infty$, as indeed they must. For purposes of comparison, b_n is tabulated as a function of the temperature of the central star for the two cases considered.

Computation (Table 1) shows the Balmer decrement to be very insensitive to the excitation temperature. It should be pointed out that the optically thin nebula here considered will have a brilliance much less than that of the nuclear star, so that the present results are not directly applicable to planetary nebulae. But the enormous effect of the nuclear radiation suggests that even in an optically thick nebula the nuclear radiation might show some influence on the Balmer decrement, which would tend to be steeper in consequence.

In the next paper the theory of radiative transfer for the Lyman levels and continuum will be discussed.

HARVARD OBSERVATORY

Reprinted from *The Astrophysical Journal,* Vol. 89, No. 5, June, 1939.

PHYSICAL PROCESSES IN GASEOUS NEBULAE

VI. THE EQUATIONS OF RADIATIVE TRANSFER

LAWRENCE H. ALLER, JAMES G. BAKER,[1] AND DONALD H. MENZEL

ABSTRACT

The general equations of transfer for bound-bound, bound-free, and free-free transitions for a hydrogenic nebula are derived from the usual macroscopic equations by the identification of the emissivity and absorptivity functions with those given in Paper I of this series. The integration of these equations is reserved for another paper.

In the previous papers[2] we have considered the problem of the decrement of the Balmer series and have shown the decrement to be very sensitive to the method of excitation. Thus far we have distinguished three cases: A, B, and C. Cases A and B are based on a model of a nebula in which the nuclear star radiates no energy in the Lyman lines. Case A is for a nebula transparent in the lines, so that all the Lyman-line radiation, produced in the nebula as a result of cyclic processes, is allowed to escape from the nebula. In case B none of the Lyman-line radiation, except Lyman α, escapes. This is the extreme case considered by Zanstra. The third model, C, is for a relatively transparent nebula (as in case A) but with a nuclear star radiating like a black body in the region of the Lyman lines. The additional stellar line radiation absorbed by the nebula alters the decrement markedly, raising the intensity of $H\alpha$ relative to $H\beta$.

In Table 1 we summarize the theoretical Balmer decrements for various values of the electron temperatures according to the three cases. The tabulated value is the intensity of $H\alpha$ referred to $H\beta$ as unity. The observational data for the planetary nebulae agree best, as we have shown, with the results of case B.

The surface brightness of a nebula that follows either case A or case C will be low. Such nebulae will, accordingly, be difficult to observe spectroscopically. Greenstein and Henyey[3] have observed,

[1] Society of Fellows, Harvard University.

[2] Paper II in *Ap. J.*, **86**, 70, 1937; Paper III in *ibid.*, **88**, 52, 1938; Paper V in *ibid.*, **88**, 422, 1938.

[3] *Ap. J.*, **87**, 79, 1938.

in certain faint extended nebulae, what appears to be a high Balmer decrement. Greenstein kindly warns us, by letter, that the observation is still open to question because of the possible blending of *Hα* with the nebular lines of *N* II. Whatever may be the explanation of this phenomenon, a high Balmer decrement leads to no theoretical difficulty, since it may be explained in terms of case C.

The great theoretical differences in decrement according to various models show that a more exact treatment of the nebular radiation field is to be desired. In the present paper we lay the basis for a more thorough analysis of the transfer problem.

TABLE 1

T_ϵ (K)	A	B	C
5,000	1.86	2.43	12.5
10,000	1.92	2.50	6.0
20,000	1.98	2.59	4.0
40,000	2.07	2.71	3.3
80,000	2.17	2.83	3.0

In the notation of paper I we may write down the equations of transfer almost at once. We consider the nebulous envelope to be stratified in parallel planes, since Milne has shown that the problem of the spherical shell can be reduced to a plane problem as far as radiative transfer is concerned. We choose an x-axis with co-ordinates increasing outward. Let a beam of energy of specific intensity $I(x, \theta)$, confined within a solid angle $d\omega$ making an angle θ with the positive direction of x, fall on a slab of 1 square centimeter in cross-section and of thickness dx. The total amount of energy removed from the beam by a transition from state n' to n will be

$$\frac{1}{4\pi} E_{n'n}\, d\omega\, dx \sec\theta ,$$

where we have integrated $I(x, \theta)$ with respect to the frequency over the line. The energy returned to the beam by the emissions in $d\omega$ is

$$\frac{1}{4\pi} E_{nn'}\, d\omega\, dx \sec\theta .$$

The induced emissions, treated as negative absorptions, are included in the term $E_{n'n}$. To derive the transfer equations we must consider the emission and absorption in the frequency range $d\nu$. The two preceding expressions may be multiplied by a function $\varphi(\nu - \nu_{nn'}) \equiv \varphi_\nu$, which represents the profile of the absorption or emission coefficient. We must have

$$\int_0^\infty \varphi(\nu - \nu_{nn'})d\nu = 1 . \qquad (1)$$

We note, also, that

$$a_{n'n}(\nu) = \frac{\pi\epsilon^2}{mc} f \varphi_\nu . \qquad (2)$$

Should the spectral distribution of the emitted quantum not be equal to that of the absorbed, we must introduce different expressions for φ_ν for the two processes.

The equation of transfer for the lines now becomes

$$\cos\theta \frac{dI_\nu}{dx} = -\frac{1}{4\pi}[E_{n'n} - E_{nn'}]\varphi_\nu . \qquad (3)$$

The equivalent form

$$\cos\theta \frac{dI_\nu}{dx} = -\frac{1}{4\pi}[F_{n'n} - F_{nn'}]h\nu\, \varphi_\nu , \qquad (4)$$

which refers to the number of transitions, is sometimes useful.

In an analogous manner the transfer equation for the bound-free transitions for a particular ν may be written

$$\cos\theta \frac{dI_\nu}{dx} = -\frac{1}{4\pi}\sum_{n'}[E_{n'\kappa} - E_{\kappa n'}] . \qquad (5)$$

The summation over n' is due to the fact that radiation of a given frequency may arise by capture on any one of a number of discrete levels from various elements $d\kappa$ of the continuum, such that

$$\nu = RZ^2\left(\frac{1}{n'^2} + \frac{1}{\kappa^2}\right) .$$

Since $E_{1\kappa} = F_{1\kappa} h\nu$, we obtain the equivalent form

$$\cos\theta \frac{dI_\nu}{dx} = -\frac{1}{4\pi} \sum_{n'} [F_{n'\kappa} - F_{\kappa n'}] h\nu \,. \tag{6}$$

Similarly, the equation of transfer for the case of free-free transitions becomes

$$\cos\theta \frac{dI_\nu}{dx} = -\frac{1}{4\pi}\int_0^\infty [E_{\kappa'\kappa} - E_{\kappa\kappa'}] d\nu_{\kappa'} = -\frac{1}{4\pi}\int_0^\infty [F_{\kappa'\kappa} - F_{\kappa\kappa'}] h\nu d\nu_{\kappa'} \,. \tag{7}$$

The integration in (7) is carried out subject to the condition

$$\nu = RZ^2 \left(\frac{1}{\kappa^2} - \frac{1}{\kappa'^2}\right) , \tag{8}$$

where ν is held constant.

Although we have represented the transfer in terms of three separate equations, one for each process, we shall have, in general, all three types of processes acting at any given frequency, and the right-hand side of the transfer equation will include all three terms. Frequently, of course, only one will predominate. We should further note that, when neighboring lines of a series overlap, as always occurs near a series limit, the right-hand side of equations (3) and (4) should include a summation over the significant values of n.

We now consider the transformation of these equations with the aid of expressions developed in earlier papers for the emissivity and absorptivity terms. We first discuss the equation of transfer for the bound-bound processes.

Recalling I-4, I-5, and I-9, and making use of (3) or (4), we obtain

$$\cos\theta \frac{dI_\nu}{dx} = -\frac{N_i N_\epsilon}{4\pi\sigma} \frac{KZ^4}{T_\epsilon^{3/2}} \frac{g}{n'^3} e^{X_{n'}} \left\{ [b_{n'} - b_n e^{-h\nu/kT_\epsilon}] I_\nu - b_n \sigma e^{-h\nu/kT_\epsilon} \right\} \frac{2hRZ^2}{n^3} \varphi_\nu \,, \tag{9}$$

where $\sigma = 2h\nu^3/c^2$. In the foregoing form the equation of transfer is useful for problems in which the variation of absorption coefficient with wave length is of prime importance.[4]

[4] E.g., this form may be convenient for a discussion of absorption-line profiles in stellar atmospheres.

For the problem at hand, considerable simplification may be effected without too great a sacrifice of accuracy if, at least in the earlier approximations, we assume an average absorption coefficient over the line. Let $\Delta\nu_{nn'}$ be the effective half-width of the absorption coefficient. Accordingly, we may define a mean φ_ν such that

$$\int_0^\infty \varphi_\nu \, d\nu = \overline{\varphi}_\nu \, \Delta\nu = 1 \,, \qquad \text{or} \qquad \overline{\varphi}_\nu = \frac{1}{\Delta\nu} \,. \tag{10}$$

Equation (9) now becomes

$$\cos\theta \frac{dI_\nu}{dx} = -\frac{N_i N_\epsilon}{4\pi\sigma} \frac{KZ^4}{T_\epsilon^{3/2}} \frac{g}{n'^3} e^{X_{n'}} \left\{ [b_{n'} - b_n e^{-h\nu/kT_\epsilon}] I_\nu - b_n \sigma e^{-h\nu/kT_\epsilon} \right\} \frac{2hRZ^2}{n^3} \frac{1}{\Delta\nu_{nn'}} \,. \tag{11}$$

It is interesting to note that the frequency difference, $\Delta\nu'$, between two successive members of a given series, is

$$\Delta\nu' = RZ^2 \left[\frac{1}{n^2} - \frac{1}{(n+1)^2} \right] \sim \frac{2RZ^2}{n^3} \,. \tag{12}$$

Hence, if the line broadening is such that successive lines just overlap,

$$\Delta\nu_{nn'} \sim \Delta\nu' \,, \tag{13}$$

and the factor

$$\frac{2RZ^2}{n^3} \frac{1}{\Delta\nu_{nn'}} \sim 1 \,. \tag{14}$$

Near the series limit, where $b_n \to 1$, we see that (9) must run continuously into the analogous equation for the bound-free transitions, which is easily found to be

$$\cos\theta \frac{dI_\nu}{dx} = -\frac{N_i N_\epsilon}{4\pi\sigma} \frac{KZ^4}{T_\epsilon^{3/2}} \sum_{n'_{\min}}^{\infty} \frac{g}{n'^3} e^{X_{n'}} \{ [b_{n'} - e^{-h\nu/kT_\epsilon}] I_\nu - \sigma e^{-h\nu/kT_\epsilon} \} \,. \tag{15}$$

The summation of (16) must obviously be extended over only those values of n' whose series limits lie to the red of the frequency in question. The quantity $n'_{\min}$ is thus seen to be the smallest value of n' that will satisfy the relation

$$\frac{RZ^2}{n'^2} \leq \nu \, . \tag{16}$$

For free-free transitions the equation becomes

$$\cos\theta \frac{dI_\nu}{dx} = -\frac{N_i N_\epsilon}{4\pi\sigma} \frac{KZ^4}{T_\epsilon^{3/2}} \frac{kT_\epsilon}{2hRZ^2} \bar{g} \{[1 - e^{-h\nu/kT_\epsilon}]I_\nu - 1\} \, . \tag{17}$$

Note that, since we are dealing here entirely with the continuum (assumed to be in thermodynamic equilibrium), the b-factors do not enter. The present transfer equations differ from those given by Milne,[5] Ambarzumian,[6] Chandrasekhar,[7] etc., in the specific character of the absorption coefficients. The number of discrete states has not been limited to two, three, or four, as is commonly done. These equations are quite general and involve no assumptions concerning the dilution of the radiation. They are equally applicable to stellar atmospheres or gaseous nebulae,[8] although in the former case a solution involves great complexities. The ratios of the successive values of the b_n's will determine whether a line appears in absorption or emission. Fluorescent cycles and interlocking effects are all taken into account by the equations of transfer used in conjunction with those of radiative and of statistical equilibrium. The latter can be generalized to include the effects of collisional excitation and de-excitation of the stationary states.

The solution of the transfer equations for the lines even in the nebular problem is very difficult, as one must consider the interlocking between the continuum and the discrete states. For the Lyman continuum the equation of transfer can be solved in a straightforward manner by means of successive approximations. The solution of the transfer equations for the free-free transitions is very simple.

[5] *Zs. f. Ap.*, **1**, 98, 1930.

[6] *M.N.*, **93**, 50, 1932.

[7] *Zs. f. Ap.*, **9**, 266, 1935.

[8] D. H. Menzel, *Pop. Astr.*, **47**, 6, 1939.

We shall see that we can set up and solve the transfer equation for the Lyman continuum quite independently of any knowledge of the population of the discrete levels (except the first). For each point in the nebula the solution will be characterized by a value of the dilution factor of the Lyman continuum radiation, the value of b_1, and the electron temperature T_ϵ. When the transfer problem for the lines is considered, we shall see that an intimate knowledge of T_ϵ, and therefore of both the dilution factor and b_1, is required, because the population of the higher states depends largely on the electron captures. Before we can attack the problem of radiation transfer for the lines, we must solve it for the Lyman continuum. In the next paper of the series we shall discuss the transfer of radiation in the Lyman continuum.

Harvard University
December 19, 1938

Reprinted from *The Astrophysical Journal,* Vol. 90, No. 2, September, 1939.

PHYSICAL PROCESSES IN THE GASEOUS NEBULAE

VII. THE TRANSFER OF RADIATION IN THE LYMAN CONTINUUM

JAMES G. BAKER,[1] LAWRENCE H. ALLER,[1] AND
DONALD H. MENZEL

ABSTRACT

The differential equations for the transfer of radiation in the Lyman continuum are derived. In order to take into account the varying quality of the radiation as it passes through the nebula and also the variation of the electron temperature, it is necessary to consider the transfer in each element of the continuum rather than to treat the continuum as a single level. The equation of transfer, as written down for a particular frequency in the continuum, therefore contains a term that describes the "interlocking" between the elements of the continuum. The application of these equations to a numerical example is deferred to a later paper.

In the preceding paper[2] we have given the equations of transfer for bound-bound, bound-free, and free-free transitions. For the case of a gaseous nebula exposed to dilute temperature radiation, these equations admit of certain simplifications and modifications. In this paper we shall consider the transfer of radiation in the Lyman continuum for a model planetary nebula consisting of a shell whose thickness is small compared with its radius. For this problem it is not necessary to suppose that the nebular shell is in static equilibrium. The rates of expansion of the planetary nebulae, as observed by Campbell and Moore,[3] introduce Doppler shifts of only negligible importance in the continuum. In the case of the Lyman lines, however, the problem of transfer as discussed in paper VI is further complicated by these internal motions in the nebular shell. Even for the simple case of a homogeneous gaseous shell expanding at a constant rate, the boundary conditions for the differential equations of transfer become involved.

[1] Society of Fellows, Harvard University.

[2] "Physical Processes in the Gaseous Nebulae," I, *Ap. J.*, **85**, 330, 1937; II, *ibid.*, **86**, 70, 1937; III, *ibid.*, **88**, 52, 1938; IV, *ibid.*, p. 313; V, *ibid.*, p. 422; VI, *ibid.*, **89**, 587, 1939. The papers of this series are referred to in the text as papers I, II, III, etc.

[3] *Pub. Lick Obs.*, **13**, 77, 1918.

The transfer equation for the bound-free transitions for a particular frequency ν may be written as given in equation VI (5):

$$\cos\theta \frac{dI_\nu}{dx} = -\frac{1}{4\pi}\sum_{n'}(E_{n'\kappa} - E_{\kappa n'}) . \tag{1}$$

Since the same frequency may be emitted by the capture of a very fast electron on the second or higher level as by the capture of a much slower electron on the first level, we have summed over n'. This overlapping effect will be of importance, however, only for extremely high temperatures. Ordinarily, for a given spectral region and almost certainly for the frequencies in the Lyman continuum, only the first term of the summation need be retained. Since the energy absorbed or emitted is related to the number of transitions by means of the formula $E_{1\kappa}d\nu = F_{1\kappa}h\nu d\nu$, the equation of radiative transfer becomes:

$$\cos\theta \frac{dI_\nu}{dx} d\nu = -\frac{1}{4\pi}[F_{1\kappa} - F_{\kappa 1}]h\nu d\nu . \tag{2}$$

In order to solve this equation we must find a relation between $F_{1\kappa}$ and $F_{\kappa 1}$. Note that equation (2) is written for an element $d\nu$ of the continuum. Previous investigators of the transfer problem have introduced the approximation of considering the continuum as a single level. The absorption coefficient was taken as constant over the level. We shall remove this severe restriction. If we are to study the variation of electron temperature and of the radiation field through the nebula, it will be necessary to consider the transfer problem in each element of the continuum rather than to treat the continuum as a single level.

In a steady state the number of electrons entering the continuum must be equal to the number of electrons leaving it. As we have shown in paper V, the number of ionizations from the second and higher levels is very small compared with the number of ionizations from the first, the reason being the small number of atoms in the higher levels as well as the much weaker radiation intensity in the visible and infrared regions of the stellar spectrum. Hence, neglecting ionizations from the higher levels, we find that the electrons are supplied to the continuum from the ground level. The converse process, however, may be a recapture on any one of the discrete levels. Therefore, for statistical equilibrium,

$$\int_{\nu_1}^{\infty} F_{1\kappa} d\nu = \sum_{n=1}^{\infty} \int_{\nu_n}^{\infty} F_{\kappa n} d\nu \,. \tag{3}$$

The summation should be extended to a certain maximum value of n depending on the density in a manner which we shall discuss in a later paper. For purposes of simplification we here make the approximation of summing to infinity. The error introduced by this approximation will be of small importance in the transfer problem of the Lyman continuum. Of the totality of captured ν-electrons, the fraction p_ν entering the first level is given by

$$p_\nu = \frac{F_{\kappa 1} d\nu}{\sum_1^{\infty} F_{\kappa n} d\nu} = \frac{\sigma_{\kappa 1}}{\sum_1^{\infty} \sigma_{\kappa n}} \,, \tag{4}$$

where $\sigma_{\kappa n}$ is the target area of the nth level. For an electron moving with a velocity v, κ is given by

$$\tfrac{1}{2} m v^2 = \frac{hRZ^2}{\kappa^2} \,. \tag{5}$$

The expression for $\sigma_{\kappa n}$ for hydrogen is

$$\sigma_{\kappa n} = \frac{2^4}{3\sqrt{3}} \frac{\epsilon^2 h}{m^2 c^3} \frac{1}{\left(\frac{1}{n'^2} + \frac{1}{\kappa^2}\right)} \frac{\kappa^2}{n'^3} g_{\text{II}} \,, \tag{6}$$

where h, m, and c have their usual meanings and g_{II} is the correction factor to Kramers' formula for bound-free transitions. The quantity κ, a real positive number not necessarily integral, is related to the frequency of the emitted radiation by

$$\nu_{n'\kappa} = RZ^2 \left(\frac{1}{n'^2} + \frac{1}{\kappa^2}\right) ,$$

where n' is integral and R is the Rydberg constant. For the first level

$$\frac{\nu_{1\kappa}}{RZ^2} = \frac{1}{\kappa^2} + 1 \,,$$

and we obtain finally

$$p_\nu = \frac{\dfrac{g_{II}}{1+\dfrac{1}{\kappa^2}}}{\displaystyle\sum_{n'=1}^{\infty}\frac{g_{IIn'}}{\left(\dfrac{1}{n'^2}+\dfrac{1}{\kappa^2}\right)n'^3}} = \frac{\dfrac{RZ^2}{\nu}}{\displaystyle\sum_{n'=1}^{\infty}\frac{\dfrac{g_{IIn'}}{g_{II}}}{\left\{n'+\left(\dfrac{\nu}{RZ^2}-1\right)n'^3\right\}}} . \tag{7}$$

The dependence of g_{II} on the frequency is known. For the purposes of the present discussion we have found it permissible to take the ratio $g_{IIn'}/g_{II}$, as constant. Values of p_ν were then adopted as follows:

ν/ν_1	p_ν	ν/ν_1	p_ν
1.001	0.21	1.40	0.64
1.01	0.31	1.60	0.68
1.03	0.38	2.00	0.72
1.06	0.44	2.50	0.75
1.10	0.48	3.00	0.77
1.20	0.57	6.00	0.80

Previous investigators have employed p_ν as a parameter in their equations. In this manner they have evaded the troublesome task of determining a properly weighted mean value of p_ν consistent with their treatment of the continuum as a single discrete level.

Equation (3) imposes the condition that the number of electrons entering the continuum per second must be equal to the number leaving it per second. We have yet to consider the requirement that each element $d\kappa$ of the continuum must be in a steady state; that is, the number of atoms entering the range $d\kappa$ per second by all possible processes must be equal to the number leaving that element per second by all processes. We may express this condition as follows:

$$\left.\begin{aligned} &F_{1\kappa}d\kappa + \int_\kappa^\infty F_{\kappa'\kappa}d\kappa'' d\kappa + C_1(\kappa)d\kappa \\ &\qquad = \sum_{n=1}^{\infty} F_{\kappa n}d\kappa + \int_0^\infty F_{\kappa\kappa'}d\kappa + C_2(\kappa)d\kappa . \end{aligned}\right\} \tag{8}$$

The left-hand side of the equation expresses the totality of ways in which an electron may enter the element $d\kappa$ of the continuum. The respective terms represent photo-ionizations from the ground state, free-free transitions from a higher state, and collisional processes involving radiationless impact. The right-hand side of the equation enumerates the totality of ways by which an electron can leave the element $d\kappa$ of the continuum, by capture on the nth level of an atom, by a free-free transition to a lower level, or by a collision with another electron. We denote the collisional terms by $C_1(\kappa)$ and $C_2(\kappa)$, respectively, and shall find that we do not need to evaluate them explicitly. From equation (4) we have

$$F_{\kappa 1} = p_\nu \sum_1^\infty F_{\kappa n} \,. \tag{9}$$

We re-write equation (8) in the form

$$\sum_1^\infty F_{\kappa n} d\kappa = F_{1\kappa} d\kappa + \eta d\kappa \,, \tag{10}$$

where

$$\eta = \left\{ \int_\kappa^\infty F_{\kappa''\kappa} d\kappa'' - \int_0^\kappa F_{\kappa\kappa'} d\kappa + C_1(\kappa) - C_2(\kappa) \right\} .$$

Equation (10) is taken as a definition of η. Now equation (3) requires that

$$\int_0^\infty \eta d\kappa = 0 \,, \tag{11}$$

an expression that proves to be a useful check on the accuracy of the calculations of the transfer problem. Physically, η is a measure of the extent to which the electron velocities are reshuffled by collisions and by free-free transitions among the various elements $d\kappa$ of the continuum. Previous writers have usually considered the continuum as a single level, in which case η has no significance. We shall find η to be of importance in the solution of the equations of transfer. From equations (9) and (10) we have

$$p_\nu [F_{1\kappa} + \eta] = F_{\kappa 1} \,. \tag{12}$$

Using equation (2), we obtain

$$\cos\theta \frac{dI_\nu}{dx} = -\frac{1}{4\pi}[1 - p_\nu]F_{1\kappa}h\nu + \frac{p_\nu \eta h\nu}{4\pi} . \tag{13}$$

The element of optical depth is now introduced as

$$d\tau_\nu = N_1 \alpha_{1\nu} dx , \tag{14}$$

where N_1 is the number of atoms in the ground state and α_ν is the absorption coefficient at the frequency ν. In the notation of previous papers we have

$$d\tau_\nu = b_1 \frac{N_i N_\epsilon}{4\pi\sigma} \frac{KZ^4}{T_\epsilon^{3/2}} hge^{X_1} dx , \tag{15}$$

where $\sigma = 2h\nu^3/c^2$. We must remark that the optical depth of any given nebula will vary with the frequency of the radiation with which it is measured. The optical depth will decrease with increasing frequency roughly as ν^{-3}. It is convenient to define a τ_{ν_1} which represents the optical depth at the Lyman limit.

Our method of discussion will be to set up the differential equation of transfer for each frequency element of the continuum. The equations of transfer will contain the terms in η representing the "interlocking" between the various elements of the continuous levels. It is in the introduction of η that our method differs from that of Ambarzumian,[4] who treated the continuum as a single level. Our equations will be similar to his in form, but we find it necessary to introduce for each ν the total optical depth at that frequency. The relation between the optical depth at a given frequency and the frequency at the Lyman limit is known. The geometrical thickness of the shell is, of course, constant.

Let us consider the radiation field to which an elementary volume of the nebula is exposed. First of all, there is the direct radiation from the star. If τ_ν is measured outward from the inner surface of the nebular shell, the incident stellar radiation at a given level will be $S_\nu e^{-\tau_\nu}$, where S_ν is the flux of direct stellar radiation incident upon the inner face of the nebula. The direct beam from the star is de-

[4] *M.N.*, **93**, 50, 1932; see also Chandrasekhar, *Zs.f. Ap.*, **9**, 266, 1935.

pleted in an exponential manner as it passes through the gaseous shell. We assume the geometrical portion of the dilution factor for the stellar radiation to be constant through the shell. Therefore, this part of the dilution factor cancels out of our equations and need not be considered. To take properly into account, in any given case, the variation of this geometrical factor, we should have to know the thickness and radius of the shell.

In the second place, we must consider the Lyman continuum radiation originating in the nebula itself. This energy has been absorbed from the direct beam and reradiated in the nebula in all directions. We denote the intensity of this diffuse radiation at frequency ν by J_ν. Our problem is to find J_ν as a function of both optical depth and frequency. The number of transitions from the first state to the element $d\kappa$ of the continuum corresponding to the frequency ν will be determined by the number of atoms in the ground level, the absorption coefficient for a frequency ν, and the total intensity of the diffuse and direct radiation of frequency ν. Thus,

$$F_{1\kappa} h\nu d\nu = N_1 a_\nu [J_\nu + S_\nu e^{-\tau_\nu}] d\nu \,. \tag{16}$$

We substitute equation (16) into equation (13), multiply by $2\pi \sin \theta d\theta$, integrate to obtain the differential equation for the total ν-flux, F'_ν, and obtain

$$\frac{dF'_\nu}{dx} = -[1 - p_\nu] N_1 a_\nu [J_\nu + S_\nu e^{-\tau_\nu}] + p_\nu \eta h\nu \,. \tag{17}$$

For the direct beam from the star, the flux and intensity are both given by $S_\nu e^{-\tau_\nu}$. The radiation absorbed and re-emitted in the Lyman continuum in the nebula, i.e., the diffuse flux, is to be found by integrating the equation of transfer. Let

$$F'_\nu = F_\nu + F^*_\nu \,, \tag{18}$$

where $F^*_\nu = S_\nu e^{-\tau_\nu}$ and F_ν is the diffuse flux. The equation for the latter is, therefore,

$$\frac{dF_\nu}{dx} = -J_\nu N_1 a_\nu + p_\nu N_1 a_\nu [J_\nu + S_\nu e^{-\tau_\nu}] + p_\nu \eta h\nu \,. \tag{19}$$

By introducing the definition of the optical depth, we get:

$$\frac{dF_\nu}{d\tau_\nu} = -J_\nu + p_\nu[J_\nu + S_\nu e^{-\tau_\nu}] + R_\nu \,, \tag{20}$$

where

$$R_\nu = \frac{p_\nu \eta h\nu}{N_1 a_\nu} \,. \tag{21}$$

The expression for R_ν is obtained with the aid of equations (12), I (23), (14), and (4) and V (2). It is:

$$R_\nu = \frac{e^{X_1} 8\pi h\nu^3 e^{-h\nu/kT_\epsilon}}{c^2 G_{T_\epsilon}} \int_{y_1}^{\infty} W_{1\kappa} \frac{dy}{y(e^y - 1)} - p_\nu W_{1\kappa} \frac{8\pi h\nu^3}{c^2} \frac{1}{e^y - 1} \,, \tag{22}$$

where $y = h\nu/kT_1$. To solve equation (20) we use the relation between F_ν and J_ν given by the Eddington approximation, namely:

$$F_\nu = -\frac{1}{3}\frac{\partial J_\nu}{\partial \tau_\nu} \,, \tag{23}$$

and our differential equation becomes

$$\frac{d^2 J_\nu}{d\tau_\nu^2} = 3(1 - p_\nu)J_\nu - 3p_\nu S_\nu e^{-\tau_\nu} - 3R_\nu \,. \tag{24}$$

To simplify matters we find it convenient to recast the equations in units of S_ν, the direct radiation from the star upon the inner surface of the nebular shell. Let $\mathfrak{J}_\nu$ denote the diffuse intensity and $\mathfrak{R}_\nu$ the interlocking term. With this new unit we obtain

$$\frac{d^2 \mathfrak{J}_\nu}{d\tau_\nu^2} = \lambda_\nu^2 \mathfrak{J}_\nu - 3p_\nu e^{-\tau_\nu} - 3\mathfrak{R}_\nu \,, \tag{25}$$

where

$$\lambda_\nu^2 = 3(1 - p_\nu) \,. \tag{26}$$

We also get from equation (22)

$$\left.\begin{aligned} \mathfrak{R}_\nu = \frac{(e^{h\nu/kT_1} - 1)}{G_{T_\epsilon}} e^{-h(\nu-\nu_1)/kT_\epsilon} \int_{y_1}^{\infty} (\mathfrak{J}_\nu + e^{-\tau_\nu}) \frac{dy}{y(e^y - 1)} \\ - p_\nu(\mathfrak{J}_\nu + e^{-\tau_\nu}) \,. \end{aligned}\right\} \tag{27}$$

Since W is the dilution factor for both diffuse flux and direct starlight, we have in the new units:

$$W_{1\kappa} = \mathfrak{J}_\nu + e^{-\tau_\nu} .$$

Now $\mathfrak{J}_\nu$ is a function of both τ_ν and ν, and therefore we should properly use partial derivatives. From equations (25) and (27) we find

$$\frac{\partial^2 \mathfrak{J}_\nu}{\partial \tau_\nu^2} = 3\mathfrak{J}_\nu - \frac{3(e^{h\nu/kT_1} - 1)}{G T_\epsilon} e^{-h(\nu-\nu_1)/kT_\epsilon} \int_{y_1}^{\infty} (\mathfrak{J}_\nu + e^{-\tau_\nu}) \frac{dy}{y(e^y - 1)} . \quad (28)$$

We have still to indicate the dependence of the optical depth upon the frequency. If we neglect the g-factor, we obtain the result:

$$\frac{\tau_\nu}{\tau_{\nu_1}} = \left(\frac{\nu_1}{\nu}\right)^3 = \left(\frac{y_1}{y}\right)^3 \quad \text{or} \quad \tau_\nu = \left(\frac{y_1}{y}\right)^3 \tau_{\nu_1} . \quad (29)$$

The equation of transfer for any frequency ν, expressed in terms of the two independent variables τ_{ν_1} and ν, becomes:

$$\left.\begin{aligned} &\frac{\partial^2 \mathfrak{J}_\nu}{\partial \tau_{\nu_1}^2} \left(\frac{y}{y_1}\right)^6 \\ &\quad = 3\mathfrak{J}_\nu - \frac{3(e^y - 1)}{G T_\epsilon} e^{(y_1 - y)T_1/T_\epsilon} \int_{y_1}^{\infty} (\mathfrak{J}_\nu + e^{-\tau_\nu}) \frac{dy}{y(e^y - 1)} . \end{aligned}\right\} \quad (30)$$

There is no rigorous solution of this equation readily obtainable in terms of elementary functions. The boundary conditions for the problem have been given by Milne.[5] Since no diffuse radiation is created within the nebular shell, the net diffuse flux must vanish at the inner boundary of the shell. In other words, the inward and outward fluxes just balance. The total optical depth of the nebula at frequency ν is expressed by $\tau_{0\nu}$. For the outer boundary condition we make use of the approximate relationship

$$J_\nu = 2F_\nu . \quad (31)$$

This boundary condition has often been used in treating problems of radiation transfer in stellar atmospheres. In the case of plane-

[5] *Zs. f. Ap.*, 1, 98, 1930.

parallel layers of radiating material, this means that there is no appreciable back radiation.

The transfer problem for the free-free transitions is extremely simple. We note, in the equation of transfer VI (17),

$$\cos\theta \frac{dI_\nu}{dx} = -\frac{N_i N_\epsilon}{4\pi\sigma}\frac{KZ^4}{T_\epsilon^{3/2}}\frac{kT_\epsilon}{2hRZ^2}\bar{g}\,\{[1 - e^{-h\nu/kT_\epsilon}]I_\nu - 1\}\,, \quad (32)$$

that the first term contains I_ν and therefore contains the dilution factor $W_{\kappa\kappa'}$ for the radiation in question. This first term is, accordingly, very small compared with unity and may be neglected. The transfer equation then becomes

$$\cos\theta \frac{dI_\nu}{d\tau_\nu} = \frac{kT_\epsilon}{2hRZ^2}\frac{\bar{g}}{g}\frac{\int_{y_1}^{\infty} W_{1\kappa}\frac{dy}{y(e^y - 1)}}{G_{T_\epsilon}}\,, \quad (33)$$

where we have used equation V (2) to express b_1 in terms of the parameters of the electron temperature and the radiation field. The equation falls into this form because of our definition of the optical depth.

The integration of equation (33) can be carried through numerically as soon as the functional dependence of T_ϵ and $W_{1\kappa}$ on optical depth is known from the solution of the equations for the Lyman continuum. The boundary conditions of the problem will be the same as for the diffuse Lyman flux.

In the next paper the solution of the foregoing equations of transfer will be carried out numerically for the case of a model nebula of total optical depth $\tau_{0\nu_1} = 3.0$, with a central star radiating as a black body at a temperature of 80,000°.

HARVARD COLLEGE OBSERVATORY
May 15, 1939

Reprinted from *The Astrophysical Journal,* Vol. 90, No. 4, November, 1939.

PHYSICAL PROCESSES IN GASEOUS NEBULAE

VIII. THE ULTRAVIOLET RADIATION FIELD AND ELECTRON TEMPERATURE OF AN OPTICALLY THICK NEBULA

LAWRENCE H. ALLER,[1] JAMES G. BAKER,[1] AND DONALD H. MENZEL

ABSTRACT

The theory of the transfer of radiation in the Lyman continuum that was developed in papers VI and VII of this series is applied to a numerical example. A central star radiating as a black body at a temperature of 80,000° C. is assumed to illuminate a nebular shell whose optical thickness at the Lyman series limit is 3.0. The numerical methods of solution are discussed in detail. The redistribution of electron velocities by collisions and free-free transitions is shown to have considerable effect upon the character of the radiation field. Near the series limit more radiation gets through the nebula than would otherwise be expected. The electron temperature at the inner boundary is almost that of an optically thin nebular shell, and at the outer boundary it approaches that of the central star.

In the preceding paper of this series[2] we have derived the differential equations of transfer for radiation in the Lyman continuum for optically thick nebulae. In order to study the variation of the radiation field and electron temperature throughout a given nebula we found it necessary to consider the transfer of radiation in each element of the continuum rather than to treat the continuum as a single level. A term was introduced to take into account the interlocking between the elements of the continuum. The equation of transfer (VII, 25) is

$$\frac{d^2\mathfrak{J}_\nu}{d\tau_\nu^2} = \lambda_\nu^2\mathfrak{J}_\nu - 3p_\nu e^{-\tau_\nu} - 3\mathfrak{R}_\nu\,, \tag{1}$$

where $\lambda_\nu^2 = 3(1 - p_\nu)$ and $\mathfrak{J}_\nu$ is the intensity of the diffuse radiation. The expression for $\mathfrak{R}_\nu$ is

$$\mathfrak{R}_\nu = \frac{(e^{y_1} - 1)}{GT_\epsilon} e^{T_1(y_1-y)/T_\epsilon}\int_{y_1}^{\infty} (\mathfrak{J}_\nu + e^{-\tau_\nu})\,\frac{dy}{y(e^y - 1)} - p_\nu(\mathfrak{J}_\nu + e^{-\tau_\nu})\,. \tag{2}$$

[1] Society of Fellows, Harvard University.

[2] (Part I) *Ap. J.*, **85**, 330, 1937; (II) **86**, 70, 1937; (III) **88**, 52, 1938; (IV) **88**, 313, 1938; (V) **88**, 422, 1938; (VI) **89**, 587, 1939; (VII) **90**, 271, 1939. The papers of this series are referred to in the text as (I), (II), (III), etc.

An alternate form of the equation of transfer, as given in (VII, 30), is

$$\frac{\partial^2 \mathfrak{J}_\nu}{\partial \tau_{\nu_1}^2}\left(\frac{y}{y_1}\right)^6 = 3\mathfrak{J}_\nu - \frac{3(e^y - 1)}{G_{T_\epsilon}} e^{T_1(y_1-y)/T_\epsilon} \int_{y_1}^{\infty} (\mathfrak{J}_\nu + e^{-\tau_\nu}) \frac{dy}{y(e^y - 1)}, \quad (2a)$$

where the g-factor has been neglected. For the numerical calculation we found it preferable to deal with equation (1), in which we have taken the Gaunt factor into account. The optical depth at any frequency is given, then, by

$$\frac{\tau_{\nu_1}}{\tau_\nu} = \frac{a_1}{a_\nu} = \left(\frac{\nu}{\nu_1}\right)^3 \frac{g_{\nu_1}}{g_\nu}. \quad (3)$$

In the solution of (1) we must remember that the variables employed are not strictly independent. The relation(3) must always be kept in mind when we manipulate (1).

Except for the term in $\mathfrak{R}_\nu$, (1) is of the same form as the equation of transfer given by Ambarzumian and Chandrasekhar. This "correction term" is a measure, for any frequency, of the effect of the interlocking between the various levels of the continuum. $\mathfrak{R}_\nu$ can scarcely be regarded, however, as a mere correction, because it often proves to be as large as, or larger than, the other terms on the right-hand side of (1). To neglect it would be to consider the continuum as a manifold of levels interacting only with discrete levels and not with each other. We could avoid some of the difficulty by treating the continuum as a single level, in which case $\mathfrak{R}_\nu$ would vanish; but we should encounter the new difficulty of choosing a proper mean value of p_ν, which would certainly vary throughout the nebula along with the variation of the radiation field and T_ϵ.

The trouble arises from the fact that $\mathfrak{R}_\nu$ contains the unknown $\mathfrak{J}_\nu$ integrated over all frequencies, whereas (1) is valid for only one particular frequency. Note also that $\mathfrak{R}_\nu$ contains the electron temperature both explicitly and implicitly through G_{T_ϵ}. No rigorous solution of the differential integral equation (2a), in terms of elementary functions, appears to exist. One is therefore compelled to seek a solution by a method of successive approximations. As a working basis, we found it expedient to start our approximation by setting $\mathfrak{R}_\nu = 0$.

By so doing, we have adopted a preliminary set of values of $\mathfrak{J}_\nu$ for which the interlocking between the elements of the continuum has been neglected. The resultant solution of (1) is well known and may be written as[3]

$$\mathfrak{J}_0 = A_0 e^{\lambda\tau} + B_0 e^{-\lambda\tau} + \frac{3p}{\lambda^2 - 1} e^{-\tau}, \tag{4}$$

where

$$A_0 = B_0 + \frac{3p}{\lambda(\lambda^2 - 1)},$$

$$B_0 = \frac{-3p}{\lambda(\lambda^2 - 1)} \frac{(1 + \frac{2}{3}\lambda)e^{\lambda\tau_0} + \frac{1}{3}\lambda e^{-\tau_0}}{(1 + \frac{2}{3}\lambda)e^{\lambda\tau_0} + (1 - \frac{2}{3}\lambda)e^{-\lambda\tau_0}}. \tag{5}$$

These values of $\mathfrak{J}_0$ lead immediately to the determination of a preliminary set of values of $W_{1\kappa}$ by the relation

$$W_{1\kappa} = \mathfrak{J}_\nu + e^{-\tau_\nu}. \tag{6}$$

With this approximate value of $W_{1\kappa}$ one can compute first the electron temperature T_ϵ by the method outlined in (V), and then $\mathfrak{R}_\nu$ by (2). This procedure completes the first step. For the purposes of the next approximation, $\mathfrak{R}_\nu$ can be taken as constant with optical depth, although its functional dependence on frequency is retained. The resultant solution of (1) is of the form

$$\mathfrak{J}_\nu = A'_\nu e^{\lambda_\nu\tau_\nu} + B'_\nu e^{-\lambda_\nu\tau_\nu} + \frac{3p_\nu e^{-\tau_\nu}}{\lambda_\nu^2 - 1} - \frac{3\mathfrak{R}_\nu}{\lambda^2}, \tag{7}$$

where the new constants of integration, A' and B', are related to the old by simple additive terms:

$$A' = A_0 + \Delta' A ; \qquad B' = B_0 + \Delta' B ,$$

where

$$\Delta' A = \Delta' B = -\frac{3\mathfrak{R}_\nu}{\lambda^2\{(1 - \frac{2}{3}\lambda)e^{-\lambda\tau_0} + (1 + \frac{2}{3}\lambda)e^{\lambda\tau_0}\}}. \tag{8}$$

With these improved values of the $\mathfrak{J}$'s, double-entry tables of $W_{1\kappa}$ and $\mathfrak{R}_\nu$ as functions of optical depth are computed. The dependence

[3] Cf. Rosseland, *Theoretical Astrophysics*, p. 328.

of T_ϵ on the optical depth is, of course, obtained as an intermediate step in passing from $W_{1\kappa}$ to $\mathfrak{R}_\nu$. With these values of $\mathfrak{R}_\nu$ we now proceed to a third approximation.

For the third and succeeding approximations we have found that, to a high order of accuracy, $\mathfrak{R}_\nu$ may be taken as a linear function of the optical depth.[4] Therefore, we adopt[5]

$$\mathfrak{R} = a_0 + a_1\tau \,, \tag{9}$$

whence

$$\mathfrak{J} = Ae^{\lambda\tau} + Be^{-\lambda\tau} + \frac{3pe^{-\tau}}{\lambda^2 - 1} + \alpha_0 + \alpha_1\tau \,, \tag{10}$$

where

$$\alpha_0 = \frac{3a_0}{\lambda^2} \,; \qquad \alpha_1 = \frac{3a_1}{\lambda^2} \,. \tag{11}$$

Then

$$\Delta B = \frac{e^{\lambda\tau_0}(1 + \frac{2}{3}\lambda)\dfrac{\alpha_1}{\lambda} - (\alpha_0 + \frac{2}{3}\alpha_1) - \alpha_1\tau_0}{\{(1 + \frac{2}{3}\lambda)e^{\lambda\tau_0} + (1 - \frac{2}{3}\lambda)e^{-\lambda\tau_0}\}} \tag{12}$$

and

$$\Delta A = \Delta B - \frac{\alpha_1}{\lambda} \tag{13}$$

and

$$A = A_0 + \Delta A \,; \qquad B = B_0 + \Delta B \,. \tag{14}$$

These expressions for the constants of integration are taken as final.

[4] It might be supposed on first sight that $\mathfrak{R}_\nu$ should be expanded as a power series in the optical depth. In assuming a series expansion, however, we encounter a rather unusual mathematical difficulty. The coefficients of the higher-order terms in the expansion of $\mathfrak{R}_\nu$ turn out to be large numerically. The contribution of each term to $\mathfrak{R}_\nu$ is small because the optical depth raised to a high power is small. If we assume that $\mathfrak{R}_\nu$ is a power series, it is clear that the coefficients in this series must be related to the coefficients in a corresponding power series in $\mathfrak{J}$. From the nature of the boundary conditions it turns out that the rth coefficient in the expansion for $\mathfrak{R}_\nu$ and also the $(r + 2)$, $(r + 4)$, $(r + 6)$, etc., coefficients occur with factors only of the order of unity. As a result, the uncertain and numerically large coefficients overpower the significant coefficients a_0 and a_1 and render the resultant $\mathfrak{J}$ meaningless.

[5] All the quantities in these equations are functions of ν. The subscript ν is dropped as superfluous.

For a numerical application of the present theory we have adopted a gaseous nebula of optical depth 3.0 at the Lyman limit surrounding a hot central star which is assumed to radiate as a black body at a temperature of 80,000°. The procedure outlined above was followed closely in the numerical calculations. For each of the steps, double-entry tables of the various functions needed in the calculation were prepared. The independent variables were taken to be ν/ν_1 and τ_{ν_1} (e.g., see Table 1).

TABLE 1

ν/ν_1	$\tau_{0\nu}$	g_ν/g_{ν_1}	S_ν
1.001	2.991	1.000	1.0665
1.01	2.937	1.010	1.0724
1.03	2.775	1.010	1.0877
1.06	2.568	1.020	1.1079
1.10	2.340	1.035	1.1331
1.20	1.836	1.060	1.1802
1.40	1.200	1.100	1.2244
1.60	0.831	1.135	1.2077
2.00	0.444	1.186	1.0542
2.50	0.234	1.220	0.7589
3.00	0.138	1.235	0.4892
6.00	0.015	1.250	0.1085

First of all, tables of τ_ν, g, and S_ν were computed. With these fundamental starting data, A_0 and B_0 were computed with the aid of (5), and $\mathfrak{J}_0$ from (4). For the calculation of the electron temperature by the method of (V) we were required to find the integrals

$$\int W_{1\kappa} \frac{dy}{e^y - 1} \quad \text{and} \quad \int W_{1\kappa} \frac{dy}{y(e^y - 1)} ,$$

where

$$W_{1\kappa} = \mathfrak{J}_0 + e^{-\tau} .$$

By the methods described in (V), values of T_ϵ have been computed for the four values of $\tau_{\nu_1} = 0$, 1, 2, and 3. For the purpose of the second approximation, values of $\mathfrak{R}_\nu$ and T_ϵ were adopted corresponding to an optical depth τ_{ν_1} of about 1.5. For this second approximation the dependence of $\mathfrak{R}_\nu$ on frequency was retained, of course, and formulas (7) and (8) were applied. For the third and higher approximations a linear variation of $\mathfrak{R}_\nu$ with τ_ν was adopted, and from the $\mathfrak{J}$-values of the second approximation it was possible to compute $\mathfrak{R}_\nu$ for the four optical depths corresponding to $\tau_{\nu_1} = 0, 1, 2,$ and 3 and to evaluate the coefficients a_0 and a_1 in (9). By means of equations (11), (12), (13), and (14), $\mathfrak{J}$ was now calculated, $\mathfrak{R}_\nu$ was derived from (2), and the procedure was repeated for the fourth approximation. The character of $W_{1\kappa}$ on the basis of the third and fourth approximations is exhibited in Figure 1, computed for $\tau = 0$, where a comparison with the Ambarzumian value (first approximation) is shown.

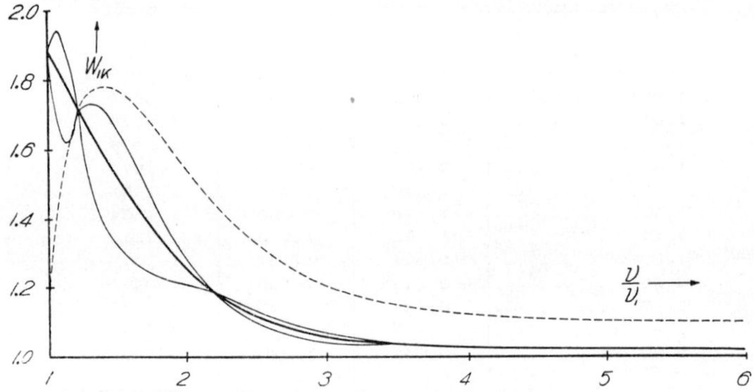

FIG. 1.—$W_{1\kappa}$ at the inner boundary

The successive approximations yield curves that pass through nodal points, analogously to the way in which a Fourier series approximates a given function. These nodal points must also lie on the final curve, and, indeed, one would obtain a very excellent approximation to the final curve merely by evaluating these nodal points. In future calculations involving nebulae of various total optical depths the position of the nodal points should be determined first. Then the functions can be evaluated for these special points. The calculations will be considerably shortened if this procedure is followed. Having noted the character of the approximations, we found it expedient to guess at the final form of the function $W_{1\kappa}$ and thereby to arrive immediately at an approximation of a high order. If successive approximations were carried through blindly, the slow convergence would render the computations of hopeless length. A mean curve was drawn through the nodal points given by the intersection of the third and fourth approximations to $W_{1\kappa}$. The Ambarzumian curve passes through the second nodal point, which lies at the frequency for which $\mathfrak{R}_\nu = 0$. The frequencies of the various nodal points are nearly independent of optical depth, at least within the errors of the calculation. This new curve was treated as the fifth approximation, and one more numerical solution of the equations was effected. There were, thus, six approximations in all.

The comparison of the final values of $\mathfrak{J}$ with those obtained in the first approximation (i.e., the $\mathfrak{J}_0$ values derived from the Chandrasek-

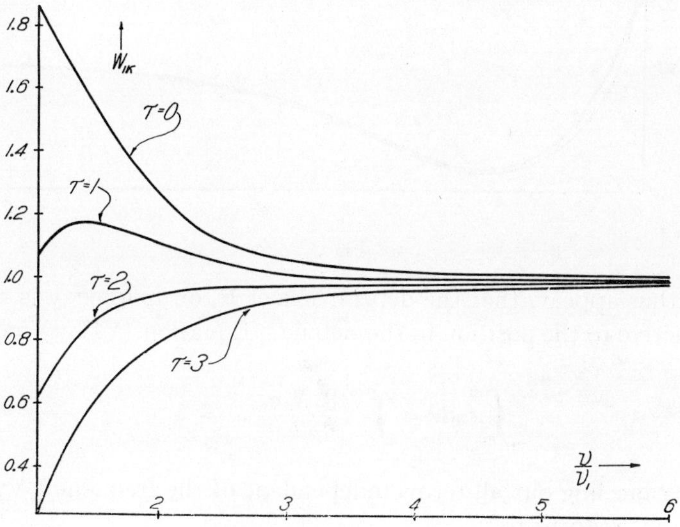

FIG. 2

har modification of the Ambarzumian equations) exhibits strikingly the effect of the interlocking between the various states of the continuum. Much energy is degraded to the lower frequencies by collisions and free-free emissions. The heavy full curve in Figure 1 is the function $W_{1\kappa}$ that satisfies the differential equation of transfer (2) and also (2*a*). The dashed curve represents the Ambarzumian values, and the full light-curves represent the third and fourth approximations. These are the curves for the inner boundary, but they are typical of the other three sets of curves for $\tau_{\nu_1} = 1$, 2, and 3. Figure 2 gives $W_{1\kappa}$ as a function of frequency for the optical depths $\tau_{\nu_1} = 0$, 1, 2, and 3. Figure 3 gives $\mathfrak{R}_\nu$ as a function of frequency at the inner and outer boundaries of the nebula. Since it was possible to take $\mathfrak{R}_\nu$ as a linear function of τ_ν, values of $\mathfrak{R}_\nu$ for the inner points of the nebula are intermediate between the two curves drawn.

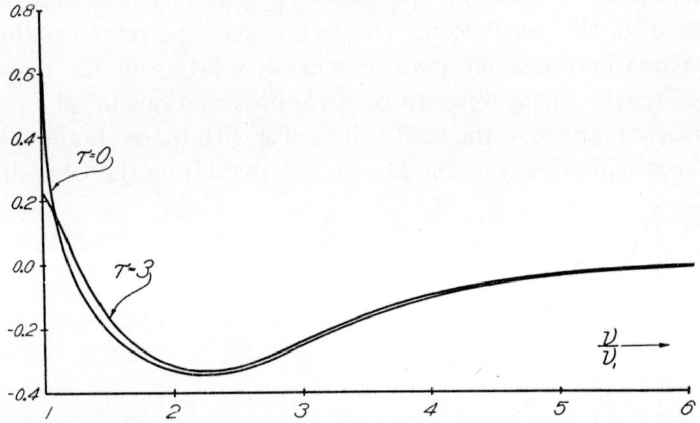

FIG. 3

It thus appears that the dependence of $\mathfrak{R}_\nu$ on frequency is rather insensitive to the position in the nebula. Equation (VII, 3) requires that

$$\int_{\nu_1}^{\infty} \eta d\nu = \int_{\nu_1}^{\infty} \frac{R_\nu N_1 a_\nu}{p_\nu h\nu} d\nu = 0 . \tag{15}$$

After canceling out all terms independent of the frequency, we find that (15) reduces to

$$\int_1^{\infty} \frac{R_\nu \dfrac{\tau_{0\nu}}{\tau_{0\nu_1}}}{p_\nu \left(\dfrac{\nu}{\nu_1}\right)} d\left(\frac{\nu}{\nu_1}\right) = 0 .$$

This check was applied to the curves in (3) and was found to hold within the error of the computations.

Another point of interest is the change of the electron temperature through the nebula (this variation is shown in Fig. 4):

τ_{ν_1}	T_ϵ
0	59,000
1	61,000
2	64,000
3	76,000

From the results of the calculations of (V) we recall that, for a thin nebular shell illuminated by a star radiating as a black body at a temperature of 80,000°, the electron temperature is 57,000°. We arrive, therefore, at the interesting conclusion that a nebula of optical thickness 3.0 at the Lyman limit will assume on its inner boundary a temperature almost corresponding to that of a thin nebular shell and that, on the outer boundary, the temperature will be close to that of the central star.

These calculations were carried out partly with the intention of providing a guide for future numerical work on nebulae of different optical thicknesses and central stars of different temperatures. The

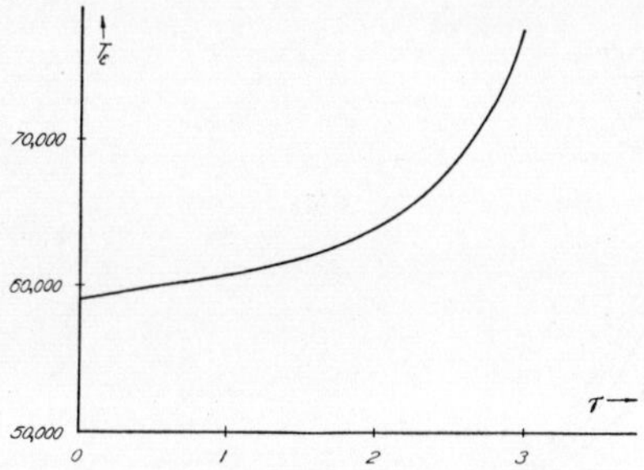

FIG. 4

slow convergence of the approximations seems very discouraging until one remembers that the initial approximation, i.e., that of no coupling between the various states of the continuum, is a poor one, as a glance at the curve for $\mathfrak{R}_\nu$ will show.

In a recent comprehensive publication[6] Y. Hagihara discusses the problem of transfer in the Lyman continuum. Hagihara takes the problem further than does Chandrasekhar or Ambarzumian in that he uses the temperature of the nuclear star as a parameter rather than mean capture-probabilities. He still considers the continuum as a single state, however, and neglects the effect of the reshuffling of electron velocities. In addition he makes the assumption that the temperature of the electron gas is constant throughout the nebula and equal to that of the nuclear star. This restriction cannot be fulfilled, as we have shown, for a pure hydrogen nebula, so that Hagihara's results do not agree too well with our final values. They

[6] *Japanese J. of Astron. and Geoph.*, **15**, 1, 1938.

correspond more closely to our first approximation, weighted according to frequency by Hagihara's method.

With this paper we shall leave, for the present, the problem of the equilibrium of a nebula consisting of pure hydrogen. The effect of collisions in populating the higher atomic levels remains still to be considered; but we shall postpone discussion of this question, although a partial analysis of the problem has been carried out. We

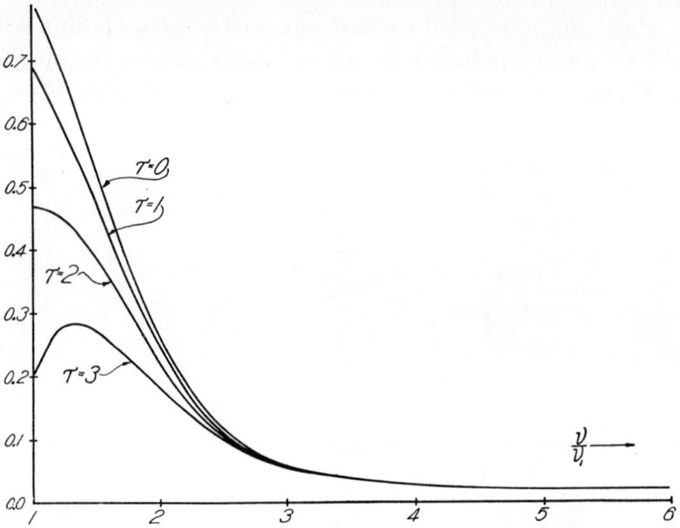

FIG. 5.—$\mathfrak{J}_\nu$ as a function of frequency and optical depth

shall also defer discussion of the problem of transfer in the hydrogen lines because of the complications introduced by expansion of the nebular shell. It is our desire to develop the theory along lines that can be directly compared with observation.

The next papers of this series will deal, accordingly, with the excitation of the lines of oxygen, helium, nitrogen, etc. The discussion will include investigation of significant atomic properties, transition probabilities, capture probabilities, and target areas for collisional excitation. The quantum mechanical calculations are essentially complete, and the relevant papers will appear shortly.

HARVARD COLLEGE OBSERVATORY
June 8, 1939

Reprinted from *The Astrophysical Journal,* Vol. 91, No. 3, April, 1940.

PHYSICAL PROCESSES IN GASEOUS NEBULAE

IX. ON THE EXCITATION OF FRACTIONAL MULTIPLETS BY ELECTRON CAPTURE

GEORGE H. SHORTLEY AND DONALD H. MENZEL

ABSTRACT

Observations show that certain permitted multiplets of *O* III and *N* III are not completely represented in the spectra of gaseous nebulae. Some mechanism apparently acts to give selective excitation of certain levels of a term, without exciting other levels. Bowen attributes the phenomenon to resonance absorption, principally of the ultimate line of *He* II, which happens to coincide with an ultimate line of *O* III. The absorptions originate from the highest, instead of the lowest, levels of the ground terms. Since these levels have mean lifetimes of the order of several hours, Bowen's theory requires the operation of some additional mechanism to maintain a sufficient population in the levels involved in the fluorescence process. Another method of excitation of fractional multiplets, direct electron capture by an ion in the ground level of a term (e.g., $^2P_{1/2}$ of *O* IV), is examined. The components predicted by quantum mechanics do not, however, agree with those observed in the nebulae. The conclusion is reached that Bowen is probably right, although a few discrepancies still remain unexplained. His method apparently requires inelastic electron impacts to be sufficiently frequent to maintain approximately a thermodynamic equilibrium population of atoms in the levels of the ground term. This paper includes the relevant quantum mechanical analysis of the relative probabilities of capture into the different levels of a configuration.

Several years ago William H. Wright[1] identified a number of lines in the far ultraviolet spectrum of the planetary nebulae with certain permitted lines of doubly ionized oxygen. A most striking feature of the identification was the fact that the complete multiplet 2p3p ^3P − 2p3d $^3P^o$ does not appear, but only the lines arising from the level $^3P^o_2$.

In Table 1 are given the lines of two significant transition arrays of *O* III. The lines observed in the gaseous nebulae are shown in boldface type. Nebular spectra to the violet of about λ 3200 are only incompletely observed. With the exception of the lines λ 3122 and λ 5592, all of the observed lines would emanate from initial excitation of the level 2p3d $^3P^o_2$, with subsequent cascadings to lower levels.

Bowen[2] has suggested a possible mechanism for the selective excitation of this single atomic level. He has noted that a very close coincidence exists between the frequencies of the ultimate lines 1s ^2S − 2p $^2P^o$ of *He* II and 2p^2 $^3P_2$ − 2p3d $^3P^o_2$ of *O* III. Most of the

[1] *Pub. A.S.P.*, **46**, 142, 149, 280, 1934. [2] *Pub. A.S.P.*, **46**, 146, 1934.

nebulae where fractional multiplets of *O* III are observed are evidently of high excitation. The great intensity of the line λ 4686 of *He* II suggests that the ultimate line of *He* II, referred to above, is also very intense. Bowen argues that resonance absorption by *O* III of the intense *He* II radiation would lead to excitation of 2p3d $^3P^o_2$

TABLE 1

TRANSITION ARRAYS 2p3d→2p3p AND 2p3p→2p3s OF *O* III

			1s² 2s² 2p3p									
			^{3}D			^{3}P			^{3}S	^{1}D	^{1}P	^{1}S
		J	3	2	1	2	1	0	1	2	1	0
1s² 2s² 2p3d	^{3}F	4	3265.41	......	......	......	......	......	......	......	......	......
		3	3284.56	3260.95	......	......	......	......	......	......	......	......
		2	3305.77	3281.92	3267.29	......	......	......	......	......	......	......
	^{3}D	3	3017.69	2997.70	......	3715.07	......	......	......	......	......	......
		2	*	3004.35	2992.10	3725.29	3707.28	......	......	......	......	......
		1	......	3008.75	2996.49	*	3714.01	3702.70	......	......	......	......
	^{3}P	2	2836.35	2818.80	*	**3444.15**	**3428.70**	......	**3132.87**	......	......	......
		1	......	2809.72	2799.01	3430.66	3415.33	3405.79	**3121.71**†	......	......	......
		0	......	......	2794.21	......	3408.20	......	3115.74	......	......	......
	^{1}F	3	......	......	......	......	......	......	......	2983.78	......	......
	^{1}D	2	......	......	......	......	......	......	......	*	2959.68	......
	^{1}P	1	......	......	......	......	......	......	......	*	2390.44	5268.06
1s² 2s² 2p3s	^{3}P	2	**3759.83**	3791.26	3810.99	**3047.14**	3059.31	......	**3340.78**	......	......	......
		1	......	3754.65	3774.01	**3023.42**	3035.92	3043.01	**3312.35**	......	......	......
		0	......	......	3757.20	......	3024.54	......	**3299.39**	......	......	......
	P	1	......	......	......	......	......	......	......	3961.58	**5592.37**‡	2454.99

* Faint unobserved or unidentified line. Triplets according to Mihul, *Ann. de phys.*, **9**, 261, 1928. Singlets according to Edlén, *Nov. Act. Reg. Soc. Sci. Upsala*, IV, **9**, No. 6, 1933. Edlén also identifies a number of faint intercombination lines not included in this table.

† Although λ 3122 does not originate in 2p3d $^3P_2$, Wright and Bowen have suggested that it may arise from resonance absorption of λ 303, by atoms moving with relative velocities of 90 km/sec.

‡ Bowen and Wyse record the presence of λ 5592 in Nebulae NGC 7027 and 7662, with approximately one-half the intensity of λ 3760. We are indebted to these scientists for the communication of this information in advance of publication.

but not to excitation of the other levels of $^3P^o$, because the wave lengths do not coincide. As a further argument in support of his theory, Bowen points out that the line 2p3s $^3P^o_1$ − 2p² $^3P_2$ of *O* III, one of the end-products of the cycle of excitation and radiation, coincides in turn with the line 2p $^2P^o_{3/2}$ − 3d $^2D_{5/2}$ of *N* III. This coincidence initiates another cycle, with further excitation of fractional multiplets of *N* III.

Bowen's ingenious explanation, if true, places some definite restrictions on the physical conditions obtaining in planetary nebulae.

It does not seem to have been emphasized previously that the levels $^3P_2$ of *O* III and $^2P_{3/2}$ of *N* III, from which the original fluorescent absorption arises, *are not the ground levels* of the terms involved. Furthermore, although the mean lives of the levels $2p^2\ ^3P_2$ and $2p\ ^2P_{3/2}$ are long,[3] they are, nevertheless, much shorter than the expected time the atom must wait before being excited by the foregoing mechanism. Therefore, if only radiation processes are involved, we should expect the large majority of the atoms to be in the true ground levels, $^3P_0$ and $^2P_{1/2}$, respectively, of *O* III and *N* III. That a condition of this sort holds in interstellar space is shown by the observations of T. Dunham,[4] who points out that the only observable interstellar absorption lines of *Ti* II are those arising from the true ground level. Lines from other levels of the same term are not observed.

In view of the fact that there is an alternative method by means of which fractional multiplets may be obtained—namely, in electron capture by ions not statistically distributed—we have thought it wise to undertake the relevant quantum mechanical analysis. Although the results are apparently negative, as far as explanation of the fractional multiplets of planetary nebulae is concerned, the calculations appear to be of sufficient interest to warrant a summary here.

The problem is extremely simple to visualize. For the sake of explicitness, we shall discuss first the case of radiative electron capture by *O* IV to give directly *O* III in the configuration $1s^2\ 2s^2\ 2p3d$. Under the conditions that obtain in the nebulae we might reasonably assume that practically all the *O* IV ions are in the very lowest level $1s^2\ 2s^2\ 2p\ ^2P_{1/2}$. If the *O* IV ions were distributed between $^2P_{3/2}$ and $^2P_{1/2}$ according to the statistical weights, we should obtain (because of the sum rules) approximately a statistical distribution of the captured electrons among the levels of *O* III 2p3d. But on the assumption that all the ions are in the level $^2P_{1/2}$, we expect and obtain, large peculiarities in the distribution in 2p3d.

To make a (dipole) radiative transition into 2p3d, the electron before capture must be in either a p or an f orbit. The *O* IV ion in the level $2p\ ^2P_{1/2}$, plus the free electron in a hyperbolic orbit of energy

[3] The mean lives are: *O* III $2p^2\ ^3P_2$, 2.9 hours; *N* III $2p\ ^2P_{3/2}$, 5.3 hours.

[4] *Nature*, **139**, 246, 1937.

E, are to be considered together as forming a term of the continuous spectrum of *O* III, which we write as

$$2p_{1/2}Ep_{1/2}\,,\ 2p_{1/2}Ep_{3/2}\,,\ 2p_{1/2}Ef_{5/2}\,,\ 2p_{1/2}Ef_{7/2}\,, \qquad (1)$$

according to whether the free electron is in a p or an f orbit and according to the total (spin+orbital) angular momentum of this free electron. These are definite terms of *O* III in *jj* coupling, capable of making radiative transitions to the levels of 2p3d, which are, however, in *LS* coupling. The problem is then one of obtaining the strengths of the lines in the transitions from the configurations 2pEp and 2pEf in *jj* coupling to the configuration 2p3d in *LS* coupling. This problem may be solved in all detail, but with considerable labor, by the standard methods of transformation theory; but a theorem we shall shortly prove will furnish very quickly the answers to the questions of interest here.

We are interested only in the *total* strength from all the levels comprehended in (1)[5] to a given level of 2p3d. *If* the configuration 2p3d were in *jj* coupling, the answer would be given very simply by the sum rule:[6] the total strength to any level ($2p_{1/2}\,3d_j\,J$) in which the 2p electron has the same j value as in (1) would be proportional to $2J+1$; the strengths to all levels for which the 2p electron has the j value $\frac{3}{2}$ would vanish. The theorem below shows that with 2p3d in any other coupling, in particular Russell-Saunders, we need only express the eigenfunction of each level in terms of *jj* coupling levels, e.g.,[7]

$$\psi(pd\,^3F_2) = \left(\frac{1}{15\sqrt{2}}\right)[4\sqrt{21}\,\psi(p_{1/2}d_{3/2}) - \sqrt{14}\,\psi(p_{1/2}d_{5/2}) + 2\sqrt{21}\,\psi(p_{3/2}d_{3/2}) - 4\psi(p_{3/2}d_{5/2})]\,,$$

$$\psi(pd\,^3P_1) = \left(\frac{1}{\sqrt{30}}\right)[-\sqrt{5}\,\psi(p_{1/2}d_{3/2}) + 4\psi(p_{3/2}d_{3/2}) + 3\psi(p_{3/2}d_{5/2})]\,,$$

etc. We then take the sum of the squares of the coefficients of the states $p_{1/2}d_{3/2}$ and $p_{1/2}d_{5/2}$ (which sum we call the *$p_{1/2}$ parentage factor*)

[5] I.e., the levels $2p_{1/2}\,Ep_{1/2}$, $J = 0, 1$; $2p_{1/2}\,Ep_{3/2}$, $J = 1, 2$; etc.

[6] Condon and Shortley, *Theory of Atomic Spectra*, p. 279.

[7] *Ibid.* All the necessary transformation coefficients are tabulated on p. 294.

and multiply this sum by $2J+1$ to get the relative total strengths of the capture transitions to the different levels of 2p3d. For example, for $^3F_2$ and $^3P_1$ the $p_{1/2}$ parentage factors are

$$(1/225 \cdot 2)(16 \cdot 21 + 14) = \tfrac{7}{9}\,,$$

$$(\tfrac{1}{30})5 = \tfrac{1}{6}\,,$$

respectively. The relative total strengths are then $\frac{7}{9} \cdot 5 = 35/9$ and $\frac{1}{6} \cdot 3 = 1/2$.

It is in this way that the $p_{1/2}$ parentage factors and the relative total strengths of Table 2 are computed for various transition arrays of interest.[8] These values are independent of the energy and the angular momentum of the captured electron, but to get transition probabilities from these strengths requires multiplication by the cube of the wave number of the radiation emitted in the capture, averaged in a complicated way over the energy of the captured electron. The transition probability falls rapidly with increasing electron energy. Consequently, for low electron temperatures, T, such that $kT < \chi$, where χ is the ionization energy, one should get satisfactory relative transition probabilities by using the cube of the wave-length limit of the continuous radiation. For very high temperatures the averaged factors for the various continua approach equality.

The method used in these computations is a special case of a rather general theorem of matrix mechanics, which is demonstrated at the end of the paper. For the modifications necessary to handle capture into a configuration, such as p^2, involving equivalent electrons, see the discussion under this theorem. We shall proceed first to the astrophysical discussion of Table 2.

In Table 3 are given the relative strengths of the lines in the configuration arrays ps − pp and pp − pd. The relative intensities are to be obtained from these strengths by multiplication by the fourth power of the wave number and by a population factor that is unity if each upper level is populated in proportion to its statistical weight

[8] If we had been interested in the case in which the ion is in $^2P_{3/2}$, we should, by the same procedure, obtain the complementary relative strengths, which with the strengths from $^2P_{1/2}$ make up the whole $2J+1$. These are also listed in Table 2.

$(2J + 1)$. The population of a level depends on the rates at which atoms enter and leave the level, by all processes. For the assumption of excitation of pp and pd, respectively, purely by electron capture,

TABLE 2

RELATIVE TOTAL STRENGTHS OF TRANSITIONS INVOLVED IN RADIATIVE ELECTRON CAPTURE BY AN ION IN LEVELS p $^2P_{1/2}$ AND p $^2P_{3/2}$

TRANSITION	TERM	ION IN LEVEL $^2P_{1/2}$	ION IN LEVEL $^2P_{3/2}$	SUM	$P_{1/2}$ PARENTAGE FACTOR
Capture of p or f electron to give pd	$^3F_4$	0	9	9	0
	$^3F_3$	28/9	35/9	7	4/9
	$^3F_2$	35/9	10/9	5	7/9
	$^1F_3$	7/3	14/3	7	1/3
	$^3D_3$	14/9	49/9	7	2/9
	$^3D_2$	35/18	55/18	5	7/18
	$^3D_1$	3/2	3/2	3	1/2
	$^1D_2$	5/3	10/3	5	1/3
	$^3P_2$	5/2	5/2	5	1/2
	$^3P_1$	1/2	5/2	3	1/6
	$^3P_0$	0	1	1	0
	$^1P_1$	1	2	3	1/3
Capture of s or d electron to give pp	$^3D_3$	0	7	7	0
	$^3D_2$	5/2	5/2	5	1/2
	$^3D_1$	5/2	1/2	3	5/6
	$^1D_2$	5/3	10/3	5	1/3
	$^3P_2$	5/6	25/6	5	1/6
	$^3P_1$	3/2	3/2	3	1/2
	$^3P_0$	2/3	1/3	1	2/3
	$^1P_1$	1	2	3	1/3
	$^3S_1$	1	2	3	1/3
	$^1S_0$	1/3	2/3	1	1/3
Capture of s or d electron to give p²	$^3P_2$	5/3	25/3	10	1/6
	$^3P_1$	3	3	6	1/2
	$^3P_0$	4/3	2/3	2	2/3
	$^1D_2$	10/3	20/3	10	1/3
	$^1S_0$	2/3	4/3	2	1/3
Capture of p electron to give ps	$^3P_2$	0	5	5	0
	$^3P_1$	2	1	3	2/3
	$^3P_0$	1	0	1	1
	$^1P_1$	1	2	3	1/3

Table 4 takes into account solely the rate of entrance of the atoms. It is obtained from Table 3 by multiplication of the files referring to the upper levels by the $p_{1/2}$ *parentage* factors of Table 2.[9]

[9] Note also that the variation in the (wave number)^3 factor necessary to convert Table 2 into transition probabilities is neglected.

Table 4 contains only part of the story, however. Since 2p3d feeds very strongly into $2p^2$ as well as into 2p3p, only a fraction, and probably a small fraction at that, of the atoms entering 2p3d will be available for transition to 2p3p. Most of the atoms in the terms ^3D, ^3P, ^1F, ^1D, or ^1P of 2p3d will have a much higher probability,

TABLE 3

LINE STRENGTHS IN NORMAL ARRAYS

			$1s^2\,2s^2\,2p3p$									
			^3D			^3P			^3S	^1D	^1P	^1S
		J	3	2	1	2	1	0	1	2	1	0
$1s^2\,2s^2\,2p3d$	^3F	4	54	0	0	0	0	0	0	0	0	0
		3	14/3	112/3	0	0	0	0	0	0	0	0
		2	2/15	14/3	126/5	0	0	0	0	0	0	0
	^3D	3	28/3	7/6	0	63/2	0	0	0	0	0	0
		2	7/6	125/24	9/8	45/8	135/8	0	0	0	0	0
		1	0	9/8	27/8	3/8	45/8	15/2	0	0	0	0
	^3P	2	7/10	1/8	1/120	75/8	25/8	0	50/3	0	0	0
		1	0	3/8	1/8	25/8	15/8	5/2	10	0	0	0
		0	0	0	1/6	0	5/2	0	10/3	0	0	0
	^1F	3	0	0	0	0	0	0	0	42	0	0
	^1D	2	0	0	0	0	0	0	0	15/2	45/2	0
	^1P	1	0	0	0	0	0	0	0	1/2	15/2	10
$1s^2\,2s^2\,2p3s$	^3P	2	7	5/4	1/12	15/4	5/4	0	5/3	0	0	0
		1	0	15/4	5/4	5/4	3/4	1	1	0	0	0
		0	0	0	5/3	0	1	0	1/3	0	0	0
	^1P	1	0	0	0	0	0	0	0	5	3	1

because of the large ν-cube factor, of relapsing directly to ^3P, ^1D, or ^1S of $2p^2$ than of going to the terms of 2p3p. Atoms in any one of the levels of ^3F of 2p3d, however, can relapse only to ^3D of 2p3p. The intensity of this multiplet remains as given in Table 4, whereas all the others are diminished by a very appreciable factor. The strengths in the second part of Table 4, from the array 2p3p − 2p3s, are unaffected by any such complications. Comparison of Tables 3 and 4, even without the depletion just discussed, shows that the relative strengths of the various multiplet lines are abnormal. Certain lines are missing or are very weak. As far as the multiplets $^3P - {}^3D$, $^3P - {}^3P$, and $^3P - {}^3S$ are concerned, the intensity distor-

TABLE 4

MODIFICATION OF TABLE 3 ON THE ASSUMPTION OF EXCITATION BY DIRECT ELECTRON CAPTURE IN THE UPPER LEVELS OF 2p3d→2p3p AND 2p3p→2p3s

			^3D			^3P			^3S	^1D	^1P	^1S
		J	3	2	1	2	1	0	1	2	1	0
			1s² 2s² 2p3p									
1s² 2s² 2p3d (excited by capture)	^3F	4	0	0	0	0	0	0	0	0	0	0
		3	56/27	448/27	0	0	0	0	0	0	0	0
		2	14/135	98/27	98/5	0	0	0	0	0	0	0
	^3D	3	56/27	7/27	0	7	0	0	0	0	0	0
		2	49/108	875/432	7/16	35/16	105/16	0	0	0	0	0
		1	0	9/16	27/16	3/16	45/16	15/4	0	0	0	0
	^3P	2	7/20	1/16	1/240	75/16	25/16	0	25/3	0	0	0
		1	0	1/16	1/48	25/48	5/16	5/12	5/3	0	0	0
		0	0	0	0	0	0	0	0	0	0	0
	^1F	3	0	0	0	0	0	0	0	14	0	0
	^1D	2	0	0	0	0	0	0	0	5/2	15/2	0
	^1P	1	0	0	0	0	0	0	0	1/6	5/2	10/3
			1s² 2s² 2p3p (excited by capture)									
1s² 2s² 2p3s	^3P	2	0	5/8	5/72	5/8	5/8	0	5/9	0	0	0
		1	0	15/8	25/24	5/24	3/8	2/3	1/3	0	0	0
		0	0	0	25/18	0	1/2	0	1/9	0	0	0
	^1P	1	0	0	0	0	0	0	0	5/3	1	1/3

tion agrees fairly well with observation. The partial excitation of $^3P_1$ is accounted for naturally, whereas Bowen was forced to consider the extremely high velocity differential of 90 km/sec to achieve excitation of 2p3d $^3P_1$ by resonance absorption of the ultimate line of *He* II. But failure of at least the lines of the $^3F - {}^3D$ and $^3D - {}^3P$ multiplets to appear is definite evidence against the validity of the mode here considered of exciting fractional multiplets. It is true that we have taken into account only the excitation of 2p3d directly by electron capture, while electrons will also be captured into higher configurations and feed down into 2p3d. But the changes occasioned by taking this into account will certainly *decrease* the intensity anomalies and will not aid matters.

Looking now at 2p3p − 2p3s in Table 4, we see that direct capture into 2p3p could not possibly account for the observed intensity anomalies. A rough computation shows that the assumption of ex-

TABLE 5

RELATIVE STRENGTHS OF OBSERVABLE LINES ON BOWEN'S HYPOTHESIS

			$1s^2\,2s^2\,2p3p$						
		J	^3D			^3P			^3S
			3	2	1	2	1	0	1
$1s^2\,2s^2\,2p3d$..	^3P	2	7/10	1/8	1/120	75/8	25/8	0	50/3
$1s^2\,2s^2\,2p3s$ ..	^3P	2	7/10	1/32	1/4320	225/32	125/96	0	250/27
		1	0	3/32	1/288	75/32	75/96	0	50/9
		0	0	0	1/216	0	25/24	0	50/27

citation of 2p3p principally by transitions from 2p3d does not improve the agreement.

Also, the observed anomalous intensities in the multiplet $3p\,^2P^o - 3d\,^2D$ of *N* III cannot be accounted for by the present scheme, because the ground term of *N* IV is $^1S_0$, a single level.

We are apparently forced to return to the Bowen mechanism for the excitation of these lines, despite the fact that it does not account for the strength of the singlet line λ 5592. In Table 5 are given the relative theoretical strengths of the lines of the significant multiplets of *O* III, calculated on the basis of initial selective excitation of $^3P_2$ alone. The strengths in the two transition arrays of this table have the correct relative values and need only be multiplied by ν^4 to give the relative intensities of all the lines. The tabulated strengths show good qualitative agreement with the lines observed in the nebulae.

As was noted earlier in this paper, acceptance of the Bowen mechanism places a definite restriction on physical conditions within a nebula. In interstellar space the high concentration of atoms occurs in the lowest level, whereas in the nebulae an appreciable population must be maintained in the levels $^3P_2$ and $^2P^\circ_{/2}$ of O III and N III, respectively. The mean lifetimes of these levels are of the order of several hours,[3] and a concentration will occur only when the rate of excitation to these levels is also about once in three hours. Radiation processes, with the high dilution existing in nebulae, cannot be called on for so frequent excitations. A study of the problem indicates that collisional excitations from the ground level must be sufficiently frequent to maintain the required population. This investigation, by Hebb and Menzel, will form the subject of the next paper of the present series.

We proceed now to the proof of a theorem concerning the transformation properties of the line-strength matrix which will justify the method used in the above computations.

THEOREM.—*Consider a configuration* α *and all the levels* βJ *of a particular angular momentum* J *belonging to a configuration* β. *In a particular coupling let the levels of* α *and of* βJ *break into two groups* α', α'' *and* $\beta' J$, $\beta'' J$ *such that* $\beta'' J$ *does not combine with* α' ($\beta' J$ *may combine with both* α' *and* α''). *In this case the matrix*[10] *of* $S^{1/2}$ *connecting the levels of configuration* α *with those of* βJ *will have the form*

$$
\begin{array}{c|cccc|cccc|}
 & & \beta' J & & & & \beta'' J & & \\
\hline
 & s_{11} & s_{12} & \cdots & s_{1\nu} & 0 & 0 & \cdots & 0 \\
\alpha' & s_{21} & s_{22} & \cdots & s_{2\nu} & 0 & 0 & \cdots & 0 \\
 & \cdots & \cdots & \cdots & \cdots & \cdots & \cdots & \cdots & \cdots \\
 & s_{n1} & s_{n2} & \cdots & s_{n\nu} & 0 & 0 & \cdots & 0 \\
\hline
 & x & x & \cdots & x & x & x & \cdots & x \\
\alpha'' & x & x & \cdots & x & x & x & \cdots & x \\
 & \cdots & \cdots & \cdots & \cdots & \cdots & \cdots & \cdots & \cdots \\
 & x & x & \cdots & x & x & x & \cdots & x \\
\hline
\end{array}
\qquad (2)
$$

[10] See Condon and Shortley, *op. cit.*, p. 277, for definition of this matrix, whose components are the square roots of the line strengths chosen with phases corresponding to those of the matrix components of electric moment.

Let the strength of each of the files connecting one of the levels of $\beta' J$ with all the levels of α' have the same value C_J, and let all pairs of these files be orthogonal. That is, let

$$\sum_{i=1}^{n} s_{i\xi}\, s_{i\eta} = C_J\, \delta_{\xi\eta}\,. \tag{3}$$

Now consider a level βJ of configuration β in any other coupling. When this level is expressed in the β', β'' scheme, let the sum of the squares of the transformation amplitudes connecting this level with the levels of $\beta' J$ be τ. Then the total strength of all transitions connecting the level βJ and the levels of the part-configuration α' will be τC_J.

To prove this statement let the level βJ be connected with those of β'J and β''J by transformation amplitudes contained in the one-column matrix

$$\begin{array}{c|c|}
 & \beta \mathrm{J} \\
\cline{2-2}
 & t_1 \\
\beta' \mathrm{J} & t_2 \\
 & \cdots \\
 & t_\nu \\
\cline{2-2}
 & \mathrm{x} \\
 & \mathrm{x} \\
\beta'' \mathrm{J} & \cdots \\
 & \mathrm{x} \\
\cline{2-2}
\end{array} \tag{4}$$

where by hypothesis

$$\sum_{\xi=1}^{\nu} t_\xi^2 = \tau \leq 1\,. \tag{5}$$

Now, since α' does not combine with β''J, we may get the $S^{1/2}$ matrix connecting α' and βJ by multiplying the upper left square of (2) with the upper part of (4) to get

$$\begin{array}{c|c|} & \beta J \\ \hline & \sum_{\xi=1}^{\nu} s_{1\xi}\, t_\xi \\ \alpha' & \sum s_{2\xi}\, t_\xi \\ & \cdots\cdots \\ & \sum s_{n\xi}\, t_\xi \\ \hline \end{array}$$

The sum of the squares of these elements is

$$\sum_{i=1}^{n}\sum_{\xi=1}^{\nu}\sum_{\eta=1}^{\nu} s_{i\xi}\, t_\xi\, s_{i\eta}\, t_\eta = C_J \sum_{\xi=1}^{\nu} t_\xi^2 = \tau C_J\,,$$

where the right-hand members are obtained by carrying out first the sum over i, then that over ξ. Note that in cases where this theorem is valid, one may obtain the desired sums without knowing the relation between the phases in (2) and the phases in (4). This is sometimes a great convenience.

We may now apply this theorem explicitly to transitions from αEl to αn′l′ where the electron n′l′ is not equivalent to any in the parent configuration α. Let α_1 represent a given *level* of the parent configuration in the ion. Then, if we characterize the levels of αn′l′ by the same parents, there will be no transitions from any level of α_1n′l′ to any levels of αEl except those of α_1El. The sum of the strengths from a given level α_1n′l′J to all levels of α_1El is

$$C_J = (2J+1)\,|(\mathrm{El} \vdots P \vdots \mathrm{n'l'})|^2\, \Xi(\mathrm{l}, \mathrm{l'})\,, \qquad (6)^{11}$$

[11] *Ibid.*, p. 281. Here $\Xi(\mathrm{l}, \mathrm{l}+1) = (\mathrm{l}+1)(2\mathrm{l}+3)$; $\Xi(\mathrm{l}, \mathrm{l}-1) = \mathrm{l}(2\mathrm{l}-1)$; $(\mathrm{El} \vdots P \vdots \mathrm{n'l'})$ is an integral of the product of r and the radial functions of the two states as defined in *ibid.*, p. 132. Since one of the levels is in the continuum, we must consider (6) as referring to unit energy range and the radial function for the continuum normalized in the proper fashion. See Kemble, *Quantum Mechanics*, for a good discussion of this normalization.

proportional to $2J + 1$. The orthogonality conditions of the foregoing theorem are also satisfied;[11] hence, in the case of a level of $\alpha n'l'J$ in any other coupling, having a total component τ in the sense of (5) in the direction of $\alpha_1 n'l'J$, the total strength to all levels of $\alpha_1 El$ will be τC_J. Note that *the relative values of the strength sums from the different levels of a given discrete* configuration $\alpha n'l'$ to continuous states $\alpha_1 El$ is independent of whether $l = l' + 1$ or $l' - 1$, and of the energy E.

O III is just a special case of the previous paragraph, with $\alpha = 1s^2\,2s^2\,2p$; $\alpha_1 = 1s^2\,2s^2\,2p\ {}^2P_{1/2}$; hence our earlier calculations are justified except for the case $p({}^2P_{1/2})Es$ or $p({}^2P_{1/2})\ Ed \rightarrow p^2$, which violates the inequivalence specification. But in this case an examination of the $S^{1/2}$ matrices in *jj* coupling shows them to be of the form (2), if for the p^2 configuration we take $\beta'0 = p_{1/2}p_{1/2}0$; $\beta'1 = p_{3/2}p_{1/2}1$; $\beta'2 = p_{3/2}p_{1/2}2$ (for the case ps, p^2 see the matrix on page 265 of Condon and Shortley). The C_J's in these cases are no longer proportional to $2J + 1$ but have the same relative values for capture of s and d electrons, namely, $C_0 : C_1 : C_2 = 2:3:5$.

MENDENHALL LABORATORY OF PHYSICS
OHIO STATE UNIVERSITY
AND
HARVARD COLLEGE OBSERVATORY

Reprinted from *The Astrophysical Journal,* Vol. 92, No. 3, November, 1940.

PHYSICAL PROCESSES IN GASEOUS NEBULAE

X. COLLISIONAL EXCITATION OF NEBULIUM

MALCOLM H. HEBB AND DONALD H. MENZEL

ABSTRACT

Target areas for collisional excitation of *O* III by electron impact are calculated by wave-mechanical methods. Coulomb wave functions are used for the colliding electron. The resulting target areas are large, particularly for excitation between the levels of ^3P. Application to the nebulae leads to the result that collisional de-excitations from $^3P_2$ predominate over the spontaneous transitions. In consequence, the levels of $^3P_2$ attain a high population, suited to the mechanism proposed by Bowen: resonance absorption of λ 303 of *He* II by the coincident line $^3P_2 - {}^3P_2$ of *O* III. The figures also suggest, for the lower electron densities of interstellar space, a high population of atoms in the ground level, a result consistent with Dunham's observations of the interstellar lines of *Ti* II.

The high intensities of the forbidden lines in gaseous nebulae lead one to conclude that the mode of excitation for these lines may differ from that for normal lines. Bowen has suggested that the free electrons, liberated from hydrogen by photo-ionization, may excite the "nebulium" lines by inelastic impact. To calculate the effectiveness of this excitation process, we must have advance knowledge of the atomic collision cross-sections. Since experimentally determined values of these parameters are not available, we must have recourse to theoretical methods, via wave mechanics. The calculations of the target areas have been carried out specifically for the various metastable levels of *O* III, the atom responsible for the most intense of the nebular lines.

THE BORN APPROXIMATION

We consider an *O* III ion initially in the state determined by the quantum numbers J, L, S, M, which give, respectively, the total, orbital, and spin angular momenta and the component of the total angular momentum of the ion along some convenient axis. Let us suppose that an electron with momentum p at infinity and spin component m_s impinges on the ion, changing its state to J', L', S', M', while at the same time the electron escapes with a momentum p'

within solid angle $d\Omega$ and spin component m_s'. The first-order differential cross-section for this process can be written[1]

$$\sigma(JLSMpm_s \rightarrow J'L'S'M'p'm_s')d\Omega = \frac{d\Omega}{4\pi^2}\frac{p'}{p}$$
$$(JLSMpm_s|V|J'L'S'M'p'm_s)(J'L'S'M'p'm_s'|V|JLSMpm_s) ,$$

or, briefly,

$$\sigma(n \rightarrow n')d\Omega = \frac{d\Omega}{4\pi^2}\frac{p'}{p}(n|V|n')(n'|V|n) . \quad (1)$$

The bracketed symbol on the right side is the matrix element of V, the mutual potential of the electron and ion connecting the two complete states n and n'. We can write, symbolically,

$$(n'|V|n) = \int \Psi_{n'}^* V \Psi_n d\tau , \quad (2)$$

where $\Psi_{n'}$ and Ψ_n represent wave functions for the complete system of free electron and ion, and the integral sign represents an integration over the space co-ordinates of the several electrons concerned. Throughout this paper an asterisk will be used to designate the complex conjugate. Disregarding the exclusion principle for the moment, we can write the complete wave function as the product $\Psi_n = \Psi_n \chi_n$, where Ψ_n refers to the ion alone and χ_n to the free electron. The usual procedure is to take plane waves for the free electron, i.e., to let $\chi_n = e^{i(p \cdot r)}$. This procedure is suitable for collision problems involving neutral atoms, but for collisions with charged ions the incoming and outgoing waves will be greatly distorted from plane, especially for low velocities in which we are particularly interested. Instead of plane waves we shall therefore use functions that are exact solutions of the wave equation for the Coulomb field and which will be given later. The simple product functions $\Psi_n \chi_n$ do not

[1] See, e.g., Dirac, *Quantum Mechanics* (2d ed.), chap. ix, or Mott and Massey, *Theory of Atomic Collisions*, chap. viii. In eq. (1) and in the following we use *atomic units* in which the units of mass, charge, and action are, respectively, the electronic mass m and charge e (in absolute value) and Dirac's action constant $\hbar = h/2\pi$. The unit of length is then the first Bohr radius, and the unit of energy is twice the ionization energy of the hydrogen atom. For a complete list see Condon and Shortley, *Theory of Atomic Spectra*, p. 432. In eq. (1) we take $(n|V|n')(n'|V|n)$ instead of $|(n|V|n')|^2$, since we wish to have $p^2\sigma(n \rightarrow n') = p'^2\sigma(n' \rightarrow n)$. Cf. Mott and Massey, *op. cit.*, p. 153.

satisfy the exclusion principle and do not allow for the possibility of exchange between the incident electron and the electrons of the ion. To satisfy these two conditions, we must write Ψ_n as a sum of terms $\Psi_n\chi_n$, in which the arguments of the electrons are permuted. In our case we shall consider only the two 2p electrons of the ion, so that $\Psi_n = \Psi_n\,(1, 2)$ contains the co-ordinates of the two electrons. We shall suppose that Ψ_n already satisfies the exclusion principle, i.e., $\Psi_n(2, 1) = -\Psi_n(1, 2)$. Then the wave function for the complete system is given by

$$\sqrt{3}\Psi_n(1, 2, 3) = \Psi_n(1, 2)\chi_n(3) + \Psi_n(2, 3)\chi_n(1) + \Psi_n(3, 1)\chi_n(2)\,, \quad (3)$$

where the arguments $1, 2, 3$ stand for the co-ordinates and spin of the three electrons.

The potential V can now be stated explicitly, namely,

$$V(a, b, c) = \frac{2}{r_c} - \frac{1}{r_{ac}} - \frac{1}{r_{bc}} = V(a, c) + V(b, c)\,, \quad (4)$$

where r_c is the distance of the colliding electron from the ion, and r_{ac}, r_{bc} are the distances from the colliding electron to the two electrons of the ion. It should be noticed that V reduces to zero faster than $-2/r_c$ as r_c goes to infinity, since the Coulomb term $-2/r_c$ has been included in the wave equation for χ_n and V represents the actual mutual energy of the colliding electron and ion less this Coulomb term. If we insert the expressions (3) and (4) into (2), we obtain after some simplification

$$\left.\begin{aligned}(n'|V|n) = {} & 2\int\Psi^*_{n'}(1, 2)\Psi_n(1, 2)V(1, 3)\chi^*_{n'}(3)\chi_n(3)d\tau \\ & + 2\int\Psi^*_{n'}(2, 3)\Psi_n(1, 2)V(1, 3)\chi^*_{n'}(1)\chi_n(3)d\tau \\ & + 2\int\Psi^*_{n'}(3, 1)\Psi_n(1, 2)V(1, 3)\chi^*_{n'}(2)\chi_n(3)d\tau\,.\end{aligned}\right\} \quad (5)$$

The first term alone is what would be obtained with the simple product functions which neglect exchange. The second and third terms are clearly exchange integrals, the third being of a rather peculiar nature, since the potential $V\,(1, 3)$ does not directly connect the incoming and outgoing electrons 2 and 3.

We are not primarily interested in the differential cross-section (1) for polarized beams of electrons and definite transitions of the magnetic quantum number M. Instead, we require the total integral cross-section $\sigma(JLSp \rightarrow J'L'S'p')$, which is obtained by integrating (1) over $d\Omega$, i.e., over all directions of p', averaging over the initial values M and m_s and summing over the final values M' and m_s'. The result is

$$\sigma(JLSp \rightarrow J'L'S'p') = \frac{1}{2(2J+1)} \sum_{MM'm_s m_s'} \int \sigma(n \rightarrow n') d\Omega \,. \qquad (6)$$

We shall now examine the wave functions for the O III ion and for the free electron which enter into the integrals in equation (5).

WAVE FUNCTIONS FOR O III

The radial wave function for a 2p electron in O III has been given in tabular form by Hartree and Black.[2] It is convenient for our purpose to obtain an analytic approximation to this function in the manner proposed by Slater.[3] We write, namely, the radial function in the form

$$R(r) = r^2(ae^{-br} + ce^{-dr}) \,, \qquad (7)$$

where r is the distance from the center of the ion, and adjust the constants a, b, c, d to give the best fit with the tabulated functions, subject to the normalization condition $\int_0^\infty R^2 dr = 1$. The adjusted values are

$$a = 10.415 \,, \qquad b = 3.604 \,, \qquad c = 4.520 \,, \qquad d = 2.014 \,.$$

The two-electron wave functions $\Psi_{JLSM}(1, 2)$ of the ion are now built up from angular-momentum considerations.[4] The atom of O III is sufficiently near LS coupling for the ground configuration that we need not concern ourselves with the departures from precise LS coupling that are so important for optical transitions. We choose a spherical co-ordinate system (r, θ, ϕ) and let the polar axis $\theta = 0$ coincide with the direction of the momentum $\boldsymbol{p}$ of the incident electron. Then, if this axis is also the direction of quantization for the magnetic quantum numbers, we find

[2] *Proc. R. Soc.*, A, **139**, 311, 1933. [3] *Phys. Rev.*, **42**, 33, 1932.

[4] M. H. Johnson, *Phys. Rev.*, **39**, 197, 1932; Condon and Shortley, *op. cit.*, p. 220.

$$
\left.
\begin{array}{cccccc}
\Psi_{JLSM}\,(1,2)r_1r_2/R(r_1)R(r_2) & J & L & S & M \\
\Phi_1\alpha_1\alpha_2 & 2 & 1 & 1 & +2 \\
\frac{1}{\sqrt{2}}\left[\Phi_1\frac{\alpha_1\beta_2+\beta_1\alpha_2}{\sqrt{2}}-i\Phi_0\alpha_1\alpha_2\right] & 2 & 1 & 1 & +1 \\
\frac{1}{\sqrt{6}}\left[\Phi_1\beta_1\beta_2+\Phi_1^*\alpha_1\alpha_2-2i\Phi_0\frac{\alpha_1\beta_2+\beta_1\alpha_2}{\sqrt{2}}\right] & 2 & 1 & 1 & 0 \\
\frac{1}{\sqrt{2}}\left[\Phi_1^*\frac{\alpha_1\beta_2+\beta_1\alpha_2}{\sqrt{2}}-i\Phi_0\beta_1\beta_2\right] & 2 & 1 & 1 & -1 \\
\Phi_1^*\beta_1\beta_2 & 2 & 1 & 1 & -2 \\
\frac{1}{\sqrt{2}}\left[\Phi_1\frac{\alpha_1\beta_2+\beta_1\alpha_2}{\sqrt{2}}+i\Phi_0\alpha_1\alpha_2\right] & 1 & 1 & 1 & +1 \\
\frac{1}{\sqrt{2}}\left[\Phi_1\beta_1\beta_2-\Phi_1^*\alpha_1\alpha_2\right] & 1 & 1 & 1 & 0 \\
\frac{1}{\sqrt{2}}\left[\Phi_1^*\frac{\alpha_1\beta_2+\beta_1\alpha_2}{\sqrt{2}}+i\Phi_0\beta_1\beta_2\right] & 1 & 1 & 1 & -1 \\
\frac{1}{\sqrt{3}}\left[\Phi_1\beta_1\beta_2+\Phi_1^*\alpha_1\alpha_2+i\Phi_0\frac{\alpha_1\beta_2+\beta_1\alpha_2}{\sqrt{2}}\right] & 0 & 1 & 1 & 0 \\
\Phi_{D2}\frac{\alpha_1\beta_2-\beta_1\alpha_2}{\sqrt{2}} & 2 & 2 & 0 & +2 \\
\Phi_{D1}\frac{\alpha_1\beta_2-\beta_1\alpha_2}{\sqrt{2}} & 2 & 2 & 0 & +1 \\
\Phi_{D0}\frac{\alpha_1\beta_2-\beta_1\alpha_2}{\sqrt{2}} & 2 & 2 & 0 & 0 \\
\Phi_{D1}^*\frac{\alpha_1\beta_2-\beta_1\alpha_2}{\sqrt{2}} & 2 & 2 & 0 & -1 \\
\Phi_{D2}^*\frac{\alpha_1\beta_2-\beta_1\alpha_2}{\sqrt{2}} & 2 & 2 & 0 & -2 \\
\Phi_S\frac{\alpha_1\beta_2-\beta_1\alpha_2}{\sqrt{2}} & 0 & 0 & 0 & 0
\end{array}
\right\}\quad(8)
$$

where

$$\Phi_1 = \frac{3}{8\pi}\,[\cos\theta_1 \sin\theta_2 e^{i\phi_2} - \cos\theta_2 \sin\theta_1 e^{i\phi_1}]$$

$$\Phi_0 = \frac{3\sqrt{2}}{8\pi}\,\sin\theta_1 \sin\theta_2 \sin(\phi_1 - \phi_2)$$

$$\Phi_{D0} = \frac{\sqrt{6}}{8\pi}\,[\sin\theta_1 \sin\theta_2 \cos(\phi_1 - \phi_2) - 2\cos\theta_1 \cos\theta_2]$$

$$\Phi_{D1} = \frac{3}{8\pi}\,[\cos\theta_1 \sin\theta_2 e^{i\phi_2} + \cos\theta_2 \sin\theta_1 e^{i\phi_1}]$$

$$\Phi_{D2} = \frac{3}{8\pi}\,\sin\theta_1 \sin\theta_2 e^{i(\phi_1+\phi_2)}$$

$$\Phi_S = \frac{\sqrt{3}}{4\pi}\,[\cos\theta_1 \cos\theta_2 + \sin\theta_1 \sin\theta_2 \cos(\phi_1 - \phi_2)]\,.$$

In equations (8), α_i, β_i represent single-electron spin symbols for which the spin component is, respectively, $\pm\frac{1}{2}$. They satisfy $\alpha_i\beta_i = \beta_i\alpha_i = 0$ and $\alpha_i^2 = \beta_i^2 = 1$.

CONTINUOUS COULOMB WAVE FUNCTIONS

As already stated, the wave functions that we use to represent the incoming and outgoing electrons are the appropriate solutions of the wave equation in the Coulomb field of the ion. For the incident electron this function is[5]

$$\chi_{p,\pm 1/2}(3) = e^{\pi Z/2p}\Gamma\left(1 - \frac{iZ}{p}\right) e^{i(\boldsymbol{p}\cdot\boldsymbol{r}_3)}{}_1F_1\left(\frac{iZ}{p};\ 1;\ ipr_3 - i(\boldsymbol{p}\cdot\boldsymbol{r}_3)\right)_{\beta_3}^{\alpha_3}, \quad (9)$$

where Z is the charge on the ion (in our case $Z = 2$), and ${}_1F_1$ is a confluent hypergeometric function. Similarly, the wave function for the scattered electron is

$$\chi_{p',\pm 1/2}(3) = e^{\pi Z/2p'}\Gamma\left(1 + \frac{iZ}{p'}\right) e^{i(\boldsymbol{p}'\cdot\boldsymbol{r}_3)}{}_1F_1\left(-\frac{iZ}{p'};\ 1;\ -ip'r_3 - i(\boldsymbol{p}'\cdot\boldsymbol{r}_3)\right)_{\beta_3}^{\alpha_3}. \quad (10)$$

The function (9) reduces approximately to a plane wave of unit amplitude at large distances in the direction opposite to $\boldsymbol{p}$, while

[5] Mott and Massey, *op. cit.*, chap. iii.

(10) is asymptotic to an approximate plane wave in the forward direction of p'. For $Z = 0$, (9) and (10) become, respectively, $e^{i(p\cdot r_3)}$ and $e^{i(p'\cdot r_3)}$, apart from the spin factors α_3 and β_3.

In the next section we shall require the expansions of (9) and (10) in surface harmonics. With the co-ordinates defined in the last section, (9) is the same as[6]

$$\chi_{p,\pm 1/2}(3) = \frac{1}{r_3}\sqrt{\frac{2}{p}}\sum_{l=0}^{\infty}\sqrt{(2l+1)}\, i^l e^{i\eta_l}F_l(r_3)P_{l0}\,(\cos\,\theta_3)^{\alpha_3}_{\beta_3}, \quad (11)$$

where

$$F_l(r_3) = \frac{e^{\pi Z/2p}}{2\sqrt{p}}\cdot\frac{\left|\Gamma\left(l+1-\frac{iZ}{p}\right)\right|}{(2l+1)!}\, e^{ipr_3}(2pr_3)^{l+1}\,{}_1F_1\left(l+1-\frac{iZ}{p};\, 2l+2;\, -2ipr_3\right), \quad (12)$$

and P_{lm} is the normalized Legendre function which satisfies $\int_0^\pi P^2_{lm}\,(\cos\,\theta)\,\sin\,\theta d\theta = 1$. The phases η_l are functions of p, but their precise form is of no particular interest in the present work. It can easily be shown that the F_l defined by (12) are real. The expansion of (10) is similar to (11), since (10) can be obtained from (9) by reversing the sign of i and replacing p by $-p'$. A difference in form arises, since p' is not directed along the polar axis as is p. If we let the direction of p' be specified by θ, ϕ, and apply the addition theorem[7] for Legendre polynomials, we find

$$\left.\begin{aligned}&\chi_{p',\pm 1/2}(3)\\ &= \frac{1}{r_3}\sqrt{\frac{2}{p'}}\sum_{l'=0}^{\infty}\sqrt{(2l'+1)}\, i^{l'}e^{i\eta'_{l'}}F'_{l'}(r_3)\sum_{m'=-l'}^{l'}P_{l'm'}\,(\cos\,\theta_3)P_{l'm'}\,(\cos\,\theta)e^{im'(\phi_3-\phi)}{}^{\alpha_3}_{\beta_3}\,.\end{aligned}\right\} \quad (13)$$

Our functions $F_l(r)$ differ slightly from those usually defined. For example, the quantity $L_l(r)$ defined on page 39 of Mott and Massey's book[1] is, except for notation and units, the same as $F_l(r)/(r\sqrt{p})$.

[6] C. G. Darwin, *Proc. R. Soc.*, A, **118**, 654, 1928; H. A. Bethe, *Handb. d. Phy.* (2), **24**, Part I, p. 554.

[7] Whittaker and Watson, *Modern Analysis*, p. 326.

The reason for our choice[8] is that for small p, $F_l(r)$ is rather insensitive to p near the origin $r = 0$, while under the same conditions $L_l(r)$ varies as $1/\sqrt{p}$. This means that, using $F_l(r)$, we can express most of the dependence of the cross-sections on the velocity explicitly, in the region of small velocities, and the implicit dependence entering through the F_l is less important.

METHOD OF PARTIAL CROSS-SECTIONS

Each of the matrix elements (5), which must be evaluated to obtain the cross-sections, contains integrals over the nine space coordinates of the three electrons.[9] In order to perform these integrations, we use the method of partial cross-sections, i.e., we expand the wave functions of the incoming and outgoing electrons and the potential function in surface harmonics and then integrate term by term. For neutral atoms it is well known that the successive terms in this method decrease rapidly for small velocities on account of the fact that for such velocities the classical distance of closest approach increases rapidly with the angular momenta l, l'. With ions instead of neutral atoms, of course this convergence will not be quite so good. The required expansions for the continuous wave functions have already been given in equations (11) and (13). For V we have

$$\left.\begin{aligned} V(r_1, r_3) &= \frac{1}{r_3} - \frac{1}{r_{13}} \\ &= -\sum_{s=0}^{\infty} \gamma_s(r_1, r_3) \sum_{t=-s}^{s} P_{st}(\cos\theta_1) P_{st}(\cos\theta_3) e^{it(\phi_1-\phi_3)} \end{aligned}\right\} \quad (14)$$

with

$$\begin{aligned} \gamma_s(r_1, r_3) &= \frac{2}{2S+1}\left(\frac{r_1^s}{r_3^{s+1}} - \frac{\delta_s^0}{r_3}\right) \qquad (r_1 < r_3) \\ &= \frac{2}{2S+1}\left(\frac{r_3^s}{r_1^{s+1}} - \frac{\delta_s^0}{r_3}\right). \qquad (r_1 > r_3) \end{aligned}$$

If we interchange the order of summation and integration, we have a multiple sum of integrals corresponding to each integral of (5), the

[8] Yost, Wheeler, and Breit, *Phys. Rev.*, **49**, 174, 1936.

[9] The spin symbols disappear on application of the rules given after eqs. (8).

indices of summation being l, l', m', s, t. This result is appreciably simplified when the angular integrations are considered. The integrations over the angles ϕ_1, ϕ_2, ϕ_3 and most of those over θ_1, θ_2, θ_3 are elementary. A few integrals occur that contain the products of three associated Legendre functions, and for these the rules of Gaunt[10] are useful. As is to be expected, the angular integrations severely restrict the indices l, l', m', s, t. Before specifying the relations between these indices, we find it convenient to distinguish the three types of integral arising from (5). It will be seen that the threefold radial integrals which remain after the angular integrations have been carried out are of the three following forms, corresponding to the three terms in (5):

$$f_{l'l}^{(s)}(p';p) = \iiint dr_1 dr_2 dr_3 R_1^2 R_2^2 \gamma_s(r_1, r_3) F'_{l'}(r_3) F_l(r_3) \,, \tag{15}$$

$$g_{l'l}^{(s)}(p';p) = \iiint dr_1 dr_2 dr_3 R_1 R_2^2 R_3 \gamma_s(r_1, r_3) F'_{l'}(r_1) F_l(r_3) \,, \tag{16}$$

$$h_{l'l}^{(s)}(p';p) = \iiint dr_1 dr_2 dr_3 R_1^2 R_2 R_3 \gamma_s(r_1, r_3) F'_{l'}(r_2) F_l(r_3) \,. \tag{17}$$

Then, on account of the angular integrations, only the following terms appear:

$$f_{ll}^{(2)},\ f_{l\,l\pm 2}^{(2)},\ g_{ll}^{(l\pm 1)},\ g_{l\,l\pm 2}^{(l\pm 1)},\ h_{11}^{0},\ h_{11}^{(2)},\ h_{13}^{(2)} \,.$$

Thus, finally, the matrix elements (5) become single sums of expressions containing the integrals (15), (16), and (17):

$$(n'|V|n) = \frac{2\pi}{\sqrt{p'p}} \sum_{l'=0}^{\infty} A_{l'}(n';n) P_{l'm'}(\cos\theta) e^{im'\phi} \,. \tag{18}$$

In general, A_l is complex and depends even in absolute magnitude on the phases η_l of equations (11) and (13). It will be noticed that the factors $1/\sqrt{p'}$ and $1/\sqrt{p}$ which arise from equations (11) and (13), respectively, appear explicitly in equation (18), so that the expressions A_l contain p, p' only through F_l and $F'_{l'}$. Hence, for small velocities in which we are primarily interested A_l should be roughly independent of velocity. The presence of the factor $e^{im'\phi}$

[10] *Phil. Trans.*, **228**, 192, 1929.

could have been inferred from the conservation of the component of momentum about the axis $\theta = 0$ (direction of incident electrons). For the component in this direction we have initially $M + m_s$, since the incoming partial waves all have $m = 0$. For the final state, on the other hand, a partial wave with angular momentum l' has a component m', while the ion and spin contribute $M' + m_s'$. Hence, $m' = M - M' + m_s - m_s'$, the same for all partial waves.

For the integral cross-section, we require

$$\int_0^{2\pi} d\phi \int_0^{\pi} \sin\theta d\theta (n|V|n')(n'|V|n) = \frac{8\pi^3}{p'p} \sum_{l'} A_{l'}(n'; n) A_{l'}(n; n') .$$

The final cross-section (6) is, thus,

$$\sigma(JLSp \to J'L'S'p') = \frac{\pi}{2J+1} \frac{1}{p^2} \sum_{l'} \sum_{MM'm_sm_s'} A_{l'}(n'; n) A_{l'}(n; n') . \quad (19)$$

In contrast to the expression for the individual A_l's, the summed expression is real and is independent of the phases $\eta_{l'}'$, η_l. Further, in line with what has already been said, the sum in equation (19) is only slightly dependent on p, p', provided they are small, and hence the cross-sections in the first approximation for small velocities are inversely proportional to the square of the initial velocity. This is a direct consequence of the Coulomb field, which is responsible for the form of the continuous wave functions (11) and (13). Instead of designating the partial cross-sections, as in (19), by the angular momentum l' of the outgoing electron, we adopt a more convenient and symmetrical enumeration in terms of the mean $\lambda = (l + l')/2$ of the initial and final electronic angular momenta. Then we can write

$$\left.\begin{aligned} &(2J+1)p^2\sigma(JLSp \to J'L'S'p') \\ &\qquad = (2J'+1)p'^2\sigma(J'L'S'p' \to JLSp) \\ &\qquad = \pi\Omega(JLSp; J'L'S'p') \\ &\qquad = \pi\sum_{\lambda} \Omega_\lambda(JLSp; J'L'S'p') , \end{aligned}\right\} \quad (20)$$

where the Ω_λ are symmetrical in their two sets of arguments, and the initial and final states of the system are indicated by the arrows. For the various transitions between the states $^3P_0$, $^3P_1$, $^3P_2$, $^1D_2$, $^1S_0$, we find

$$\left.\begin{aligned}
\Omega_0(^3P_0p;\,^3P_1p) &= 8(g_{11}^{(0)})^2 \\
\Omega_0(^3P_1p;\,^3P_2p) &= 10(g_{11}^{(0)})^2 \\
\Omega_1(^3P_0p_1;\,^1Sp_2) &= \tfrac{1}{3}\Omega_1(^3P_1p_1;\,^1Sp_2) = \tfrac{1}{5}\Omega_1(^3P_2p_1;\,^1Sp_2) \\
&= \tfrac{4}{3}[g_{11}^{(0)} - g_{11}^{(2)} + h_{11}^{(0)} - h_{11}^{(2)}](p_1;\,p_2) \\
&\quad \times [g_{11}^{(0)} - g_{11}^{(2)} + h_{11}^{(0)} + 2h_{11}^{(2)}](p_2;\,p_1) \\
\Omega_1(^3P_0p_1;\,^1Dp_2) &= \tfrac{1}{3}\Omega_1(^3P_1p_1;\,^1Dp_2) = \tfrac{1}{5}\Omega_1(^3P_2p_1;\,^1Dp_2) \\
&= 6[g_{11}^{(0)} + h_{11}^{(0)} - h_{11}^{(2)}][g_{11}^{(0)} + h_{11}^{(0)} + \tfrac{1}{5}h_{11}^{(2)}] \\
&\quad + \tfrac{2}{3}[g_{11}^{(0)} - g_{11}^{(2)} + h_{11}^{(0)} - h_{11}^{(2)}] \\
&\quad \times [g_{11}^{(0)} - g_{11}^{(2)} + h_{11}^{(0)} + \tfrac{1}{5}h_{11}^{(2)}] + \tfrac{6}{5}g_{11}^{(2)}g_{11}^{(2)} \\
&\quad + 6g_{02}^{(1)}g_{20}^{(1)} + 6g_{20}^{(1)}g_{02}^{(1)} \\
\Omega_1(^1Dp_1;\,^1Sp_2) &= \tfrac{4}{15}[-12f_{11}^{(2)} + 5g_{11}^{(0)} + g_{11}^{(2)} + 5h_{11}^{(0)} + h_{11}^{(2)}] \\
&\quad \times [-12f_{11}^{(2)} + 5g_{11}^{(0)} + g_{11}^{(2)} + 5h_{11}^{(0)} + 10h_{11}^{(2)}] \\
&\quad + 8[2f_{02}^{(2)} - g_{02}^{(1)}][2f_{20}^{(2)} - g_{20}^{(1)}] \\
&\quad + 8[2f_{20}^{(2)} - g_{20}^{(1)}][2f_{02}^{(2)} - g_{02}^{(1)}] \\
\Omega_1(^3P_0p;\,^3P_1p) &= 5[g_{11}^{(0)} + \tfrac{4}{5}g_{11}^{(2)} + h_{11}^{(0)} - h_{11}^{(2)}]^2 \\
&\quad + 9(g_{11}^{(2)})^2 + 2(g_{02}^{(1)})^2 \\
\Omega_1(^3P_0p;\,^3P_2p) &= 5[-\tfrac{8}{5}f_{11}^{(2)} + g_{11}^{(0)} + h_{11}^{(0)} - h_{11}^{(2)}]^2 \\
&\quad + [\tfrac{8}{5}f_{11}^{(2)} - g_{11}^{(2)}]^2 + \tfrac{576}{25}(f_{11}^{(2)})^2 \\
&\quad + 64(f_{02}^{(2)})^2 - 32f_{02}^{(2)}g_{02}^{(1)} + 10(g_{02}^{(1)})^2 \\
\Omega_1(^3P_1p;\,^3P_2p) &= \tfrac{35}{2}[-\tfrac{36}{35}f_{11}^{(2)} + g_{11}^{(0)} + \tfrac{2}{7}g_{11}^{(2)} + h_{11}^{(0)} - h_{11}^{(2)}]^2 \\
&\quad + \tfrac{2376}{35}(f_{11}^{(2)})^2 + \tfrac{108}{35}f_{11}^{(2)}g_{11}^{(2)} + \tfrac{225}{14}(g_{11}^{(2)})^2 \\
&\quad + 144(f_{02}^{(2)})^2 - 72f_{02}^{(2)}g_{02}^{(1)} + 25(g_{02}^{(1)})^2
\end{aligned}\right\} \quad (21)$$

EVALUATION OF THE RADIAL INTEGRALS

The integrals (15), (16), and (17) can immediately be simplified. In the first two the integration over r_2 can be performed with the help of the normalizing condition $\int_0^\infty R_2^2 dr_2 = 1$. Also, in (15) the integral over r_1 is elementary, with the form (7) for R_1. If we let

$$I_s(r_3) = \int_0^\infty dr_1 R_1^2 \gamma_s(r_1;\,r_3)\,, \tag{22}$$

then (15) becomes

$$f_{l'l}^{(s)}(p';\, p) = \int_0^\infty dr_3 I_s(r_3) F'_{l'}(r_3) F_l(r_3)\,, \tag{23}$$

and (16) is

$$g_{l'l}^{(s)}(p';\, p) = \int_0^\infty \int_0^\infty dr_1 dr_3 R_1 R_2 \gamma_s(r_1,\, r_3) F'_{l'}(r_1) F_l(r_3)\,. \tag{24}$$

Similarly,

$$h_{l'l}^{(s)}(p';\, p) = \int_0^\infty dr_2 R_2 F'_{l'}(r_2) \cdot \int_0^\infty dr_3 I_s(r_3) R_3 F_l(r_3)\,. \tag{25}$$

The remaining integrations have to be carried out numerically, and for this purpose tables of the Coulomb functions $F_l(r)$ are necessary. Each of the integrals is a function of the two momenta p, p', with the difference between their squares fixed by the excitation energy of the ion for the transition under consideration. To shorten the work the smaller of p, p' was taken equal to zero, and the remaining p was taken to correspond to the transition energy. Thus, for each transition each integral was computed only once and not at several points, to give the dependence on velocity. Hence, only the first approximation, in which the cross-sections vary as the inverse square of the incident velocity, was obtained. The values of the integrals computed are given in Table 1. In Table 2 are shown the corresponding numerical values of Ω computed from equations (21).

Extensive tables of the Coulomb functions $F_l(r)$ were prepared for $l = 0, 1, 2$ over the ranges of p and r appropriate for our problem. Lack of space forbids their reproduction here.

COLLISIONAL EXCITATION IN NEBULAE

In the present paper we shall confine our attention to a single illustrative problem, reserving further applications for a later communication. We now transform equation (20) from atomic to general and c.g.s. units, for which

$$\sigma_{AB} = \frac{1}{2J_A + 1} \frac{h^2}{4\pi m^2} \frac{\Omega(A,\, B)}{v^2} = \frac{4.17}{2J_A + 1} \frac{\Omega(A,\, B)}{v^2}\, cm^2\,. \tag{26}$$

In this equation, A symbolizes the initial and B the final atomic level. The formula may be applied to calculations of collisions of either the first or second kind.

TABLE 1

VALUES OF f, g, AND h INTEGRALS

$(p_1; p_2)$ $(p_2; p_1)$	$^3P-{}^1S$ $p_1^2-p_2^2=0.394$	$^1D-{}^1S$ $p_1^2-p_2^2=0.209$	$^3P-{}^1D$ $p_1^2-p_2^2=0.185$	$^3P-{}^3P$ $p_1^2-p_2^2\cong 0$
$f_{11}^{(2)}$	....	+0.098	....	+0.126
$f_{02}^{(2)}$	....	{− .045 − .050}	....	− .045
$f_{20}^{(2)}$	....	{− .050 − .045}	....	− .045
$g_{00}^{(1)}$	....	....	....	+ .072
$g_{11}^{(0)}$	{−0.203 − .297}	− .240 − .293	{−0.246 − .292}	− .288
$g_{11}^{(2)}$	+ .098	+ .103	+ .103	+ .104
$g_{02}^{(1)}$	....	{− .046 − .048}	{− .046 − .048}	− .047
$g_{20}^{(1)}$	....	{− .048 − .046}	{− .048 − .046}	− .047
$h_{11}^{(0)}$	{− .230 − .268}	− .246 − .265	{− .248 − .265}	− .262
$h_{11}^{(2)}$	{+ .084 +0.092}	+ .090 +0.094	{+ .090 +0.094}	+0.096

TABLE 2

TARGET AREAS, Ω, FOR COLLISIONAL EXCITATION OF O III BY ELECTRON IMPACT

	$^3P_0$	$^3P_1$	$^3P_2$	^1D
^1S	0.39	1.18	1.96	2.71
^1D	2.22	6.66	11.1	
$^3P_2$	4.06	11.2		
$^3P_1$	1.72			

For electrons with initial velocities corresponding to the transition energy, the σ's are as follows:

$$\left.\begin{aligned} \sigma(^3P_0\rightarrow{}^3P_1) &= 1.69\times 10^3\sigma_H\,, & \sigma(^3P_0\rightarrow{}^3P_2) &= 1.45\times 10^3\sigma_H\,,\\ \sigma(^3P_0\rightarrow{}^1D_2) &= 12.0\sigma_H\,, & \sigma(^3P_0\rightarrow{}^1S_0) &= 9.91\sigma_H\,, \end{aligned}\right\}\quad(27)$$

where σ_H is the theoretical area, πa^2, of the 1s orbit of hydrogen,

$$\sigma_H = 8.765 \times 10^{-17} \text{ cm}^2 . \tag{28}$$

The target areas are large, especially for excitation within the levels of ^{3}P.

We assume that the velocity distribution of the nebular electrons is Maxwellian, defined by a kinetic temperature T_ϵ, so that the fraction $f(v)$ of electrons having velocities between v and $v + dv$ is

$$f(v)dv = 4\pi \left(\frac{m}{2\pi k T_\epsilon} \right)^{3/2} v^2 e^{-\frac{1}{2}mv^2/kT_\epsilon} dv . \tag{29}$$

Then the total number of collisional excitations (or de-excitations) from level A to B, per cubic centimeter per second,[11] is

$$\mathfrak{F}_{AB} = N_A N_\epsilon \int_{v_0}^{\infty} \sigma_{AB} v f(v) dv , \tag{30}$$

where v_0 is the lowest velocity capable of inducing the transition; v_0 is zero for superelastic collisions and is defined by

$$\tfrac{1}{2} m v_0^2 = \chi_{AB} \tag{31}$$

for inelastic collisions. χ_{AB} is the energy difference between the levels A and B. Carrying out the integration, we find for inelastic impacts,

$$\left.\begin{aligned} \mathfrak{F}_{AB} &= N_A N_\epsilon \frac{\Omega(AB)}{\pi_A} \frac{h^2}{2\pi m^2} \left(\frac{m}{2\pi k T_\epsilon} \right)^{1/2} e^{-\chi_{AB}/kT_\epsilon} \\ &= 8.54 \times 10^{-6} \frac{N_A N_\epsilon}{T_\epsilon^{1/2}} \frac{\Omega(AB)}{\tilde{\omega}_A} e^{-\chi_{AB}/kT_\epsilon} . \end{aligned}\right\} \tag{32}$$

For superelastic impacts, the equation is

$$\mathfrak{F}_{AB} = 8.54 \times 10^{-6} \frac{N_B N_\epsilon}{T_\epsilon^{1/2}} \frac{\Omega(AB)}{\tilde{\omega}_B} . \tag{33}$$

[11] R. H. Fowler, *Statistical Mechanics.*

We have written ϖ_A for $2J_A + 1$, the statistical weight, in the above equations.

When the gaseous assembly permits detailed balancing, so that

$$\mathfrak{F}_{AB} = \mathfrak{F}_{BA} , \tag{34}$$

we have the condition

$$\frac{N_B}{N_A} = \frac{\varpi_B}{\varpi_A} e^{-\chi_{AB}/kT_\epsilon} , \tag{35}$$

which is the Boltzmann relationship for thermodynamic equilibrium. If the only means of removing atoms from levels B and A are collisional, equation (35) will hold exactly, even for gaseous nebulae. Leakage of atoms from level B by spontaneous emissions will cause departure from the Boltzmann formula. It is easy to show, however, that (35) will still be a fair approximation in any nebula where the probability of de-excitation of B by superelastic impact exceeds that for spontaneous transition, i.e., when

$$\sum_A \mathfrak{F}_{BA} > N_B \sum_A A_{BA} , \tag{36}$$

where A_{BA} is the Einstein coefficient of spontaneous emission.

Of particular interest is the relative population in the levels of ^3P. In Paper IX of this series Shortley and Menzel discussed the fractional multiplets observed in gaseous nebulae. The spectra suggest an initial excitation of the single level 2p3d $^3P^0_2$, with a subsequent cascade to the levels of the lower configurations 2p3p and 2p3s. The source of excitation, as Bowen has pointed out, is probably the ultimate line λ 303 of *He* II, which must be very intense in certain nebulae. The line $2p^2\ ^3P_2$ — 2p3d $^3P^0_2$ of *O* III coincides very closely with this helium line, the resonance absorption of which by *O* III will lead to selective excitation of $^3P^0_2$, as observed.

For the Bowen fluorescent process to work effectively, however, requires that the population of *O* III atoms in the level $2p^2\ ^3P_2$ be large. Since $^3P_2$ has a mean lifetime of 2.9 hours,[12] there will be an

[12] *Ap. J.*, **91**, 307, 1940.

effective population in that level only if the inequality (36) is fulfilled.

$$\sum A_{BA} = \frac{1}{2.9} \times 3600 \sim 10^{-4} . \qquad (37)$$

$$\sum \mathfrak{F}_{BA} = 8.54 \times 10^{-6} \frac{N_B N_\epsilon}{10^2} \frac{\Omega(^3P_2, {}^3P_1) + \Omega(^3P_2, {}^3P_0)}{5} , \qquad (38)$$

for $T_\epsilon = 10^4$ deg. Hence, from (36) we obtain the condition that

$$N_\epsilon > 4 \times 10^2 . \qquad (39)$$

An independent investigation, the details of which will be given in a later paper, shows that the brighter nebulae, in particular the ones showing the lines of the fluorescent mechanism, have electron densities somewhat in excess of this figure. We conclude, therefore, that collisions occur with sufficient frequency to maintain approximately a Boltzmann distribution among the levels of ^3P.

Dunham observed that interstellar lines of *Ti* II appear from only the lowest atomic level. The excitation parameters for *O* III are, of course, not directly applicable to *Ti* II. But the figures, which indicate a high accumulation of atoms in the lowest level as a result of the very low electron density in interstellar space, are consistent with the observational data.

Extension of the calculations to the $^1D_2$ and $^1S_0$ levels of *O* III leads to an independent determination of the electron temperature and of the abundance of *O* III. These results will be given in a later paper.

We wish to acknowledge the advice and assistance of Dr. J. H. Van Vleck. A grant to Dr. Van Vleck from the Milton Fund and Clark Bequest of Harvard University enabled Mr. Hebb to pursue the present investigation.

DUKE UNIVERSITY
AND
HARVARD UNIVERSITY
April 1940

Reprinted from *The Astrophysical Journal,* Vol. 93, No. 1, January, 1941.

PHYSICAL PROCESSES IN GASEOUS NEBULAE

XI. STRENGTHS OF FORBIDDEN LINES IN p^2, p^3, AND p^4 AS A FUNCTION OF COUPLING

G. H. SHORTLEY, L. H. ALLER, J. G. BAKER, AND D. H. MENZEL

ABSTRACT

The coupling of configurations p^2, p^3, and p^4 is determined by a single parameter χ which ranges from 0 for *LS* to ∞ for *jj* coupling. This paper contains a tabulation of the magnetic-dipole strengths and the coefficient of the radial integral in the electric-quadrupole strengths for all the forbidden lines in these configurations for all values of χ. The transformation coefficients which express the actual eigenfunctions in terms of the *LS*-coupling eigenfunctions are also tabulated.

For a study of physical conditions in celestial objects, the night sky, and laboratory sources of "forbidden" radiation, it is necessary to have detailed theoretical information concerning the strengths of the lines. The first steps in this direction were made by Stevenson[1] and by Condon,[2] who considered certain lines of p^2 for *C* I, *N* II, and *O* III; of p^3 for *O* II and *S* II; and of p^4 for *O* I. But, since the properties of all possible coupling cases[3] for these configurations can be tabulated against a single parameter χ and since careful estimates of probable χ values for the various ions exist,[4] it seems desirable once and for all to tabulate the strengths of the lines as a function of χ with no approximations except the sometimes dubious one of using a pure configuration assignment. The parameter $\chi = \frac{1}{5}\zeta_p/F_2$ is essentially a measure of the departure from *LS* coupling, being 0 for pure *LS* coupling and ∞ for pure *jj* coupling.

The *absolute* values of the strengths of the magnetic-dipole lines are tabulated against χ; but for the quadrupole lines we may tabulate only the coefficients of the squares of the integrals

$$s_q = s_q(np, np) = \tfrac{2}{5}\epsilon \int_0^\infty r^2 R^2(np)\, dr \,, \tag{1}$$

which involve the radial factor of the p-orbital and must be considered separately for each atom.[5] In atomic units ϵa^2, s_q is of the order of magnitude of unity. It will be noticed that we include the strengths of the very low-frequency "lines" connecting two levels of the same term. These values, which have been neglected in the past, are important in considerations of the populations of the levels in rarefied atmospheres. We also tabulate, for reference, the often needed transformation coefficients which express the intermediate-coupling eigenfunctions in terms of *LS*-coupling eigenfunctions.

[1] *Proc. Roy. Soc.*, Ser. A, **137**, 298, 1932.

[2] *Ap. J.*, **79**, 217, 1934. We disagree with certain of the formulas represented by equations (27), (31), (32), (41), (44), and (46) of this paper and, hence, with the numerical values obtained therefrom.

[3] See Condon and Shortley, *Theory of Atomic Spectra*, pp. 274–276, 300–301; and see, in particular, note 4 of this article.

[4] Robinson and Shortley, *Phys. Rev.*, **52**, 713, 1937, Tables V, VI, and VII. The $2p^3$ χ-values which are missing from Table VI may be obtained by interpolation in the data of Fig. 5 of this reference, which gives the following values for $\chi(2p^3)$: *N* I, 0.0086; *O* II, 0.0153; *F* III, 0.0251; *Ne* IV, 0.0365; *Na* V, 0.0549; *Mg* VI, 0.0772; *Al* VII, 0.103; *Si* VIII, 0.134; *P* IX, 0.168.

[5] This integral has been estimated by Condon[2] for certain cases, recently for many others by Pasternack, *Ap. J.*, **92**, 129, 1940.

In natural excitation the spontaneous-transition probability associated with the emission of *magnetic-dipole* radiation is given by

$$A_{\mathrm{m}}(J, J') = 35{,}320 \frac{\sigma^3 S_{\mathrm{m}}(J, J')}{2J + 1} \sec^{-1} . \tag{2}$$

Here J and J' are the quantum numbers of the upper and lower levels; σ is the wave-number of the line expressed in Rydbergs (1 Ryd. = 109,737 cm^{-1}); and $S_{\mathrm{m}}(J, J')$ is the magnetic-dipole strength[6] in atomic units $e^2\hbar^2/4\mu^2c^2$, here given by Tables 1, 4, and 7.

The *electric-quadrupole* transition probability is

$$A_{\mathrm{q}}(J, J') = 2648 \frac{\sigma^5 S_{\mathrm{q}}(J, J')}{2J + 1} \sec^{-1} . \tag{3}$$

The quadrupole strength $S_{\mathrm{q}}(J, J')$ is here to be expressed in atomic units $\epsilon^2 a^4$ (where a is the Bohr radius $\hbar^2/\mu\epsilon^2$). This strength may be written as the product of a numerical coefficient $C_{\mathrm{q}}(J, J')$, which depends only on the coupling parameter χ, by the square of the integral in equation (1),

$$S_{\mathrm{q}}(J, J') = C_{\mathrm{q}}(J, J')\, s_{\mathrm{q}}^2(np, np) . \tag{4}$$

It is the coefficient C_{q} which is given in Tables 2, 5, and 8.

THE CONFIGURATION p^2

The actual eigenfunctions for the levels of p^2 are functions of a parameter χ which ranges from 0 to ∞ as the coupling goes from extreme *LS* to extreme *jj*. If these functions (distinguished by primes) are expressed in terms of the pure Russell-Saunders functions (phases chosen as in n. 3) by

$$\left.\begin{aligned} \psi(^1D_2') &= a\psi(^1D_2) + b\psi(^3P_2) , \qquad \psi(^3P_0') = c\psi(^3P_0) + d\psi(^1S_0) , \\ \psi(^3P_2') &= a\psi(^3P_2) - b\psi(^1D_2) , \qquad \psi(^1S_0') = c\psi(^1S_0) - d\psi(^3P_0) , \\ & \qquad \psi(^3P_1') = \psi(^3P_1) , \end{aligned}\right\} \tag{5}$$

we find the following series for the coefficients near *LS* coupling, i.e., for small values of χ:

$$\left.\begin{aligned} a &= 1 - \tfrac{25}{144}\chi^2 - \tfrac{125}{864}\chi^3 + \ldots , \\ b &= \tfrac{5}{12}\sqrt{2}\,\chi(1 + \tfrac{5}{12}\chi - \tfrac{25}{72}\chi^2 + \ldots) , \\ c &= 1 - \tfrac{1}{9}\chi^2 + \tfrac{2}{27}\chi^3 + \ldots , \\ d &= \tfrac{1}{3}\sqrt{2}\,\chi(1 - \tfrac{1}{3}\chi - \tfrac{2}{9}\chi^2 + \ldots) . \end{aligned}\right\} \tag{6}$$

[6] See Shortley, *Phys. Rev.*, 57, 225, 1940, for definitions. This reference contains the values of the magnetic-dipole and electric-quadrupole strengths in *LS* coupling and explains the method of transforming to intermediate coupling. We here give just the numerical values and the formulas from which they are obtained.

The magnetic-dipole strengths, in units $(-\,\epsilon h/4\pi\mu c)^2$, have, in general, the following values in terms of these coefficients and the following asymptotic expansions near LS coupling:[7]

$$\left.\begin{aligned}
&S_{\mathrm{m}}({}^1\mathrm{S}'_0,\,{}^1\mathrm{D}'_2) = 0\,,\\
&S_{\mathrm{m}}({}^1\mathrm{S}'_0,\,{}^3\mathrm{P}'_2) = 0\,,\\
&S_{\mathrm{m}}({}^1\mathrm{S}'_0,\,{}^3\mathrm{P}_1) = (-)^2\,2d^2 \quad = (-)^2\,\tfrac{4}{9}\chi^2(1 - \tfrac{2}{3}\chi - \tfrac{1}{3}\chi^2 + \ldots .)\,,\\
&S_{\mathrm{m}}({}^1\mathrm{S}'_0,\,{}^3\mathrm{P}'_0) = 0\,,\\
&S_{\mathrm{m}}({}^1\mathrm{D}'_2,\,{}^3\mathrm{P}'_2) = (+)^2\,\tfrac{15}{2}a^2b^2 = (+)^2\,\tfrac{375}{144}\chi^2(1 + \tfrac{5}{6}\chi - \tfrac{125}{144}\chi^2 + \ldots .)\,,\\
&S_{\mathrm{m}}({}^1\mathrm{D}'_2,\,{}^3\mathrm{P}_1) = (+)^2\,\tfrac{5}{2}b^2 \quad = (+)^2\,\tfrac{125}{144}\chi^2(1 + \tfrac{5}{6}\chi - \tfrac{25}{48}\chi^2 + \ldots .)\,,\\
&S_{\mathrm{m}}({}^1\mathrm{D}'_2,\,{}^3\mathrm{P}'_0) = 0\,,\\
&S_{\mathrm{m}}({}^3\mathrm{P}'_2,\,{}^3\mathrm{P}_1) = (+^2\,\tfrac{5}{2}a^2 \quad = (+)^2\,\tfrac{5}{2}(1 - \tfrac{25}{72}\chi^2 + \ldots .)\,,\\
&S_{\mathrm{m}}({}^3\mathrm{P}'_2,\,{}^3\mathrm{P}'_0) = 0\,,\\
&S_{\mathrm{m}}({}^3\mathrm{P}_1,\,{}^3\mathrm{P}'_0) = (+)^2\,2c^2 \quad = (+)^2\,2(1 - \tfrac{2}{9}\chi^2 + \ldots .)\,.
\end{aligned}\right\} \tag{7}$$

The electric-quadrupole strengths, in terms of the one-electron components in equation (9), are the following:

$$\left.\begin{aligned}
&S_{\mathrm{q}}({}^1\mathrm{S}'_0,\,{}^1\mathrm{D}'_2) = (+)^2\,\tfrac{5}{3}s_{\mathrm{q}}^2(bd + 2ac)^2 = (+)^2\,\tfrac{20}{3}s_{\mathrm{q}}^2(1 - \tfrac{7}{24}\chi^2 + \ldots .)\,,\\
&S_{\mathrm{q}}({}^1\mathrm{S}'_0,\,{}^3\mathrm{P}'_2) = (+)^2\,\tfrac{5}{3}s_{\mathrm{q}}^2(ad - 2bc)^2 = (-)^2\,\tfrac{5}{6}s_{\mathrm{q}}^2\chi^2(1 + \tfrac{11}{6}\chi + \ldots .)\,,\\
&S_{\mathrm{q}}({}^1\mathrm{S}'_0,\,{}^3\mathrm{P}_1) = 0\,,\\
&S_{\mathrm{q}}({}^1\mathrm{S}'_0,\,{}^3\mathrm{P}'_0) = 0\,,\\
&S_{\mathrm{q}}({}^1\mathrm{D}'_2,\,{}^3\mathrm{P}'_2) = (-)^2\,\tfrac{105}{4}s_{\mathrm{q}}^2a^2b^2 \qquad = (-)^2\,\tfrac{875}{96}s_{\mathrm{q}}^2\chi^2(1 + \tfrac{5}{6}\chi + \ldots .)\,,\\
&S_{\mathrm{q}}({}^1\mathrm{D}'_2,\,{}^3\mathrm{P}_1) = (+)^2\,\tfrac{15}{4}s_{\mathrm{q}}^2b^2 \qquad = (+)^2\,\tfrac{125}{96}s_{\mathrm{q}}^2\chi^2(1 + \tfrac{5}{6}\chi + \ldots .)\,,\\
&S_{\mathrm{q}}({}^1\mathrm{D}'_2,\,{}^3\mathrm{P}'_0) = (+)^2\,\tfrac{5}{3}s_{\mathrm{q}}^2(2ad - bc)^2 = (+)^2\,\tfrac{5}{24}s_{\mathrm{q}}^2\chi^2(1 - \tfrac{19}{6}\chi + \ldots .)\,,\\
&S_{\mathrm{q}}({}^3\mathrm{P}'_2,\,{}^3\mathrm{P}_1) = (+)^2\,\tfrac{15}{4}s_{\mathrm{q}}^2a^2 \qquad = (+)^2\,\tfrac{15}{4}s_{\mathrm{q}}^2(1 - \tfrac{25}{72}\chi^2 + \ldots .)\,,\\
&S_{\mathrm{q}}({}^3\mathrm{P}'_2,\,{}^3\mathrm{P}'_0) = (-)^2\,\tfrac{5}{3}s_{\mathrm{q}}^2(2bd + ac)^2 = (-)^2\,\tfrac{5}{3}s_{\mathrm{q}}^2(1 + \tfrac{13}{24}\chi^2 + \ldots .)\,,\\
&S_{\mathrm{q}}({}^3\mathrm{P}_1,\,{}^3\mathrm{P}'_0) = 0\,.
\end{aligned}\right\} \tag{8}$$

THE CONFIGURATION p^4

For the p^4 eigenfunctions we may conveniently use the same notation (eq. [5]) as for p^2. In this case the asymptotic values of a, b, c, and d may be obtained from equation (6) by writing $-\chi$ in place of χ.

The magnetic-dipole strengths are the same functions of a, b, c, d as for p^2, and their asymptotic forms are to be obtained from those of p^2 by reversing the sign of χ.

[7] The phases of $S^{1/2}$ to be used in transformations of coupling are noted for completeness. They are of no immediate significance (see n. 6 for details).

As shown in note 6, the quadrupole-moment matrix components in *LS* coupling are equal but of opposite sign for p^2 and p^4. Hence, the intermediate-coupling strengths are the same functions of a, b, c, and d in the two cases, and the asymptotic forms for p^4 are obtained from those of p^2 by reversing the sign of χ.

THE CONFIGURATION p^3

In this configuration we designate the eigenfunctions by

$$\left.\begin{aligned}
\psi({}^2D'_{3/2}) &= a\psi({}^2P_{3/2}) + b\psi({}^4S_{3/2}) + c\psi({}^2D_{3/2}) \,, \\
\psi({}^4S'_{3/2}) &= a'\psi({}^2P_{3/2}) + b'\psi({}^4S_{3/2}) + c'\psi({}^2D_{3/2}) \,, \\
\psi({}^2P'_{3/2}) &= a''\psi({}^2P_{3/2}) + b''\psi({}^4S_{3/2}) + c''\psi({}^2D_{3/2}) \,, \\
\psi({}^2P'_{1/2}) &= \psi({}^2P_{1/2}) \,, \qquad \psi({}^2D'_{5/2}) = \psi({}^2D_{5/2}) \,.
\end{aligned}\right\} \quad (9)$$

We choose the phases of the Russell-Saunders eigenfunctions as in Condon and Shortley.[8] Near *LS* coupling these coefficients have the form

$$\left.\begin{aligned}
a &= -\tfrac{5}{12}\sqrt{5}\,\chi(1 - \tfrac{1525}{864}\chi^2 + \ldots.) \,, & a' &= -\tfrac{1}{3}\chi\,(1 + \tfrac{7}{108}\chi^2 + \ldots.) \,, \\
b &= -\tfrac{25}{108}\sqrt{5}\,\chi^2(1 - \tfrac{1025}{864}\chi^2 + \ldots.) \,, & b' &= 1 - \tfrac{1}{18}\chi^2 + \ldots. \,, \\
c &= 1 - \tfrac{125}{288}\chi^2 + \ldots. \,, & c' &= \tfrac{5}{54}\sqrt{5}\,\chi^2(1 - \tfrac{13}{108}\chi^2 + \ldots.) \,, \\
& & a'' &= 1 - \tfrac{47}{96}\chi^2 + \ldots. \,, \\
& & b'' &= \tfrac{1}{3}\chi(1 - \tfrac{91}{96}\chi^2 + \ldots.) \,, \\
& & c'' &= \tfrac{5}{12}\sqrt{5}\,\chi(1 - \tfrac{157}{96}\chi^2 + \ldots.) \,.
\end{aligned}\right\} \quad (10)$$

The magnetic-dipole strengths are

$$\left.\begin{aligned}
({}^2P'_{3/2}, {}^2P_{1/2}) &= (+)^2\,\tfrac{4}{3}a''^2 &&= (+)^2\,\tfrac{4}{3}(1 - \tfrac{47}{48}\chi^2 + \ldots.) \,, \\
({}^2P'_{3/2}, {}^2D_{5/2}) &= (+)^2\,\tfrac{12}{5}c''^2 &&= (+)^2\,\tfrac{25}{12}\chi^2(1 - \tfrac{157}{48}\chi^2 + \ldots.) \,, \\
({}^2P'_{3/2}, {}^2D'_{3/2}) &= (+)^2\,\tfrac{4}{15}(10aa'' + 15bb'' + 6cc'')^2 &&= (-)^2\,\tfrac{100}{27}\chi^2(1 - \tfrac{397}{108}\chi^2 + \ldots.) \,, \\
({}^2P_{1/2}, {}^2D_{5/2}) &= 0 \,, \\
({}^2P_{1/2}, {}^2D'_{3/2}) &= (+)^2\,\tfrac{4}{3}a^2 &&= (-)^2\,\tfrac{125}{108}\chi^2(1 - \tfrac{1525}{432}\chi^2 + \ldots.) \,, \\
({}^2P'_{3/2}, {}^4S'_{3/2}) &= (+)^2\,\tfrac{4}{15}(10a'a'' + 15b'bb'' + 6c'c'')^2 &&= (+)^2\,\tfrac{20}{27}\chi^2(1 - \tfrac{1267}{432}\chi^2 + \ldots.) \,, \\
({}^2P_{1/2}, {}^4S'_{3/2}) &= (+)^2\,\tfrac{4}{3}a'^2 &&= (-)^2\,\tfrac{4}{27}\chi^2(1 + \tfrac{7}{54}\chi^2 + \ldots.) \,, \\
({}^2D_{5/2}, {}^2D'_{3/2}) &= (+)^2\,\tfrac{12}{5}c^2 &&= (+)^2\,\tfrac{12}{5}(1 - \tfrac{125}{144}\chi^2 + \ldots.) \,, \\
({}^2D_{5/2}, {}^4S'_{3/2}) &= (+)^2\,\tfrac{12}{5}c'^2 &&= (+)^2\,\tfrac{25}{243}\chi^4(1 - \tfrac{13}{54}\chi^2 + \ldots.) \,, \\
({}^2D'_{3/2}, {}^4S'_{3/2}) &= (+)^2\,\tfrac{4}{15}\,(10aa' + 15bb' + 6cc')^2 &&= (-)^2\,\tfrac{3025}{972}\chi^4(1 - \tfrac{929}{432}\chi^2 + \ldots.) \,.
\end{aligned}\right\} \quad (11)$$

[8] *Op. cit.*, pp. 225 and 268.

The quadrupole strengths are

$$
\left.\begin{aligned}
S_q({}^2P'_{3/2}, {}^2P_{1/2}) &= (-)^2\, 6s_q^2\, c''^2 &&= (-)^2\, \tfrac{125}{24} s_q^2\, \chi^2(1 - \tfrac{157}{48}\chi^2 + \ldots) ,\\
S_q({}^2P'_{3/2}, {}^2D_{5/2}) &= (-)^2\, 14 s_q^2\, a''^2 &&= (-)^2\, 14 s_q^2(1 - \tfrac{47}{48}\chi^2 + \ldots),\\
S_q({}^2P'_{3/2}, {}^2D'_{3/2}) &= (-)^2\, 6s_q^2(a''c + ac'')^2 &&= (-)^2\, 6s_q^2(1 - \tfrac{43}{12}\chi^2 + \ldots) ,\\
S_q({}^2P_{1/2}, {}^2D_{5/2}) &= (+)^2\, 4s_q^2 , &&\\
S_q({}^2P_{1/2}, {}^2D'_{3/2}) &= (-)^2\, 6s_q^2\, c^2 &&= (-)^2\, 6s_q^2(1 - \tfrac{125}{144}\chi^2 + \ldots) ,\\
S_q({}^2P'_{3/2}, {}^4S'_{3/2}) &= (-)^2\, 6s_q^2(a''c' + a'c'')^2 &&= (+)^2\, \tfrac{125}{1944} s_q^2\, \chi^4(1 - \tfrac{3017}{432}\chi^2 + \ldots) ,\\
S_q({}^2P_{1/2}, {}^4S'_{3/2}) &= (-)^2\, 6s_q^2\, c'^2 &&= (-)^2\, \tfrac{125}{486} s_q^2\, \chi^4(1 - \tfrac{13}{54}\chi^2 + \ldots) ,\\
S_q({}^2D_{5/2}, {}^2D'_{3/2}) &= (-)^2\, 14 s_q^2\, a^2 &&= (+)^2\, \tfrac{875}{72} s_q^2\, \chi^2(1 - \tfrac{1525}{432}\chi^2 + \ldots) ,\\
S_q({}^2D_{5/2}, {}^4S'_{3/2}) &= (-)^2\, 14 s_q^2\, a'^2 &&= (+)^2\, \tfrac{14}{9} s_q^2\, \chi^2(1 + \tfrac{7}{54}\chi^2 + \ldots) ,\\
S_q({}^2D'_{3/2}, {}^4S'_{3/2}) &= (-)^2\, 6s_q^2(ac' + a'c)^2 &&= (+)^2\, \tfrac{2}{3} s_q^2\, \chi^2(1 + \tfrac{181}{432}\chi^2 + \ldots) .
\end{aligned}\right\} \qquad (12)
$$

NUMERICAL MAGNITUDES

We may add a few general remarks concerning orders of magnitude. If S is of the order of unity, as will be true if the line is permitted by all selection rules, and $\sigma \sim 0.2$ Ryd. (corresponding to $\lambda = 4580$ A), we find from equations (2), (3), and note 3 (p. 134),

$$
\left.\begin{aligned}
A_p &\sim 2 \times 10^7 \text{ sec}^{-1} ,\\
A_m &\sim 3 \times 10^2 \text{ sec}^{-1} ,\\
A_q &\sim 8 \times 10^{-1} \text{ sec}^{-1} ,
\end{aligned}\right\} \qquad (13)
$$

where A_p refers to the ordinary electric-dipole transition probability. Hence, other things being equal, the magnetic-dipole transitions will be of more importance than the electric-quadrupole.

However, all lines connecting different terms violate the magnetic-dipole selection rule which prohibits combinations between different terms; the multiplets p^2, p^4, ${}^1S \rightarrow {}^3P$, ${}^1D \rightarrow {}^3P$; p^3, ${}^2P \rightarrow {}^4S$, ${}^2D \rightarrow {}^4S$ violate the quadrupole prohibition of intersystem combinations. Such lines owe their existence to departure from Russell-Saunders coupling, and their strengths appear in the tables with order of magnitude χ^2 or χ^4, which, for example, have values for *O* III $2p^2$ of 1.4×10^{-4} and 2×10^{-8}, respectively. The transition probabilities are, hence, smaller than those of equation (13) by these factors. As one progresses in the periodic system, χ becomes of the order of unity, and we recover values of the order of equation (13).

For the transition between two levels of a term, with $\sigma \sim 109 \text{ cm}^{-1} = 10^{-3}$ Ryd., as for the ground term of *O* III, we find

$$
\left.\begin{aligned}
A_m &\sim 4 \times 10^{-5} \text{ sec}^{-1} ,\\
A_q &\sim 2 \times 10^{-12} \text{ sec}^{-1} .
\end{aligned}\right\} \qquad (14)
$$

Although such transitions are rigorously forbidden for electric-dipole radiation, they are always allowed for magnetic-dipole for J changes of one unit. The quadrupole contributions, because of the σ^5 factor, are unimportant.

INTERFERENCE OF MAGNETIC-DIPOLE AND ELECTRIC-QUADRUPOLE RADIATION

In cases where the electric-quadrupole and magnetic-dipole transition probabilities are of comparable magnitude one must ask whether these *probabilities* are to be directly added to get the whole transition probability or whether *amplitudes* are to be added in some more complicated fashion.

This question may be answered by looking into the derivation of formulas (2) and (3) from the Dirac radiation theory. It turns out that the intensity of a line as a whole, which is of course isotropic and unpolarized in natural excitation, is the sum of the intensities of the quadrupole and magnetic-dipole radiation as separately computed, so that $A_m(J, J')$ *and* $A_q(J, J')$ *are to be directly added to obtain the total transition probability*. But the individual Zeeman components of the line interfere in a complicated fashion, which we sketch below.[9]

There is no interference among the Zeeman components except for spontaneous transitions of the type

$$J, M \rightarrow J', M \pm 1 ,$$

to which we confine attention.

We define a quantity $\mathcal{N}$ in such a way that $|\mathcal{N}|^2$ is the quadrupole strength $|(J, M|\mathfrak{N}'|J', M \pm 1)|^2$ of the component in question, and we take for the sign of $\mathcal{N}$ the sign of the coefficient of $\mathfrak{K}$ in formulas (3) of reference 6 for these strengths. Thus

$$\text{sign of } \mathcal{N} = \text{sign of } S_q^{1/2}(J, J') \times \text{sign of } \begin{cases} J \mp 2M & \text{if } J' = J + 1 \\ 2M \pm 1 & \text{if } J' = J \\ J \pm 2M + 1 & \text{if } J' = J - 1 . \end{cases}$$

We define a similar quantity $\mathcal{M}$ so that $|\mathcal{M}|^2$ is the magnetic-dipole strength $|(J, M|\boldsymbol{M}|J', M \pm 1)|^2$, with the sign given by the coefficient of $(\boldsymbol{i} \pm i\boldsymbol{j})$ in equation 9^311 (p. 63, in reference 3). Thus

$$\text{sign of } \mathcal{M} = \text{sign of } S_m^{1/2}(J, J') \times \begin{cases} \mp 1 & \text{if } J' = J + 1 , \\ +1 & \text{if } J' = J , \\ \pm 1 & \text{if } J' = J - 1 . \end{cases}$$

The signs of $S_q^{1/2}$ and $S_m^{1/2}$ needed for the above definitions are given for p^2 and p^3 (s_q is essentially positive) by equations (7), (8), (11), and (12). For p^4, the quadrupole signs are *opposite*, the magnetic-dipole signs the *same*, as for p^2 (cf. reference 6).

With this notation the intensity of radiation for $M \rightarrow M \pm 1$, at angle θ with the magnetic field, is proportional to

$$(\mathcal{N}\sigma\pi \cos 2\theta \pm \sqrt{2}\,\mathcal{M})^2 \text{ polarized in } \theta \text{ direction } (\pi)$$

$$(\pm\mathcal{N}\sigma\pi + \sqrt{2}\,\mathcal{M})^2 \cos^2\theta \text{ polarized in } \varphi \text{ direction } (\sigma) .$$

[9] Similar considerations have been independently made by E. Gerjuoy, as noted in the abstract of paper No. 16 presented at the Pasadena meeting of the American Physical Society, December 20, 1940, in connection with actual observations of the Zeeman effect of the $6p^2$ forbidden lines of *Pb* I.

That is, the two θ *amplitudes add* if the signs of $\mathcal{N} \cos 2\theta$ and $\mathcal{M}$ are the $\left\{\begin{matrix}\text{same}\\ \text{opposite}\end{matrix}\right\}$; they *subtract* otherwise. The two φ *amplitudes add* if the signs of $\mathcal{N}$ and $\mathcal{M}$ are the $\left\{\begin{matrix}\text{same}\\ \text{opposite}\end{matrix}\right\}$; they *subtract* otherwise.

The polarization is right elliptic if

$$\mathcal{N}\sigma\pi \cos 2\theta \pm \sqrt{2}\,\mathcal{M} \quad \text{and} \quad (\pm\mathcal{N}\sigma\pi + \sqrt{2}\mathcal{M}) \cos \theta$$

have the same sign, left elliptic if they have opposite signs.

The *total intensity* of any Zeeman component, integrated over the sphere, is just the sum of the separate magnetic-dipole and electric-quadrupole intensities.*

The authors wish to express their appreciation to Mrs. Martha E. Shapley for assistance in carrying out the calculations.

MENDENHALL LABORATORY OF PHYSICS
OHIO STATE UNIVERSITY
AND
HARVARD COLLEGE OBSERVATORY
April 1940

*For further discussion of this problem, see the last paper in this volume.

TABLE 1*

MAGNETIC DIPOLE STRENGTHS FOR p^2

	$S_m(^1D'_2, ^3P'_2)$	$S_m(^1D'_2, ^3P_1)$	$S_m(^1S'_0, ^3P_1)$	$S_m(^3P'_2, ^3P_1)$	$S_m(^3P_1, ^3P'_0)$
χ 0.00	0.0000	0.0000	0.0000	2.5000	2.0000
.04	$0.^{1}0452$	$0.^{1}0151$	$0.^{2}0692$	2.4990	1.9996
.08	0.0177	$0.^{1}0590$	$0.^{1}0269$	2.4940	1.9976
.12	.0406	$0.^{1}1362$	$0.^{1}0588$	2.4860	1.9940
.16	.0735	0.0248	$0.^{1}1011$	2.4751	1.9896
.20	.1162	.0394	$0.^{1}1528$	2.4602	1.9848
.24	.1687	.0575	$0.^{1}2126$	2.4438	1.9784
.28	.2296	.0790	$0.^{1}2794$	2.4206	1.9721
.32	.2986	.1038	$0.^{1}3522$	2.3966	1.9646
.36	.3739	.1316	$0.^{1}4298$	2.3683	1.9570
.40	.4548	.1621	$0.^{1}5114$	2.3387	1.9491
.48	.6250	.2294	$0.^{1}6838$	2.2705	1.9314
.56	.7977	.3026	$0.^{1}8636$	2.1968	1.9137
.64	0.9641	.3787	0.1046	2.1215	1.8954
.72	1.1156	.4545	.1228	2.0453	1.8771
.80	1.2498	.5283	.1408	1.9714	1.8594
0.88	1.3657	.5985	.1583	1.9014	1.8417
χ 1.00	1.5064	.6956	.1835	1.8046	1.8164
1/χ 0.95	1.5563	.7347	.1941	1.7653	1.8061
.90	1.6055	.7759	.2056	1.7243	1.7944
.85	1.6506	.8182	.2179	1.6810	1.7823
.80	1.6952	.8632	.2312	1.6366	1.7687
.75	1.7366	.9099	.2457	1.5904	1.7544
.70	1.7727	0.9579	.2612	1.5421	1.7387
.65	1.8045	1.0074	.2783	1.4927	1.7220
.60	1.8307	1.0579	.2965	1.4421	1.7035
.55	1.8515	1.1096	.3163	1.3905	1.6836
.50	1.8661	1.1621	.3377	1.3381	1.6620
.45	1.8742	1.2149	.3609	1.2856	1.6391
.40	1.8747	1.2677	.3861	1.2324	1.6139
.35	1.8688	1.3206	.4131	1.1792	1.5867
.30	1.8569	1.3731	.4424	1.1269	1.5576
.25	1.8376	1.4247	.4739	1.0749	1.5260
.20	1.8134	1.4757	.5076	1.0240	1.4923
.15	1.7844	1.5261	.5437	0.9744	1.4562
.10	1.7489	1.5749	.5823	0.9254	1.4176
.05	1.7098	1.6213	.6234	0.8788	1.3768
1/χ 0.00	1.6664	1.6668	0.6666	0.8332	1.3333

*In this and the succeeding tables the exponent following the decimal indicates the number of ciphers actually omitted. Thus: $0.^{1}2454$ signifies 0.02454; $0.^{2}0692$ signifies 0.000692, etc.

TABLE 2. ELECTRIC QUADRUPOLE STRENGTHS FOR p^2 (cf. Eq. 4)

	$C_q({}^1S_0', {}^1D_2')$	$C_q({}^1S_0', {}^3P_2')$	$C_q({}^1D_2', {}^3P_2')$	$C_q({}^1D_2', {}^3P_1)$	$C_q({}^1D_2', {}^3P_0')$	$C_q({}^3P_2', {}^3P_1)$	$C_q({}^3P_2', {}^3P_0')$
χ 0.00	6.6666	0.0000	0.0000	0.0000	0.0000	3.7500	1.6666
.04	6.6657	$0.^{2}1548$	$0.^{1}1580$	$0.^{1}0226$	$0.^{3}2668$	3.7485	1.6687
.08	6.6545	$0.^{2}6095$	$0.^{1}6182$	$0.^{2}8854$	$0.^{2}1020$	3.7410	1.6726
.12	6.6360	$0.^{1}1453$	0.1422	$0.^{1}2043$	$0.^{2}1972$	3.7290	1.6790
.16	6.6126	$0.^{1}2720$	0.2573	$0.^{1}3713$	$0.^{2}2973$	3.7126	1.6886
.20	6.5827	$0.^{1}4447$	0.4069	$0.^{1}5906$	$0.^{2}3902$	3.6902	1.7006
.24	6.5501	$0.^{1}6655$	0.5906	$0.^{1}8630$	$0.^{2}4678$	3.6657	1.7160
.28	6.5025	$0.^{1}9353$	0.8035	0.1185	$0.^{2}5227$	3.6610	1.7311
.32	6.4540	0.1252	1.0453	.1558	$0.^{2}5568$	3.5949	1.7493
.36	6.3976	.1613	1.3086	.1973	$0.^{2}5703$	3.5524	1.7684
.40	6.3395	.2017	1.5918	.2431	$0.^{2}5627$	3.5081	1.7896
.48	6.2018	.2929	2.1872	.3441	$0.^{2}4987$	3.4058	1.8323
.56	6.0566	.3933	2.8442	.4539	$0.^{2}4050$	3.2952	1.8782
.64	5.9070	.4988	3.3742	.5681	$0.^{2}3010$	3.1823	1.9254
.72	5.7560	.6058	3.9113	.6831	$0.^{2}2007$	3.0680	1.9722
.80	5.6084	.7058	4.3751	.7925	$0.^{2}1307$	2.9570	2.0174
0.88	5.4686	.8020	4.7801	0.8978	$0.^{2}0742$	2.8521	2.0613
χ 1.00	5.2759	.9328	5.2710	1.0435	$0.^{2}0662$	2.7068	2.1255
1/χ 0.95	5.1973	0.9845	5.4469	1.1020	$0.^{2}0118$	2.6479	2.1523
.90	5.1147	1.0378	5.6191	1.1639	$0.^{2}0040$	2.5865	2.1805
.85	5.0287	1.0922	5.7776	1.2274	$0.^{3}0028$	2.5215	2.2111
.80	4.9398	1.1485	5.9325	1.2948	$0.^{3}0097$	2.4549	2.2446
.75	4.8479	1.2058	6.0769	1.3649	$0.^{3}0580$	2.3856	2.2815
.70	4.7506	1.2632	6.2055	1.4369	$0.^{3}1537$	2.3132	2.3195
.65	4.6526	1.3195	6.3158	1.5111	$0.^{3}2646$	2.2390	2.3626
.60	4.5496	1.3747	6.4076	1.5868	$0.^{3}4005$	2.1632	2.4068
.55	4.4472	1.4285	6.4811	1.6643	$0.^{3}5400$	2.0858	2.4567
.50	4.3417	1.4802	6.5310	1.7432	$0.^{3}6868$	2.0072	2.5105
.45	4.2373	1.5283	6.5599	1.8223	$0.^{3}7993$	1.9284	2.5689
.40	4.1307	1.5710	6.5625	1.9016	$0.^{3}8588$	1.8485	2.6313
.35	4.0233	1.6092	6.5415	1.9809	$0.^{3}8817$	1.7688	2.6979
.30	3.9194	1.6408	6.4995	2.0596	$0.^{3}8363$	1.6904	2.7714
.25	3.8158	1.6653	6.4312	2.1370	$0.^{3}7350$	1.6123	2.8506
.20	3.7137	1.6833	6.3472	2.2136	$0.^{3}5953$	1.5360	2.9349
.15	3.6152	1.6932	6.2449	2.2891	$0.^{3}4108$	1.4616	3.0263
.10	3.5172	1.6938	6.1215	2.3624	$0.^{3}2243$	1.3881	3.1227
.05	3.4243	1.6847	5.9850	2.4319	$0.^{3}0640$	1.3182	3.2253
1/χ 0.00	3.3333	1.6667	5.8328	2.5000	$0.^{3}0000$	1.2500	3.3333

TABLE 3

TRANSFORMATION COEFFICIENTS FOR p^2 (cf. Eq. 5)

	a	b	c	d
χ 0.00	1.0000	0.0000	1.0000	0.0000
.04	0.9998	0.^{1}2454	0.9999	.0186
.08	.9988	0.^{1}4859	.9994	.0367
.12	.9972	0.^{1}7382	.9985	.0542
.16	.9950	0.^{1}9950	.9974	.0711
.20	.9920	0.1255	.9962	.0874
.24	.9887	.1517	.9946	.1031
.28	.9840	.1778	.9930	.1182
.32	.9791	.2038	.9911	.1327
.36	.9733	.2294	.9892	.1466
.40	.9672	.2546	.9872	.1599
.48	.9530	.3029	.9827	.1849
.56	.9374	.3479	.9782	.2078
.64	.9212	.3892	.9735	.2287
.72	.9045	.4264	.9688	.2478
.80	.8880	.4597	.9642	.2653
0.88	.8721	.4893	.9596	.2813
χ 1.00	.8496	.5275	.9530	.3029
1/χ 0.95	.8403	.5421	.9503	.3115
.90	.8305	.5571	.9472	.3206
.85	.8200	.5721	.9440	.3301
.80	.8091	.5876	.9404	.3400
.75	.7976	.6033	.9366	.3505
.70	.7854	.6190	.9324	.3614
.65	.7727	.6348	.9279	.3730
.60	.7595	.6505	.9229	.3850
.55	.7458	.6662	.9175	.3977
.50	.7316	.6818	.9116	.4109
.45	.7171	.6971	.9053	.4248
.40	.7021	.7121	.8983	.4394
.35	.6868	.7268	.8907	.4545
.30	.6714	.7411	.8825	.4703
.25	.6557	.7549	.8735	.4868
.20	.6400	.7683	.8638	.5038
.15	.6243	.7813	.8533	.5214
.10	.6084	.7937	.8419	.5396
.05	.5929	.8053	.8297	.5583
1/χ 0.00	0.5773	0.8165	0.8165	0.5773

TABLE 4. MAGNETIC DIPOLE STRENGTHS FOR p^3

	$S_m(^2P_{1/2}, ^4S'_{3/2})$	$S_m(^2P'_{3/2}, ^2P_{1/2})$	$S_m(^2P_{1/2}, ^2D'_{3/2})$	$S_m(^2D_{5/2}, ^4S'_{3/2})$	$S_m(^2P'_{3/2}, ^2D_{5/2})$	$S_m(^2D_{5/2}, ^2D'_{3/2})$	$S_m(^2P'_{3/2}, ^2D'_{3/2})$	$S_m(^2P'_{3/2}, ^4S'_{3/2})$	$S_m(^2D'_{3/2}, ^4S'_{3/2})$
χ 0.00	0.0000	1.3333	0.0000	0.0000	0.0000	2.4000	0.0000	0.0000	0.0000
.05	$0.^{3}3705$	1.3301	$0.^{2}2868$	$0.^{6}6426$	$0.^{2}5166$	2.3948	$0.^{2}9175$	$0.^{2}1838$	$0.^{4}1935$
.10	$0.^{2}1483$	1.3207	$0.^{1}1118$	$0.^{4}1026$	$0.^{1}2017$	2.3798	$0.^{1}3573$	$0.^{2}7196$	$0.^{3}3047$
.15	$0.^{2}3343$	1.3059	$0.^{1}2412$	$0.^{4}5180$	$0.^{1}4368$	2.3563	$0.^{1}7703$	$0.^{1}1563$	$0.^{2}1504$
.20	$0.^{2}5955$	1.2968	$0.^{1}4055$	$0.^{3}1630$	$0.^{1}7377$	2.3261	0.1295	$0.^{1}2651$	$0.^{2}4596$
.25	$0.^{2}9329$	1.2648	$0.^{1}5919$	$0.^{3}3956$	0.1083	2.2913	0.1892	$0.^{1}3907$	$0.^{1}1078$
.30	$0.^{1}1347$	1.2410	$0.^{1}7884$	$0.^{3}8144$	.1454	2.2537	.2530	$0.^{1}5259$	$0.^{1}2134$
.35	$0.^{1}1839$	1.2164	$0.^{1}9849$	$0.^{2}1495$	.1834	2.2151	.3181	$0.^{1}6636$	$0.^{1}3759$
.40	$0.^{1}2410$	1.1919	0.1174	$0.^{2}2524$	.2209	2.1765	.3831	$0.^{1}7980$	$0.^{1}6082$
.45	$0.^{1}3060$	1.1679	.1348	$0.^{2}3991$	.2572	2.1387	.4461	$0.^{1}9244$	$0.^{1}9204$
.50	$0.^{1}3789$	1.1449	.1506	$0.^{2}5994$	.2915	2.1026	.5073	0.1039	0.1323
.55	$0.^{1}4596$	1.1230	.1644	$0.^{2}8627$	.3237	2.0677	.5665	.1139	.1823
.60	$0.^{1}5479$	1.1024	.1762	$0.^{1}1198$	.3535	2.0345	.6228	.1225	.2428
.65	$0.^{1}6438$	1.0831	.1859	$0.^{1}1614$	.3810	2.0028	.6773	.1293	.3138
.70	$0.^{1}7466$	1.0651	.1936	$0.^{1}2118$	.4062	1.9726	.7299	.1343	.3953
.75	$0.^{1}8562$	1.0483	.1995	$0.^{1}2716$	.4293	1.9436	.7808	.1377	.4872
.80	$0.^{1}9718$	1.0337	.2035	$0.^{1}3410$	.4503	1.9156	.8314	.1391	.5887
.85	0.1093	1.0181	.2059	$0.^{1}4201$	.4695	1.8882	.8777	.1398	.6992
.90	.1219	1.0046	.2070	$0.^{1}5091$	.4870	1.8619	.9250	.1389	.8169
0.95	.1349	0.9920	.2066	$0.^{1}6074$	.5034	1.8362	0.9686	.1368	0.9410
χ 1.00	.1481	.9802	.2051	$0.^{1}7147$	.5175	1.8111	1.0139	0.1336	1.0695
1/χ 0.90	.1783	.9567	.1983	$0.^{1}9808$	.5455	1.7564	1.3638	1.1067	0.1239
.80	.2161	.9318	.1854	0.1354	.5735	1.6912	1.7267	1.2147	0.1085
.70	.2626	.9054	.1654	.1863	.6008	1.6129	2.1458	1.3390	$0.^{1}8783$
.60	.3178	.8774	.1382	.2526	.6268	1.5206	2.5786	1.4796	$0.^{1}6354$
.55	.3481	.8628	.1224	.2912	.6390	1.4699	2.7772	1.5560	$0.^{1}5109$
.50	.3797	.8499	0.1058	.3326	.6520	1.4170	2.9502	1.6400	$0.^{1}3936$
.45	.4121	.8324	$0.^{1}8884$	.3759	.6608	1.3633	3.0863	1.7196	$0.^{1}2859$
.40	.4447	.8164	$0.^{1}7219$	.4199	.6700	1.3101	3.1798	1.8070	$0.^{1}1954$
.35	.4769	.8000	$0.^{1}5640$	.4631	.6779	1.2590	3.2227	1.8977	$0.^{1}1236$
.30	.5082	.7831	$0.^{1}4201$	.5044	.6840	1.2116	3.2283	1.9928	$0.^{2}7109$
.25	.5383	.7656	$0.^{1}2942$	.5425	.6882	1.1694	3.1901	2.0924	$0.^{2}3599$
.20	.5670	.7475	$0.^{1}1892$	.5766	.6900	1.1334	3.1188	2.1971	$0.^{2}1537$
.15	.5941	.7286	$0.^{1}1066$	.6062	.6892	1.1046	3.0232	2.3072	$0.^{2}0499$
.10	.6200	.7089	$0.^{2}4741$	.6314	.6853	1.0836	2.9132	2.4228	$0.^{3}1007$
.05	.6438	.6883	$0.^{2}1185$	.6512	.6779	1.0709	2.7826	2.5370	$0.^{4}0637$
1/χ 0.00	0.6667	0.6667	0.0000	0.6666	0.6666	1.0667	2.6667	2.6667	0.0000

TABLE 5. ELECTRIC QUADRUPOLE STRENGTHS FOR p^3 (cf. Eq. 4)

	$C_q(^2P_{\frac{3}{2}}, ^2P_{\frac{1}{2}})$	$C_q(^2P_{\frac{3}{2}}, ^2D'_{\frac{5}{2}})$	$C_q(^2P'_{\frac{3}{2}}, ^2D'_{\frac{3}{2}})$	$C_q(^2P_{\frac{1}{2}}, ^2D'_{\frac{3}{2}})$	$C_q(^2P_{\frac{3}{2}}, ^4S'_{\frac{3}{2}})$	$C_q(^2P_{\frac{1}{2}}, ^4S'_{\frac{3}{2}})$	$C_q(^2D_{\frac{5}{2}}, ^2D'_{\frac{3}{2}})$	$C_q(^2D_{\frac{5}{2}}, ^4S'_{\frac{3}{2}})$	$C_q(^2D'_{\frac{3}{2}}, ^4S'_{\frac{3}{2}})$
x 0.00	0.0000	14.0000	6.0000	6.0000	0.0000	0.0000	0.0000	0.0000	0.0000
.05	$0.^{1}1291$	13.9660	5.9468	5.9871	$0.^{6}3949$	$0.^{5}1606$	$0.^{1}3012$	$0.^{2}3890$	$0.^{2}1668$
.10	$0.^{1}5044$	13.8670	5.7936	5.9495	$0.^{5}6001$	$0.^{4}2566$	0.1174	$0.^{1}1557$	$0.^{2}6695$
.15	0.1092	13.7116	5.5578	5.8907	$0.^{4}2793$	$0.^{3}1295$	.2533	$0.^{1}3510$	$0.^{1}1514$
.20	.1844	13.5117	5.2531	5.8152	$0.^{4}7866$	$0.^{3}4075$	.4258	$0.^{1}6253$	$0.^{1}2712$
.25	.2709	13.2805	4.9346	5.7281	$0.^{3}1664$	$0.^{3}9890$	.6215	$0.^{1}9795$	$0.^{1}4277$
.30	.3636	13.0307	4.5952	5.6344	$0.^{3}2918$	$0.^{2}2036$	0.8279	0.1415	$0.^{1}6229$
.35	.4585	12.7727	4.2589	5.5378	$0.^{3}4466$	$0.^{2}3738$	1.0341	.1931	$0.^{1}8586$
.40	.5523	12.5150	3.9401	5.4413	$0.^{3}6156$	$0.^{2}6309$	1.2324	.2531	0.1137
.45	.6429	12.2633	3.6450	5.3468	$0.^{3}7830$	$0.^{2}9979$	1.4157	.3213	.1460
.50	.7288	12.0212	3.3769	5.2566	$0.^{3}9288$	$0.^{1}1498$	1.5814	.3978	.1829
.55	.8092	11.7915	3.1353	5.1694	$0.^{2}1039$	$0.^{1}2157$	1.7268	.4826	.2244
.60	.8837	11.5753	2.9206	5.0861	$0.^{2}1102$	$0.^{1}2996$	1.8498	.5753	.2705
.65	0.9524	11.3725	2.7302	5.0069	$0.^{2}1113$	$0.^{1}4036$	1.9520	.6759	.3211
.70	1.0155	11.1833	2.5624	4.9315	$0.^{2}1075$	$0.^{1}5296$	2.0334	.7840	.3760
.75	1.0732	11.0068	2.4142	4.8589	$0.^{3}9918$	$0.^{1}6739$	2.0945	0.8990	.4348
.80	1.1258	10.8535	2.2870	4.7890	$0.^{3}8640$	$0.^{1}8524$	2.1371	1.0204	.4969
.85	1.1738	10.6901	2.1692	4.7206	$0.^{3}7272$	0.1050	2.1623	1.1476	.5617
.90	1.2175	10.5479	2.0677	4.6549	$0.^{3}5736$	.1273	2.1734	1.2797	.6286
0.95	1.2586	10.4154	1.9781	4.5906	$0.^{3}4274$	.1519	2.1689	1.4160	.6969
x 1.00	1.2938	10.2918	1.8997	4.5278	$0.^{3}2799$	.1787	2.1535	1.5554	.7657
1/x 0.90	1.3637	10.0456	1.7548	4.3911	$0.^{3}0577$	.2452	2.0822	1.8721	0.9163
.80	1.4336	9.7840	1.6191	4.2280	$0.^{3}0182$	.3384	1.9468	2.2691	1.0915
.70	1.5021	9.5066	1.4960	4.0322	$0.^{3}3384$	.4657	1.7361	2.7574	1.2803
.60	1.5671	9.2128	1.3894	3.8015	$0.^{2}1118$	.6314	1.4505	3.3366	1.4584
.55	1.5975	9.0597	1.3436	3.6747	$0.^{2}1642$	.7279	1.2852	3.6551	1.5326
.50	1.6299	8.9237	1.3066	3.5425	$0.^{2}2177$	.8315	1.1107	3.9872	1.5908
.45	1.6519	8.7398	1.2696	3.4083	$0.^{2}2641$	0.9398	0.9328	4.3263	1.6298
.40	1.6750	8.5728	1.2434	3.2753	$0.^{2}2947$	1.0496	.7580	4.6693	1.6475
.35	1.6946	8.4006	1.2245	3.1476	$0.^{2}3024$	1.1578	.5922	5.0072	1.6444
.30	1.7101	8.2228	1.2136	3.0291	$0.^{2}2833$	1.2609	.4411	5.3360	1.6226
.25	1.7204	8.0391	1.2117	2.9234	$0.^{2}2405$	1.3561	.3089	5.6521	1.5860
.20	1.7251	7.8484	1.2185	2.8334	$0.^{2}1804$	1.4415	.1986	5.9529	1.5391
.15	1.7230	7.6503	1.2344	2.7615	$0.^{2}1146$	1.5155	0.1119	6.2378	1.4867
.10	1.7132	7.4437	1.2592	2.7091	$0.^{2}0555$	1.5785	$0.^{1}4978$	6.5100	1.4336
.05	1.6948	7.2274	1.2924	2.6773	$0.^{3}1453$	1.6279	$0.^{1}1244$	6.7605	1.3808
1/x 0.00	1.6667	7.0000	1.3333	2.6667	0.0000	1.6666	0.0000	7.0000	1.3333

TABLE 6. TRANSFORMATION COEFFICIENTS FOR p^3 (cf. Eq. 9)

	a	b	c	a'	b'	c'	a"	b"	c"
χ 0.00	-0.0000	-0.0000	1.0000	-0.0000	1.0000	0.0000	1.0000	0.0000	0.0000
.05	$0.^{1}4638$	$0.^{2}1290$	0.9989	$0.^{1}1667$	0.9999	$0.^{3}5174$	0.9988	$0.^{1}1663$	$0.^{1}4640$
.10	$0.^{1}9157$	$0.^{2}5116$	.9958	$0.^{1}3335$	.9994	$0.^{2}2068$	.9952	$0.^{1}3302$	$0.^{1}9169$
.15	0.1345	$0.^{1}1135$	.9908	$0.^{1}5007$	.9987	$0.^{2}4645$	.9896	$0.^{1}4899$	0.1349
.20	.1744	$0.^{1}1981$	.9845	$0.^{1}6683$	.9977	$0.^{2}8241$	.9824	$0.^{1}6436$	.1753
.25	.2107	$0.^{1}3028$	.9771	$0.^{1}8365$	.9964	$0.^{1}1284$	.9740	$0.^{1}7902$	.2124
.30	.2432	$0.^{1}4252$	.9690	0.1005	.9948	$0.^{1}1842$	.9648	$0.^{1}9293$	.2462
.35	.2718	$0.^{1}5631$	.9607	.1174	.9928	$0.^{1}2496$	.9552	0.1061	.2764
.40	.2967	$0.^{1}7147$	.9523	.1344	.9904	$0.^{1}3243$	.9455	.1184	.3034
.45	.3180	$0.^{1}8777$	.9440	.1515	.9876	$0.^{1}4078$	.9359	.1300	.3273
.50	.3361	0.1051	.9360	.16857	.9844	$0.^{1}4998$	.9266	.1410	.3485
.55	.3512	.1232	.9282	.18566	.9808	$0.^{1}5996$	.9177	.1513	.3672
.60	.3634	.1421	.9207	.2027	.9767	$0.^{1}7066$	.9093	.1610	.3838
.65	.3734	.1615	.9135	.2197	.9721	$0.^{1}8202$	.9013	.1701	.3984
.70	.3811	.1813	.9066	.2366	.9670	$0.^{1}9395$	.8938	.1787	.4114
.75	.3868	.2015	.8999	.2534	.9615	0.1064	.8867	.1869	.4229
.80	.3907	.2218	.8934	.2700	.9555	.1192	.8805	.1946	.4332
.85	.3930	.2423	.8870	.2863	.9490	.1323	.8738	.2020	.4423
.90	.3940	.2626	.8808	.3023	.9420	.1456	.8680	.2089	.4505
0.95	.3936	.2828	.8747	.3180	.9396	.1591	.8625	.2156	.4580
χ 1.00	.3922	.3028	.8687	.3333	.9269	.1726	.8574	.2219	.4644
$1/\chi$ 0.90	.3857	.3456	.8555	.3657	.9085	.2021	.8471	.2349	.4767
.80	.3729	.3953	.8394	.4026	.8840	.2375	.8360	.2494	.4888
.70	.3521	.4516	.8198	.4438	.8517	.2786	.8240	.2657	.5004
.60	.3219	.5126	.7960	.4882	.8102	.3244	.8112	.2842	.5111
.55	.3030	.5438	.7826	.5110	.7859	.3483	.8044	.2943	.5160
.50	.2817	.5747	.7684	.5337	.7594	.3723	.7984	.3056	.5212
.45	.2581	.6044	.7537	.5560	.7309	.3958	.7901	.3169	.5247
.40	.2327	.6324	.7388	.5775	.7011	.4183	.7825	.3294	.5284
.35	.2057	.6581	.7243	.5980	.6704	.4393	.7746	.3428	.5315
.30	.1775	.6809	.7105	.6174	.6393	.4584	.7664	.3573	.5339
.25	.1485	.7005	.6980	.6354	.6085	.4754	.7578	.3729	.5355
.20	0.1191	.7166	.6872	.6521	.5784	.4901	.7487	.3897	.5362
.15	$0.^{1}8942$	.7292	.6784	.6675	.5494	.5026	.7392	.4079	.5359
.10	$0.^{1}5963$	.7382	.6720	.6819	.5220	.5129	.7292	.4275	.5344
.05	$0.^{1}2981$	.7436	.6680	.6949	.4958	.5209	.7185	.4487	.5315
$1/\chi$ 0.00	-0.0000	-0.7454	0.6667	-0.7071	0.4714	0.5270	0.7071	0.4714	0.5270

TABLE 7

MAGNETIC DIPOLE STRENGTHS FOR p^4

	$S_m(^1D_2', ^3P_2')$	$S_m(^1D_2', ^3P_1)$	$S_m(^1S_0', ^3P_1)$	$S_m(^3P_2', ^3P_1)$	$S_m(^3P_1, ^3P_0')$
x 0.00	0.0000	0.0000	0.0000	2.5000	2.0000
.04	$0.^{1}0402$	$0.^{1}0134$	$0.^{1}0730$	2.4975	1.9992
.08	$0.^{1}1547$	$0.^{1}0517$	$0.^{1}0299$	2.4935	1.9968
.12	$0.^{1}3340$	$0.^{1}1119$	$0.^{1}0687$	2.4885	1.9928
.16	$0.^{1}5681$	$0.^{1}1908$	$0.^{1}1246$	2.4805	1.9876
.20	$0.^{1}8470$	$0.^{1}2857$	$0.^{1}1980$	2.4706	1.9800
.24	0.1162	$0.^{1}3938$	$0.^{1}2894$	2.4602	1.9713
.28	.1507	$0.^{1}5127$	$0.^{1}3987$	2.4498	1.9602
.32	.1868	$0.^{1}6392$	$0.^{1}5257$	2.4357	1.9468
.36	.2246	$0.^{1}7726$	$0.^{1}6699$	2.4226	1.9326
.40	.2630	$0.^{1}9101$	$0.^{1}8305$	2.4084	1.9173
.48	.3410	0.1194	0.1196	2.3810	1.8806
.56	.4174	.1479	.1612	2.3522	1.8382
.64	.4907	.1760	.2065	2.3237	1.7940
.72	.5600	.2032	.2542	2.2963	1.7458
.80	.6250	.2294	.3029	2.2705	1.6972
.88	.6852	.2542	.3516	2.2458	1.6486
x 1.00	.7666	.2890	.4226	2.2104	1.5774
1/x 0.95	.8000	.3035	.4526	2.1968	1.5470
.90	.8340	.3186	.4848	2.1814	1.5148
.85	.8702	.3349	.5194	2.1655	1.4802
.80	.9076	.3521	.5564	2.1479	1.4433
.75	.9469	.3706	.5958	2.1294	1.4042
.70	0.9879	.3903	.6375	2.1096	1.3626
.65	1.0310	.4113	.6815	2.0889	1.3187
.60	1.0747	.4335	.7279	2.0662	1.2723
.55	1.1212	.4573	.7761	2.0430	1.2237
.50	1.1677	.4825	.8259	2.0169	1.1741
.45	1.2165	.5094	.8772	1.9901	1.1229
.40	1.2664	.5380	.9294	1.9616	1.0705
.35	1.3171	.5683	0.9833	1.9312	1.0179
.30	1.3694	.6007	1.0354	1.8997	0.9644
.25	1.4204	.6345	1.0882	1.8654	.9118
.20	1.4723	.6706	1.1400	1.8297	.8599
.15	1.5231	.7084	1.1908	1.7918	.8092
.10	1.5721	.7480	1.2403	1.7514	.7597
.05	1.6207	.7899	1.2879	1.7098	.7119
1/x 0.00	1.6660	0.8333	1.3333	1.6663	0.6666

TABLE 8. ELECTRIC QUADRUPOLE STRENGTHS FOR p^4 (cf. Eq. 4)

	$C_q(^1S_0', ^1D_2')$	$C_q(^1S_0', ^3P_2')$	$C_q(^1D_2, ^3P_2')$	$C_q(^1D_2', ^3P_1')$	$C_q(^1D_2', ^3P_0')$	$C_q(^3P_2', ^3P_1)$	$C_q(^3P_2', ^3P_0')$
χ 0.00	6.6666	0.0000	0.0000	0.0000	0.0000	3.7500	1.6666
.04	6.6600	$0.^{2}1235$	$0.^{1}1407$	$0.^{2}2012$	$0.^{3}3765$	3.7463	1.6673
.08	6.6507	$0.^{2}4552$	$0.^{1}5413$	$0.^{2}7751$	$0.^{2}1684$	3.7403	1.6714
.12	6.6380	$0.^{2}9390$	0.1169	$0.^{1}1678$	$0.^{2}4200$	3.7328	1.6792
.16	6.6195	$0.^{1}1523$	.1988	$0.^{1}2863$	$0.^{2}8200$	3.7208	1.6893
.20	6.5922	$0.^{1}2159$	.2966	$0.^{1}4286$	$0.^{1}1395$	3.7058	1.7015
.24	6.5663	$0.^{1}2811$	.4069	$0.^{1}5906$	$0.^{1}2169$	5.6904	1.7173
.28	6.5335	$0.^{1}3445$	.5276	$0.^{1}7691$	$0.^{1}3163$	3.6746	1.7353
.32	6.4912	$0.^{1}4032$	.6539	$0.^{1}9589$	$0.^{1}4387$	3.6532	1.7532
.36	6.4523	$0.^{1}4563$	.7862	0.1159	$0.^{1}5859$	3.6338	1.7750
.40	6.4085	$0.^{1}5023$	0.9203	.1365	$0.^{1}7580$	3.6124	1.7985
.48	6.3115	$0.^{1}5708$	1.1936	.1790	0.1175	3.5715	1.8487
.56	6.2005	$0.^{1}6077$	1.4608	.2218	.1681	3.5284	1.9010
.64	6.0897	$0.^{1}6188$	1.7173	.2639	.2263	3.4856	1.9570
.72	5.9687	$0.^{1}6085$	1.9598	.3048	.2898	3.4444	2.0115
.80	5.8513	$0.^{1}5838$	2.1874	.3441	.3570	3.4058	2.0672
.88	5.7332	$0.^{1}5500$	2.3982	.3814	.4255	3.3686	2.1203
χ 1.00	5.5602	$0.^{1}4910$	2.6830	.4335	.5275	3.3158	2.1953
1/χ 0.95	5.4650	$0.^{1}4643$	2.8001	.4552	.5712	3.2951	2.1973
.90	5.4102	$0.^{1}4350$	2.9180	.4778	.6182	3.2719	2.2597
.85	5.3277	$0.^{1}4028$	3.0468	.5025	.6693	3.2483	2.2960
.80	5.2375	$0.^{1}3692$	3.1762	.5282	.7234	3.2216	2.3337
.75	5.1433	$0.^{1}3335$	3.3138	.5558	.7817	3.1939	2.3743
.70	5.0432	$0.^{1}2975$	3.4576	.5854	.8429	3.1642	2.4172
.65	4.9375	$0.^{1}2606$	3.6083	.6169	.9076	3.1335	2.4637
.60	4.8247	$0.^{1}2234$	3.7622	.6502	0.9748	3.0994	2.5113
.55	4.7073	$0.^{1}1873$	3.9241	.6860	1.0443	3.0645	2.5623
.50	4.5855	$0.^{1}1536$	4.0871	.7237	1.1147	3.0255	2.6158
.45	4.4608	$0.^{1}1220$	4.2575	.7641	1.1862	2.9850	2.6730
.40	4.3330	$0.^{2}9358$	4.4323	.8070	1.2565	2.9422	2.7332
.35	4.2035	$0.^{2}6905$	4.6045	.8524	1.3252	2.8969	2.7963
.30	4.0732	$0.^{2}4787$	4.7932	.9011	1.3923	2.8496	2.8632
.25	3.9423	$0.^{2}3110$	4.9718	0.9518	1.4542	2.7982	2.9322
.20	3.8153	$0.^{2}1857$	5.1529	1.0058	1.5105	2.7446	2.1770
.15	3.6888	$0.^{2}0957$	5.3288	1.0624	1.5613	2.6876	3.0822
.10	3.5643	$0.^{2}0380$	5.5020	1.1220	1.6046	2.6272	3.1620
.05	3.4455	$0.^{3}0845$	5.6726	1.1850	1.6395	2.5646	3.2457
1/χ 0.00	3.3333	0.0000	5.8328	1.2500	1.6667	2.5000	3.3333

TABLE 9

TRANSFORMATION COEFFICIENTS FOR p^4 (cf. Eq. 5)

	a	b	c	d
x 0.00	+1.0000	-0.0000	+1.0000	-0.0000
.04	0.9995	$0.^{1}2316$	0.9998	$0.^{1}1910$
.08	.9990	$0.^{1}4547$	.9992	$0.^{1}3866$
.12	.9977	$0.^{1}6689$	.9982	$0.^{1}5862$
.16	.9961	$0.^{1}8737$	.9969	$0.^{1}7893$
.20	.9941	0.1069	.9950	$0.^{1}9951$
.24	.9920	.1255	.9928	0.1203
.28	.9899	.1432	.9900	.1412
.32	.9870	.1599	.9866	.1621
.36	.9844	.1758	.9830	.1830
.40	.9815	.1908	.9791	.2038
.48	.9759	.2185	.9697	.2446
.56	.9700	.2432	.9587	.2839
.64	.9641	.2653	.9471	.3214
.72	.9584	.2851	.9343	.3565
.80	.9530	.3029	.9212	.3892
.88	.9478	.3189	.9079	.4193
x 1.00	.9403	.3400	.8881	.4597
$1/x$ 0.95	.9374	.3484	.8795	.4757
.90	.9341	.3570	.8703	.4923
.85	.9307	.3660	.8603	.5096
.80	.9269	.3753	.8495	.5274
.75	.9229	.3850	.8379	.5458
.70	.9186	.3951	.8254	.5646
.65	.9141	.4056	.8120	.5838
.60	.9091	.4164	.7976	.6033
.55	.9040	.4277	.7822	.6229
.50	.8982	.4393	.7662	.6426
.45	.8922	.4514	.7943	.6623
.40	.8858	.4639	.7316	.6817
.35	.8789	.4768	.7134	.7008
.30	.8717	.4902	.6944	.7195
.25	.8638	.5038	.6752	.7376
.20	.8555	.5179	.6557	.7549
.15	.8466	.5323	.6361	.7716
.10	.8370	.5470	.6163	.7875
.05	.8270	.5621	.5966	.8024
$1/x$ 0.00	+0.8164	+0.5773	+0.5773	-0.8165

Reprinted from *The Astrophysical Journal,* Vol. 93, No. 1, January, 1941.

PHYSICAL PROCESSES IN GASEOUS NEBULAE

XII. THE ELECTRON DENSITIES OF SOME BRIGHT PLANETARY NEBULAE

DONALD H. MENZEL AND LAWRENCE H. ALLER

ABSTRACT

The electron densities of some typical, bright, planetary nebulae have been estimated from the intensity of the continuous emission beyond the limit of the Balmer series. The densities so obtained are of the order of 10^4 electrons/cm^3 for the objects studied and are probably somewhat lower in the fainter planetaries.

Before the predictions in theories of gaseous nebulae can be compared with the observations, one must first determine the values of two fundamental parameters: the electron density and the electron temperature. In this paper we shall show how the electron density may be estimated from the intensity of the emission beyond the limit of the Balmer series, and in the next paper of this series estimates of the electron temperatures will be derived from a comparison of the green nebular lines of *O* III (λ 4959 and λ 5007) with the λ 4363 line of the same ion. The electron density was derived some time ago by one of us[1] by a similar method, but it is now possible to make closer estimates because of more accurate observational data. The earlier determinations gave the correct order of magnitude.

From equation (23) of the first paper of this series[2] we find that the emission at the head of the Balmer series, $n' = 2$ and $\nu = R/4$, per unit volume in the frequency interval $d\nu$, is

$$E_{\kappa n} d\nu = N_i N_\epsilon \frac{K}{T_\epsilon^{3/2}} \frac{g}{8} h d\nu \, . \tag{1}$$

For convenience, we re-write this formula in terms of a wave-length interval rather than a frequency interval, $d\nu = (c/\lambda^2) d\lambda$, where $c = 3 \times 10^{10}$ cm/sec and $\lambda = 3.65 \times 10^{-5}$ cm at the Balmer limit. If $\Delta\lambda$ is expressed in angstrom units, $d\nu = 2.25 \times 10^{11} \Delta\lambda$. From (I, 6)[2] we have $Kh = 6.99 \times 10^{-33}$; also $g = 0.88$ at the Balmer limit. Therefore,

$$E_{\kappa n} d\nu = N_i N_\epsilon \frac{1.73 \times 10^{-22}}{T_\epsilon^{3/2}} \Delta\lambda \, . \tag{2}$$

For the observational data used in the present paper, an interval of 20 A at the Balmer limit was employed. Our equation then becomes

$$E_{\kappa n} \Delta\nu = 3.46 \times 10^{-21} \frac{N_i N_\epsilon}{T_\epsilon^{3/2}} \, . \tag{3}$$

In a homogeneous spherical nebula the total emission will be

$$\tfrac{4}{3}\pi E_{\kappa n} \Delta\nu A^3$$

[1] Menzel, *Pub. A.S.P.*, **43**, 337, 1931.

[2] Menzel, *Ap. J.*, **85**, 330, 1937. The Roman numeral refers to the number of the paper in the series, the Arabic number to the number of the equation. In the original (I, 6) the *h* was inadvertently omitted.

and in a homogeneous shell of thickness d and outer radius A, the total emission will be

$$\tfrac{4}{3}\pi E_{\kappa n}\Delta\nu[3A^2d - 3Ad^2 + d^3] ,$$

or

$$4\pi E_{\kappa n}\Delta\nu A^2 d ,$$

when d is much smaller than A. The surface through which this energy must flow has an area $4\pi A^2$, so that the average surface brightness, S_n, for this 20 A range of Balmer continuum will be, for the three separate cases

$$S_n = \tfrac{1}{3}E_{\kappa n}\Delta\nu A = \frac{1.15 \times 10^{-21}}{T_\epsilon^{3/2}} N_i N_\epsilon A \qquad \text{(homogeneous sphere)} , \qquad (4)$$

$$S_n = E_{\kappa n}\Delta\nu d = \frac{3.46 \times 10^{-21}}{T_\epsilon^{3/2}} N_i N_\epsilon d \qquad \text{(thin homogeneous shell)} , \qquad (5)$$

$$\left. S_n = \frac{1.15 \times 10^{-21} N_i N_\epsilon}{T_\epsilon^{3/2}} 3d \left[1 - \frac{d}{A} + \frac{d^2}{3A^2}\right] = \frac{1.15 \times 10^{-21}}{T_\epsilon^{3/2}} N_i N_\epsilon D , \quad \text{(thick homogeneous shell)} , \right\} \qquad (6)$$

where

$$D = 3d\left[1 - \frac{d}{A} + \frac{d^2}{3A^2}\right] . \qquad (7)$$

The photometrically determined surface brightness of a planetary is usually expressed in magnitudes per square minute of arc. Our next problem is to convert the surface brightness of the nebula at the Balmer limit, as measured in magnitudes, to the physical units of brightness in ergs emitted per square centimeter. Since surface brightness is independent of distance, no knowledge of the parallax is required.

To effect the conversion we find it convenient to employ the sun as a reference standard. According to Kuiper,[3] the apparent bolometric magnitude of the sun is -26.95. The surface area of the sun is 806 square minutes of arc. Therefore, the surface brightness in magnitudes per square minute of arc is

$$H_\odot = -19.68 . \qquad (8)$$

The amount of solar energy $S_\odot$ radiated per second[4] is 6.25×10^{10} erg/cm^{-2}. Let H_n denote the surface brightness of the nebula at the Balmer limit, in terms of magnitudes per square minute of arc. Then we easily obtain the elementary conversion formula

$$\log S_\odot - \log S_n = 0.4 \log (H_n - H_\odot)$$

$$S_n = 0.840 \times 10^3 (2.512)^{-H_n} \qquad \text{in ergs sec}^{-1}\text{ cm}^{-2} . \qquad (9)$$

We assume that the number of hydrogen ions is equal to the number of electrons, which is equivalent to assuming that hydrogen is the predominant element. This assumption

[3] *Ap. J.*, **88**, 438, 1938.

[4] Russell, Dugan, and Stewart, *Astronomy*, **2**, 534.

leads to a minimum value for N_ϵ and a maximum for N_i. If we put $N_i = N_\epsilon$ we obtain from equation (4)

$$N_\epsilon = 2.94 \times \frac{10^{10} T_\epsilon^{3/4} S_n^{1/2}}{A^{1/2}} = 8.54 \times 10^{11} \frac{T_\epsilon^{3/4}(2.512)^{-H_n/2}}{A^{1/2}}, \qquad (10)$$

and a similar equation in which A is replaced by D as defined above. Equation (10) suffices to determine N_ϵ when A, H_n, and T_ϵ are known.

The radius of a planetary may be determined from its apparent size if the distance is known. The distances here considered were taken from the data of Berman;[5] the angular sizes, from the estimates of H. D. Curtis;[6] and the apparent photographic magnitudes from the work of Vorontsov-Velyaminov and Parenago and Hubble as tabulated by Louis Berman.[5]

The remaining necessary data are, of course, the intensity of the Balmer continuum at the series limit and the electron temperature. The latter, as will be discussed in the next paper of this series, is evaluated from the relative intensities of λλ 5007, 4959, and 4363 or *O* III. For the planetaries in question, the electron temperatures lie between 6000° K and 10,000° K. The intensity of the continuum must be obtained from spectrophotometric observations. Fortunately, the results are insensitive to the exact temperature.

A spectrophotometric study of a number of the brighter and more interesting planetaries has been undertaken by one of us. The plates were secured with the quartz slitless spectrograph attached to the Crossley reflector of the Lick Observatory. Although a detailed account of this investigation is to be reported in a subsequent paper of the series, it seems worth while to summarize here the portion of the study relating to the intensity of the Balmer continuum at the series limit.

Of the eleven planetary nebulae observed during the summers of 1938 and 1939, suitable plates for a study of the Balmer continuum were obtained of the following objects: NGC 6543, 6572, 6826, 7009, 7027, 7662, and IC 418. In addition, plates of NGC 1535 taken by Mr. R. H. Stoy were kindly made available for this photometric study. Standards were impressed by means of a tube sensitometer. For the region of the Balmer continuum, a Wratten 18A filter was utilized; in the visual region, a 50 filter. The star π_1 Cygni was employed as the comparison star. The energy distribution for this star was that assumed by Popper[7] in his study of Nova Lacertae and depends ultimately on the spectrophotometric studies by Barbier and his colleagues.[8]

The spectrograms were measured with the microdensitometer of the Harvard Observatory, by a method similar to that employed by Berman;[9] that is, successive tracings were made parallel to the dispersion until the total surface of the planetary had been covered.

For all these objects there seemed to be a general continuum in addition to that of the Balmer series. In some cases it was rather difficult to estimate this continuum because of the numerous overlapping images of the faint lines; near the Balmer limit the faint lines on slitless spectra overlap, and, in consequence, the intensity of the general continuum has to be estimated upon the basis of some likely wave-length variation. In practice, the intensity of the general continuum in the neighborhood of the Balmer limit was estimated by eye from the reduced tracings. Thus, the estimates of the intensity of the Balmer continuum may be systematically in error because of unknown variations in the intensity of the general continuum. The values tabulated here are, consequently,

[5] *Lick Obs. Bull.*, **18**, 73, 1937.

[6] *Lick Obs. Pub.*, **13**, 55, 1918.

[7] *Ap. J.*, **92**, 262, 1940.

[8] Barbier *et al.*, *J. d. obs.*, **19**, 149, 1936; *C.R.*, **210**, 99, 1940.

[9] *Lick Obs. Bull.*, **15**, 86, 1930.

to be considered as tentative estimates which may easily be in error by as much as 25 per cent or perhaps even 40 per cent. The order of magnitude of the calculated electron densities is insensitive, however, to errors of this sort.

It may be remarked in passing that in the case of NGC 7662, which is a particularly suitable object for such a study because of the small amount of space absorption in this region of the sky, the wave-length variation of the intensity of the Balmer continuum was investigated in an effort to estimate the electron temperature, but no satisfactory value could be found. Photometric difficulties, the effect of space absorption, and the unknown wave-length variation of the general continuum render this method of estimating the electron temperature uncertain.

The intensity of the 20 A interval of the Balmer continuum was estimated from the reduced tracings. With the intensity of $H\beta$ taken as 10, we obtain the results shown in the accompanying table. The means of two plates were used for each of these objects except for NGC 1535, for which only one suitable plate was available.

NGC 7027	0.27	IC 418	0.69
6572	0.68	NGC 7662	0.62
6826	0.75	7009	0.86
6543	0.84	1535	1.57

We have yet to reduce these intensities to the apparent-magnitude scale in order to be able to use our formulae. The determination of the magnitudes of gaseous nebulae is very different from the determination of stellar magnitudes, since in the former practically all the light is concentrated within a few bright lines. The strongest lines in most of the gaseous nebulae are the green lines of *O* III, but the plate sensitivity of most blue-sensitive emulsions is low in their neighborhood. For a tentative estimate of the magnitude of the planetaries at the Balmer limit the following procedure was adopted. It was assumed that the magnitude scale taken from Berman's paper[5] corresponded to a plate-*plus*-instrument sensitivity combination similar to that found by Yü[10] for a Seed 23 plate and a silver-coated mirror. The relative intensities of all the strong nebular lines (on the scale $H\beta = 10$) were then multiplied by this plate-*plus*-instrument sensitivity factor and added together. The intensity of the Balmer continuum at the series head was then compared with this sum to give the amount by which the nebular magnitude had to be corrected to give the magnitude of the 20 A interval of Balmer continuum.

The details of the calculation are summarized in Table 1. Successive columns give the designations of the nebula according to its NGC or IC number, the magnitude of the planetary as quoted by Berman,[5] the dimensions in seconds of arc as derived from the data by Curtis,[11] and the value of r in parsecs according to Berman. The columns labeled A'' and A (see Table 1) give, respectively, the assumed radii of these planetaries in seconds of arc and in centimeters. Uncertainties introduced in estimating these radii provide a fundamental limitation upon the accuracy of our results. As soon as the area in seconds of arc has been calculated, we may compute the mean surface brightness of the planetary in terms of magnitudes per square minute of arc. The latter quantity is then corrected for the ratio of brightness of the continuum at the Balmer limit to the brightness of the whole nebula multiplied by the assumed plate-*plus*-instrument sensitivity. The column labeled ΔH_n gives the amount by which the tabulated magnitude m_n is to be corrected to reduce it to the surface brightness of the nebula in magnitudes, in terms of a 20 A interval at the Balmer limit. The choice of this wave-length interval, as previously explained, depended upon convenience. We could have expressed the bright-

[10] *Lick Obs. Bull.*, **12**, 104, 1926. [11] *Op. cit.*

ness in terms of unit-frequency interval, but there seemed to be little point in multiplying by a large factor and dividing by it again. The quantity S_n may be immediately computed. For each planetary a value of T_ϵ was assumed on the basis of calculations which are yet to be described. With these data we may compute N_ϵ by means of equation (10).

For the ring nebulae, which are apparently shell structures, estimates of the thickness, d, of the shell were taken from the paper by Curtis and the corresponding corrections to the densities were evaluated.

The electron densities of these brighter planetaries seem to range from about 9×10^3 electrons/cm³ for NGC 1535 to about 4×10^4 electrons/cm³ for NGC 7009. It is to be

TABLE 1

ELECTRON DENSITIES FROM INTENSITY OF BALMER CONTINUUM

Nebula	m_n	Dimensions	r (Parsecs)	d/A	A''	$A \times 10^{-17}$ (Cm)	Δm/Sq. Min.
NGC 7027	9.8	12″×8″	2130	1.0	5	1.59	−3.94
6572	9.2	15 ×11	1230	1.0	7	1.29	3.35
6826	9.2	27 ×24	1050	1.0	13	2.04	1.85
6543	8.7	22 ×16	1080	0.2	10	1.61	2.52
IC 418	10.6	14 ×11	1800	0.2	6	1.61	3.42
NGC 7662	8.7	17 ×14*	1200	0.2	8	1.43	2.95
7009	7.9	25 ×12*	930	0.3	9	1.25	2.70
1535	9.9	20 ×17	1720	0.3	9	2.32	−2.56

Nebula	Δm (Balmer Continuum)	ΔH_n	H_n	$\sqrt{D} \times 10^{-8}$	T_ϵ	N_ϵ
NGC 7027	+5.85	+1.91	+11.71	4.10	9,500	9 3×10³
6572	4.71	1.36	10.56	3.59	9,200	17
6826	4.39	2.54	11.74	4.52	7,200	7
6543	4.16	1.64	10.34	2.80	6,000	18
IC 418	4.17	0.75	11.35	2.80	6,800	12
NGC 7662	5.27	2.32	11.04	2.64	10,300	20
7009	4.52	1.82	9.72	2.87	9,500	33
1535	+3.82	+1.26	+11.16	3.91	10,000	13

* These are the dimensions of the inner and brighter shell of the planetary.

kept in mind that these quantities are only orders of magnitude, as the irregular structures of the planetaries preclude the specification of more accurate values. More accurate parallaxes and magnitudes would improve these mean values. It is to be remarked that the electron densities of the fainter planetaries may be somewhat lower.

We next examine the possibility of estimating nebular densities from the intensities of the hydrogen lines, in particular $H\beta$. This method should prove useful when the continuum is too weak to be measured with any degree of certainty. The intensity of any line of the Balmer series may be written down from (I, 9) as

$$E_{n2} = N_i N_\epsilon \frac{6.99 \times 10^{-33}}{T_\epsilon^{3/2}} b_n Z^4 \frac{g}{8} \frac{2RZ^2}{n^3} e^{X_n} \quad \left(X_n = \frac{hRZ^2}{n^2 kT_\epsilon} \right) . \quad (11)$$

For $H\beta$ we have, since $g = 0.822$,

$$E_{H\beta} = 7.37 \times 10^{-20} \frac{N_i N_\epsilon}{T_\epsilon^{3/2}} b_4 e^{X_4} . \qquad (12)$$

Therefore,

$$S_4 = \frac{2.46 \times 10^{-20}}{T_\epsilon^{3/2}} b_4 N_i N_\epsilon D e^{X_4} , \qquad (13)$$

and, by setting $N_i = N_\epsilon$ as before, we obtain

$$N_\epsilon = 1.84 \times 10^{11} \frac{T_\epsilon^{3/4} e^{-X_4/2}}{\sqrt{D b_4}} (2.512)^{-H_n/2} , \qquad (14)$$

where H_n is now taken to mean the brightness of $H\beta$ in terms of magnitudes per square minute of arc. When T_ϵ is adopted from other data, as before, we may work out the

TABLE 2

b_4 FOR TYPICAL PLANETARIES

	$\Delta m(H\beta)$	$H(H\beta)$	e^{X_n}	$N_\epsilon\sqrt{b_4}$	b_4*
NGC 7027........	+1.95	+7.80	2.79	7.2×10^3	0.61
6572........	1.80	7.65	2.91	8.3	.23
6826........	1.57	8.92	3.87	2.7	.16
6543........	1.46	7.64	5.14	5.9	.11
IC 418........	1.27	8.45	4.23	4.8	.15
NGC 7662........	1.24	7.00	2.59	17.3	.75
7009........	1.86	7.05	2.79	14.2	.19
1535........	+1.84	+9.18	2.66	4.2	0.11

* Mean value of b_4 (excluding NGC 7662 and 7027) is 0.16 ± 0.01.

TABLE 3

ELECTRON DENSITY FROM INTENSITY OF $H\beta$

	m_n	Dimensions	A''	r (Parsecs)	$A\times 10^{17}$ (Cm)	d/A	Δm/Sq.Min. ($H\beta$)
NGC 40........	9.5	38″×35″	18	1220	3.29	0.2	−1.08
NGC 2022........	12.4	22 ×17	10	4570	6.84	0.2	−2.46

	$m(H\beta)$	$H(H\beta)$	T_ϵ(Adopted)	$\sqrt{D}\times 10^{-8}$	$N_\epsilon\sqrt{b_4}$	N_ϵ
NGC 40........	0.96	9.4	6,000	4.01	1.84×10^3	5×10^3
NGC 2022........	2.03	12.0	10,000	5.78	0.78	2×10^3

values of $N_\epsilon\sqrt{b_4}$ (see Table 2). We obtain the values of b_4 from a comparison of $N_\epsilon\sqrt{b_4}$ with the previous values of N_ϵ. Except for NGC 7027 and NGC 7662 the values of b_4 are remarkably constant from one nebula to another. It seems fairly safe, therefore, to assume $b_4 \sim 0.16$, and to use the observed intensities of $H\beta$ to deduce the densities of nebulae for which the Balmer continuum is too faint to be observed.

We may apply this method to NGC 40 and NGC 2022, both of which have low surface brightnesses. The relevant data are given in Table 3. To estimate N_ϵ, we assumed that $b_4 = 0.16$. The electron temperature of NGC 40 is assumed to be 6000°, that of NGC 2022, 10,000°. Both nebulae are assumed to possess shell structures. If the adopted value of b_4 is approximately correct, it is apparent that the density of the electrons is somewhat lower in these objects than in the others, as one would expect on the basis of their lower surface brightnesses.

On the strength of these results, we shall adopt electron densities of the order of 10^4 electrons/cm^3 in future theoretical studies of the planetary nebulae. This value seems to be about correct for the brighter planetaries. For the fainter objects, values of the order of 2×10^3 electrons/cm^3 may be assumed. In the next paper of this series we shall show how the electron temperature may be derived.

Harvard College Observatory
June 8, 1940

Reprinted from *The Astrophysical Journal,* Vol. 93, No. 2, March, 1941.

PHYSICAL PROCESSES IN GASEOUS NEBULAE

XIII. THE ELECTRON TEMPERATURES OF SOME TYPICAL PLANETARY NEBULAE

DONALD H. MENZEL, LAWRENCE H. ALLER, AND MALCOLM H. HEBB

ABSTRACT

The electron temperatures of several typical planetary nebulae have been determined from the relative intensities of the line λ 4363 and the green nebular lines of *O* III. The values obtained range from about 6000° K to 10,000° K, and seem to be insensitive to the degree of excitation of the nebular spectrum and probably also of the temperature of the central star.

In a previous paper[1] we have shown how the electron densities for a number of planetary nebulae may be found when the intensity of the continuum at the Balmer limit is known and when T_ϵ is known. In this paper we shall show how the electron temperature of a planetary nebula may be determined from the intensities of certain lines in the nebular spectrum and from the electron density.

The levels of the p^2 ground configuration of *O* III are, respectively, $^3P_{0,1,2}$, $^1D_2$, and $^1S_0$. The ground term is ^3P. Transitions from $^1D_2$ to $^3P_2$ and $^3P_1$ give the green nebular lines at λ 5007 and λ 4959, respectively. The transition from the highest level, $^1S_0$, to the $^1D_2$ level gives the λ 4363 auroral-type line. It was suggested by Bowen[2] that collisions of electrons with the *O* III ions in the ^3P term raised these ions to the metastable $^1S_0$ and $^1D_2$ levels. As the ions returned to the ground levels the nebular lines λλ 4363, 4959, and 5007 were emitted. Now, if the collisional cross-sections of *O* III are known, we may calculate the relative numbers entering the $^1S_0$ and $^1D_2$ levels as a function of the electron density and the temperature. Since the transition probabilities for λλ 4363, 4959, and 5007 are known, we may compute the electron temperature from the measured intensities of the relevant nebular lines.

If we assume that the electron velocities possess a Maxwellian distribution corresponding to some temperature T_ϵ, the total number of transitions from the lower level A to the higher level B will be given by (X, 32)[1]

$$\mathfrak{F}_{AB} = 8.54 \times 10^{-6} \frac{N_A N_\epsilon \Omega_{AB}}{\bar{\omega}_A T_\epsilon^{1/2}} e^{-\chi_{AB}/kT_\epsilon} , \qquad (1)$$

where $\frac{1}{2} m v_0^2 = \chi_{AB}$, the excitation potential in volts of level B with respect to level A, and $\bar{\omega}_A$ is the statistical weight of level A, while Ω_{AB} is a parameter calculated by wave mechanics and tabulated in Paper X for *O* III. Although only electrons with velocities greater than v_0 are effective in exciting atoms from level A to level B, it is apparent that electrons with any velocity whatever may be effective in de-exciting the atoms in the higher levels. Thus, from (X, 33) the number of de-excitations is given by

$$\mathfrak{F}_{BA} = 8.54 \times 10^{-6} \frac{N_B N_\epsilon \Omega_{AB}}{\bar{\omega}_B T_\epsilon^{1/2}} . \qquad (2)$$

[1] "Physical Processes in Gaseous Nebulae. XII," *Ap. J.*, **93**, 195, 1941; X, *Ap. J.*, **92**, 408, 1940.

[2] *Ap. J.*, **81**, 1, 1935.

We set up the equations of statistical equilibrium for the levels of the p^2 configuration by equating the number entering each level by collisional and radiative transitions to the number leaving in a similar fashion. The problem is greatly simplified if we are permitted to suppose that the population of each of the levels of the ^3P term is proportional to its statistical weight. This will be the case, provided the collisions are sufficiently numerous to maintain a Boltzmann distribution, i.e., the probability of the collisional de-excitation of a level of the ^3P term is much greater than the probability of a radiative transition downward. As was shown in (X, 39), if the electron density is greater than about 4×10^3 electrons/cm³ a Boltzmann distribution will obtain among the levels of the ^3P term.

Before setting up the general equations of statistical equilibrium we shall sketch a useful approximate method of treating the same problem. The number of atoms entering the $^1S_0$ level by collisions from the ground level will be

$$\mathfrak{F}_{PS} = 8.54 \times 10^{-6} N_{^3P_0} N_\epsilon \frac{\sum_i \Omega_{P_i S}}{T_\epsilon^{1/2}} e^{-\chi_{PS}/kT_\epsilon} \qquad \sum_i \Omega_{P_i S} = 3.53 \,. \quad (3)$$

By the use of the number of atoms in the $^3P_0$ level of unit weight we have avoided the necessity of introducing the weight factors in the other terms, since $N_{^3P_0} = N_{^3P_1}/3 = N_{^3P_2}/5$. Of the number arriving in $^1S_0$, the fraction

$$\frac{A_{SD}}{\sum_J A_{SJ}} ,$$

when J refers to the inner quantum numbers of both ^1D and ^2P, goes to form the line λ 4363.[3] According to the data of Pasternack,[3] $A_{SD} = 2.7$, $A_{SJ} = 2.84$, so that if there were no collisional de-excitation of the $^1S_0$ level, 95 per cent of the atoms in $^1S_0$ would go to produce the λ 4363 line. Similarly, the number of atoms arriving in $^1D_2$ by collision from ^3P will be

$$\mathfrak{F}_{PD} = 8.54 \times 10^{-6} N_{^3P_0} N_\epsilon \frac{\sum \Omega_{P_i D}}{T_\epsilon^{1/2}} e^{-\chi_{PD}/kT_\epsilon} \qquad \sum_i \Omega_{P_i D} = 20 \,. \quad (4)$$

We want to know what fraction of these will return to the ground state by collisions. According to Pasternack, $A_{^1D_2-^3P_2} + A_{^1D_2-^3P_1} = 0.0215$. The number of superelastic collisions will be

$$\mathfrak{F}_{DP} = 8.54 \times 10^{-6} \frac{N_{^1D_2} N_\epsilon}{5} \frac{20}{T_\epsilon^{1/2}} = 34.2 \times 10^{-6} \frac{N_\epsilon}{T_\epsilon^{1/2}} N_{^1D} \,.$$

If we take T_ϵ as 10,000° K, N_ϵ as 10^4 electrons/cm³, we obtain $\mathfrak{F}_{DP} = 3.42\times 10^{-3} N_D$, and the fraction of atoms collisionally de-excited will be only 0.14. Similarly, we find that an entirely negligible fraction of atoms in $^1S_0$ are collisionally de-excited at nebular densities. Thus, the intensity of λ 4363 will be proportional to the number of atoms entering the $^1S_0$ level multiplied by 0.95. Likewise, the intensity of the green nebular pair will be proportional to the number entering $^1D_2$ by collision from ^3P *plus* the number cascading from ^1S multiplied by $(1 - \beta)$, where β is the fraction of atoms in the $^1D_2$ level col-

[3] *Pub. A.S.P.*, **51**, 160, 1939; see also *Ap. J.*, **92**, 129, 1940.

lisionally de-excited. Therefore, the ratio of the intensity of the green nebular lines to λ 4363 will be given by

$$\left.\begin{aligned}\frac{I_{N_1+N_2}}{I_{4363}} &= \frac{[1-\beta]\{20e^{-\chi_{DP}/kT_\epsilon}+e^{-\chi_{PS}/kT_\epsilon}\,3.53\times 0.95\}h\nu_{5000}}{e^{-\chi_{PS}/kT_\epsilon}\,3.53\times 0.95\times h\nu_{4363}}\\ &= [1-\beta]\{5.97e^{\chi_{DS}/kT_\epsilon}+1\}0.875\,,\end{aligned}\right\}\tag{5}$$

where

$$\chi_{DS} = h\nu_{DS} = h\nu_{4363}\,,$$

$$\chi_{PD} = h\nu_{DP} = h\nu_{5000}\,,$$

$$\chi_{PS} = h\nu_{PS}\,.$$

If $\beta = 0.14$

$$\frac{I_{N_1+N_2}}{I_{4363}} = 4.4e^{33,000/T_\epsilon} + 0.75\,.\tag{6}$$

In this equation we have neglected the effects of a number of processes that are not of great importance in the present case because of the particular values which the parameters assume for likely values of the electron density and temperature.

We shall now set up the general equations of statistical equilibrium for the terms of the p^2 configuration of *O* III. Ambarzumian has treated the problem of statistical equilibrium with four discrete states. We shall assume that the electron density is sufficiently high for the relative populations of the levels of the ^3P term to be proportional to their statistical weights.

As previously mentioned, we equate the number of atoms entering the $^1S_0$ level by collisions to the number leaving by collisions and radiative transitions,

$$\left.\begin{aligned}N_P N_\epsilon \frac{\Omega_{PS}}{\bar{\omega}_P} e^{-\chi_S/kT_\epsilon} C + N_D N_\epsilon \frac{\Omega_{DS}}{\bar{\omega}_D} e^{-(\chi_S-\chi_D)/kT_\epsilon} C\\ = N_\epsilon N_S \frac{\Omega_{SP}\Omega_{SD}}{\bar{\omega}_S} C + N_S(A_{SD}+A_{SP})\,.\end{aligned}\right\}\tag{7}$$

The number of atoms entering the $^1D_2$ term by collisions from the ^3P and ^1S terms and by radiative transitions from the ^1S term must equal the number collisionally de-excited to ^3P or collisionally excited to $^1S_0$ plus the number of radiative transitions from $^1D_2$ to ^3P.

$$\left.\begin{aligned}N_P N_\epsilon \frac{\Omega_{PD}}{\bar{\omega}_P} e^{-\chi_D/kT_\epsilon} C + N_S N_\epsilon \frac{\Omega_{SD}}{\bar{\omega}_S} C + N_S A_{SD}\\ = N_D N_\epsilon \frac{\Omega_{DS}}{\bar{\omega}_D} e^{-(\chi_S-\chi_D)/kT_\epsilon} C + N_D N_\epsilon \frac{\Omega_{DP}}{\bar{\omega}_D} C + N_D A_{DP}\,,\end{aligned}\right\}\tag{8}$$

where

$$C = \frac{8.54\times 10^{-6}}{T_\epsilon^{1/2}}\,,$$

$$\bar{\omega}_S = 2J_S + 1 = 1\,,$$

$$\bar{\omega}_D = 2J_D + 1 = 5\,.$$

If we eliminate N_P between these equations, we obtain for the relative populations of the S and D levels

$$\frac{N_S}{N_D} = \frac{\frac{\bar{\omega}_S}{\bar{\omega}_D} e^{-(\chi_S - \chi_D)/kT_\epsilon} \left\{ 1 + \frac{\Omega_{DS} e^{-(\chi_S-\chi_D)/kT_\epsilon}}{\Omega_{DP}\left(1 + \frac{\Omega_{SD}}{\Omega_{SP}}\right)} + \frac{A_{DP}\bar{\omega}_D}{\Omega_{PD} N_\epsilon \left(1 + \frac{\Omega_{SD}}{\Omega_{SP}}\right) C} \right\}}{1 + \left(\frac{A_{SD} + A_{SP}}{\Omega_{SP}}\right) \frac{\bar{\omega}_S}{\left(1 + \frac{\Omega_{SD}}{\Omega_{SP}}\right) N_\epsilon C} + \frac{\Omega_{SD} e^{-(\chi_S-\chi_D)/kT_\epsilon}}{\Omega_{PD}\left(1 + \frac{\Omega_{SD}}{\Omega_{SP}}\right)} + \frac{A_{SD}\bar{\omega}_S e^{-(\chi_S-\chi_D)/kT}}{A_{PD}\left(1 + \frac{\Omega_{SD}}{\Omega_{SP}}\right) N_\epsilon C}} . \quad (9)$$

When numerical values of the relevant parameters are substituted, we obtain

$$\frac{N_S}{N_D} = \frac{\frac{1}{5} e^{-Y}\left[1 + 0.0765 e^{-Y} + \frac{A_{DP}}{N_\epsilon} T_\epsilon^{1/2} 1.65 \times 10^4\right]}{1 + \left[0.0765 + T_\epsilon^{1/2} \frac{8.93 \times 10^3}{N_\epsilon}\right] e^{-Y} + \frac{A_{SD} + A_{SP}}{N_\epsilon} T^{1/2} 1.88 \times 10^4} , \quad (10)$$

where

$$Y = \frac{\chi_S - \chi_D}{kT_\epsilon} .$$

To evaluate the quantities within the parentheses, we make use of formula (6) to estimate T_ϵ, and we use the electron densities determined in XI. For a good approximation it is sufficient to take $T_\epsilon = 9000°$ and $N_\epsilon \sim 10^4$. Then $e^Y \sim 39$. The values of the A's are taken from the calculations of Pasternack and the collisional parameters from X. Then

$$\frac{N_S}{N_D} = 0.2 e^{-Y} \frac{4.4}{508} ,$$

and the relative intensities of λ 4363 and the green nebular lines will be given by

$$\frac{N_D A_{DP} h\nu_{DP}}{N_S A_{SD} h\nu_{SD}} = 4.07 e^{33,000/T_\epsilon} , \quad (11)$$

which is in good agreement with our formula (6) derived from elementary considerations.

As a check on equation (9) we note that as N_ϵ becomes large, equation (9) approaches the form of the Boltzmann equation. This result is to be expected, since whenever the higher levels are excited and de-excited, principally by collisions, a Boltzmann distribution of the atoms among the various levels obtains. When N_ϵ is low, we see that the Einstein A's play a very significant role.

An estimate of the electron temperature of any given planetary may be obtained from the use of equation (6). This estimated value of T_ϵ may then be utilized in the calculation of the denominator of equation (10) and the terms inclosed in brackets in the numerator. We ultimately obtain an equation similar to equation (11) but with a different numerical coefficient.

The intensities of all of the stronger lines in a number of the brighter and more interesting planetaries have been determined by one of us from plates taken with the quartz slitless spectrograph attached to the Crossley Reflector of the Lick Observatory. In a number of objects the sum of the intensities of the green nebular lines has been determined, and the intensity of the λ 4363 line has been measured whenever possible upon the slitless spectrograms. When λ 4363 was too faint to be measured we have used the intensity estimates of Wright[1] reduced to our system of intensities. In a number of the planetaries studied the λ 4363 line was blended with $H\gamma$, and it was necessary to disentangle the effects of the overlapping images. A wave-length and intensity calibration of Wright's intensity scale was effected from the relative intensities of lines in objects common to our two lists. In this way approximate intensities for λλ 5007, 4959, and 4363 for nineteen objects observed by Wright but not by any of us were obtained, and electron temperatures were estimated by the application of equation (6).

TABLE 1

ELECTRON TEMPERATURES OF PLANETARY NEBULAE

Object	$N_1 + N_2$	4363	Ratio	$T_\epsilon(6)$	$T_\epsilon(11)$
NGC 6543	81	(0.08)	1000	6,100	6,000
6826	88	(0.22)	400	7,300	7,200
6572	129	0.90	143	9,500	9,200
7009	130	1.03	126	9,900	9,500
7027	187	1.46	128	9,800	9,500
7662	211	2.14	98.5	10,600	10,300
IC 418	16.9	(0.034)	497	7,000	6,800
NGC 1535	109	(1.0)	109	10,300	10,000
J 320	151	1.48	102	10,500	10,100
NGC 6210	126	(0.3)	420	7,300	7,100
IC 4997	72.3	5.3	13.6	31,800	27,000
IC 351	139	0.58	240	8,300	8,100
NGC 2440	236	0.83	284	7,900	7,700
2165	145	1.35	107	10,400	10,000
2392	130	2.65	49	13,900	13,200
3242	154	0.48	317	7,700	7,500
IC 4634	140	0.22	637	6,600	6,500
4776	57	0.19	300	7,800	7,600
NGC 6741	322	0.52	620	6,700	6,500
IC 4846	102	1.85	55.3	13,300	12,600
NGC 6790	263	1.24	212	8,500	8,300
6803	108	0.34	317	7,700	7,500
6807	106	0.50	212	8,500	8,300
6818	202	0.46	440	7,200	7,000
6833	86	0.55	157	9,300	9,000
6879	88	0.25	350	7,600	7,400
6884	204	0.57	355	7,500	7,400
6886	230	1.02	225	8,400	8,200
IC 5117	214	0.95	226	8,400	8,200
5217	106	0.48	218	8,500	8,200

The intensities of the nebular green lines in the upper part of the table are taken from the results of a photometric study of these planetaries. The intensities of λ 4363, when not inclosed in parentheses, are also obtained directly from the photometric observations. The values in parentheses are obtained from Wright's (*Lick Obs. Pub.*, **13**, 193, 1918) estimates reduced to our system of intensities.

The intensities of all the lines in the lower part of the table are taken from Wright's measures reduced to our system. They are uncertain.

The calculation of the electron temperatures is summarized in Table 1. The first column gives the NGC or IC number of the object. The second and third columns give the intensities of the green nebular pair and λ 4363. The column headed R gives the ratio of intensities, and that headed T_ϵ, (6) the electron temperatures computed by equation (6). The column headed T_ϵ (11) gives the electron temperature as computed by the application of equation (11). The order of magnitude is unchanged if the temperature is computed by equation (10) with improved values taken from the results of the application of equation (6). At 6000° an error of 400 per cent in the ratio of the intensity of λ 4363 to that of the green nebular lines makes an error of about 30 per cent in T_ϵ. At the higher temperatures, the line λ 4363 is stronger relative to the green nebular lines, and the ratio may be measured fairly accurately. The tabulated temperatures are probably good to within 10 or 20 per cent in even the worst cases.

The electron temperatures range from about 6000° for NGC 6543 to about 10,000° for NGC 7027. NGC 2392, observed by Wright, gives a temperature of about 14,000°, but this may easily arise from errors in the intensities of the relevant nebular lines. The case of the stellar planetary IC 4997 deserves special attention. With an assumed electron density of 10^4 electrons/cm^3, we obtain an electron temperature of about 27,000°. If we assume an electron density of 10^5 electrons/cm^3, the temperature is depressed to about 19,000°. If we assume an electron density of 10^6 electrons/cm^3, an electron temperature of 7500° is obtained. It would seem reasonable to suppose that the electron density in IC 4997 is relatively high, very likely between 10^5 and 10^6 electrons/cm.^{3} Widened spectra show a strong hydrogen continuum at the head of the Balmer series. Unfortunately, the angular size of this object is very small, and therefore we cannot obtain a good estimate of the electron density by use of the method described in XI.

The electron temperature seems to be relatively insensitive to the degree of excitation of the nebular lines or to the nuclear temperatures found by Zanstra. A glance at the table will show that the electron temperature of highly excited objects such as NGC 7027 is not much greater than that of objects such as NGC 6572. This discrepancy between the observed temperatures and those which one would be led to expect on the basis of the calculations for a purely hydrogenic nebula, as outlined in V, seems to arise from loss of energy by the electrons in the collisional excitation of the metastable levels of ions of *O* III, *O* II, *Ne* III, etc. The cooling effect of the presence of oxygen and other impurities upon the electrons photoelectrically ejected from hydrogen will be considered in a subsequent paper. It will there be shown that the density of *O* III in the planetaries is of the right amount to depress the electron temperatures to the observed values.

Miyamoto has attempted the calculation of the electron temperature by the same method as outlined here, but his collisional cross-sections were calculated upon the basis of plane wave functions for the impinging electron. As shown in X, it is necessary to take into account the distortion of the wave front by the field of the ion. His values of the electron temperature differ from ours and are somewhat higher. Page[4] attempted the determination of the electron temperature from the distribution of energy in the continuum beyond the limit of the Balmer series and obtained very low values. This method of the determination of the electron temperature is subject to very great systematic errors on account of photometric difficulties and space reddening. By the same method, one of us obtained an electron temperature for NGC 7662 much in excess of 10,000°.

The next paper of this series will be devoted to a discussion of the observational data upon which these and other theoretical calculations rest.

Harvard College Observatory
June 18, 1940

[4] *M.N.*, **96**, 604, 1936.

Reprinted from *The Astrophysical Journal,* Vol. 93, No. 2, March, 1941.

PHYSICAL PROCESSES IN GASEOUS NEBULAE

XIV. SPECTROPHOTOMETRY OF SOME TYPICAL PLANETARY NEBULAE

LAWRENCE H. ALLER

ABSTRACT

Total intensities of emission lines in the spectra of a number of typical planetaries have been determined from spectrograms made with the quartz slitless spectrograph attached to the Crossley reflector of the Lick Observatory. The plates have been measured with the microdensitometer of the Harvard Observatory. Intensities of all the strong lines between λ 5007 and λ 3133 are tabulated. Intensities of lines in the visual region of NGC 7027, 7662, and 7009 have also been measured.

The present paper deals with the results of a study undertaken to supply observational data for the theory, now being developed at the Harvard Observatory, of the gaseous nebulae. The objective is to establish a system of total intensities[1] for the stronger lines in the spectra of a number of the brighter and more interesting planetaries.

Eleven planetaries were observed with the quartz slitless spectrograph attached to the Crossley reflector of the Lick Observatory during the summers of 1938 and 1939. Of the 150 plates taken, the best 53 have been reduced. Particular attention was paid to the high-excitation objects NGC 7009, 7027, and 7662. For a detailed study the advantageous location of NGC 7662 in declination and galactic latitude minimizes atmospheric extinction and space reddening. Its angular dimensions—$32'' \times 28''$—enable us to investigate the distribution of the various radiations. The irregular nebula NGC 7027 contains many strong lines and is also favorably placed for observation. NGC 7009 was included because of its brightness and interesting structural details.

Observations in the photographic region were secured for the objects of intermediate excitation—NGC 6210, 6543, 6572, and 6826—and for the objects of low excitation—BD $+ 30^\circ$ 3639, NGC 40, and IC 418. The stellar nebula IC 4997 was studied because of the abnormal intensity of λ 4363. Plates taken by R. H. Stoy, of NGC 1535, Jonckheere 320, NGC 2022, and $18^h13^m + 10.^\circ6$ were kindly made available to me by Director Wright of the Lick Observatory.

Eastman 33's were generally used for the photographic region, although for the ultraviolet lines and faint images faster emulsions were employed. I used Eastman Super Pancro Press plates for the visual regions of NGC 7009, 7027, and 7662. Photometric standards for various wave-length regions were impressed on the plates by means of a tube sensitometer. For the visual region Wratten No. 26 and No. 40 filters were used; for the photographic region a Wratten No. 50 filter was employed, and for the ultraviolet calibrations I found a Wratten 18A filter satisfactory. Early in the series, standards were impressed upon a plate taken from the same box as the one exposed on the planetary, and the two were developed together. Later, standards were placed on all plates of the nebulae and the comparison star. The plates were developed in a tank under constant agitation for seven minutes at 67° in D11.

The B3 comparison star π_1 Cygni (HD 206672), magnitude 4.78, used by Popper[2] in his study of Nova Lacertae, was likewise used here. Stebbins[3] finds for the former a

[1] Ideally, we should give total intensities and isophotic contours for each image in order to reduce our values to the rate of energy emission per unit volume. The unpublished material permits this to be done in some cases.

[2] "The Spectrum of Nova Lacertae," *Ap. J.*, **92**, 262, 1940.

[3] Unpublished.

color excess of 0.03 mag. from λ 4250 to λ 4750, but I have made no attempt to correct for space reddening. The spectral energy distribution was suggested by inference from the work of Barbier[4] and his colleagues. From λ 6600 to λ 3700, a Planckian energy distribution corresponding to 17,000° K is assumed. Because of the overlapping hydrogen lines, a discontinuity at λ 3700 amounting to a factor of 1.55 has been adopted. Beyond this point the energy-curve is supposed to correspond to a temperature of 20,000° K. Thus, on the redward side of λ 3700 the intensity of the continuum is taken as $A\,f(17{,}000^\circ\ \mathrm{K})$ and on the violet side as $A/1.55\,f(20{,}000^\circ\ \mathrm{K})$.

On each exposure on the comparison star the image was run back and forth twenty times by the slow-motion motor to widen the spectrum. Between successive exposures the brightness of the star was diminished by 1 mag. with the end of a diaphragm with circular holes placed over the end of the telescope. We readily see that it is then possible to obtain the plate-*plus*-instrument sensitivity and also to check the photometric calibration-curves derived from laboratory data.

To correct for atmospheric extinction we make use of the expression

$$\frac{I_\lambda}{I_{0\lambda}} = 10^{-f_\lambda \sec z} ,$$

where I_λ is the intensity at zenith distance z, $I_{0\lambda}$ is the intensity outside the earth's atmosphere, and f_λ is the atmospheric extinction factor given by

$$f_\lambda = a_\lambda + b ,$$

where a_λ depends on the wave length and b is a constant. The values of these quantities were taken from the unpublished work of Popper.

To calculate the relative amounts of energy falling upon the plate per millimeter along the dispersion, I assumed the aforementioned spectral energy distribution of the comparison star and computed the dispersion by Hartmann's formula from measures of the Balmer lines. By comparing these values with the observed intensity distribution corrected for atmospheric extinction, we may find the sensitivity of the plate-*plus*-instrument combination at once.

All spectrograms were analyzed with the new microdensitometer of the Harvard Observatory.[5] The procedure adopted for all except a few of the widened spectra of some of the smaller planetaries was to run the plates parallel to the dispersion and to set the beam over 0.05 mm or 0.10 mm (depending on the size of the planetary) between successive runs. The tracing of each plate consisted of from four to eight separate runs along the spectrum. Plates run in this manner, with a magnification of 25, resulted in about 2000 feet of tracings. The comparison-star spectra were analyzed with the same arrangements of slits, magnifications, and sensitivity adjustments.

The D–log I-curves for the different wave lengths were joined, on the assumption that the amount of light required to produce a density of 0.6 was the same for all colors. As both the comparison star and the nebula were reduced in a similar manner, I believe the errors tend to cancel out. A considerable change of contrast was noted between the visual and the photographic regions of the spectrum. The check provided by the diaphragmed exposures of the comparison star yielded reassuring results for the photographic region but did not turn out well for the visual region. In any case the intensities of the lines in the visual region are less accurate than those determined for the photographic region.

[4] *J. d. obs.*, **19**, 149, 1936, and *C.R.*, **210**, 99, 1940.

[5] Dimitroff and Menzel, *Harvard Ann.*, **105**, 99, 1937.

The total intensities were found in an obvious manner by the addition of the image sections supplied by successive tracings. Also, as was pointed out by Page,[6] we must subtract from each image the contribution of the continuous background, whose intensity may be inferred from the stronger exposures. Errors in the estimation of the continuous background are more serious for the weaker lines.

Vignetting, usually dependent on wave length in the slitless spectrograph, requires that the nebula and comparison star be photographed in the same part of the field. Variations of atmospheric extinction, especially in the far ultraviolet, and uncertainties in the energy-curve of the comparison star are also causes of concern. Likewise, nonuniform sky fog is hard to correct for. Because of the rapid variations in the sensitivity of the Eastman Super Pancro Press plates with wave length, especially near $H\alpha$, numerous uncertainties are introduced into the intensities of the visual lines. Finally, since the planetaries are Milky Way objects, they are affected by unknown amounts of selectively absorbing material in the line of sight.

The objects observed and plates reduced are listed in Table 1. Successive columns give the NGC or IC number, the α and δ for 1900, the plate number, the observation date, the exposure time in minutes, the spectral region, and the zenith distance. In Table 2 are listed the line intensities observed in the five planetaries NGC 2022, 7009, 7027, 7662, and 6572. Table 3 gives the total intensities of the lines observed in the other objects. Blends are inclosed in brackets, and the procedure here has been to measure and tabulate the sum of the intensities of the blended lines, unless the images were far enough apart for reliable estimates of the separate contributions to be made. For term designations and estimates of the relative intensities of the lines contributing to a blend, reference to the paper of Bowen and Wyse[7] is recommended. Ozone absorption introduces considerable uncertainty in the ultraviolet. The intensities of the lines beyond λ 3133 in the spectrum of NGC 7662 were obtained from only one plate and are to be regarded as very approximate.

A comparison of these results with those obtained by Plaskett,[8] by Page,[6] and by Berman[9] has been made for the objects common to our lists. The intensities found by Plaskett and Page from slit spectra are not immediately comparable. In some cases we might expect a better agreement, particularly in the visual region where the intensities tabulated here seem too low. It is more appropriate to compare our results with those of Berman, after his measures have been corrected for the error in dispersion pointed out by Page. Five objects—NGC 7027, 7009, 6572, 6826, and 6543—were common to the two lists. Usually, I find greater intensities for the green nebular lines than Berman found. Since the plate sensitivity is low in this region, slight errors in its determination are capable of producing relatively large errors in the measured intensities. The fainter lines seem to be measured systematically stronger by Berman than by the writer. This effect very likely arises from the method used in allowing for the general continuum. For intermediate intensities, ranging from 1 to 10 on the scale $H\beta = 10$, the two systems show no systematic differences depending on intensity or wave length, and the scatter of individual points amounts to less than 0.1 in log I. For the weaker lines, even when the systematic difference is allowed for, the scatter is much greater because of random errors.

Since both Berman and I used the same telescope and spectrograph, the differences between our results show the size of the errors of observation. Although the same general method of photometry was used in both cases, the reductions were done by quite independent methods.

The eye-estimates of Bowen and Wyse for the fainter lines suggest intensities greater than those found here. Their intensity scale was calibrated with respect to the published

[6] *M.N.*, **96**, 604, 1936.

[7] *Lick Obs. Bull.*, **19**, 1, 1939.

[8] *Pub. Dom. Ap. Obs.*, **4**, 187, 1928.

[9] *Lick Obs. Bull.*, **15**, 86, 1930.

TABLE 1

LIST OF OBSERVATIONS

Object	α	δ	Plate	Date	Exp. (Min.)	Region	Z
NGC 6210	$16^h40^m.3$	23° 59′	171	1938 Sept. 15	12	Pg.	51°
6543	17 58.6	66 38	8	July 8	15	Pg.	36
			12	July 8	60	Pg.	45
6572	18 7.2	6 50	48	Aug. 5	30	Pg.	37
			49	Aug. 5	10	Pg.	41
			82	Aug. 15	5	Pg.	36
			90	Aug. 15	30	Pg.	51
			91	Aug. 15	10	Pg.	55
BD 30°3639	19 30.8	30 18	114	Sept. 6	15	Pg.	14
			122	Sept. 6	8	Pg.	44
			143	Sept. 13	12	Pg.	11
			149	Sept. 13	11	Pg.	39
NGC 6826	19 42.1	50 17	107	Sept. 5	20	Pg.	30
			108	Sept. 5	40	Pg.	35
IC 4997	20 15.6	16 25	57	Aug. 6	25	Pg.	23
			59	Aug. 6	25	Pg.	21
			61	Aug. 6	12	Pg.	23
			92	Aug. 15	17	Pg.	44
			94	Aug. 15	17	Pg.	47
			98	Aug. 15	13	Pg.	58
			99	Aug. 15	38	Pg.	64
NGC 7009	20 58.7	−11 46	65	Aug. 6	87	Pg.	59
			204	1939 Sept. 6	61	Vis.	54
			206	Sept. 6	40	Vis.	51
			208	Sept. 6	30	Pg.	57
			209	Sept. 6	20	Pg.	60
			228	Sept. 8	152	U.V.	49
			231	Oct. 5	60	U.V.	55
NGC 7027	21 3.3	41 50	16	1938 July 9	10	Pg.	34
			17	July 9	10	Pg.	32
			18	July 9	15	Pg.	27
			21	July 9	20	Pg.	16
			156	Sept. 13	25	Pg.	13
			157	Sept. 13	20	Pg.	9
			220	1939 Sept. 7	60	Vis.	7
			221	Sept. 7	20	Vis.	14
			244	Oct. 8	85	U.V.	34
NGC 7662	23 21.1	41 59	23	1938 July 9	10	Pg.	24
			24	July 9	18	Pg.	21
			26	July 9	10	Pg.	14
			50	Aug. 5	20	Pg.	13

TABLE 1—*Continued*

Object	α	δ	Plate	Date	Exp. (Min.)	Region	Z
				1938			
			51	Aug. 5	80	Pg.	5
			110	Sept. 5	150	U.V.	30
				1939			
NGC 7662	$23^h21^m.1$	41° 59′	194	Aug. 25	138	U.V.	34
			225	Sept. 7	130	Vis.	28
			237	Oct. 5	57	U.V.	45
			249	Oct. 8	177	U.V.	49
				1938			
NGC 40	0 7.6	71 32	125	Sept. 6	73	Pg.	39
			137	Sept. 7	11	Pg.	69
			138	Sept. 7	31	Pg.	65
IC 418	5 22.8	−12 46	140	Sept. 7	15	Pg.	58
			141	Sept. 7	5	Pg.	56
			168	Sept. 14	17	Pg.	60
				1934			
NGC 1535*	4 9.6	−13 0	S167	Nov. 2	70	Pg.	52
			S177	Nov. 11	120	Pg.	53
J 320*	5 0.0	10 36	....	Dec. 14	120	Pg.	66
NGC 2022*	5 36.6	9 2	....	....	....	Pg.	32
1813 10°.6*	....	....	S133	June 16	105	Pg.	47

* The asterisks refer to plates taken by R. H. Stoy in 1934.

intensities of Plaskett and Berman.[10] The intensities published by Wright[11] have been reduced to the present system by the following procedure: First, the wave-length variation was found from a comparison of the intensities of the stronger lines. Next, the lines in Wright's tables were corrected for the wave-length effect and plotted against the present measures. To effect this calibration the planetaries NGC 6543, 6572, 6826, 7009, 7027, 7662, and IC 418, common to our lists, were employed. The scatter of individual points was great, and the weaker lines were relatively stronger as measured by Wright. Since his work was done before the present techniques in photometry were available, it is not surprising that some discrepancies exist.

Objects observed by Stoy are indicated by an asterisk. Photometric standards were impressed on his plates in only the blue region, and different development was used for the individual plates of his series. I considered it impracticable to use his comparison stars in the reduction of his plates, since a different star was photographed on each plate and it would consequently have been hard to place the intensities on a uniform scale. Also, as the comparison star has been photographed on the same field as the nebula, pre-exposure effects may exist. Therefore, mean plate-*plus*-instrument sensitivity-curves for the types of plates used by Stoy have been employed. We may, therefore, expect larger errors in the line intensities derived from his plates.

[10] Apparently Bowen and Wyse have used the uncorrected Berman values for NGC 7027, so that their intensities for this object have a systematic error depending on wave length in the blue and violet spectral regions.

[11] *Lick Obs. Pub.*, **13**, 191, 1918.

TABLE 2

Line	Element	NGC 2022*	NGC 6572	NGC 7009	NGC 7027	NGC 7662
6583	*N* II					
6563	*H* I	....	....	14.5(?)	45(?)	17.5(?)
6548	*N* II					
6364	*O* I					
6312	*S* III	....	....	....	5.7	....
6300	*O* I					
5876	*He* I	....	....	1.35	2.4	1.1
5755	*N* II				1.0	
5412	*He* II				0.4	0.4
5007	*O* III	74	94	96	138	156
4959	*O* III	27	35	34	49	56
4861	*H* I	10.00	10.0	10.0	10.0	10.0
4740	*A* IV	....	0.12	0.44	1.02	0.67
4712	*A* IV		0.13			
4686	*He* II	19.0	0.081	1.35	3.89	5.26
4659	C IV					
4650	C III	....	0.16	0.38	0.58	0.32
4641	*N* III					
4634	*N* III					
4571	Mg I?	....	0.044	....	0.06	....
4542	*He* II	....	....	....	0.17	0.165
4471	*He* I	....	0.58	0.43	0.27	0.25
4363	*O* III	8.7	0.90	1.03	1.46	2.14
4340	*H* I		3.60	3.42	2.71	4.29
4267	C II	....	0.05	0.13	0.04	....
4200	*He* II	....	....	....	0.03	0.08
4144	*He* I	....	0.034	....	0.024	....
4103	*N* III					
4101	*H* I	4.4	2.11	2.49	1.30	2.63
4097	*N* III					
4076	*S* II					
4069	*S* II	....	0.24	....	0.48	....
4026	*He* I	....	0.20	0.14	0.10	....
3970	*H* I					
3967	*Ne* III	4.5	3.50	3.95	2.16	4.12
3924	*He* II	....	....	....	0.02	....
3889	*He* I, *H* I	9.8	1.43	9.85	0.80	11.4
3869	*Ne* III		5.85		4.43	
3835	*H* I	....	0.60	....	0.31	....
3798	*H* I	....	0.34	....	0.22	....
3771	*H* I	....	0.23	....	0.12	....
3760	*O* III	....	0.18	....	0.15	....
3750	*H* I	....	....	....	0.12	....
3728.6	*O* II					
3726.1	*O* II	....	3.53	1.66	1.19	2.4
3712	*H* I	....	....	....	0.05	....
3444	*O* III	11.3	....	0.47	0.72	2.0
3425	*Ne* V			0.11	2.98	1.2
3346	*Ne* V					
3343	*Ne* III	4.4	....	0.36	1.36	1.57
3340	*O* III					
3312	*O* III	....	....	0.35	0.33	1.05
3299	*O* III	....	....	....	0.07	0.26
3203	*He* II	4.6	....	0.31	0.46	1.89
3178	*He* I	....	....	0.21	....	....
3133	*O* III	7.9	....	1.23	1.35	7.10
3121	*O* III	....	....	....	....	0.66
3047	*O* III	....	....	....	....	2.08
3023	*O* III					very weak

As the theoretical discussion is taken up elsewhere in this series, no attempt at an ordered arrangement of the planetaries will be made. An empirical classification on the basis of their spectra was made by Wright, and there seems to be nothing further on that subject to add here. However, although he described most of their characteristics, a few remarks concerning certain of these planetaries may not be out of place.

NGC 1535 consists of an intense inner ring surrounded by a faint outer one. λ 4686 is represented only in the inner ring. λ 3727 has a fairly large but faint image. The lines of the Bowen fluorescent mechanism are present but very weak.

TABLE 3

Line	NGC 40	IC 418	NGC 1535	J 320	NGC 6210	NGC 6543	18^h13^m 10.6	BD $30^\circ3639$	NGC 6826	IC 4997
5007	......	13.9	80.0	120.	92	60	......	......	80	54.3
4959	......	5.0	28.6	31.0	34	21	......	......	27	18.9
4861	10.0	10.0	10.0	10.0	10.0	10.0	10.0	10.0	10.0	10.0
4686	......	......	1.3	......	......	......	......	......	0.56	......
4471	......	0.4	0.21	......	0.4	0.36	......	......	0.57	0.92
4363 }	......	...... }	[5.5]	1.5 }	[4.0]	[4.13]	......	...... }	[4.0]	{ 5.6
4340 }	4.8	4.5 }		5.7 }			4.0	3.6 }		{ 3.5
4101	2.5	2.4	3.1	3.4	1.8	2.14	2.1	2.0	1.9	1.78
4069, 76	......	0.21	......	......	......	......	0.31	......	......	0.32
4026	......	0.11	......	......	......	0.21	......	......	......	0.34
3967	1.3	1.50	4.5	7.9	3.0	2.33	1.2	1.26	2.3	4.50
3889	0.9	1.26 }	[11.8]	2.7	1.1 }	[5.3]	{ 0.9 }	[1.51]	[5.1]	{ 1.53
3868	......	...... }		15	5.6 }		{ ...... }			{ 7.6
3835	......	0.64	......	......	......	......	0.35	0.55	......	0.92
3820	......	......	......	......	......	......	......	......	......	0.29
3798	......	0.34	......	......	......	......	0.34	......	......	0.53
3770 }	......	......	......	......	......	......	0.46	......	......	{ 0.45
3750 }										{ 0.38
3727	13.5	15.7	1.37	2.7	2.00	3.22	10.2	5.4	2.6	2.20

TABLE 4

(NGC 7009)

λ 5007	106	λ 4340	4	λ 3889	2
4959	40	4101	2	3868	11
4861	10	3967	4	3727	18

NGC 2022, a low-density object, seems to be the highest excitation planetary among the objects studied. It is a rather faint ring nebula, characterized by strong λ 4686 and lines of the Bowen fluorescent mechanism.

NGC 7009, a beautiful and interesting planetary, exhibits a spindle shape and a pair of faint extensions terminated by bright ansae. The ultraviolet hydrogen continuous spectrum is strong, and the Bowen fluorescent mechanism lines are not conspicuous.[12] A helium line at λ 3178 gives an image that seems to be diffuse and larger than the neighboring image of the λ 3203 line of *He* II. The mean relative intensities of the lines in the two ansae have been determined from plate 228. The degree of excitation in the two ansae is about the same and is much lower than in the main body of the planetary (cf. Table 2). The λ 4686 line and the lines of the Bowen fluorescent mechanism do not appear in the ansae.

[12] On the strongest plate the spectrum of the central star runs down to about λ 3000, and there is a faint suggestion of yet another line of the Bowen fluorescent mechanism.

The spectral characteristics of NGC 7662 and NGC 7027 will be discussed in some detail in later papers. The data on the continuous spectrum at the Balmer limit have been discussed in a previous paper[13] in connection with the determination of the electron density. In the later papers of this series these observational data will be subjected to further theoretical analysis. We shall discuss, among other topics, the ionization formula for gaseous nebulae, stratification, the Bowen fluorescent mechanism, and abundances.

I am very grateful to Director Wright of the Lick Observatory who make it possible for me to secure the observations for this investigation; and to Dr. A. B. Wyse I am grateful for many helpful suggestions. I am deeply indebted to Dr. Menzel whose interest in this investigation, constant encouragement, and valuable suggestions have made the tedious reductions much easier.

Harvard College Observatory
July 13, 1940

[13] Menzel and Aller, "Physical Processes in Gaseous Nebulae," XII, *Ap. J.*, **93**, 195, 1941.

Reprinted from *The Astrophysical Journal,* Vol. 93, No. 2, March, 1941.

PHYSICAL PROCESSES IN GASEOUS NEBULAE

XV. THE STATISTICAL EQUILIBRIUM OF NEUTRAL HELIUM

LEO GOLDBERG

ABSTRACT

The statistical equilibrium of an assembly of neutral helium atoms has been investigated with reference to the problem of gaseous nebulae. The helium atom was assumed to contain an infinite number of energy levels, but a single equation of equilibrium was written down for all triplet levels above 2 $^3P^o$ and, similarly, for all singlets above 2 $^1P^o$. Mean values are thus obtained for the populations of the high singlets and the high triplets. The population is given in terms of a quantity b_n, the ratio between the true population and the value for thermodynamic equilibrium at a temperature T_ϵ. The temperature T_ϵ refers to an assumed Maxwellian distribution of the free electrons.

The nebular atoms are assumed to be opaque to the stellar radiation field only in the frequencies corresponding to absorptions from levels 1 ^1S, 2 ^1S, and 2 ^3S. Numerical solutions of the equations are tabulated for stellar temperatures, T_1, of 50,000° and 100,000° and electron temperatures of 5,000°, 10,000°, 20,000°, 40,000°, and 80,000°, and also for the special case $T_1 = T_\epsilon = 20{,}000°$.

The procedure for calculating the distribution of atoms among their various energy levels when a gaseous assembly is not in thermodynamic equilibrium has been indicated by Rosseland.[1] In a nonvariable spectrum the condition of statistical equilibrium must be preserved. In order that a constant number of atoms may be maintained in a given level, the total number of atoms per second entering by radiative or by collisional processes must balance the total number leaving. The condition of statistical equilibrium for a given atom in a particular stage of ionization may be expressed as follows for a given level *n:*

$$\sum_{n''} F_{n''n} + \sum_{n'} F_{n'n} = \sum_{n'} F_{nn'} + \sum_{n''} F_{nn''}\,, \tag{1}$$

where the F's denote the number of transitions per second between the involved levels. The subscript n' refers to levels with energy less than n, whereas n'' refers to higher levels, and the summations are to be taken over all possible levels n' and n'', including the continuous states. The F's are expressible in terms of the populations of the levels and the atomic and physical parameters that govern the assembly. Each atomic level yields an equation of the form of equation (1); and since, if we assume that the atomic and physical parameters are known, there is but one unknown per level—the population—the problem is theoretically soluble. In practice, however, difficulties arise from the fact that each atom has an infinite number of levels. The solution of the infinite determinant has been accomplished for hydrogen by Menzel and Baker,[2] but their solution cannot readily be adapted to more complex atoms. Other investigators[3] have avoided the difficulty by neglecting the higher atomic levels and treating the continuum as a single discrete state. Struve and Wurm, for example, have dealt with a six-state neutral helium atom comprising the five lowest levels and a discrete continuum.

The presence of two low-lying metastable levels makes the statistical equilibrium problem for *He* I of considerable interest. The influence of diluted radiation in over-

[1] *Ap. J.*, **63**, 218, 1926. [2] *Ap. J.*, **86**, 70, 1937.

[3] Cf. Plaskett, *Harvard Circ.*, No. 335, 1928; Carroll, *M.N.*, **90**, 558, 1930; Cillie, *M.N.*, **92**, 820, 1932; **96**, 771, 1936; Struve and Wurm, *Ap. J.*, **88**, 84, 1938.

populating the metastable levels and, indeed, all the triplet levels with respect to the singlets should be very much more pronounced in the nebulae than in the distended stellar atmospheres treated by Struve and Wurm.

In the present investigation we shall attempt to solve the equations of statistical equilibrium for an eight-state helium atom and at the same time allow in part both for the variation of the continuous absorption coefficient with frequency, and for the infinity in the number of discrete levels. We may write equation (1) for each of the five lowest levels: 1 ¹S, 2 ¹S, 2 ¹P°, 2 ³S, 2 ³P°, and for the continuum. The contribution of the higher levels may then be accounted for by the inclusion of two additional equations, one of which specifies the condition that the total number of atoms per second entering all the singlet levels above 2 ¹P° equals the number leaving, and similarly for the triplet levels above 2 ³P°. For the present only radiative processes will be considered.

We consider a source of black-body radiation at temperature T_1 shining on a distant unit volume of nebular material, whose temperature T_ϵ is defined by a Maxwellian distribution of the free electrons. If the dilution factor is W, the intensity of radiation of a given frequency striking the nebular material will be $4\pi I_\nu W$. If we denote the lower of two atomic levels by n' and the upper by n, the total number of absorptions per second will be

$$F_{nn'} = 4\pi \int_0^\infty I_\nu W N_{n'} a_{n'n} \frac{d\nu}{h\nu}, \tag{2}$$

where $N_{n'}$ is the number of atoms in level n', and $a_{n'n}$ is the atomic absorption coefficient. The variation of I_ν/ν over the line width may be assumed to be so small that it may be removed from the integration sign. In terms of the oscillator strength $f_{n'n}$,

$$\int_0^\infty a_{n'n} d\nu = \frac{\pi\epsilon^2}{mc} f_{n'n}. \tag{3}$$

The number of emissions from n to n', including stimulated transitions, will be

$$F_{nn'} = N_n A_{nn'} + 4\pi \int_0^\infty I_\nu W N_n a_{nn'} d\nu, \tag{4}$$

where $A_{nn'}$ is the Einstein transition probability, and

$$a_{nn'} = \frac{\tilde{\omega}_{n'}}{\tilde{\omega}_n} a_{n'n}. \tag{5}$$

Following the procedure of Menzel,[4] we define a parameter b_n, which, for a given level n, is an index to the departure from thermodynamic equilibrium with respect to the ionized portion of the gas,

$$N_n = b_n N_i N_\epsilon \frac{h^3}{(2\pi m k T_\epsilon)^{3/2}} \frac{\tilde{\omega}_n}{4} e^{I_n/kT_\epsilon}. \tag{6}$$

When $b_n = 1$, equation (6) reduces to the ionization equation for thermodynamic equilibrium. The factor 4 in the denominator is the product of the weights of the electron spin and of the ground level of ionized helium.

[4] *Ap. J.*, 85, 330, 1937.

If we substitute equations (3) and (6) in equations (2) and (4) and replace I_ν by the Planck function, we get, for the number of absorptions,

$$F_{n'n} = K\,\frac{N_i N_\epsilon}{T_\epsilon^{3/2}}\,\tilde{\omega}_{n'} f_{n'n} b_{n'} \nu^2\,\frac{W}{e^{h\nu/kT_1}-1}\,e^{I_{n'}/kT_\epsilon}\,, \tag{7}$$

and for the number of emissions,

$$F_{nn'} = K\,\frac{N_i N_\epsilon}{T_\epsilon^{3/2}}\,\tilde{\omega}_{n'} f_{n'n} b_n \nu^2 e^{I_n/kT_\epsilon}\left(1+\frac{W}{e^{h\nu/kT_1}-1}\right), \tag{8}$$

where

$$K = \frac{h^3}{(2\pi m k)^{3/2}}\cdot\frac{2\pi^2\epsilon^2}{mc^3}\,. \tag{9}$$

We number the eight states as follows: (1) 1 ^{1}S, (2) 2 ^{1}S, (3) 2 ^{1}P$^\circ$, (4) high singlets, (5) 2 ^{3}S, (6) 2 ^{3}P$^\circ$, (7) high triplets, and (8) continuum. The equations of statistical equilibrium may now be set down with the aid of equations (7) and (8) and solved for $b_1, b_2, \ldots, b_8$. Since the continuum is assumed to be in thermodynamic equilibrium, we set $b_8 = 1$. The values of b_4 and b_7 are mean values for the high levels. The presence of the Boltzmann factor in equation (6) tends to weight the lower levels in each group, for it is N_n that is actually averaged. The condition of statistical equilibrium leads to a seventh-order determinant that must be equated to zero:

$$\left.\begin{array}{cccccccc}
b_1 & b_2 & b_3 & b_4 & b_5 & b_6 & b_7 & 1 \\
\hline
D_{11} & D_{21} & D_{31} & \ldots & \ldots & \ldots & \ldots & D_{81} \\
D_{12} & D_{22} & \ldots & \ldots & \ldots & \ldots & \ldots & D_{82} \\
D_{13} & \ldots & \ldots & \ldots & \ldots & \ldots & \ldots & D_{83} \\
\ldots & \ldots & \ldots & \ldots & \ldots & \ldots & \ldots & \ldots \\
\ldots & \ldots & \ldots & \ldots & \ldots & \ldots & \ldots & \ldots \\
D_{18} & \ldots & \ldots & \ldots & \ldots & \ldots & \ldots & D_{88}
\end{array}\right\} \tag{10}$$

The coefficients D_{ij} and D_{ji}, where $i < j$, are given, except for transitions involving states 4, 7, and 8, by

$$D_{ij} = \tilde{\omega}_i f_{ij} \nu^2\,\frac{W}{e^{h\nu/kT_1}-1}\,e^{I_i/kT_\epsilon}\,; \tag{11}$$

$$D_{ji} = \tilde{\omega}_i f_{ij} \nu^2 e^{I_j/kT_\epsilon}\left(1+\frac{W}{e^{h\nu/kT_1}-1}\right). \tag{12}$$

When $j = 4$ or 7,

$$D_{i4} = W\tilde{\omega}_i e^{I_i/kT_\epsilon}\sum_{n=3}^{\infty} f_{in}\,\frac{\nu^2}{e^{h\nu/kT_1}-1}\,; \tag{13}$$

$$D_{4i} = \tilde{\omega}_i e^{I_i/kT_\epsilon}\sum_{n=3}^{\infty} f_{in} e^{-h\nu/kT_\epsilon}\left(1+\frac{W}{e^{h\nu/kT_1}-1}\right). \tag{14}$$

The index n refers to the total quantum number. When $j = 8$, the D's involve integrations over the continuous states:[4]

$$D_{i8} = \frac{1}{2R}\,\tilde{\omega}_i e^{I_i/kT_\epsilon}\int_{\nu_i}^{\infty}\frac{\kappa^3 f_{i\kappa}\nu^2}{e^{h\nu/kT_1}-1}\,d\nu\;; \tag{15}$$

$$D_{8i} = \frac{1}{2R}\,\tilde{\omega}_i e^{I_i/kT_\epsilon}\int_{0}^{\infty}\kappa^3 f_{i\kappa}\nu^2 e^{-h\nu/kT_\epsilon}\left(1+\frac{W}{e^{h\nu/kT_1}-1}\right)d\nu\,. \tag{16}$$

In equations (15) and (16) κ is the continuous quantum number, defined by

$$E_\kappa = +\frac{R}{\kappa^2} = \tfrac{1}{2}mv^2\,, \tag{17}$$

where E_κ is the positive energy of a continuous state, in frequency units, and R is Rydberg's constant. When $i = 4$ or 7, equations (15) and (16) must be summed over all values of the total quantum number n from 3 to infinity. The coefficients D_{ii} are given by

$$\left.\begin{aligned} D_{11} &= -\,(D_{12}+D_{13}+\,\ldots\,+D_{18})\,,\\ D_{22} &= -\,(D_{21}+D_{23}+\,\ldots\,+D_{28})\,,\\ &\ldots\ldots\ldots\ldots\ldots\ldots\ldots\ldots\ldots\ldots \end{aligned}\right\} \tag{18}$$

Several simplifications may now be introduced. Because of the close adherence of *He* I to *LS* coupling, intersystem combinations are forbidden. Hence all values of D referring to transitions between singlets and triplets may be set equal to zero. In equations (12), (14), and (16) the right-hand term in the parenthesis represents the contribution of stimulated emissions. This term may obviously be neglected if it is small compared with unity, although it becomes important for small ν and large T_1. For the low-frequency transition $2\,^3P^o - 2\,^3S$, for example,

$$\frac{W}{e^{h\nu/kT_1}-1} \sim 3W,\,7W\,, \tag{19}$$

when $T_1 = 50{,}000^\circ$ and $100{,}000^\circ$, respectively. The small values of W in the gaseous nebulae thus justify the neglect of stimulated emissions, even when W is as large as 0.001. We also neglect all absorptions other than those from $1\,^1S$, $2\,^1S$, and $2\,^3S$.

The evaluation of the D_{ij} and D_{ji} requires a knowledge of the atomic transition probabilities. The exact values for helium are not known, but sufficiently good approximations have been derived by various investigators.[5] For the principal series a general expression derived by Vinti has been employed, with a very slight modification. Vinti made use of exact hydrogenic wave functions, with $Z = 1$, for the $n\,^1P^o$ levels, whereas calculations based on the variation method suggest that $Z = 0.97$ is more ap-

[5] Cf. Hylleraas, *Zs. f. Phys.*, **106**, 395, 1937; Goldberg, *Ap. J.*, **90**, 414, 1939; Vinti, *Phys. Rev.*, **42**, 632, 1932.

propriate.[6] For the 2p − nd and 2s − np series, formulae given by the writer were used. For the 2p − ns series the following general expression was derived:

$$\left.\begin{aligned} f(2p - ns) = {} & \frac{2^{18}\pi^2 \nu m a_0^2}{27h} n^9 Z_1^3 Z_2^5 \frac{(nZ_2 - 2Z_1)^{2n-8}}{(nZ_2 + 2Z_1)^{2n+8}} \\ & \times [n^2(3Z_2^3 - 18Z_1Z_2^2 + 24Z_1^2Z_2 - 8Z_1^3) + 4Z_1^2(3Z_2 + 4Z_1)]^2 . \end{aligned}\right\} \quad (20)$$

The formula (20) was obtained from hydrogenic wave functions with appropriate values of Z_1 and Z_2, the average effective charges:

$$\left.\begin{aligned} Z_1(2\,^3S) &= 1.22\,; & Z_2(n\,^3P^0) &= 1.10\,; \\ Z_1(2\,^1S) &= 1.22\,; & Z_2(n\,^1P^0) &= 0.97\,. \end{aligned}\right\} \quad (21)$$

The following asymptotic formulae for the f-values were found useful in carrying out the summations indicated in equations (13) and (14):

$$\left.\begin{aligned} f(1\,^1S - n\,^1P^0) &\sim [31.8146] n^{-3}\nu^{-2} \\ f(2\,^1S - n\,^1P^0) &\sim [30.3603] n^{-3}\nu^{-2} \\ f(2\,^3S - n\,^3P^0) &\sim [30.2191] n^{-3}\nu^{-2} \\ f(2\,^1P^0 - n\,^1D) &\sim [45.1979] n^{-3}\nu^{-3} \\ f(2\,^3P^0 - n\,^3D) &\sim [45.4570] n^{-3}\nu^{-3} \\ f(2\,^1P^0 - n\,^1S) &\sim [43.9323] n^{-3}\nu^{-3} \\ f(2\,^3P^0 - n\,^3S) &\sim [43.9698] n^{-3}\nu^{-3} \end{aligned}\right\} \quad (22)$$

The summations were performed accurately from $n = 3$ to $n = 10$; beyond that point the summation was replaced by an integral from $n = 11$ to infinity, and the asymptotic formulae (22) were employed. When n is replaced by the continuous quantum number κ, equation (22) applies to the bound-free transitions near the series limit and may be used to evaluate equations (15) and (16). Although the asymptotic formulae hold only close to the series limit, the integrations may be safely performed to infinity, because the exponential term in ν makes the contribution to the integral small when ν is large.

The coefficients D_{87} and D_{84}, which represent the rate of electron capture in the high triplet and singlet levels, may be obtained if we set $i = n'$ in equation (16), perform the indicated integration over the continuous states, and then sum over n' from 3 to infinity. For this purpose the levels of different l that belong to a given n' were assumed to be degenerate, and the hydrogenic formula for $f_{n'\kappa}$[7] was employed. The procedure is justified by the fact that the levels with large values of l are favored in the capture process and helium levels with $l = 1$, or greater, are practically hydrogenic. Since the integral in equation (16) reduces to an E_i function, which cannot be expressed analytically in simple form, the summations were performed numerically with the aid of the tables of E_i functions given by Menzel and Baker.[2]

The determinant, equation (10), was solved for two values of T_1: 50,000° and 100,000°, and for five values of T_ϵ: 5,000°, 10,000°, 20,000°, 40,000°, and 80,000°. In addition, the special case $T_1 = T_\epsilon = 20{,}000°$ was treated, to afford at least a rough comparison be-

[6] Goldberg and Clogston, *Phys. Rev.*, **56**, 696, 1939.

[7] Menzel and Pekeris, *M.N.*, **96**, 77, 1935.

tween the present results and those obtained by Struve and Wurm. The results are summarized in Table 1. Since the stellar radiation is absorbed only by the ground and metastable levels, the populations of these levels depend only upon the dilution factor. The preference of atoms for the metastable levels is clearly shown, as is the preponderance of triplets over singlets. The large values of b for 2 $^3P^o$ are notable. Once an atom

TABLE 1

VALUES OF b FOR LEVELS OF *He* I AS A FUNCTION OF W, T_1, AND T_ϵ

T_ϵ	5,000°	10,000°	20,000°	40,000°	80,000°
			$T_1 = 100,000°$		
b_1W...	2.27×10^{-25}	1.05×10^{-12}	2.92×10^{-6}	6.66×10^{-3}	4.41×10^{-1}
b_2W...	4.25×10^{-6}	9.28×10^{-4}	1.76×10^{-2}	1.06×10^{-1}	3.64×10^{-1}
b_3....	5.05×10^{-5}	5.35×10^{-3}	6.93×10^{-2}	3.37×10^{-1}	1.02
b_4....	5.34×10^{-3}	6.96×10^{-2}	2.34×10^{-1}	5.35×10^{-1}	1.08
b_5W...	1.41×10^{-5}	8.14×10^{-3}	1.83×10^{-1}	1.24	3.67
b_6....	4.17×10^{-3}	6.42×10^{-1}	7.36	35.9	88.4
b_7....	2.02×10^{-2}	2.60×10^{-1}	7.96×10^{-1}	1.69	2.80
			$T_1 = 50,000°$		$T_1 = T_\epsilon = 20,000°$
b_1W...	8.04×10^{-24}	3.78×10^{-11}	1.04×10^{-4}	2.36×10^{-1}	5.67×10^{-1}
b_2W...	1.59×10^{-5}	3.58×10^{-3}	6.82×10^{-2}	4.05×10^{-1}	5.02×10^{-1}
b_3....	1.29×10^{-4}	1.39×10^{-2}	1.80×10^{-1}	8.82×10^{-1}	5.89×10^{-1}
b_4....	9.36×10^{-3}	1.18×10^{-1}	4.08×10^{-1}	9.74×10^{-1}	6.07×10^{-1}
b_5W...	6.02×10^{-5}	3.46×10^{-2}	7.80×10^{-1}	5.29	11.9
b_6....	6.20×10^{-3}	9.52×10^{-1}	10.95	53.4	35.5
b_7....	2.67×10^{-2}	3.35×10^{-1}	1.02	2.23	2.01

arrives in this level, it may exit only by a spontaneous transition to 2 ^3S. The probability of such a transition is small, because the frequency is low; hence this level becomes abundant at the expense of the higher triplet levels.

In a later paper the results derived here will be discussed in connection with observations of helium lines in the nebulae. The writer is indebted to Professor D. H. Menzel for many interesting discussions relating to the problem of statistical equilibrium.

HARVARD OBSERVATORY
December 5, 1940

Reprinted from *The Astrophysical Journal*, Vol. 94, No. 1, July, 1941.

PHYSICAL PROCESSES IN GASEOUS NEBULAE

XVI. THE ABUNDANCE OF *O* III

DONALD H. MENZEL AND LAWRENCE H. ALLER[1]

ABSTRACT

From the intensities of the green nebular lines the density of *O* III ions proves to be roughly 1 per cm^3, corresponding to an abundance about ten thousand times less than that of the hydrogen ions. The numbers, nevertheless, are sufficient to produce a marked depression of the electron temperature, through the medium of collisional excitation. The observed electron temperatures, given in Paper XIII, are consistent with the theory developed in Paper V, modified for the effects of collisions.

In earlier papers of this series we have shown how electron densities and temperatures of planetary nebulae may be determined from the absolute intensity of the Balmer continuum and the relative intensities of the lines of *O* III, λλ 5007, 4959, and 4363. We now proceed to a determination of the abundance of *O* III from the absolute intensities of these same lines.

The number of inelastic collisions from level *A* to level *B* per unit volume per second is given by equation (X, 32)[2] as

$$\mathfrak{F}_{AB} = 8.54 \times 10^{-6} \frac{N_A N_\epsilon}{T_\epsilon^{1/2}} \frac{\Omega(A, B)}{\tilde{\omega}_A} e^{-\chi_{AB}/kT_\epsilon} . \tag{1}$$

For the green nebular lines *A* refers to one of the levels of ^3P, and *B* to the single level $^1D_2$. For all three transitions, χ_{AB} is essentially identical. Further, for all three transitions,

$$\frac{\Omega_{AB}}{\tilde{\omega}_A} = 2.22 ,$$

as may be seen from reference to Table 2 of Paper X. Hence, setting

$$\Sigma N_A = N_P ,$$

where N_P is the number of atoms in the combined levels of ^3P, we find for $\mathfrak{F}_{PD}$, the number of collisional excitations of $^1D_2$ from ^3P per cubic centimeter per second,

$$\mathfrak{F}_{PD} = 1.90 \times 10^{-5} \frac{N_P N_\epsilon}{T_\epsilon^{1/2}} e^{-\chi_{PD}/kT_\epsilon} . \tag{2}$$

Not all the atoms excited to the $^1D_2$ level will return to the levels of the ground term by radiative transitions with consequent emission of the nebular lines. We suppose that a fraction, β, is collisionally de-excited. Thus the contribution per unit volume to the intensity of the two nebular lines is $(1 - \beta)\mathfrak{F}_{PD} h\nu_{PD}$.

[1] Society of Fellows, Harvard University.

[2] *Ap. J.*, **92**, 421, 1940.

As in Paper XIII, we assume a planetary model of outer radius A and thickness d. The mean surface brightness, S_n, in the green nebular lines becomes

$$S_n = \tfrac{1}{3}(1 - \beta)\mathfrak{F}_{PD}h\nu_{PD}3d\left[1 - \frac{d}{A} + \frac{d^2}{3A^2}\right] = \tfrac{1}{3}(1 - \beta)\mathrm{D}\mathfrak{F}_{PD}h\nu_{PD}\,. \qquad (3)$$

For a homogeneous sphere $d = A$, and $\mathrm{D} = A$.

From (XI, 9) we have the surface brightness expressed in terms of the magnitude per square minute of arc,

$$S_n = 8.40 \times 10^2(2.512)^{-H_n}\,. \qquad (4)$$

From equation (3) we get

$$S_n = 2.48 \times 10^{-17}\frac{e^{-\chi/kT_\epsilon}}{T_\epsilon^{1/2}}(1 - \beta)\mathrm{D}N_P N_\epsilon\,. \qquad (5)$$

Solving equations (4) and (5) for N_P, the number of *O* III atoms in the ground ^{3}P term, we find

$$N_P = 3.38 \times 10^{19}\frac{T_\epsilon^{1/2}e^{+\chi/kT_\epsilon}}{N_\epsilon \mathrm{D}(1 - \beta)}(2.512)^{-H_n}\,. \qquad (6)$$

The needed observational quantities are H_n, the surface brightness of the planetary in the green nebular lines expressed in magnitudes per square minute of arc, the electron temperature, T_ϵ, and the electron density, N_ϵ. The quantity H_n may be obtained from the data in Paper XIV; T_ϵ and N_ϵ have already been determined for a number of representative nebulae. These two latter quantities are needed for a calculation of β, which is, by definition,

$$\beta = \frac{\mathfrak{F}_{DP}}{\mathfrak{F}_{DP} + F_{DP}} = \frac{1}{1 + \dfrac{6.35 \times 10^2 T_\epsilon^{1/2}}{N_\epsilon}}, \qquad (7)$$

since

$$\mathfrak{F}_{DP} = 3.42 \times 10^{-5}\frac{N_\epsilon}{T_\epsilon^{1/2}}N_D\,, \qquad (8)$$

from (X, 33) and

$$F_{BA} = N_D \Sigma A_{DP_i} = N_D 2.15 \times 10^{-2}\,, \qquad (9)$$

the latter expressing the number of radiative transitions per cubic centimeter per second.

We have computed $(1 - \beta)$ as a function of T_ϵ and N_ϵ (see Table 1). We note that $(1 - \beta)$ is much more sensitive to N_ϵ than to T_ϵ. As one would expect, the efficiency of the electrons in de-exciting the $^1D_2$ level is much more dependent upon the number of electrons available than upon their temperature and velocity distribution.

We have estimated the surface brightnesses in the green nebular lines, in magnitudes per square minute of arc, by the same procedure as was used earlier in determining the surface brightness of the planetaries in the continuous spectrum at the Balmer limit. In most cases the correction to the magnitude is negative because the green nebular lines are usually much the strongest in the spectrum and lie in a region where the plate sensitivity is low.

We give the relevant data in Table 2. The successive columns refer respectively to the NGC number of the planetary, the electron temperature, the electron density, D, as computed from the data of Curtis[3] for NGC 1535, 6543, 7009, 7662, and IC 418 (for the other objects a uniform sphere model is assumed), H_n the surface brightness in the green nebular lines in magnitudes per square minute of arc, and finally N_P.

Nearly all the *O* III ions are in the ground ^3P level. We see that the number of *O* III ions ranges from about one in every 10 cm³ for IC 418 or NGC 1535 to about two per

TABLE 1

T_ϵ	$N_\epsilon \times 10^{-4}$			
	0.5	1.0	2.0	4.0
6,000	0.90	0.83	0.71	0.55
8,000	.92	.85	.74	.59
10,000	.92	.86	.76	.61
12,000	.93	.87	.77	.63
14,000	.93	.88	.79	.65
16,000	.94	.89	.80	.67
20,000	0.95	0.90	0.82	0.69

TABLE 2

Object	T_ϵ	N_ϵ	D	H_n	N_P
NGC 7027	9,500	0.93×10^4	1.59×10^{17}	4.60	0.76
6572	9,200	1.71	1.29	4.86	0.49
6826	7,200	0.7	2.04	6.36	0.37
6543	6,000	1.8	0.78	5.38	2.14
IC 418	6,800	1.2	0.78	7.75	0.19
NGC 7662	10,300	2.0	0.70	4.70	0.70
7009	9,500	3.3	0.82	4.27	0.73
1535	10,000	1.3	1.53	6.56	0.09

cubic centimeter for NGC 6543. The *O* III ions are thus about ten thousand times less abundant than the electrons or hydrogen ions.

We evaluate the effective f value for $^3P_2 - {}^1D_2$, the strongest nebular line, from the relation[4]

$$f = \frac{\tilde{\omega}_n}{\tilde{\omega}_{n'}} \frac{mc^3}{8\pi^2 e^2 \nu^2} A_{nn'} = \tfrac{5}{5} \times 3.78 \times 10^{-9} \times 0.016 = 6.05 \times 10^{-11} , \qquad (10)$$

with Pasternack's value of $A_{nn'}$.[5] The average absorption coefficient over the center of the line will be

$$\bar{a}_\nu \sim \frac{\pi e^2}{mc} f \frac{1}{\Delta\nu} , \qquad (11)$$

[3] *Pub. Lick Obs.*, **13,** 1918. [4] *Ap. J.*, **85,** 330, 1937, eq. (I, 1).

[5] *Ap. J.*, **92,** 129, 1940; also *Pub. A.S.P.*, **51,** 160, 1939.

where $\Delta\nu$ is the effective half-width of the line as determined by random Doppler motions. The minimum value for $\Delta\nu$ is

$$\Delta\nu = \sqrt{\frac{2kT_a}{m_a}}\frac{1}{\lambda} \sim 1.06 \times 10^{11}\,, \tag{12}$$

where m_a is the atomic mass and T_a the gas-kinetic temperature of the atoms. We set $T_a = T_\epsilon$, for we may reasonably expect that momentum exchange between the free electrons and atoms will lead to equipartition of energy. Thus

$$\bar{a}_\nu \leq 1.5 \times 10^{-24}\,, \tag{13}$$

and the total optical depth in N_1 becomes

$$T \leq N_P 2 D a_\nu \sim 1.5 \times 10^{-6} \quad \text{with} \quad N_P = 1.0 \quad \text{and} \quad 2D = 1 \times 10^{17}\ \text{cm}\,. \tag{14}$$

The assumption that the nebula is transparent in the nebular lines is, therefore, fully justified.

The radiation in the forbidden nebular lines derives from the kinetic energy of the electrons. The permanent gases—oxygen, nitrogen, argon, neon, etc.—act as temperature regulators. When the temperature of the electron gas is sufficiently high, the electrons lose a large fraction of their energy in the excitation of metastable levels. Hence the equilibrium temperatures are considerably lower than those calculated in Paper V for an optically thin nebula composed of pure hydrogen, with only radiative and capture processes active. We must now modify the equation of radiative equilibrium previously given for the continuum to allow for the collisional excitation of the nebular lines. The energy removed from the continuum per second is merely that radiated in the nebular lines. For the moment let us consider a nebula composed of hydrogen and *O* III; the argument will give the correct order of magnitude for any nebula in which the green lines are strong.

We have two fundamental equations. Equation IV(2), which specifies that the number of ionizations from the ground level of hydrogen equals the number of captures on all states, remains unchanged in form. The second equation, IV(4), guaranteeing radiative equilibrium, must be altered in form to allow for the fact that much of the electronic energy is radiated in the nebular lines. In its original form, equation IV(4) states that all the energy absorbed by hydrogen atoms is re-radiated in free-free, bound-free, and bound-bound transitions. We first replace the factor $2N_iN_\epsilon KZ^6hR/T_\epsilon^{3/2}$, previously canceled from both sides of the equation, and add the term $(1-\beta)\mathfrak{F}_{PD}h\nu_{PD}$ to the right-hand side, representative of the nebular line emission.

We now use equation IV (2) to eliminate the unknown parameter, b_1. The resulting equation, with the common factors again divided out, is identical with equation IV (5), except for an additional correction term, $C(T_\epsilon)$ on the right side. From equation (2) we have

$$C(T_\epsilon) = 1.90 \times 10^{-5}\frac{N_P}{N_i}(1-\beta)\frac{h\nu_{PD}}{KZ^4 2hRZ^2 G_{T_\epsilon}}e^{-\chi/kT_\epsilon}$$

$$= 1.63\frac{T_\epsilon}{G_{T_\epsilon}}e^{-\chi/kT_\epsilon}\frac{N_P}{N_i}(1-\beta)\,, \tag{15}$$

where G_{T_ϵ} is the function defined in V(2). In the symbolic notation of equation V(5), our equation of radiative equilibrium becomes

$$f_1(T_1) = f_2(T_\epsilon) + C(T_\epsilon) = f_2'(T_\epsilon) . \tag{16}$$

The left-hand side of the equation is a function only of the temperature of the nuclear star and may be readily calculated; $f_2(T_\epsilon)$, similarly, is a function only of the electron

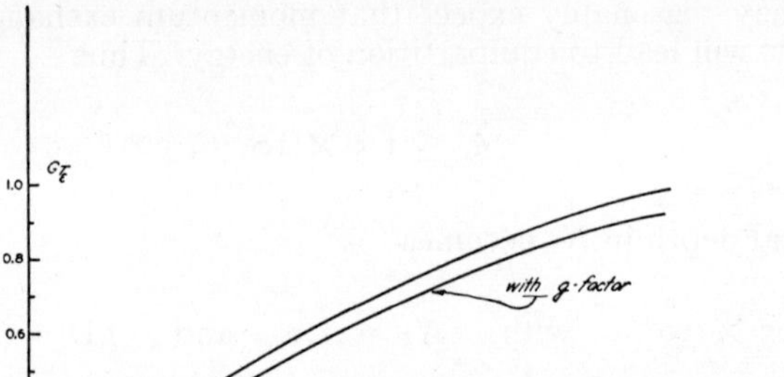

FIG. 1.—G_{T_ϵ} as a function of the electron temperature, T_ϵ. The lower curve is calculated with the Gaunt *g*-factor included. In the upper curve it is omitted. The latter curve is used in the present calculations, as the *g*-factor has been omitted in the numerical calculations of Paper V, with which our present results are compared.

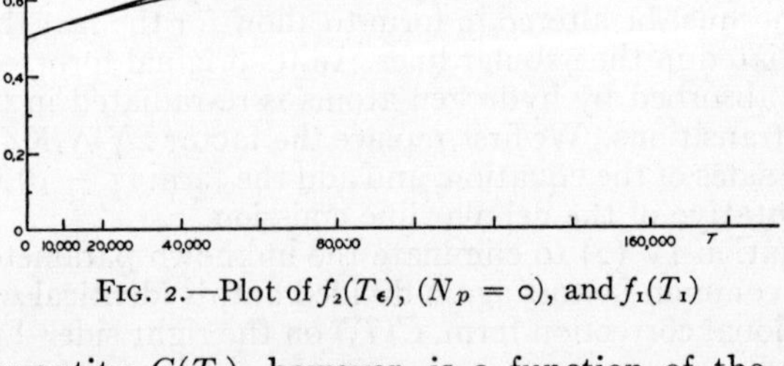

FIG. 2.—Plot of $f_2(T_\epsilon)$, $(N_P = 0)$, and $f_1(T_1)$

temperature. The quantity $C(T_\epsilon)$, however, is a function of the abundance of O III atoms in the ^3P term relative to the hydrogen ions, as well as of the temperature.

For future reference, we give G_{T_ϵ} in Figure 1 as a function of T_ϵ. Figure 2 gives $f_1(T_1)$ and also $f_2(T_\epsilon)$. By noting the two abscissae for a given value of the ordinate, we recover the relation between T_1 and T_ϵ tabulated in Paper V, corresponding to the case $N_P = 0$.

The new function $f_2'(T_\epsilon)$ is now a function of N_ϵ $(= N_i)$ and N_P as well as of T_ϵ. Hence we require several sets of curves, which are reproduced in Figure 3. Unlike $f_2(T_\epsilon)$, the new function curves sharply upward and, beginning at T_ϵ about 8000° or 10,000°, de-

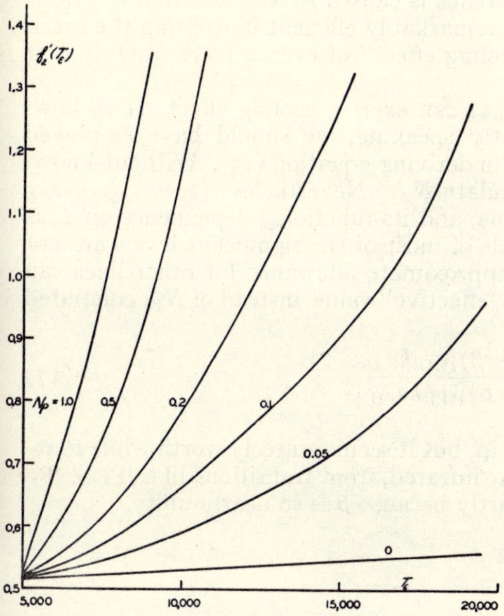

a) $N_\epsilon = 0.5 \times 10^4$; $N_p = 0.05, 0.1, 0.2, 0.5, 1.0$

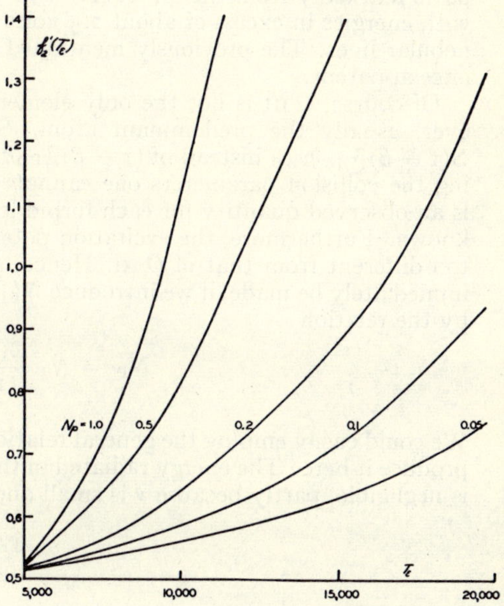

b) $N_\epsilon = 1.0 \times 10^4$; $N_p = 0.05, 0.1, 0.2, 0.5, 1.0$

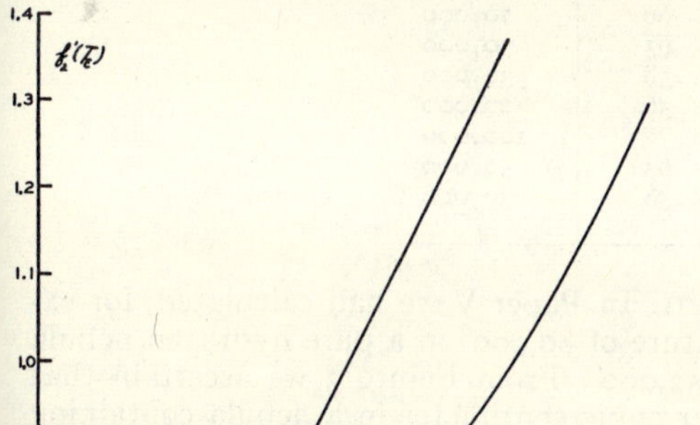

c) $N_\epsilon = 2.0 \times 10^4$; $N_p = 0.05, 0.1, 0.2, 0.5, 1.0$

d) $N_\epsilon = 4.0 \times 10^4$; curves for $N_p = 0.05, 0.1, 0.2$, etc.

FIG. 3

parts markedly from the old curve. The difference is caused by the fact that electrons with energies in excess of about 2.5 volts are remarkably efficient in exciting the green nebular lines. The previously mentioned "cooling effect" of even a trace of *O* III is at once apparent.

Of course, *O* III is not the only element that can exert a cooling effect. It is, however, usually the predominant atom. Strictly speaking, we should have employed $\Sigma(1-\beta)\mathfrak{F}_{AB}h\nu_{AB}$ instead of $(1-\beta)\mathfrak{F}_{\mathrm{PD}}h\nu_{\mathrm{PD}}$ in deriving equation (15). Without knowing the collision parameters one cannot calculate N_A. Nevertheless $(1-\beta)\mathfrak{F}_{AB}h\nu_{AB}$ is an observed quantity for each forbidden line, and its functional dependence on T_ϵ is known. Furthermore, the excitation potentials of most of the significant levels are not too different from that of *O* III. Hence, an approximate allowance for other lines can immediately be made, if we introduce N_{P}', an "effective" value, instead of N_{P}, computed by the relation

$$N_{\mathrm{P}'} = N_{\mathrm{P}} \frac{\Sigma(1-\beta)\mathfrak{F}_{AB}h\nu_{AB}}{(1-\beta)\mathfrak{F}_{\mathrm{PD}}h\nu_{\mathrm{PD}}}\,. \tag{17}$$

We could easily employ the general relationship, but it seems scarcely worth while to reproduce it here. The energy radiated in the far infrared, from transitions like $^3P_2 - {}^3P_1$, is negligible, partly because ν is small and partly because β is so nearly unity.

TABLE 3

Object	$f_1(T_1)$	T_1
NGC 7027	0.88	300,000
6572	.66	70,000
6826	.63	50,000
6543	.58	30,000
IC 418	.56	20,000
NGC 7662	.72	100,000
7009	.63	50,000
1535	0.58	30,000

We refer again to the cooling effect of *O* III. In Paper V we had calculated, for example, that a nuclear star having a temperature of 80,000° in a pure hydrogen nebula would produce an electron temperature of 52,000°. From Figure 3 we ascertain that an abundance of only 0.05 atom of *O* III per cubic centimeter in a nebula containing 10^4 ions per cubic centimeter—a relative abundance of 1:200,000—will lower the temperature to 18,000°, while an abundance of 1 *O* III atom per cubic centimeter will lower the value to 8000°. The low values we have found for the electron temperatures are, therefore, consistent with theory.

With accurate values of N_ϵ, T_ϵ, and N_{P} we could, of course, proceed in the reverse direction and estimate the temperature of the nucleus. Unfortunately, small inaccuracies in these observationally determined quantities would yield large errors in T_1. We have proceeded, nevertheless, to attempt the calculation, with the results given in Table 3. We place no emphasis on the figures, however, except for noting a general tendency for the objects whose spectra exhibit greater excitation to be associated with higher values of T_1. The only serious discrepancy is NGC 1535, and here a lowering of the electron density by a factor of only 1.4 will raise T_1 to 48,000°; a factor of 2.0, in the same direction, will raise T_1 to 150,000°, which value would be more in keeping with the excitation.

In conclusion one may say that Zanstra's assumption that most of the electronic energy goes into the excitation of nebular lines appears to be a good one.

Harvard College Observatory
January 15, 1941

Reprinted from *The Astrophysical Journal,* Vol. 94, No. 3, November, 1941.

PHYSICAL PROCESSES IN GASEOUS NEBULAE

XVII. FLUORESCENCE IN HIGH-EXCITATION PLANETARIES

DONALD H. MENZEL AND LAWRENCE H. ALLER[1]

ABSTRACT

Some years ago Bowen attributed the appearance of "fractional multiplets" of *O* III in the spectra of high-excitation planetaries to absorption by *O* III atoms in the $^2p^2\ ^3p_2$ level of the λ 303.80 "Lyman alpha" radiation of helium. The subsequent cascading of these atoms from 2p3d $^3D_2$ of *O* III back to the ground level via the 2p3p and 2p3s configurations are supposed to produce the observed lines. We have attempted to put this theory on a quantitative basis by setting up the appropriate equations of statistical equilibrium for the excited levels and predicting the relative intensities of the observed ultraviolet lines. A comparison of observed relative intensities with predicted relative intensities yields a consistency check on the theory. The populations of the excited levels are found to be 10^{-14}–10^{-16} atoms cm^{-3}. The intensity of the radiation in the *O* III 303.80 line is from 10^2 to 10^3 times greater than that radiated by the nuclear star, which illustrates why permitted lines of *O* III, other than those originating by cascade from 3d$^3P_2$, are not observed. The excess is attributed to the concentration of radiation in these lines, by cyclic processes resulting from ionization of He^+. We have also shown that theoretically the images of the Bowen lines should be about 5 per cent larger than those of the *He* II lines, which is in rough agreement with the observational data. The paper contains tables of the relevant *f*'s and *A*'s *computed* from approximate wave functions based on Slater's rules.

In the spectra of certain planetaries, e.g., NGC 7662 and NGC 7027, a number of strong permitted lines appear in the astronomically accessible ultraviolet. Wright,[2] who obtained the observational data, pointed out that other lines, equally strong in laboratory spectra, are entirely absent from the nebular radiation. Bowen[3] has noted the fact that all the observed lines emanate either directly or indirectly from excitation of a single upper level, 3d $^3P_2$. He further noticed that the λ 303.799 line corresponding to the transition 2p$^2\ ^3P_2$ — 2p3d $^3P_2$ almost coincides with the Lyman α line of ionized helium, whose wave length is λ 303.780 A.

Bowen suggested that the excitation of the *O* III level was caused by resonance absorption of the helium radiation. An *O* III atom, thus raised to the 2p3d $^3P_2$ level, would either return directly to the ground level or cascade back via the terms of the configurations 2p3p and 2p3s, emitting the observed lines in transit. The λ 304 radiation of *He* II provides a very rich reservoir of energy for the *O* III atoms, upon which transitions other than 2p$^2\ ^3P_2$ — 2p3d $^3P_2$ cannot draw. The circumstance that the ultraviolet *O* III radiations appear only in planetaries where λ 4686 of *He* II is strong strengthens this suggestion. The images of the ultraviolet *O* III lines are never significantly larger than those of *He* II (e.g., λ 3203). An excellent example is NGC 7009. Although the Bowen fluorescent lines are conspicuous in the main body of the planetary, they are quite missing in the ansae. The spectra of the latter are definitely of a low-excitation type in which the *He* II lines do not appear.[4] The occurrence of the green nebular lines, even though they are faint, shows that some *O* III is present in the outer portions of the nebula, in a state where it could absorb the λ 304 radiation if it existed.

Bowen's suggestion provides an excellent qualitative explanation for the observed

[1] Society of Fellows, Harvard University.

[2] *Lick Obs. Bull.*, **17**, 1, 1934.   [3] *Ap. J.*, **81**, 1, 1935.

[4] For the observational data see Paper XIV of this series (*Ap. J.*, **93**, 236, 1941). See also Papers XII (*Ap. J.*, **93**, 195, 1941) and XVI (*Ap. J.*, **94**, 30, 1941) for the electron density and *O* III density determinations, respectively.

phenomena. We propose, in the present paper, to investigate the theory of the Bowen fluorescent mechanism from the quantitative viewpoint.

Previous studies[4] have yielded estimates of the densities of atoms of *O* III, electrons, and hydrogen ions in typical planetaries. Rough intensities of the ultraviolet lines of helium, neon, and oxygen are available. We shall proceed by setting up the equations of statistical equilibrium for each of the excited levels. The number of atoms entering any excited level per second must equal the number leaving that same level each second. Since all the excited levels are fed from the 3d $^3P_2$ level, we should be able to predict the relative intensities of the ultraviolet lines of *O* III.

Atomic theory yields the Einstein *A*'s or oscillator strengths (*f*-values) for the transitions involved. The *A*'s or *f*'s for the transition $J - J'$ are as follows:

$$A(J, J') = \frac{1}{2J+1} \frac{64\pi^4\nu^3}{3hc^3} \frac{\epsilon^2 a^2}{4l^2 - 1} S\rho^2 , \tag{1}$$

$$f(J', J) = \frac{1}{2J'+1} \frac{8\pi^2 m\nu}{3h} \frac{a^2}{4l^2 - 1} S\rho^2 ; \tag{2}$$

accordingly,

$$A(J, J') = \frac{2J'+1}{2J+1} \frac{8\pi^2\epsilon^2\nu^2}{mc^3} f(J, J') , \tag{2a}$$

where J' and J are the inner quantum numbers of the lower and upper levels, respectively; $a = 0.528 \times 10^{-8}$ cm, the radius of the first Bohr orbit; S is the numerical "strength" of the line; and l refers to the larger of the initial or final values for the two configurations. The symbols, ϵ, m, c, h, and ν have their usual meanings.

Calculation of the radial quantum integral,

$$\rho = \int_0^\infty R_{nl} R_{n'l'} r^3 dr , \tag{3}$$

involves a knowledge of the radial wave functions R_{nl} and $R_{n'l'}$.[5] The detailed and accurate calculation of these functions is a difficult problem. For our present purposes we shall use approximate hydrogen-like radial wave functions whose parameters are fixed according to a rule given by Slater.[6]

The procedure is as follows. We first introduce effective quantum numbers according to the following scheme:

$$\text{for} \quad n = 1, 2, 3, 4, 5, 6 ,$$

$$\text{let} \quad n^* = 1, 2, 3, 3.7, 4.0, 4.2 .$$

Then we divide the orbits up according to shells: 1s; 2s2p; 3s3p3d; 4s4p, ... ; etc. In place of the nuclear charge Z, we use the effective charge, $Z - \sigma$, where σ, the screening constant, is fixed by the following rules. Each electron in the same shell with the

[5] In the present paper we adopt wave functions normalized to give

$$\int_0^\infty R_{nl}^2 r^2 dr = 1$$

instead of the form used in Paper X.

[6] *Phys. Rev.*, **57**, 36, 1930.

"jumping" electron contributes 0.35 to the total σ. For s and p electrons the contribution is 0.85 for each electron with a total quantum number less by 1, and for other electrons the contribution is 1.00; i.e., the screening is supposed to be complete. For d electrons we adopt the value of 1.00 for all electrons in the inner shell.

These rules lead to the following approximate wave function:

$$R_{nl} = \left[\frac{2(Z-\sigma_{nl})}{n^*}\right]^{n^*+1/2} \frac{1}{[\Gamma(2n^*+1)]^{1/2}} r^{n^*-1} e^{[-(Z-\sigma_{nl})/n^*]r} . \tag{4}$$

From equation (3) we easily obtain

$$\left.\begin{aligned} \rho = \frac{\Gamma(n^*+n'^*+2)}{[\Gamma(2n^*+1)\Gamma(2n'^*+1)]^{1/2}} \left[\frac{2(Z-\sigma_{nl})}{n^*}\right]^{n^*+1/2} \\ \times \left[\frac{2(Z-\sigma_{n'l'})}{n'^*}\right]^{n'^*+1/2} \frac{1}{\left[\dfrac{Z-\sigma_{nl}}{n^*}+\dfrac{Z-\sigma_{n'l'}}{n'^*}\right]^{n^*+n'^*+2}} . \end{aligned}\right\} \tag{5}$$

For O III the relevant values of the significant parameters, calculated for the various configurations of O III ($Z = 8$) by Slater's rules, are shown in the accompanying table.

Configuration	n	σ	$Z-\sigma$
$1s^2 2s^2 2p^2$	2	2.75	5.25
2p3s	3	4.55	3.45
2p3p	3	4.55	3.45
2p3d	3	5.00	3.00

For the ground configuration of O III, however, a Hartree wave function is available, for which Hebb and Menzel[7] have given an analytic approximation. We may use this function in the calculations of radial quantum integrals connecting the higher configurations with the ground configuration. To calculate ρ for the transition $2p^2 - 2p3d$, we have (cf. n. 4, regarding normalization convention)

$$R_{2p} = r(ae^{-br} + ce^{-dr}) , \tag{6}$$

where

$$a = 10.415 , \quad c = 4.520 ,$$

$$b = 3.604 , \quad d = 2.014 ,$$

and

$$R_{3d} = \left[\frac{2(Z-\sigma)}{3}\right]^{7/2} (6!)^{-(1/2)} r^2 e^{-(z-\sigma_{3d})r/3} = 2^{7/2}(6!)^{-(1/2)} r^2 e^{-r} . \tag{7}$$

From equation (3) we now obtain

$$2p - 3d \qquad \rho_1 = (2^7 6!)^{1/2} \left[\frac{a}{(1+b)^7} + \frac{c}{(1+d)^7}\right] = 0.68 \tag{8a}$$

[7] *Ap. J.*, **92**, 408, 1940.

and similarly

$$2p - 3s \qquad \rho_4 = 18.50(6!)^{1/2}\left[\frac{a}{(1.15 + b)^7} + \frac{c}{(1.15 + d)^7}\right] = 0.81\,. \tag{8b}$$

The modified Slater type of wave function for the ground configuration, with ρ calculated according to equation (5), gives, instead,

$$\rho_1 = 0.47\,, \tag{8a'}$$

$$\rho_4 = 0.58\,. \tag{8b'}$$

The agreement is satisfying in view of the approximate nature of the calculations. We shall use equations (8*a*) and (8*b*) in preference to equations (8*a*′) and (8*b*′). From equation (5) we find for the transitions between excited levels

$$3p - 3d \qquad \rho_2 = 3.20\,, \tag{8c}$$

$$3p - 3s \qquad \rho_3 = 3.04\,. \tag{8d}$$

The strengths, *S*, which depend on the angular factors in the wave functions, have been tabulated in a previous paper.[8]

The error in ρ arising from inaccurate wave functions is difficult to estimate. Modified hydrogenic wave functions calculated from Slater's rules are certainly better than pure hydrogenic functions. The variation method might yield better results, but the labor of applying it for excited configurations in such atoms as *O* III would be considerable. We believe that the integrals calculated from the much simpler Slater's rules will be amply accurate for the present purpose.

To obtain a brief and explicit, although somewhat awkward, notation, we have adopted the designations shown in the accompanying table for the levels involved.

Configuration and Level	Present Notation	Configuration and Level	Present Notation
$1s^2 2s^2\ 2p^2\ {}^3P_1$	*a*′	${}^3D_3$	*g*
${}^3P_2$	*a*	${}^3D_2$	*h*
$2p3d\ {}^3P_2$	*b*	${}^3D_1$	*j*
$2p3p\ {}^3P_2$	*c*	$2p3s\ {}^3P_2$	*k*
${}^3P_1$	*d*	${}^3P_1$	*m*
${}^3P_0$	*e*	${}^3P_0$	*n*
${}^3S_1$	*f*		

In Table 1 we give the wave lengths, term designations, strengths, *f*-values, and *A*'s for the transitions involved in the Bowen fluorescent mechanism. We found it convenient to recast equations (1), (2), and (2*a*) as follows:

$$\left.\begin{aligned} f &= 3.04 \times 10^{-6}\,\frac{1}{2J' + 1}\,\frac{\rho^2 S}{\lambda(4l^2 - 1)}\,, \\ A &= 2.01 \times 10^{-6}\,\frac{1}{2J + 1}\,\frac{\rho^2 S}{(4l^2 - 1)\lambda^3}\,, \end{aligned}\right\} \tag{9}$$

[8] *Ap. J.*, **91**, 307, 1940.

and

$$f = 1.509 \frac{2J + 1}{2J' + 1} \lambda^2 A \, , \tag{10}$$

where λ is expressed in centimeters.

Atoms in the $2p^2\ ^3P_2$ (*a*) level absorb the λ 304 radiation of *He* II and are raised to the $3d\ ^3P_2$ (*b*) level. The levels—*c*, *d*, *f*, *g*, etc.—are populated exclusively by atoms cascading from the *b* level.

TABLE 1

TRANSITION PROBABILITIES FOR *O* III

Transition	Wave Length	*S*	*f*	*A*
$1s^22s^22p3d\ ^3P_2 - 2p3p\ ^3D_3$	2836.35	7/10	0.0073	8.43×10^6
$^3D_2$	2818.80	1/8	.00184	1.53
$^3D_1$	2808	1/120	.00020	0.103
$^3P_2$	3444.15	75/8	.113	6.31×10^7
$^3P_1$	3428.70	25/8	.063	2.13
$^3S_1$	3132.87	50/3	.368	1.49×10^8
$1s^22s^22p3d\ ^3P_1 - 2p3p\ ^3S_1$	3121.71	10	.222	1.50×10^8
$1s^22s^22p3p\ ^3D_3 - 2p3s\ ^3P_2$	3759.83	7	.346	1.16×10^8
$^3D_2 -\ ^3P_2$	3791.26	5/4	.062	2.84×10^7
$^3D_2 -\ ^3P_1$	3754.65	15/4	.313	8.80
$^3D_1 -\ ^3P_2$	3810.99	1/12	.0041	0.312
$^3P_1$	3774.01	5/4	.103	4.80
$^3P_0$	3757.20	5/3	.416	6.50
$^3P_2 -\ ^3P_2$	3047.14	15/4	.230	1.64×10^8
$^3P_1$	3023.42	5/4	.129	0.561
$^3P_1 -\ ^3P_2$	3059.31	5/4	.0765	9.02×10^7
$^3P_1$	3035.92	3/4	.077	5.55
$^3P_0$	3024.54	1	.310	7.46
$^3S_1 -\ ^3P_2$	3340.78	5/3	.0935	9.25×10^7
$^3P_1$	3312.35	1	.0940	5.68
$^3P_0$	3299.39	1/3	.0946	1.92
$1s^22s^22p3s\ ^3P_2 - 2p^2\ ^3P_2$	374.00	15/2	.266	1.26×10^{10}
$^3P_1$	....	5/2	.148	0.42
$^3P_1 -\ ^3P_0$	....	5/2	.090	0.701
$^3P_1$	....	3/2	.089	0.421
$^3P_0$	374.00	2	.035	0.558×10^{10}
$^3P_0 -\ ^3P_1$	....	2	.119	1.68
$1s^22s^22p^2\ ^3P_2 - 2p3d\ ^3P_2$	303.80	18.75	.116	8.31×10^9
$^3P_1$	303.80	6.25	.039	4.63
$^3P_1 -\ ^3P_2$	303.62	6.25	0.0645	2.78×10^9

Our first problem will be to use the computed *A*'s to set up the equations of statistical equilibrium for each of these excited levels in turn. We shall be able to predict the populations of each of the *c*, *d*, *f* levels in terms of that of the *b* level. Since the intensity of an emission line is proportional to the product of the population of the upper level of the transition of the *A*- or *f*-value, we may use the observed intensities of the *O* III fluorescent lines to evaluate N_b. The values of N_b obtained from the various observable permitted lines should agree with one another to within the errors of observation.

The equation of statistical equilibrium for the $2p3d\ ^3P_2$ level, (*b*) follows from elementary considerations. We must have

$$F_{ab} = F_{ba} + F_{ba'} + F_{bc} + F_{bd} + F_{bf} + F_{bg} + F_{bh} + F_{bj} \, , \tag{11}$$

where F, in accordance with the earlier notation of this series, is the number of transitions per unit volume per second from one level to another as indicated by the subscripts. The number of downward transitions is $N_b \sum_i A_{bi}$, where A is the Einstein coefficient of spontaneous emission. The summation is to be carried out over all lower levels connecting with level b. We sum the appropriate A's in Table 1 to obtain the right-hand side of equation (11),

$$N_b \Sigma A_{bi} = 1.13 \times 10^{10} N_b \,. \tag{12}$$

Now F_{ab} is the number of transitions from $2p^2\,{}^3P_2$ to $2p3d\,{}^3P_2$. Practically all the O III atoms are in the ground term. In an earlier paper Hebb and Menzel[6] have shown that collisions are sufficiently active to give a statistical distribution among the three levels of ${}^3P$. Hence we take the fraction

$$\frac{(2J+1)}{\Sigma(2J+1)} = \frac{5}{9}$$

of O III atoms to be in the level $2p^2\,{}^3P_2$, i.e., $N_a = \frac{5}{9} N_P$. Thus the number of transitions from the ground level will be

$$F_{ab} = \tfrac{5}{9} N_P \frac{4\pi}{h\nu_{ab}} \int W_\nu I_\nu^0 a_{ab}(\nu) d\nu \,, \tag{13}$$

where I_ν^0 is the frequency distribution of the intensity of the radiation in a black body of an arbitrary temperature T_0 and W_ν is the dilution factor averaged over all angles for the frequency ν. We cannot formulate this function for the general case because of the complications introduced by the structure and large-scale mass motions in the planetary nebula. For a similar reason we are unable to specify $a(\nu)$. We know, however, that

$$\int a(\nu) d\nu = \frac{\pi \epsilon^2}{mc} f \,,$$

and we shall find it convenient to use a mean dilution factor, W, defined by

$$\int I_\nu^0 W_\nu a_{ab}(\nu) d\nu = \frac{\pi \epsilon^2}{mc} f_{ab} I^0 \overline{W} \,, \tag{14}$$

where I^0 is supposed to vary slowly across the line at λ 304, which corresponds to the transition (ab). Then equation (13) becomes

$$F_{ab} = \tfrac{5}{9} N_P \frac{4\pi^2 \epsilon^2}{h\nu mc} f_{ab} \overline{I} \overline{W} = 3.27 \times 10^8 \overline{I} \overline{W} N_P \,, \tag{15}$$

and, by equation (12),

$$\overline{W} \overline{I} = 34.6 \frac{N_b}{N_P} \,. \tag{16}$$

We obtain the equations of statistical equilibrium for each of the levels, c, d, f, etc., by equating the number entering a level by cascade to the number leaving. Thus,

$$N_b A_{bc} = N_c(A_{ck} + A_{cm}) , \qquad N_c = 0.286 N_b , \tag{17}$$

$$N_b A_{bd} = N_d(A_{dk} + A_{dm} + A_{dn}) , \qquad N_d = 0.097 N_b , \tag{18}$$

$$N_b A_{bf} = N_f(A_{fk} + A_{fm} + A_{fn}) , \qquad N_f = 0.885 N_b , \tag{19}$$

$$N_b A_{bg} = N_g A_{gk} , \qquad N_g = 0.072 N_b , \tag{20}$$

$$N_b A_{bh} = N_h(A_{hk} + A_{hm}) , \qquad N_h = 0.0131 N_b , \tag{21}$$

$$N_b A_{bj} = N_j(A_{jk} + A_{jm} + A_{jn}) , \qquad N_j = 8.9 \times 10^{-4} N_b . \tag{22}$$

From

$$N_c A_{ck} + N_d A_{dk} + N_f A_{fk} + N_g A_{gk} + N_h A_{hk} = N_k(A_{ka} + A_{ka'}) , \tag{23}$$

$$N_c A_{cm} + N_d A_{dm} + N_f A_{fm} + N_h A_{hm} = N_m(A_{ma} + A_{ma'}) , \tag{24}$$

$$N_d A_{dn} + N_f A_{fn} = N_n(A_{na} + A_{na'}) , \tag{25}$$

we obtain, by the use of equations (17), (18), and (19),

$$N_k = 8.3 \times 10^{-3} N_b , \tag{26}$$

$$N_m = 4.34 \times 10^{-3} N_b , \tag{27}$$

$$N_n = 1.4 \times 10^{-3} N_f . \tag{28}$$

The small values of the coefficients in equations (26), (27), and (28) arise from the large values of $A_{k,\,m,\,n-a,\,a'}$. In the derivation of these equations we have assumed that transitions from the ground level to 2p3d $^3P_1$ are relatively infrequent.

Our observational data now afford a consistency check on the theory. From the intensities of the *O* III lines, originating from the b, f, or c levels we have calculated N_b, directly or by use of equation (17) or equation (19), from N_c or N_f. If the theory is correct and our assumed A values are not too much in error, the various values of N_b should agree with one another to within the limits of observational error.

The observed surface brightness, S_λ, of any line expressed in terms of magnitudes per square minute of arc, H_λ, is related to the number of atoms in the upper level, N'', by means of the expression (cf. XI, 9)

$$S_\lambda = \tfrac{1}{3} D N'' A h\nu = 8.40 \times 10^2 (2.512)^{-H_\lambda} \quad \text{or} \quad N'' = \frac{25.2 \times 10^2}{D A h\nu} (2.512)^{-H_\lambda} . \tag{29}$$

To derive N'' from the observed intensity, I_0, we have proceeded as in Paper XI and have adopted for the dimensions of the *O* III fluorescent images of NGC 7009, 7662, and 3242, the values given by Curtis[9] for the inner shells of these planetaries. Table 2 sum-

[9] *Pub. Lick Obs.*, **13**, 55, 1918.

marizes in successive columns the observed lines,[10] the upper levels, H_λ, N'', and N_b computed from equation (17) or equation (19). The straight mean values of the numbers of atoms in the 3d $^3P_2$ level are given in the accompanying table. An approximate

NGC 2022....	1.0×10^{-16}	NGC 7027....	$(1.6\pm0.2)\times10^{-15}$
3242....	3.0×10^{-15}	7662....	$(1.2\pm0.2)\times10^{-14}$
7009....	$(5.9\pm0.2)\times10^{-15}$		

reduction of an unstandardized plate by Stoy yielded the estimate of the value of N_b in NGC 3242. The probable errors quoted are a measure of the internal agreement of the values deduced from the different lines with the aid of the equations of statistical equilibrium.

TABLE 2

POPULATION OF THE 3d $^3P_2$ LEVEL OF *O* III

λ	Upper Level	H_λ				$N''\times10^{15}$				$N_b\times10^{15}$			
		2022	7009	7027	7662	2022	7009	7027	7662	2022	7009	7027	7662
3444.15......	3d $^3P_2$	11.87	10.38	10.64	10.80	0.39	6.05	2.41	12.8	.....	6.05	2.41	12.7
3312.35......	3p $^3S_1$	......	10.67	11.50	10.45	.....	4.87	1.18	7.2	.....	5.5	1.33	8.1
3299.39......	3p $^3S_1$	......	......	13.20	11.96	.....	.....	0.73	5.2	.....	.....	0.83	4.9
3132.87......	3d $^3P_2$	12.25	9.42	9.98	8.38	0.10	6.11	1.73	17.2	0.10	6.1	1.73	17.2
3121.71......	3d $^3P_1$	......	......	......	10.96	.....	.....	.....	1.6	.....	.....	.....	1.6
3047.14......	3p $^3P_2$	......	......	......	9.70	.....	.....	.....	3.9	.....	.....	.....	3.9

Within the errors of observation, theory and observation seem to agree. The lines λ 3312 and λ 3299 appear to be too weak. The fainter lines seem to yield too small a population, as if there is present in the observations a systematic error depending on intensity.

These values of N_b at first sight may seem very small compared with the populations of the metastable $^1D_2$ and ^1S levels of the p^2 configuration. The strongest Bowen fluorescent lines have an intensity approximately one-tenth that of the green nebular lines. The ratio of transition probabilities of the permitted and forbidden lines is of the order of 10^{10}. Hence one readily sees that the effect of a small population in the excited levels is more than offset by a high transition probability.

The λ 304 radiation of *He* II corresponds to the Lyman α radiation of hydrogen. Accordingly, if the optical thickness of the nebula is sufficient, most of the quanta absorbed in the far ultraviolet—i.e., at frequencies greater than that corresponding to the series limit of *He* II—will be degraded into the λ 304 radiation in accordance with the mechanism proposed for hydrogen by Zanstra and by Menzel. Hence this radiation may be expected to be intense, especially in a nebula possessing strong lines of *He* II. The energy available in λ 304 should be far greater than the nuclear star would provide of itself.

From our observational data and equation (16) we may compute the intensity of the radiation in the *O* III 304 radiation in units of ergs/cm²/sec. If we make some reasonable assumption about the size and ultraviolet color temperatures of the central stars of the planetaries, we can estimate the intensity of the λ 304 radiation from the star itself and

[10] Since the intensities of the *O* III 3428 and 3341 lines are somewhat uncertain because of the continguous *Ne* V lines, we have omitted them from this discussion. The evidence suggests that *Ne* V is very strong in NGC 7027 and perhaps also in NGC 2022, whereas it is much weaker in NGC 7662 and NGC 7009. Slit spectra are necessary to separate the *Ne* V from the *O* III lines.

of the geometrical dilution factor. Then a comparison of the observed and computed dilution will indicate a lower limit to the degree of concentration of energy in the λ 304 radiation produced by the atoms of *He* II.

Accordingly, we have calculated $(I^\circ \overline{W})$ for the five planetaries for which we have estimates of N_b and N_a. Also we have estimated the sizes of the central stars and the geometrical dilution factor for various assumed ultraviolet color temperatures.

Berman[11] tabulates the apparent photographic magnitudes and distances of these planetaries. In the expression relating distance and absolute and apparent magnitude,

$$m = M + 5 \log \frac{r}{10} + A(r) ,$$

we have taken the correction for space absorption as[12] 0.7 mag. per kiloparsec. The calculated absolute photographic magnitudes for the nuclei of the five planetary nebulae are

TABLE 3

ESTIMATED ABSOLUTE PHOTOGRAPHIC MAGNITUDES OF THE PLANETARY NUCLEI

Nebula	m_s	r	$A(r)$	M
NGC 2022	14.7	4570	3.2	−1.8
3242	11.9	1640	1.2	−0.3
7009	11.9	930	0.6	+1.4
7027	(15.6)	2130	1.5	+2.5
7662	12.7	1200	0.8	+1.5

given in Table 3. The radius of the central star is related to its absolute magnitude by the expression[13]

$$\log R_c = \frac{7340}{T} - 0.14 - 0.2M_p + 0.5 \log (1 - 10^{-1.47 \times 10^4/T}) . \tag{30}$$

Let

$$W_g \sim \frac{1}{4} \frac{R^2}{r^2} \tag{31}$$

denote the geometrical dilution factor; R_c is the radius of the central star in terms of the sun's radius as unity and r the mean radius of the radiating layers of the planetary.

The radii of the central stars and the dilution factors have been calculated for the three assumed values of the color temperature—80,000, 100,000, and 150,000; hence the three computed values for each pair of tabular arguments in Table 4. The successive columns of Table 4 give the designation of the planetary, the computed radii, dilution factors, and $(WI)_c$ for the three assumed temperatures. The column headed $(WI)_{obs}$ is obtained from the use of equation (16), and the final column from the ratio of observed and computed dilution factors. As we expected, the observed dilution factor is larger, by

[11] *Lick Obs. Bull.*, **18**, 73, 1937.

[12] This value, adopted from Trumpler's investigation of galactic star clusters, represents some kind of an average over different galactic longitudes. A correction based on the assumption of uniform space absorption is admittedly rough but adequate for our present purposes.

[13] Russell, Dugan, and Stewart, *Astronomy*, p. 732, Boston: Ginn & Co.

a factor of 10^2 or 10^3, than the dilution factor predicted on the basis of geometrical considerations alone. Evidently the concentration of energy in the λ 304 line of *He* II is considerable. The absence of the other high-level permitted lines of *O* III is easy to understand; they cannot draw on the rich sources of energy available for excitation of the 3d $^3P_2$ level.

Photoionization and subsequent recapture of electrons in responsible for the radiations emitted by *He* II. It is this energy, radiated in the strongest *He* II line, that can be absorbed by the *O* III atoms. The observations show that the lines of the Bowen fluorescent mechanism are confined closely to the region in which the *He* II lines occur. We can reconcile this observational result with theory only if the opacity of the *O* III atoms is sufficiently great to use up all the *He* II 304 radiation within a small fraction of the total nebular radius.

TABLE 4

DILUTION OF λ 304 RADIATION IN PLANETARY NEBULAE

NGC	R_c	$W_g \times 10^{15}$	$(WI)_c \times 10^{16}$	$\times 10^{13}$ $(WI)_{obs}$	W_{obs}/W_{comp}
2022	1.20	4.6	1.82	14.	74. $\times 10^2$
	1.05	3.5	4.05	....	33.
	0.83	2.2	14.	....	9.6
3242	0.62	9.6	3.8	3.4	8.8
	0.54	7.2	9.3	....	3.6
	0.43	4.6	29.	....	1.2
7009	0.28	8.2	3.2	1.6	5.0
	0.24	6.2	8.0	....	2.0
	0.19	3.9	25.	....	0.65
7027	0.17	2.2	0.85	0.42	4.9
	0.15	1.6	2.1	....	2.0
	0.12	1.0	6.6	....	0.63
7662	0.27	5.3	2.1	3.3	16.
	0.23	4.0	5.2	....	6.4
	0.19	2.5	16.	....	2.1

We now investigate this question quantitatively. An *O* III atom that has just absorbed a λ 304 quantum and is raised to the 3d $^3P_2$ level may do one of three things: return to the initial level 2p² $^3P_2$, drop to 2p² $^3P_1$, or emit the lines of the Bowen fluorescent mechanism. The first two of these processes produce quanta that are immediately available for starting the cycle again. Only when the atom has started the cascade of the Bowen mechanism is the energy lost as far as concerns further excitation of *O* III.

In the zone of *He* II the major excitation of *O* III arises from the energy in the λ 304 line of that atom. Outside the *He* II zone, i.e., when all the ultraviolet radiation has been degraded in λ 304, a greater percentage of the radiation will lie in λ 303.62 of 2p² $^3P_1$ − 2p3d $^3P_2$. As we have stated above, this energy also can be used by *O* III to initiate the Bowen cycle. The Einstein *A*'s for the respective downward transitions *b* to *a* and *b* to *a′* have the ratio 3:1. The corresponding products of the populations in the lower levels and the absorption coefficients, $N_a a_\nu(ab)$ and $N_{a'} a_\nu(a'b)$, possess the same ratio. Hence, if the beams in the two wave lengths possess the same intensity, the absorptions and emissions will balance, except for the small fraction removed through the operation of the cyclic process. This loss of energy will in no way disturb the initial equality of the two beams. Should one or the other of the beams be initially of different intensity, absorption and emission of radiation will tend to bring them toward equality. The phenomenon is closely related to that of "interlocking," as discussed by Woolley.

We shall investigate the case when the two beams are equal. Then, according to the above argument, gains and losses from one beam into the other are exactly offset. Now, of the number of upward transitions caused by both beams, the fraction,

$$\frac{A_{ba} + A_{ba'}}{\Sigma A_{bi}} = 0.978 ,$$

reappears as downward transitions. Because of the much greater Einstein A's associated with the far ultraviolet lines, only 2.2 per cent of the transitions initiate the Bowen cycle. Thus, as far as the disappearance of radiation is concerned, the effective absorption coefficient a_{eff} is only about 0.022 times that of the true absorption coefficient. An element $d\tau_{\text{eff}}$ of effective optical depth is

$$d\tau_{\text{eff}} \sim 0.022 N_a a_\nu(ab) = 0.022 N_{a'} a_\nu(a'b) . \tag{32}$$

We do not know a_ν exactly, but we may adopt a mean value over the effective width of the line according to the formula

$$\bar{a}_\nu = \frac{\pi \epsilon^2}{mc} f \frac{1}{\Delta\nu} , \tag{33}$$

where we set

$$\frac{\Delta\nu}{\nu} = \frac{2v}{c} . \tag{34}$$

The quantity v is defined by the root-mean-square formula

$$v = \sqrt{\frac{2kT}{m_a}} = 0.321 \times 10^4 T^{1/2} . \tag{35}$$

We suppose the atoms to have the same kinetic temperature as the electrons, 10,000°. Therefore $\Delta\nu \sim 2.11 \times 10^{11}$ and

$$d\tau_{\text{eff}} \sim 3.2 \times 10^{-16} N_a dx . \tag{36}$$

Taking $N_a \sim 0.5$ and assuming that within a range of $\tau = 1 = 1.6 \times 10^{-16} x$, whence $x \sim 6 \times 10^{15}$ cm. or 400 astronomical units, we calculate that the intensity will fall to $1/e$ of its original value. The images of the O III lines should be about 5 per cent larger than those of the helium lines in a typical planetary.

To treat the problem more exactly, let us set up the equations of transfer for the $\nu(ab)$ and $\nu(a'b)$ transitions. If $I^\circ W_1'$ and $I^\circ W_2'$ are the intensities of the $\nu(ab)$ and $\nu(a'b)$ radiations, respectively, we have

$$\cos\theta \frac{d(I^\circ W_1')}{dx} = -\bar{a}_1 N_a (I^\circ W_1') + \bar{a}_1 N_b \frac{2h\nu^3}{c^2} \frac{\varpi_a}{\varpi_b} , \tag{37a}$$

$$\cos\theta \frac{d(I^\circ W_2')}{dx} = -\bar{a}_2 N_{a'} (I^\circ W_2') + \bar{a}_2 N_b \frac{2h\nu^3}{c^2} \frac{\varpi_a}{\varpi_b} , \tag{37b}$$

where

$$\bar{a}_1 = \frac{\pi\epsilon^2}{mc} f_{ab} \frac{1}{\Delta\nu}, \qquad \bar{a}_2 = \frac{\pi\epsilon^2}{mc} f_{a'b} \frac{1}{\Delta\nu}. \tag{38}$$

We also write, for the equation of statistical equilibrium,

$$\left.\begin{aligned} &\frac{4\pi}{h\nu_{ab}} \frac{\pi\epsilon^2}{mc} [N_a I^\circ W_1' f_{ab} + N_{a'} I^\circ W_2' f_{a'b}] \\ &\qquad = N_b \frac{8\pi^2\epsilon^2\nu_{ab}^2}{mc^3} \left\{ \left(\frac{\varpi_a}{\varpi_b} f_{ab} + \frac{\varpi_{a'}}{\varpi_b} f_{a'b} \right) + \sum_{i \neq a, a'} \frac{\varpi_i}{\varpi_b} \frac{\nu_i^2}{\nu_{ab}^2} f_{ib} \right\} = N_b \sum_j A_{jb}, \end{aligned}\right\} \tag{39}$$

where the summations are to be carried out over all the lower levels connecting with b. If we add equations (37a) and (37b) and substitute for N_b, we obtain

$$\left.\begin{aligned} \cos\theta \frac{d[I^\circ(W_1' + W_2')]}{dx} &= -I^\circ(N_a \bar{a}_1 W_1' + N_a \bar{a}_2 W_2') \\ &\quad + \left(\bar{a}_1 \frac{\varpi_a}{\varpi_b} + \bar{a}_2 \frac{\varpi_{a'}}{\varpi_b} \right) \frac{8\pi^2\epsilon^2\nu_{ab}^2}{mc^3} [N_a W_1' f_{ab} + N_{a'} W_2' f_{a'b}] \frac{I^\circ}{\Sigma A}. \end{aligned}\right\} \tag{40}$$

Now $N_{a'} = \frac{3}{5} N_a$ and

$$a_2 = \tfrac{5}{9}\bar{a}_1, \qquad \varpi_a = \varpi_b = 5, \qquad \varpi_{a'} = 3. \tag{41}$$

If we further suppose that everywhere

$$W_2'/W_1' = \text{constant} = 1, \tag{42}$$

in accord with the argument presented above for equality of the two beams, equation (40) becomes

$$\left.\begin{aligned} \cos\theta \frac{d(I^\circ W_1')}{dx} &= -N_a \bar{a}_1 \tfrac{2}{3} I^\circ W_1' \left[1 - \frac{4}{3} \frac{A_{ab}}{\Sigma A} \right] \\ &= -N_a \bar{a}_1 \tfrac{2}{3} I^\circ W_1' \left[\frac{\Sigma A - (A_{a'b} + A_{ab})}{\Sigma A} \right] = -0.0215 \bar{a}_1 \tfrac{2}{3} N_a I^\circ W_1', \end{aligned}\right\} \tag{43}$$

by the use of equation (2a) and the substitution of numerical values. Let

$$d\tau = 0.0215\, \bar{a}_1 \tfrac{2}{3} N_a dx = 0.0215 \frac{\pi\epsilon^2}{mc} f_{ab} \frac{1}{\Delta\nu} \tfrac{2}{3} N_a dx = 2.07 \times 10^{-16} N_a dx. \tag{44}$$

Then equation (43) becomes

$$\cos\theta \frac{d(I^\circ w_1')}{d\tau} = -W_1' I^\circ. \tag{45}$$

Integrating over all solid angles, we obtain the result

$$\frac{\partial F}{\partial\tau} = \int \cos\theta \frac{dI^\circ W_1'}{d\tau} d\omega = -J, \tag{46}$$

where

$$J = \int I^{\circ} W_1' d\omega = 4\pi \overline{W}_1 I^{\circ} . \tag{47}$$

Now, according to the Eddington approximation,

$$F = -\frac{1}{3}\frac{\partial J}{\partial \tau} , \tag{48}$$

whence

$$-\frac{1}{3}\frac{d^2 J}{d\tau^2} = -J \quad \text{or} \quad \frac{d^2 J}{d\tau^2} = +3J . \tag{49}$$

Integration gives the following result:

$$\overline{W} = \overline{W}_0 e^{-\sqrt{3}\tau} . \tag{50}$$

For the intensity to fall to $1/e$ of its original value, we must have

$$1 \sim 3.6 \times 10^{-16} N_a . \tag{51}$$

If $N_a \sim 0.5$, we find that $x \sim 5.5 \times 10^{15}$ cm or 370 astronomical units. The radii of the planetaries are about 10^{17} cm. Hence the images of the lines of the Bowen fluorescent mechanism should be only about 5 per cent larger than those of the helium images.

Bowen has mentioned the similarity in size between the λ 4686 image and the images of the ultraviolet radiations of *O* III. Measures of the sizes of the nebular images on microphotometer tracings of slitless spectrograms of NGC 7662 yield results which are in accordance with this suggestion; they indicate that the λ 3133 images of *O* III may be slightly larger than the λ 3203 image of ionized helium.

We shall discuss the fluorescent cycle in *N* III after treating the question of abundances.

HARVARD COLLEGE OBSERVATORY
August 2, 1941

Reprinted from *The Astrophysical Journal,* Vol. 102, No. 2, September, 1945.

PHYSICAL PROCESSES IN GASEOUS NEBULAE

XVIII. THE CHEMICAL COMPOSITION OF THE PLANETARY NEBULAE

LAWRENCE H. ALLER[1] AND DONALD H. MENZEL[2]
Harvard College Observatory

Received June 4, 1945

ABSTRACT

Bowen and Wyse's conclusion that the chemical composition of the planetary nebulae is essentially the same as that of the sun is substantiated by a more detailed treatment of the same problem in which an attempt has been made to estimate the relative abundances of the lighter elements. For hydrogen, helium, and carbon, which are represented only by permitted lines of the recombination spectrum, we have estimated mean relative abundances of 1000, 100, and 0.6. The abundance of hydrogen follows from the intensity of the continuous spectrum at the limit of the Balmer series. An estimate of the abundance of ionized helium is provided by the observed line intensities and the theory given by Goldberg. The singlets are stronger than they should be, according to Goldberg's theory; and we may interpret the discrepancies between observations and theory by supposing that collisional excitations and de-excitations of the atoms in the high 2^3S and 2^1S metastable levels play a role in fixing the population of the high-energy levels. The abundances of the *C* III and *C* IV ions have been evaluated from the recombination lines of *C* II and *C* III with the aid of approximate wave functions calculated by Slater's rules and of *b*-factors estimated from the hydrogenic case. The *O* III abundance in NGC 7009, as estimated from the *O* II recombination lines, appears to agree with that found from the green nebular lines, thus removing the discrepancy suggested by Wyse.

We have estimated the abundances of the other ions from their forbidden lines and, with the exception of *O* III, from approximate collisional cross-sections. To calculate the numbers of atoms of any kind, we must make some estimate of their distribution among various stages of ionization. This step, which has to be made empirically, seems to introduce the greatest share of the uncertainty in the final results. The average abundances of nitrogen, oxygen, fluorine, neon, sulphur, chlorine, and argon are found to be 0.2, 0.25, 0.0001, 0.01, 0.036, 0.002, and 0.0015, respectively—all on the basis of hydrogen as 1000. Comparisons with the compositions of the solar atmosphere and that of τ Scorpii are discussed. It was not possible to estimate the contribution of the metals. The possible effects of stratification and of the bright-line radiation of the central star are briefly discussed.

I. INTRODUCTION

Perhaps one of the most difficult of astrophysical problems is the determination of the absolute abundances of the various elements composing a planetary nebula. To some extent this difficulty arises from the different mechanisms by which the nebular radiations are produced. Thus, atoms represented by forbidden lines are favored by a factor of a thousand fold over those that show only permitted (recombination) lines. To a greater extent uncertainties arise from the fact that only one ion of a given element may be observable while the nonobservable ions may be in the majority.

In previous papers of this series[3] we have shown how to derive the abundances of hydrogen ions and *O* III ions in planetary nebulae from intensity measures of the Balmer and green nebular lines. The results indicate about ten thousand times as many hydrogen ions as *O* III ions in the planetary nebulae, despite the fact that the green nebular lines (N_1 and N_2) are often of the order of ten times as strong as $H\beta$.

[1] Now of Indiana University, Bloomington, Indiana.

[2] Now serving as lieutenant commander, United States Navy, on leave of absence from Harvard University.

[3] Series of papers entitled "Physical Processes in Gaseous Nebulae" appearing in the *Astrophysical Journal:* I, **85**, 330, 1937; II, **86**, 70, 1937; III, **88**, 52, 1938; IV, **88**, 313, 1938; V, **88**, 423, 1938; VI, **89** 587, 1939; VII, **90**, 271, 1939; VIII, **90**, 601, 1939; IX, **91**, 307, 1940; X, **92**, 408, 1940; XI, **93**, 178, 1941; XII, **93**, 195, 1941; XIII, **93**, 230, 1941; XIV, **93**, 236, 1941; XV, **93**, 244, 1941; XVI, **94**, 30, 1941; XVII, **94**, 436, 1941. The papers of this series are referred to in the text as papers I, II, III, etc.

The estimation of the abundances of the other ions from their spectral lines is a much more difficult problem. In the first place, with the exception of neutral and ionized helium, we have only rough guesses of the transition probabilities of the permitted lines of the various ions observed in the gaseous nebulae. Reasonably good transition probabilities are available from recent calculations[4] for most of the forbidden lines. So far, cross-sections for the collisional excitation of forbidden lines are available only for *O* III. The theory of the statistical equilibrium of low-lying metastable levels presents no essential difficulties. The number of such levels is small; and one can solve for the populations in the various levels in terms of the collisional cross-sections, the transition probabilities, and the temperatures and densities of the electron gas. We possess fair estimates of all the quantities but the cross-sections. If we can get some idea of the latter, we can work out the relative abundances of the various types of ions.

The next section of this paper will be devoted to estimates of the target areas for the excitation of metastable levels and to the application of these target areas in connection with transition probabilities of the forbidden lines and the appropriate equations of statistical equilibrium to calculations of ionic abundances. Once the target areas have been agreed upon, the calculation of ionic abundances proceeds as in the case of *O* III.

The third section considers the recombination spectra of ionized carbon, ionized oxygen, and helium. Here, again, necessary atomic parameters are not known except for helium, and the transition probabilities of permitted lines have been estimated with the aid of the *LS*-coupling hypothesis and radial quantum integrals derived with the aid of Slater's rules.[5] The abundance of *O* III in NGC 7009, estimated from the recombination *O* II lines, agrees well with that estimated from the green nebular lines, thus removing the discrepancy suggested by Wyse.[6]

The fourth section of the paper is concerned with the problem of ionization in a planetary nebula subjected to radiation from a hot central star. The simultaneous appearance of ions representing a considerable range in ionization potential shows that it is not possible to explain the planetary nebulae as shells of gas exposed to black-body radiation. Stratification of the nebular layers, the filamentary structure exhibited by many planetaries and suspected in most of them, combined with the probable bright-line character of the ultraviolet spectrum of the central star—all conspire to introduce considerable complexities into the problem.

Finally, from the calculated abundances of the various ions we estimate the actual abundances of the various elements that make up the planetary nebula. Since we observe only one or two stages of ionization for most elements, we have to estimate the contribution to the total from the unobserved lines. An empirical procedure seems best here; and the method actually employed introduces the assumption that the ratio of the abundance of the ions in successive stages of ionization—e.g., the ratio of *O* II/*O* III, *Ne* III/*Ne* IV, etc.—depends only on the ionization potentials involved and not at all on the type of ion. From such ratios, for a given nebula, curves giving the estimated distribution of the atoms in various stages of ionization may be constructed; and it is with the aid of these curves that the final abundances are estimated.

That the above procedure introduces an appreciable uncertainty in the estimated composition of the nebula is obvious enough, but the difficulty that one encounters here is fundamental. Only a few of the ions of each kind of atom are observed, and we must extrapolate somehow to take account of the others. Further observations and identifications of fainter lines will help us here. Especially desired are observations or identifications of ions of higher ionization potential than *Ne* V, so that we may estimate, with a higher degree of certainty than is possible at present, the proportion of very highly ionized atoms.

[4] Pasternack, *Ap. J.*, **92**, 129, 1940; see also Paper XI of series.

[5] Slater, *Phys. Rev.*, **57**, 36, 1930.

[6] *Ap. J.*, **95**, 356, 1942.

In the end it turns out, as Bowen and Wyse have suggested,[7] that the composition of the planetary nebulae is essentially the same as that of the sun. More accurately, it would be better to word the result as follows: A mass of gas, similar in composition to the sun and expanded to the size and density of a planetary nebula and exposed to dilute high-temperature radiation, would give exactly the same kind of a spectrum as the planetaries are observed to give, in so far as our present knowledge of atomic structure and absorption coefficients permit us to predict. Wyse and Bowen's conclusion is confirmed by this independent and more detailed theoretical treatment of the problem.

II. ABUNDANCES OF IONS ESTIMATED FROM THEIR FORBIDDEN LINES

The method of evaluating ionic abundances from the intensities of their forbidden lines is well illustrated by the case of *O* III, which has been treated in a previous paper.[8] The additional data needed were the electron temperature and density. If the latter could be calculated from the intensity of the Balmer continuum at the series limit, the former could be found from the ratio of λ 4363 to the green nebular lines.

The application of these theoretical considerations to any ion possessing a ground configuration with metastable terms is straightforward. Let *A* denote the ground level of the ion and *B* the upper level of the observed transition *AB*. The number of collisional excitations from *A* to *B* will be given by

$$\mathfrak{F}_{AB} = 8.54 \times 10^{-6} \frac{N_A N_\epsilon}{T_\epsilon^{1/2}} \frac{\Omega(AB)}{\varpi_A} e^{-\chi_{AB}/kT}, \tag{1}$$

and the number of collisions of the second kind (superelastic collisions) will be

$$\mathfrak{F}_{BA} = 8.54 \times 10^{-6} \frac{N_B N_\epsilon}{T_\epsilon^{1/2}} \frac{\Omega(AB)}{\varpi_B}, \tag{2}$$

where χ is the excitation potential of *B* with respect to *A*, where $\tilde{\omega}_A$ and $\tilde{\omega}_B$ are, respectively, the statistical weights of levels *A* and *B*, and where $\Omega(AB)$ is the target area factor to be calculated by wave mechanics.

Our present discussion is confined to transitions arising between the terms of the ground p^2, p^3, and p^4 configurations. The formulae of papers XIII and XVI of this series may be applied immediately to a discussion of the p^2 and p^4 configurations. For the statistical equilibrium of the p^3 configuration we shall derive slightly modified expressions.

Unfortunately, target areas for the collisional excitation of ions other than *O* III are not available, so that we have had to resort to an approximate treatment of the problem. Accordingly, rough target areas have been estimated according to the following scheme:

1. The target areas for a p^4 configuration are taken to be the same as for a p^2 configuration. In this approximation *Ne* III has the same cross-section as *O* III.

2. For a p^3 configuration the ratio $\Omega_{AB}/\tilde{\omega}_A\tilde{\omega}_B$ is taken as a constant for all transitions. (Within a factor of 2 this condition is fulfilled for *O* III.) For *O* II we have arbitrarily taken $\Omega = 10$ for the $^4S - {^2D}$ transition and $\Omega = 15$ for the $^2D - {^2P}$ auroral transition.

Thus the target areas for the p^2 and p^4 configurations are all taken to be the same as those of *O* III. Admittedly, this approximation is a crude one, but we suspect it will give the correct order of magnitude. The target areas are taken to be independent of *Z* in this treatment, although we have also carried out a calculation in which we have supposed the target area to be proportional to Z^2 for the larger values of *Z*. It seems very unlikely that the dependence on *Z* can be as severe as Z^2, but the effect of *Z* on the target area is

[7] *Lick Obs. Bull.*, 19, 1, 1939.

[8] See Paper XVI. Equations are referred to by number and paper; thus XII-7 means eq. (7) of Paper XII of the nebular series.

very difficult to see until further calculations have been made. For the main discussion we shall neglect any variation of Ω with Z but shall occasionally refer to the results of the earlier calculations in which it was supposed that Ω was proportional to Z^2. The values of Ω computed from the above approximation are collected in Table 1.

TABLE 1*

ATOMIC PARAMETERS FOR THE CALCULATION OF ABUNDANCES FROM FORBIDDEN LINES

NEBULAR TRANSITIONS

Ion	Config.	χ	ν_0/ν	ΣA	j
N II.........	p^2	1.89	1.31	0.0030	0.086×10^3
F II.........	p^4	2.58	0.956	.064	1.87
Ne III........	p^4	3.19	0.776	.275	8.05
S III.........	p^2	1.33	1.88	.092	2.69
F IV.........	p^2	3.07	0.808	.109	3.19
Cl IV........	p^2	1.59	1.56	.28	8.20
Ne V.........	p^2	3.69	0.672	.42	12.3
A V..........	p^2	1.92	1.29	0.73	21.4

AURORAL TRANSITIONS

Ion	Config.	χ	$\lambda/4363$	A_{SD}	f
N II.........	p^2	4.04	1.315	2.2	0.99
S III.........	p^2	3.32	1.411	5.6	.86
Cl IV........	p^2	4.00	1.216	6.6	0.71

NEBULAR TRANSITIONS

Ion	Config.	χ	ν_0/ν	$\bar{A}$	ξ
O II.........	p^3	3.31	0.75	0.00008	9.13
Cl III........	p^3	2.23	1.11	.0042	491
S II.........	p^3	1.84	1.35	.0014	164
A IV.........	p^3	2.62	0.95	0.0136	1590

TRANSAURORAL TRANSITION

Ion	Config.	χ	$\bar{A}$	ξ
S II..........	p^3	3.04	0.26	5×10^4

* The target area Ω = 10 for nebular transitions in p^3 and Ω = 6 for the transauroral transition.

Let us first turn our attention to the nebular $^3P - {}^1D$ transitions in p^2 or p^4 configurations. The relation between the number of atoms in the ^3P term and the surface brightness H_n (in magnitudes per square minute of arc) of the nebular images may be written in the form analogous to XVI-6:

$$N_P = 3.36\times10^{19}\frac{T_\epsilon^{1/2}e^{\chi/kT_\epsilon}}{N_\epsilon D(1-\beta)}(2.512)^{-H_n}\frac{\Omega_0}{\Omega}\frac{\nu_0}{\nu} = C\frac{T_\epsilon^{1/2}\quad e^{\chi_{PD}/kT_\epsilon}}{N_\epsilon\; D(1-\beta)}(2.512)^{-H_n}, \quad (3)$$

where C is a constant that depends on the ion involved, the target area, and the frequency; where χ_{PD} is the excitation potential of the ^{1}D level, D depends on the thickness of the nebular shell and its outer radius A, in accordance with the expression

$$D = 3d\left(1 - \frac{d}{A} + \frac{d^2}{3A^2}\right) \tag{4}$$

(cf. XII-7); and where β is the fraction of atoms excited to the ^{1}D level that are collisionally de-excited. It is given by

$$\beta = \frac{\mathfrak{F}_{DP}}{\mathfrak{F}_{DP} + F_{DP}} = \frac{1}{1 + \dfrac{\Sigma A_{DPi}}{3.42 \times 10^{-5}} \dfrac{T_\epsilon^{1/2}}{N_\epsilon} \dfrac{\Omega_0}{\Omega}} = \frac{1}{1 + \dfrac{jT_\epsilon^{1/2}}{N_\epsilon}}, \tag{5}$$

defining j. Here $\mathfrak{F}_{DP}$ represents the number of collisional de-excitations of the ^{1}D level (superelastic collisions) and F_{DP} the number of radiative transitions; Ω_0 denotes the target area for the nebular transition in the O III atom. If we write the abundance of the O III ion as (cf. XVI-6)

$$N_0 = 3.38 \times 10^{19} \frac{T_\epsilon^{1/2} e^{\chi_0/kT_\epsilon}}{N_\epsilon D(1-\beta_0)} (2.512)^{-H_0} \tag{6}$$

and recall that the last factor is simply proportional to the sum of the intensities of the two forbidden lines, we obtain for the ratio of the abundance of any ion, p, to that of doubly ionized oxygen

$$\frac{N_p}{N_0} = \frac{I_p \nu_0}{I_0 \nu} e^{(\chi-\chi_0)/kT_\epsilon} \frac{(1-\beta_0)}{(1-\beta)} \frac{\Omega_0}{\Omega}, \tag{7}$$

where the subscript zero refers to the data for the O III atom and the green nebular lines.

Likewise for the auroral transition in the O III atom, one may write a similar expression (cf. Paper XIII) by noting that in this case the ^{1}S level is fed almost entirely by collisional excitations from the ground level. Of the atoms entering the ^{1}S level, the fraction $f = A_{SD}/(A_{SD} + A_{SP})$ goes to produce the auroral line, provided the collisional de-excitation of the ^{1}S level is negligible in comparison with the radiative de-excitations. Hence we may write an expression of the form

$$N_P = 1.65 \times 10^{20} \frac{e^{\chi_{PS}/kT_\epsilon}}{Df} \frac{\Omega_0}{\Omega} \frac{\lambda}{4363} \frac{T_\epsilon^{1/2}}{N_\epsilon} (2.512)^{-H_n}, \tag{8}$$

where the wave length λ is expressed in angstrom units. We now find the ratio of the number of ions to the number of O III ions to be given by

$$\frac{N_P}{N_0} = \frac{4.88}{f} \frac{\Omega_0}{\Omega} \frac{\lambda}{4363} \frac{I_p}{I_0} e^{(\chi_{PS}-\chi_0)/kT_\epsilon} (1-\beta_0). \tag{9}$$

Table 1 gives the necessary atomic data for the calculation of N_P from H_n, T_ϵ, and N_ϵ. The values for the Einstein A's have been taken from the work of Pasternack, checked in many cases with the help of the tables or formulae of Shortley's work (Paper XI).

For some atoms, e.g., S II or N II, both the auroral (λ 5755 of N II) and the nebular (λ 6548 and λ 6584 of N II) lines are strong; and it is possible to estimate N_P in two ways. If the velocity distribution is reasonably Maxwellian, a comparison of these values should provide an estimate of the error of our approximate collisional cross-sections. The auroral lines tend to give somewhat higher abundances than the nebular lines, but the ratios are not constant and fluctuate widely from nebula to nebula. Thus the values of Table 2, based principally on the intensity measures of Wyse, with additive measures of brighter lines by one of us, show discrepancies ranging up to a factor of 3, apparently not correlated with the electron temperature.

Let us now turn to the case of the p^3 configuration. There are three terms to be considered here: a ground $^4S_{3/2}$ term and higher $^2D_{3/2,5/2}$ and $^2P_{1/2,\,3/2}$ levels. We shall have to

TABLE 2*

IONIC ABUNDANCES DETERMINED FROM FORBIDDEN LINES

NEBULA	*N* I	*N* II Neb-ular	*N* II Auro-ral	*O* II	*O* III	*Ne* III	*Ne* V	*S* II Neb-ular	*S* II Trans-auroral	*S* III	*Cl* III	*Cl* IV	*A* IV	*A* V
NGC 7027	0.037	0.224	0.58	0.1	1.11	0.055	0.063	0.0093	0.0121	0.177	0.0028	0.0281	0.0054	0.0094
NGC 6572	.046	.108	.32	0.44	0.78	.076	......	.0022	.008	.072	.0020	......	.0038	......
NGC 6826	......	......	......	0.05	0.34	.031	......	......	......	......	......	......	......	......
NGC 6543	......	.67	......	1.14	2.16	.35	......	......	......	......	......	......	......	......
IC 418	.046	.17	.16	2.1	0.072	.0024	......	.0065	.006	.028	.0017	......	......	......
NGC 7662	......	.011	.023	0.25	1.05	.0075	.0171	.0017	.0032	.035	.0011	.030	.011	.0014
NGC 7009	......	.063	.057	0.48	1.28	.0145	0.0034	.0052	.0072	.052	.0037	0.044	.0066	......
NGC 6741	.026	.207	.43	......	0.69	.053	......	......	.011	.15	.0048	......	.006	.003
NGC 2440	0.012	.054	0.11	0.02	0.10	.0095	......	0.0005	.0018	.015	0.0005	......	......	......
NGC 2165	......	.046	......	......	0.73	.059	......	......	.0084	.048	......	......	.0064	0.0015
IC 5217	......	0.032	......	......	0.67	0.064	......	......	0.0072	0.045	......	......	0.0054	......

* From the fluorine lines observed by Wyse in NGC 7027 and NGC 7662 we estimated the abundance of the *F* II ion to be 3×10^{-4} in NGC 7027 and 1.6×10^{-4} to be the abundance of the *F* IV ion in NGC 7662. All abundances refer to numbers of ions per cm^3 in the nebula. The *O* II abundances are computed on the basis of a weighted mean *A*-value of 7.2×10^{-5}, as determined from the Hartree field for *O* II (called "case *B*" in text).

consider the statistical equilibrium of these possible levels of the atom. For convenience we shall indicate the various terms and levels by the accompanying notation.

Abbreviation	Term	Abbreviation	Level
S	^4S	S	$^4S_{3/2}$
D	^2D	d	$^2D_{3/2}$
P	^2P	d'	$^2D_{5/2}$
		p	$^2P_{1/2}$
		p'	$^2P_{3/2}$

The number of atoms entering the d level by collisional excitation from the ground level and by radiative cascade from the P term will equal the number leaving by collisional and radiative processes:

$$\left.\begin{aligned} &\frac{8.54 \times 10^{-6}}{T_\epsilon^{1/2}} \frac{N_s N_\epsilon}{\varpi_s} \Omega_{sd} e^{-\chi_{sD}/kT_\epsilon} + N_p A_{pd} + N_{p'} A_{p'd} \\ &\qquad = \frac{8.54 \times 10^{-6}}{T_\epsilon^{1/2}} \frac{N_d N_\epsilon}{\varpi_d} \Omega_{sd} + N_d A_{ds} \end{aligned}\right\} \quad (10)$$

with a similar equation for d'. In these expressions we neglect the superelastic collisions which transfer atoms from P to D, since we shall suppose that the population of the P term is very small.

The number of atoms entering the P term by collision from the ground term must equal the number lost by superelastic collisions to the D or S terms and the radiative transitions. Thus for the p level we have

$$\left.\begin{aligned} &\frac{8.54 \times 10^{-6}}{T_\epsilon^{1/2}} \frac{N_\epsilon N_s}{\varpi_s} \Omega_{sp} e^{-\chi_{sp}/kT_\epsilon} = \frac{8.54 \times 10^{-6}}{T_\epsilon^{1/2}} \frac{N_p N_\epsilon}{\varpi_p} [\Omega_{sp} + \Omega_{Dp}] \\ &\qquad + N_P (A_{pd} + A_{pd'} + A_{ps}) , \end{aligned}\right\} \quad (11)$$

with a similar expression for p'. We have neglected collisional and radiative transitions between p and p' and between d and d'. If the electron density is such that we can neglect the collisional de-excitations of the p levels in comparison with the radiative de-excitations, we may write for the level d:

$$\left.\begin{aligned}&\frac{8.54\times 10^{-6}}{T_\epsilon^{1/2}}\frac{N_s N_\epsilon}{\varpi_s}\Big\{\Omega_{sd}e^{-\chi^{SD}/kT_\epsilon}+e^{-\chi_{sp}/kT_\epsilon}\Big[\Omega_{sp}\Big(\frac{A_{pd}}{A_{pd}+A_{pd'}+A_{ps}}\Big)\\&\quad+\Omega_{sp}\Big(\frac{A_{p'd}}{A_{p'd}+A_{p'd}+A_{p's}}\Big)\Big]\Big\}=\frac{8.54\times 10^{-6}}{T_\epsilon^{1/2}}\frac{N_d N_\epsilon}{\varpi_d}\Omega_{sd}+N_d A_{ds}\,,\end{aligned}\right\} \quad (12)$$

with a similar expression for d'. We may neglect the second term on the left side unless $\Omega_{Sp} >> \Omega_{Sd}$ or χ_{SP} is relatively small and T_ϵ is large. Thus our expressions become even simpler, for it turns out that we can neglect the existence of the ^{2}P term altogether in calculating the equilibrium of the ^{2}D term. Of the N_D atoms in the ^{2}D term, let us *now* use d to denote the fraction in the $^2D_{3/2}$ level and d' the fraction in the $^2D_{5/2}$ level, with $d + d' = 1$. The equation of statistical equilibrium for the ^{2}D term is then

$$\left.\begin{aligned}&\frac{8.54\times 10^{-6}}{T_\epsilon^{1/2}}\frac{N_s N_\epsilon}{\varpi_s}[\Omega_{sd}+\Omega_{sd'}]\,e^{\chi/kT_\epsilon}=\frac{8.54\times 10^{-6}}{T_\epsilon^{1/2}}N_D N_\epsilon\\&\qquad\times\Big\{\frac{d}{\varpi_d}\Omega_{sd}+\frac{d'}{\varpi_{d'}}\Omega_{sd'}\Big\}+N_D\,[d\,A_{ds}+d'A_{d's}]\,.\end{aligned}\right\} \quad (13)$$

If collisional de-excitations are more important than radiative transitions in depopulating the ^{2}D level, we may suppose that the populations of the two levels of the ^{2}D term are proportional to their statistical weights. In that case, $d = 0.4$ and $d' = 0.6$. If we let

$$\Omega_{Ab}+\Omega_{Ab'}=\Omega_{AB}\,, \quad (14)$$

and recall that $\tilde{\omega}_s = 4$, $\tilde{\omega}_d = 4$, and $\tilde{\omega}_{d'} = 6$, we obtain

$$\left.\begin{aligned}N_S&=0.4N_D\Big(1+\frac{0.4A_{ds}+0.6A_{d's}}{\Omega_{sD}}\,1.17\times 10^6\,\frac{T_\epsilon^{1/2}}{N_\epsilon}\Big)\,e^{\chi/kT_\epsilon}\\&=0.4N_D\Big[1+\xi\,\frac{T_\epsilon^{1/2}}{N_\epsilon}\Big]e^{\chi/kT_\epsilon}\,,\end{aligned}\right\} \quad (15)$$

where ξ depends on atomic constants only. The factor in parentheses is essentially b_S/b_D; it is a measure of the deviation from thermodynamic equilibrium. As N_ϵ increases, $b_S/b_D \to 1$, and

$$\frac{N_D}{N_s}=\frac{\varpi_D}{\varpi_s}\frac{b_D}{b_s}\,e^{-\chi/kT_\epsilon} \quad (16)$$

reduces to Boltzmann's law. The uncertainty in Ω will introduce but a small error in the computed value of N_S if N_ϵ is sufficiently large and the A's are sufficiently small and adequately known. Otherwise the error will be directly proportional to the error in Ω and that in A.

The surface brightness of the planetary in the light of a forbidden line will be given by

$$S_\lambda = 840\,(2.512)^{-H_\lambda}=\tfrac{1}{3}DN_D\,(d\,A_{dS}+d'A_{d'S})\,h\nu=\tfrac{1}{3}DN_D\bar{A}\,h\nu\,, \quad (17)$$

where H_λ is the surface brightness in magnitudes per square minute of arc. We thus obtain

$$N_S=\frac{1.01\times 10^3}{A\,h\nu}\frac{b_s}{b_D}\frac{e^{\chi/kT_\epsilon}}{D}(2.512)^{-H_\lambda}=C_3\frac{b_s}{b_D}\frac{e^{\chi/kT_\epsilon}}{D}(2.512)^{-H_\lambda}\,, \quad (18)$$

where C_3 depends only on atomic parameters. Table 1 gives the ion, the adopted A-values, the excitation potential χ, ξ, and f.

The transauroral $^4S - {}^2P$ transition is observed in S II; and for this line a similar theory yields

$$N_s = 1.28 \times 10^{15} \frac{e^{\chi_{sp}/kT_\epsilon}}{D} \left(\frac{b_s}{b_p}\right) (2.512)^{-H_\lambda} . \tag{19}$$

Except for *NGG* 2440 and 6572 the values of N_S for S II computed from the transauroral lines agree with those found from the nebular lines to within a factor of about 2.

The close O II pair at 3726.16 and 3728.91, corresponding to the $^4S_{3/2} - {}^2D$ transition, are second in importance only to the green nebular lines. These lines are observed in most planetary nebulae, in the diffuse nebulae of the Milky Way, and in emission nebulosities in external galaxies, but not in the novae.

Pasternack gives Einstein A-values for the $^4S_{3/2} - {}^2D_{3/2}$ and $^4S_{3/2} - {}^2D_{5/2}$ lines as 0.00011 and 0.0014, respectively. Since the statistical weights of the two ^2D levels are respectively 6 and 4, we should expect the relative intensities of these lines to be

$$\frac{6 \times A_{3729}}{4 \times A_{3726}} = 1.9 ,$$

in favor of the λ 3729 line. Actually, just the opposite is observed. The $^4S - {}^2D$ transition comprises both magnetic dipole and electric quadrupole components, whose strengths may be computed by the methods of Paper XI. For O II Shortley's paremeter, $\chi = 0.0153$, and the magnetic dipole strengths are given by

$$\left.\begin{aligned} S_m ({}^2D_{5/2} - {}^4S_{3/2}) &= \tfrac{25}{243}\chi^4 = 5.63 \times 10^{-9} \\ &\text{and} \\ A_m ({}^2D_{5/2} - {}^4S_{3/2}) &= 4.82 \times 10^{-7} \\ S_m ({}^2D_{3/2} - {}^4S_{3/2}) &= \tfrac{3025}{972}\chi^4 = 17.05 \times 10^{-8} \\ &\text{and} \\ A_m ({}^2D_{3/2} - {}^4S) &= 2.19 \times 10^{-5} . \end{aligned}\right\} \tag{20}$$

Similarly for the quadrupole transitions, we have

$$A_q ({}^2D_{5/2} - {}^4S_{3/2}) = 14.05 \times 10^{-5} s_q^2 , \quad A ({}^4S_{3/2} - {}^2D_{3/2}) = 9.08 \times 10^{-5} s_q^2 , \tag{21}$$

where s_q depends upon the radial quantum wave functions according to the expression

$$s_q = \tfrac{2}{5} \int r^2 R(nl)\, dr , \tag{22}$$

where r is in atomic units. The total Einstein A coefficient is

$$A = A_q + A_m \tag{23}$$

for any given transition; and if we let $s_q = 1$, we obtain Pasternack's value. If we use the Hartree self-consistent field radial wave function, we find $s_q = 0.73$; and the resultant A-values are

$$A ({}^2D_{5/2}, {}^4S_{3/2}) = 7.45 \times 10^{-5} \quad \text{and} \quad A ({}^2D_{3/2}, {}^4S_{3/2}) = 6.8 \times 10^{-5} , \tag{24}$$

which give relative intensities of 1.64—again in contradiction to the observations.

A third of the possible procedures is to adopt the observed intensity ratio of the λ 3729 to λ 3726 lines (2 is a sufficiently accurate value for illustrative purposes) and calculate the required values of the A's with the utilization of s_q as an adjustable parameter. Thus,

$$A\,(^2D_{3/2}, {}^4S_{3/2}) = 3A\,(^2D_{5/2}, {}^4S_{3/2})\,;$$

and, since $A = A_q + A_m$, we obtain $s_q = 0.257$ and finally

$$A\,(^2D_{3/2}, {}^4S_{3/2}) = 2.78 \times 10^{-5} \quad \text{and} \quad A\,(^2D_{5/2}, {}^4S_{3/2}) = 0.93 \times 10^{-5}\,, \quad (25)$$

values considerably smaller than those found by Pasternack. These considerations illustrate the extreme sensitivity of the A-values to the outer portions of the radial wave functions. It is quite possible that the wave function may be inaccurate enough in this region of large r to account for the discrepancy.[9]

In view of the approximate character of the available target areas and transition probabilities, our results for *O* II are of a decidedly tentative character. We have computed the abundance of singly ionized oxygen on the basis of two assumptions concerning the weighted mean value of A, viz., assumption A, $s_q = 0.257$, and $A = 1.67 \times 10^{-5}$, which gives the correct intensity ratio but has no justification in quantum mechanical calculations; and assumption B, with $s_q = 0.73$, and $A = 7.19 \times 10^{-5}$, as calculated from the best available Hartree function but which, unfortunately, does not give the correct intensity ratio of the two components of the line.

Assumption A gives us a lower limit to the transition probability and therefore an upper limit to the abundance of ionized oxygen. Probably the results found from assumption B are more nearly of the correct order of magnitude.

Many years ago Wright[10] found an infra-red line at λ 7325 in NGC 7027. This line is the unresolved $^2P - {}^2D$ (auroral) pair of [*O* II] at λ 7319 and λ 7330. Wyse, however, did not observe these lines, and no intensity measures are available; so we cannot include them in our present study. On the basis of the excitation potentials and transition probabilities we would expect the intensity of this pair to be of the same order of magnitude as that of λ 4363 of [*O* III] for comparable abundances of *O* II and *O* III.

Unless the target areas for collisional excitation of neutral nitrogen are very small, i.e., $\Omega << 10$, we can probably get a good estimate of the abundance of this atom by simply supposing that the population of the ^2D term is given by Boltzmann's law and by calculating the number of metastable atoms from the intensities of their forbidden lines. Here again, unfortunately, the same objections that apply to the transition probabilities in *O* II are valid; if spin-orbit interactions are important in ionized oxygen, they are also important in neutral nitrogen. The scatter in the abundance of nitrogen as derived from the forbidden lines may represent actual variations in nitrogen content rather than effects of ionization or excitation. Where the [*N* II] lines are strong, they are nearly always accompanied by neutral nitrogen. Conversely, when they are weak, as in NGC 7662, NGC 7009, or IC 5217, the [*N* I] lines are missing.

In most elements we observe two successive stages of ionization; but neon deserves special comment, since, although the lines of *Ne* III and *Ne* V have been found in great strength in a number of planetaries, the auroral lines of *Ne* IV at λ 4714 and λ 4720 have not been found (the nebular lines fall in the inaccessible ultraviolet). Our equations of equilibrium reveal that with an electron temperature of 10,000° K, an electron density of 10^4, and an excitation potential of 7.7 volts, these lines should be about ten thousand times fainter than the *Ne* III lines. Their absence is thus easy to understand. Intensities of the *Ne* V lines in NGC 6741, 2440, 2165, or IC 5217 were not available.

The precision of our abundance determination could be improved were better esti-

[9] A refined treatment of the λ 3727 transition probabilities is being undertaken by C. W. Ufford and one of us. In this treatment the spin interaction of one electron with the orbit of another electron is being taken into account following a suggestion by J. H. Van Vleck. Calculations are at present in progress and should yield improved values of the transition probabilities eventually.

[10] *Pub. A.S.P.*, 32, 64, 1920.

mates of the collisional cross-sections and transition probabilities available. Probably our results are correct in so far as orders of magnitude are concerned. As we shall see, the chief uncertainty of our attempted quantitative analysis comes not so much from poor knowledge of the relevant physical parameters, in many cases, as from the distribution of the atoms of a given element among the various stages of ionization. In the calculation of abundances, it is therefore not correct simply to add together the observed numbers of ions of each element. Such a procedure will give the right answer only for hydrogen and helium, since, as we shall see presently, the number of neutral hydrogen and helium atoms in the luminous body of a planetary nebula is very small. We would obtain too small an abundance of the other atoms, such as argon or oxygen, which can exist in six or eight stages of ionization. If we count only the number of observed ions of these atoms, we may get more or less correct relative abundances, but our absolute values will be in error by factors as high as 10 or more.

III. ABUNDANCES ESTIMATED FROM RECOMBINATION LINES

Permitted lines of helium, carbon, and oxygen in various stages of ionization are observed in the gaseous nebulae. Presumably these are recombination lines. Electrons are recaptured on high levels; and, as they cascade downward in energy, the atoms emit the ordinary permitted lines of the spectrum. Only for hydrogen and helium are the transition probabilities known, so that one may work out a theory of statistical equilibrium for these atoms. Hydrogen has been discussed in detail earlier in this series, and Goldberg has worked out the theory of statistical equilibrium for neutral helium.

The treatment of ionized helium follows at once from that of hydrogen; we may carry Menzel and Baker's tables from Paper III of the nebular series[3] directly over to the problem at hand. Since

$$X_n = \frac{hRZ^2}{n^2kT_\epsilon} = \frac{hR}{n^2k\left(\dfrac{T_\epsilon}{Z^2}\right)}, \tag{26}$$

we need only call the arguments of the tables $4T_\epsilon$ instead of T_ϵ; e.g., the solution for *He* II at $T_\epsilon = 80{,}000°$ K is equivalent to that of hydrogen at 20,000°. Unfortunately, the tables do not contain entries for the physically interesting case of $T_\epsilon = 10{,}000°$ K. The strongest line of ionized helium is λ 4686, the first line of the Paschen series for which $n' = 3$, $n'' = 4$, $g = 0.768$, and $Z = 2$ in the usual expression (I-9)

$$E_{n'n} = N_i N_\epsilon \frac{KZ^4 b_n}{T_\epsilon^{3/2}} g \frac{2hRZ^2}{n'^3} e^{hRZ^2/n^2kT_\epsilon} \tag{27}$$

for the amount of energy radiated per cubic centimeter per second. Here[11] $K = 3.20 \times 10^{-6}$, R is the Rydberg constant, and b_n is a number that measures the departure of the nth state from its population in thermodynamic equilibrium. Hence,

$$E_{4,3} = 3.92 \times 10^{-18} \frac{N_i N_\epsilon}{T_\epsilon^{3/2}} b_4 e^{X_4}; \tag{28}$$

and the number of doubly ionized helium atoms is, accordingly,

$$b_4 N_i = 6.43 \times 10^{20} \frac{T_\epsilon^{3/2}}{N_\epsilon D} e^{-X_4} (2.512)^{-H}. \tag{29}$$

If we knew b_4, we could find N_i at once; but, unfortunately, this quantity is not immediately available. One procedure is to assume Baker and Menzel's case B and to extra-

[11] The value of K in Paper I is incorrectly given.

polate b_4 from their table. In this way we have obtained $b_4 = \frac{1}{200}$, which is probably correct in so far as the order of magnitude is concerned. An alternative approach is to employ some of the higher lines of the Pickering series, alternate members of which are blended with the Balmer series and others of which are confused with lines of other elements. Convenient Pickering lines are λ 4542 and λ 5411, which Wyse has measured in a number of nebulae. The upper level of the first transition is $n = 9$; that of the second is $n = 7$. We obtain

$$b_9 N_i = 1.48 \times 10^{22} \frac{T_\epsilon^{3/2}}{N_\epsilon D} e^{-X_9} (2.512)^{-H_\lambda}$$

and

$$b_7 N_i = 7.19 \times 10^{21} \frac{T_\epsilon^{3/2}}{N_\epsilon D} e^{-X_7} (2.512)^{-H_\lambda}. \qquad (30)$$

In all cases we have extrapolated b_n from Menzel and Baker's table for case B, obtaining $b_4 = 0.005$, $b_7 = 0.08$, and $b_9 = 0.125$. The resultant N_i's from λ 4686[12] and λ 4542 are usually in good agreement, in view of the precision of the measurements and the assumption that the b's are the same in all the nebulae. The N_i's obtained from λ 5411 are usually larger than those obtained from the other two lines. The assigned values of b may be in error, or Wyse's intensity estimates in the visual region may be on a slightly different scale from those in the photographic region.

The recombination lines of neutral helium are frequently strong in the planetary nebulae, indicating that helium is the most abundant element there, next to hydrogen. Ambarzumian[13] first treated the problem of the helium nebula and noted the interesting effects due to the high metastable levels 2^1S and 2^3S.

Since an ionized helium atom may recapture an electron in either a singlet or a triplet term the downward cascading electron will remain in the singlet or in the triplet system. Intercombination lines are virtually forbidden, since almost strict LS coupling obtains in helium.

Atoms in the high normal n^1P levels may cascade directly to the ground level with the emission of lines of the principal series, or they may linger on the way, dropping to n^1S_0, etc. If they reach the 2^1S_0 level, they find themselves in a metastable state. If there are no collisional de-excitations, the mean lifetime of this state is determined by the probability of the simultaneous emission of two photons. Breit and Teller,[14] on this basis, estimate that the lifetime of the 2^1S level is about the same as that for the 2s level of hydrogen, for which they find a lifetime of 0.14 second. The two quanta formed in this process escape from the nebula as part of the continuous radiation field. Accordingly, quanta of the principal series in helium suffer a fate somewhat different from that of the Lyman quanta in hydrogen. Consider, for example, the $1^1S - 3^1P$ transition. From 3^1P the atom may return to 1^1S, restoring the original quantum, or to 3^1S, thence to 2^1P, and on to 2^1S or 1^1S; or it may go to 2^1S directly. Quanta of the types $3^1S - 3^1P$, $2^1S - 3^1P$, and $2^1P - 3^1S$ escape from the nebula without further reabsorptions unless the metastable level is very heavily populated. By such a process, quanta of the principle series are degraded into low-frequency line quanta and continuous radiation from the double photon $2^1S - 1^1S$ transition.

On the other hand, all electrons recaptured on triplet levels cascade ultimately to the 2^3S level. According to Breit and Teller, the probability of the spontaneous emission of two photons is very much smaller (perhaps by a million fold) than in the case of 2^1S, so

[12] There is no evidence for any self-reversal of λ 4686 in the nebulae, although such self-reversal is quite apparent in such objects as Wolf-Rayet stars.

[13] *Pulkovo Bull.*, **13**, No. 3, 1933; *M.N.*, **93**, 50, 1932.

[14] *Ap. J.*, **91**, 215, 1940.

that it seems likely that in the case of a planetary nebula the bulk of the de-excitations of the 2^3S level take place by collision.

Ambarzumian treated the 2^3S level as rigorously metastable, in which case the $2^3S - 2^3P$, λ 10,830 transition plays the role of Lyman alpha and the $2^3S - 3^3P$, λ 3889 line the role of Lyman beta. Monochromatic nebular images in the λ 10,830 line would be very interesting. Unfortunately, the λ 3889 line is confused with $H\zeta$ of hydrogen, from which it cannot be separated.

More recently, Goldberg has treated the problem of the statistical equilibrium of neutral helium. Although he formally investigates an eight-state helium atom, he was able to allow somewhat for the variation of the continuous absorption coefficient with frequency and for the infinite number of discrete levels. In solving the equations of statistical equilibrium for the b's he used the results of his previous calculations of the f-values and treated the 2^1S and 2^3S levels as rigorously metastable. His treatment is based on purely radiative processes and takes no account of collisional effects or the possibility of the $2^1S - 1^1S$ transition.

In Table 3 we have summarized the results of the calculations of the helium abundances from two singlet lines, λ 4143 and λ 4388, and two triplet lines, λ 4471 and λ 5876. Goldberg's theory predicts a very great weakening of the singlet lines with respect to the triplets. For example, the relative intensities of λ 4143 and λ 4471 for an electron temperature of 10,000° K and a central star temperature of 50,000° K is predicted to be 0.013, while the observed intensity ratio is 0.16 for NGC 7662, according to the observations of Wyse. We find, on reference to the table, that the N_i's calculated from the singlets are some eight or ten times larger than the N_i's calculated from the triplets. So large a discrepancy can scarcely be attributed to any scale error in Wyse's intensities. We are inclined to look for an explanation in collisional and radiative effects not taken into account by Goldberg's treatment.

Collisions of electrons with atoms in the metastable ^1S and ^3S levels affect the populations of these and the higher levels in the following ways. In the first place, there may be superelastic collisions resulting in the de-excitation of ^1S and ^3S. For slow electrons, i.e., those with energies in the range from 0 to 5 volts, the cross-section for the de-excitation of the triplet is much greater than that of the singlet level, because the $^1S - {}^3S$ transition represents an effect of electron exchange and will be particularly important at low electronic energies. In the second place, inelastic collisions may excite atoms to the higher permitted levels from the metastable ones, and this effect would tend to make the b's larger than predicted by Goldberg's theory. Thirdly, although no radiative transitions between the singlet and triplet systems ever take place, collisional transitions are likely to be quite frequent, because the slow electronic velocities favor electron-exchange processes.

Accordingly, changes in density throughout the nebula ought to produce marked changes in the populations of the helium levels and in the resultant spectrum. Thus, in helium the effects of density fluctuations should be apparent in one stage of ionization, whereas, in general, density effects are apparent only in considering more than one stage.

We are inclined to attribute the strengthening of the singlets with respect to the triplets primarily to radiative processes. The singlet levels can be populated by direct transitions from the ground level; and if the radiation from the central star is rich in the resonance helium lines or if the nebula is of sufficient optical thickness to permit the building-up of an appreciable radiation density in the helium lines, radiative excitation may play the dominating role in producing the strong singlet lines. In our abundance calculations we have supposed this to be the case and have estimated the quantity of helium from the triplet lines. The resulting abundances may be too low if the collisional de-excitation of the 2^3S level is important.

When we turn to the radiative processes involved in recombination in more complex atoms, we are confronted with two very great difficulties. First, we must know the transition probabilities for jumps between all the upper levels and between the higher levels and the ground level. Second, we must use these parameters to calculate the statistical equilibrium for each level. Only then can we take the intensity of a given recombination line and infer the abundance of the ion which is gaining an electron.

Actually, we have not been able to carry this program through for any atom, and the best we can utilize is a rather approximate substitute. Wave functions for the excited and

TABLE 3

ABUNDANCES OF SINGLY AND DOUBLY IONIZED HELIUM FROM RECOMBINATION LINE INTENSITIES

ABUNDANCES OF SINGLY IONIZED HELIUM FROM He I LINES

Nebula	Singlets				Triplets			
	λ 4143		λ 4388		λ 4471		λ 5876	
	b_nN_i	N_i	b_nN_i	N_i	b_nN_i	N_i	b_nN_i	N_i
NGC 7027.....	256	3700	427	6150	184	750	306	1250
NGC 6572.....	463	5050	576	6300	261	1000	360	1450
IC 418.....	225	2250	462	4620	85	210	93	370
NGC 7662.....	525	4200	500	4000	197	580	143	420
NGC 7009.....	490	4900	585	5800	202	700	370	730
NGC 6741.....	....	....	....	....	49	220	260	1200
NGC 2440.....	134	1340	110	1100	42	120	68	190
NGC 2165.....	....	....	206	3420	76	370	68	330
IC 5217.....	1200	11000	800	7000	220	670	270	840

ABUNDANCES OF DOUBLY IONIZED HELIUM FROM He II LINES

Nebula	λ 4686		λ 5411		λ 4542	
	b_4N_i	N_i	b_7N_i	N_i	b_9N_i	N_i
NGC 7027.....	3.30	660	180	2300	80	640
NGC 6572.....	0.05	10	8	100	....	....
NGC 7662.....	6.2	1200	200	2500	170	1300
NGC 7009.....	1.3	260	130	1700	60	460
NGC 6741.....	1.1	210	100	1000	46	370
NGC 2440.....	12.0	230	90	1100	47	370
NGC 2165.....	1.0	210	30	380	78	600
IC 5217.....	0.75	150	35	440	....	....

ground levels of carbon and oxygen in various stages of ionization could be derived by Hartree's method or by some variation method, but such calculations are likely to be time-consuming. As a first rough approach to the problem, we have employed, for the evaluation of the radial quantum integrals, the wave functions calculated in the manner suggested by Slater. The Slater functions often give good approximations to the energies of the excited levels, but they neglect the loops and nodes of the wave functions, and the calculation of transition probabilities may suffer as a consequence. It seems likely, however, that they are definitely better than the hydrogenic functions, since they do take some account of the screening by the inner electrons.

The number of atoms in a level j is given by

$$N_j = b_j \frac{N_i N_\epsilon}{T_\epsilon^{3/2}} \frac{h^3}{(2\pi m k)^{3/2}} \frac{\varpi_j}{\varpi_i} e^{I_j/kT_\epsilon}, \tag{31}$$

where $\tilde{\omega}_j$ is the statistical weight of level j; $\tilde{\omega}_i$ is the statistical weight of the ion; and $I_j = I_0 - \chi_j$, where I_0 is the ionization potential and χ_j is the excitation potential of the level j. Now

$$N_j = \frac{2.52 \times 10^3}{DA\,h\nu} (2.512)^{-H_\lambda},$$

so that we obtain the general formula

$$b_j N_j = \frac{(2\pi m k)^{3/2}}{h} \frac{2.52 \times 10^3}{A\,h\nu} \frac{\varpi_i}{\varpi_j} \frac{T_\epsilon^{3/2}}{N_\epsilon D} (2.512)^{-H_\lambda} e^{-I_j/kT_\epsilon}. \tag{32}$$

Consequently, if we can evaluate b_j, we can find N_j. In our preliminary reconnaissance, about the best we can do is to adopt the b's from the analogous hydrogenic problem. Accordingly, it is to be kept in mind that the abundances estimated from one or two lines are likely to be very approximate, clearly because of the necessity of making shaky assumptions about the b's. One cannot simply adopt the hydrogenic b's for the same n with any degree of assurance, because in complex atoms the l degeneracy is removed and there may be terms of several multiplicities and L-values arising from each configuration. Finally, configura tioninteraction may enter, and there may be deviations from LS coupling.

Carbon is represented only by its recombination lines. In the case of *O* II there is only the λ 4267 line corresponding to the 3d²D — 4f²F transition, for which we have derived $A = 3.18 \times 10^8$. The intensity of this line gives us an estimate of b_4 multiplied by the number of *C* III ions; and if we adopt hydrogenic b_4's, the number of *C* III atoms turns out to be of the same order of magnitude as the number of *O* III atoms. In a similar fashion we have estimated the abundance of the *C* IV ion from the λ 4187 recombination lines of *C* III with $b_5 = 0.10$.

Wyse has observed lines of *C* IV in a number of planetaries, indicating that four-times ionized carbon atoms must exist in such objects as NGC 7662 and NGC 7027. Unfortunately, for such highly ionized atoms we have no reliable means of estimating the b's, and consequently it is not possible to estimate the abundances of these ions. In particular, the λλ 5800–5813 *C* IV pair arises from a level 25 volts below the ionization limit, and $b_3 N_i$ turns out to be of the order of 10^{-12}; but b_3 is probably also very small, so we cannot guess the number of *C* V ions.

The permitted *O* II lines observed by Wyse in NGC 7009 are of particular interest because they allow an independent estimate of the *O* III abundance. Wyse believed he found a serious discrepancy between the abundance of *O* III ions calculated from the forbidden lines and that obtained from the recombination spectrum. He assumed the rate of capture to depend on Z^2 (it should have been Z^4, according to I-24) and compared the spectrum with that of hydrogen. It is not clear just what approximations one should make to treat carbon as a hydrogenic atom in view of the high order of degeneracy among the energy-levels of the latter. The radial wave functions must differ considerably, although by adjusting the effective Z-values some improvement may be effected.

As an alternative procedure we have computed the transition probabilities of the *O* II ions observed in NGC 7009 with the aid of Slater wave functions and the assumption of LS coupling. These computed A-values and Wyse's intensity measures enable one to estimate $b_n N_{O\,\text{III}}$ (see Table 4). Since we had no knowledge of b_n, we plotted $b_n N_{O\,\text{III}}$ against the excitation potential (see Fig. 1). Notice that, as the excitation potential increases, b_n increases. The circle in the upper right-hand corner indicates the estimate of

O III determined from the green nebular lines; it was placed on the diagram at the ionization limit, where $b_n = 1$. These data suggest to us that there is no discrepancy between the number of *O* III ions calculated from the green nebular lines and the number found

TABLE 4*

THE RECOMBINATION SPECTRUM OF O II IN NGC 7009

Transition	*s*	*S*	log *A*	χ	*I*	log b_n *N*(*O* III)
$p^23p-p^23d(^3P)$						
4072 $^4D-^4F$	0.24	168	8.27	28.57	0.7	−1.61
4132 $^4P-^4P$	.14	30	7.55	28.71	0.3	−0.98
4153 $^4P-^4P$	.15	....	7.41	28.70	0.4	−0.90
4156 $^4P-^4P$	.15	....	7.59	28.71	0.3	−1.03
4397 $^2D-^2D$	.56	15	7.61	28.94	0.3	−1.07
$p^23p-p^23d(^1D)$						
4185 $^2F-^2G$	.43	108	8.28	31.18	0.3	−0.60
4190 $^2F-^2G$	.56	....	8.30	31.18	0.3	−0.82
4873 $^2P-^2D$	.60	21	7.64	31.24	1.0	0.70
p^23p-p^24f						
4284 $^4D-^4F$	.04	374	8.50	31.62	0.3	0.37
4295 $^4P-^4D$	.11	227	8.75	31.58	0.3	−0.65
4304 $^2P-^4D$	.40	....	9.02	31.57	0.4	−1.09
4609 $^2D-^2F$	.58	168	8.85	31.00	0.4	−0.90
$p^23s-p^23p\ (^3P)$						
4318 $^4P-^4P$	.29	12	7.67	25.74	0.4n	−2.94
4590 $^2D-^2F$	.58	14	8.07	28.24	0.3n	−1.90
4598 $^2D-^2F$	0.40	....	8.02	28.24	0.30	−1.84

* The first column gives the various lines and term designations grouped according to configuration and parentage; *s* is the fraction of the strength of a multiplet radiated by a given line; *S* is the strength of the multiplet; log *A* is based on approximated wave functions; χ is the excitation potential; and *I* is taken from the work of Wyse.

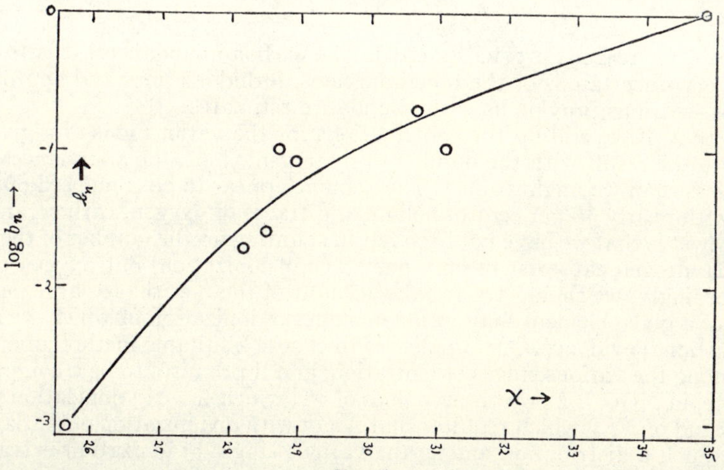

FIG. 1

from the recombination spectrum—a result especially satisfying in view of the approximate character of the *A*-values, the weakness of the lines, and the fact that target areas for *O* III may be in error by a factor of 2 or 3.

IV. IONIZATION AND STRATIFICATION

The calculated atomic abundances of the observable ions of hydrogen, helium, carbon, nitrogen, oxygen, fluorine, neon, sulphur, chlorine, and argon are summarized in Tables 2 and 5. The observational data for the weaker lines and for the lines in the visual region have been taken from the work of Wyse; for the stronger lines in the ordinary and photographic regions of the spectrum, from the investigations by one of us.[15] These tables also include the adopted electron densities and temperatures, and it will be noticed that they are somewhat different from the previously published values in some cases. For the objects observed by Wyse, for which no data on the Balmer continuum were available, we have estimated the electron density from the intensity of $H\beta$ in the manner indicated in Paper XII.

Most of the hydrogen in the luminous regions of the planetaries is ionized, so that the electron density which is equal to the ion density is essentially the density of hydrogen itself, since this gas is the most important constituent of gaseous nebulae. The estimates of ionized helium and doubly ionized helium follow from their recombination lines and

TABLE 5

ABUNDANCES ESTIMATED FROM RECOMBINATION LINES

Nebula	Hydrogen	*He* II	*He* III	C III	C IV
NGC 7027	5.9×10^3	750	650	1.08	0.3
NGC 6572	10.5	1030	10	1.2	.2
NGC 6826	4.5	181	....	....	....
NGC 6543	11.7	415	....	....	....
IC 418	11.0	210	....	0.84	....
NGC 7662	13.2	580	1200	2.12	0.2
NGC 7009	20.4	700	350	2.6	....
NGC 6741	2.2	222	300	....	....
NGC 2440	1.5	120	300	0.3	....
NGC 2165	2.4	366	300	0.7	....
IC 5217	4.4	674	150	1.0	....

are uncertain, for reasons previously cited. The carbon abundances, it is to be emphasized, are very uncertain. For the remaining ions, forbidden lines and approximate collisional cross-sections provide us with abundance estimates.

We find that, if we add up the contributions for the various ions of a given element and compare the result with the number of hydrogenic ions, the abundances tend to be somewhat less than we might expect. The nebulae appear to be composed of 90 per cent hydrogen with nearly 10 per cent of helium and traces of oxygen, carbon, and nitrogen, etc. The answer is that we have not been fair in counting up the number of the more complex ions. Hydrogen can exist in only one stage of ionization, but oxygen can exist in eight. Accordingly, we should try to take account of the fact that a large percentage of the atoms of a given element may reside in stages of ionization of which we have no direct observational evidence. Our problem is to choose a suitable method of apportioning the ions among the various stages of ionization, and it promises to be a complicated one. To illustrate, in NGC 7027 we observe lines of *N* I which have an ionization potential of 14.46 volts and of *Ne* V which requires that *Ne* IV with an ionization potential of 97 volts be ionized. Such a distribution among the various stages of ionization is scarcely to be expected in a nebula of uniform density, with no marked radial stratification and exposed to high-temperature black-body radiation. Several possible explanations suggest themselves.

[15] See Paper XIV of this series.

The planetaries are certainly not of uniform density or symmetrical structure. It seems very likely that local regions of high or low relative density may exist here and there throughout the nebula, thereby favoring fluctuations of the ionization. A nebula possessing a distinct filamentary structure may exhibit quite a different spectrum from a uniform nebula of the same mean density. Within the filaments the degree of ionization may tend to be smaller because of the higher density, the auroral lines may become more prominent with respect to the nebular lines, and the recombination lines may be especially favored. In the low-density regions, on the other hand, most radiations will be weakened; but those of highly ionized atoms may actually become enhanced, and the larger volume may compensate for the effects of a weaker emission per unit volume. In addition, the various filaments may be exposed to different radiation fields, so that the effects can become quite complicated.

Further, as Swings[16] has pointed out, the central stars of the planetaries probably do not radiate like black bodies in the ultraviolet; rather intense emission lines may be scattered throughout this region of the spectrum. We would certainly expect such ultraviolet emission lines in the Wolf-Rayet stars, and such lines may even exist in the nuclei showing continuous spectra. Within the nebula itself originate intense emission lines of Ly α, the λ 303 resonance line of *He* II, etc.; these may also play roles in ionization processes. The quantitative character of this radiation field is quite unknown; we shall be lucky if we can infer its general properties qualitatively.

The nebulae which show a symmetric type of stratification—e.g., small images of ionized helium and large *O* II images—present no essential difficulty. Many years ago Bowen[17] called attention to the fact that the high-frequency quanta would be rapidly absorbed by the nebula and that the radiation reaching the outer portions of the shells would be incapable of doubly ionizing helium or quadruply ionizing neon, for example. The general qualitative features are well understood; the details may offer some interesting problems.

Let us now develop the ionization formula for a complex atom exposed to dilute high-temperature radiation. First we shall obtain the relation between the capture coefficient for an electron of velocity v and the continuous absorption coefficient a_ν involved in the photoelectric ejection of an electron with velocity v. Since these are atomic parameters, we may evaluate them with the aid of the principle of detailed balancing in thermodynamic equilibrium; i.e., the number of ionizations in the frequency interval $d\nu$ will equal the number of captures from a velocity interval dv, viz.,

$$\frac{N_1 4\pi I_\nu a_\nu (1-e^{-h\nu/kT})\, d\nu}{h\nu} = f(v)\, v\sigma dv N_i N_\epsilon\,, \tag{33}$$

where the negative absorptions (stimulated emissions) are taken into account by the factor in parentheses on the left. Here a_ν is the absorption coefficient; σ is the target area for capture; and ν is related to the velocity of the free electron by $h\nu = \chi + \frac{1}{2}mv^2$; $hd\nu = mvdv$. The velocity distribution $f(v)$ of the free electrons is given by Maxwell's law:

$$f(v) = 4\pi\left(\frac{m}{2\pi kT}\right)^{3/2} v^2 e^{-(mv^2/2)/kT}\,, \tag{34}$$

while for I_ν Planck's law obtains. Then we have

$$\frac{N_i N_\epsilon}{N_1} = \frac{8\pi a_\nu}{c^2}\, e^{-h\nu/kT}\, \frac{\nu^2 m}{h}\left(\frac{2\pi kT}{m}\right)^{3/2} \frac{1}{4\pi}\, \frac{e^{[1/kT](h\nu-\chi)}}{\sigma v^3}\,. \tag{35}$$

[16] *Ap. J.*, **95**, 112, 1942.

[17] *Ap. J.*, **67**, 1, 1928.

If we compare this expression with the ionization formula for thermodynamic equilibrium, we find that

$$\frac{2m\nu^2}{c^2 h\nu^2}\frac{\alpha_\nu}{\sigma}\left(\frac{2\pi kT}{m}\right)^{3/2} e^{-\chi/kT} = \frac{(2\pi mkT)^{3/2}}{h^3}\frac{\varpi_\epsilon \varpi_i}{\varpi_1} e^{-\chi/kT}, \tag{36}$$

whence we derive the relation between α_ν and σ

$$\frac{\alpha_\nu}{\sigma} = \frac{m^2 v^2 c^2}{\nu^2 h^2}\frac{\varpi_\epsilon \varpi_i}{2\varpi_1}, \tag{37}$$

which holds quite generally and is not influenced by the state of the assembly or by how far it deviates from thermodynamic equilibrium.

In a planetary nebula nearly all the ionization will take place from the ground level, so that the number of ionizations

$$\int_{\chi_1}^{\infty} N_1 \frac{4\pi I_\nu \alpha_\nu}{h\nu} d\nu = \frac{8\pi N_1}{c^2}\int_{\nu_1}^{\infty}\frac{\nu^2 W_{1\kappa}\alpha_\nu}{\nu_1 (e^{h\nu/kT_1}-1)} d\nu \tag{38}$$

will equal the number of captures upon all of the bound levels, viz.,

$$\sum_j \int_0^{\infty} N_{ij} N f(v)\, v\sigma_j dv = \sum_j \int_0^{\infty} \frac{N_{ij}N_\epsilon}{T_\epsilon^{3/2}} 4\pi\left(\frac{m}{2\pi k}\right)^{1/2} e^{-(mv^2/2)/kT}\sigma_j v^2 dv\,. \tag{39}$$

Except at the very lowest velocities, most of the recaptures will take place upon the ground level. Accordingly, we shall neglect the captures upon the higher levels in our preliminary study of the problem; this corresponds to taking only the first term in $\Sigma S_n/n^3 = G_{T\epsilon}$ (cf. Paper V). We may change our variable of integration from v to ν, noting that $\chi_1 = h\nu_1$ and $v = 0$ for the ionization limit. If we substitute for σ in terms of α_ν and equate the resultant expression to the number of ionizations from the ground level, there results

$$\frac{N_i N_\epsilon}{N_1} = \frac{(2\pi mkT_\epsilon)^{3/2}}{h^3}\tilde{\omega}_\epsilon\tilde{\omega}_i \frac{\displaystyle\int_{\nu_1}^{\infty}\frac{W_{1k}\nu^2\alpha_\nu d\nu}{(e^{h\nu/kT_1}-1)}}{\displaystyle\sum_{j=1}^{\infty}\tilde{\omega}_j\left(\frac{m}{h}\right)\int_0^{\infty} e^{-mv^2/2kT_\epsilon}\nu^2\alpha_{\nu j} v dv} \tag{40}$$

If $\alpha_\nu \sim 1/\nu^2$, the ratio of the integrals reduces simply to

$$\overline{W}\frac{T_1}{T_\epsilon}\frac{e^{h\nu/kT}}{(e^{h\nu_1/kT_1}-1)},$$

where $\overline{W}$ is the mean dilution factor.

Bates[18] has calculated α_ν for the neutral atoms *C*, *N*, *O*, *F*, and *Ne* from the Hartree wave functions. Goldberg and Aller have calculated α_ν for oxygen in various stages of ionization with the aid of an analytic representation of the Hartree wave functions. For other light atoms Hartree fields were not available, and approximate Slater functions have been utilized for the ground levels. The resultant absorption coefficients are probably no more than qualitatively correct. All these calculations show that for complex atoms the absorption coefficient does not fall off hydrogenically as ν^{-3}, as many writers have assumed, but may even rise for a distance beyond the series limit. Thus the absorption coefficient of oxygen reaches a maximum a little above the series limit at $\tilde{\nu} = 160{,}000$ and then falls off gradually with frequency. On the other hand, the absorption

[18] *M.N.*, **100**, 25, 1940.

coefficient of ionized and doubly ionized oxygen falls off monotonically with increasing frequency. In general, the continuous absorption coefficient tends to become more nearly hydrogenic as one goes to higher and higher stages of ionization.[19]

If the central stars of the planetary nebulae radiated essentially as black bodies and if the nebulae were so tenuous that the atoms throughout were exposed to dilute black-body radiation, the precise character of the absorption coefficient would not play too great a role in determining the ionization; variations of a_ν according to ν^{-1}, ν^{-2}, or ν^{-3} would make but little difference in the results. If, however, the radiation impinging upon a typical volume element of the nebula is not simply dilute temperature radiation but is distorted by stellar emission lines and by emission lines arising in the nebula, the character of the absorption coefficient may assume considerable importance. For example, the absorption coefficient of oxygen falls off much more slowly with frequency than that of hydrogen. Consequently, line radiation far beyond the Lyman limit will be much more effective in ionizing oxygen than in ionizing hydrogen.

Aside from the purely geometrical aspects of the problem, W will be determined by (1) the nature of the initial radiation from the star and (2) the effect of absorption in the nebula itself upon the radiation from the central star.

As previously remarked, the central stars of the planetary nebulae are frequently stars of the Wolf-Rayet type, showing strong lines of helium, carbon, and oxygen, or helium and nitrogen, or sometimes lines of all of these elements. Swings[16] has suggested that the ultraviolet spectra of these stars in the region λλ 30–912 contain great numbers of intense lines of the aforementioned atoms. He calls attention to the fact that helium radiation from the stars at λλ 256.31, 243.02, and 237.33 may be absorbed by *C* III and *N* IV atoms, while, on the other hand, the coincidence of the ionization potentials of *O* III and *He* II cuts down the ionization of *O* III, owing to the fact that helium is much more abundant than oxygen. Thus, in NGC 6543 Swings finds fairly strong permitted lines of *C* III and *N* III in recombination, while those of *O* III are missing.[20]

In other planetaries, particularly certain high-excitation objects such as NGC 7009, NGC 3242, or NGC 7662, the nuclei exhibit apparently continuous spectra with no certain evidence of emission lines. We are ignorant of the ultraviolet spectra of these objects; but we reasonably might suppose that, if they contained strong ultraviolet emission lines, some emission lines ought to be found in the astronomically accessible regions. In any case, it does not seem safe to postulate the existence of strong emission lines in the ultraviolet spectra of all planetary nebulae.

We may illustrate the effects of the continuous absorption of radiation from the central star radiating as a black body by considering a hydrogen shell which, however, contains enough other atoms to lower the electron temperature to about 10,000° K in the manner indicated in Paper XVI. We have treated the problem of the transfer of radiation in the Lyman continuum in Paper II of this series. The method there employed essentially amounted to using a solution of the type suggested by Ambarzumian with an additional correction function, $R(\nu)$, which depended on the dilution factor W and the electron temperature T_ϵ. It turned out that the correction function was too large to be treated as a perturbation, and the problem had to be solved by successive approximations. This treatment took no account of possible inelastic collisions with foreign atoms; the electron temperature was supposed to depend on the character of the radiation field

[19] The reason for the nonhydrogenic behavior of the absorption coefficient is that an outer electron in a partially filled 2p or 3p shell is incompletely screened by its neighbors. Such an electron will move in a field, $Z - \sigma$, where σ, which is a measure of the screening, is less than $(N - 1)$, where N is the number of electrons in the ion. On the other hand, if a 2p electron is raised to a d level, it will move in a nearly hydrogenic field of charge $Z - (N - 1)$. Consequently, in calculating the transition from the bound 2p to the free κd level, one cannot employ the same value of $Z - \sigma$ in the initial and final wave functions. For a fuller discussion of this and related points see Bates (*op. cit.*).

[20] *Ap. J.*, **92**, 289, 1940.

to which the volume element was subjected, in accordance with the conditions described in Paper V. Actually, as we have seen, the electron temperature of a planetary nebula is nearly constant at about 10,000° K. Accordingly, the theory of transfer of radiation in the Lyman continuum will have to be examined from a new point of view.

Consider an interval $d\nu$ at ν in the ultraviolet continuum. The absorption of energy at this frequency will depend roughly on $1/\nu^3$, since $a_\nu \sim 1/\nu^3$; but the emission of energy in this interval will depend on the number of electrons of velocity v (such that $h\nu = \chi + \frac{1}{2}mv^2$) that are captured per second. The number with the required velocity v will be determined by Maxwell's law and electron temperature T_ϵ. Since the latter is in the neighborhood of 10,000° K, there will be very few high-velocity electrons with energies of the order of 4 or 5 volts; and consequently the captures from this velocity region will be very few. Thus, energy will be returned to the continuum only in frequencies near the series limit; and at frequencies corresponding to but a few volts beyond the series limit, the radiation will simply be extinguished by the nebula. In other words, radiation is piled up at the low-frequency end of the Lyman continuum at the expense of energy absorbed at the higher frequencies. Since the absorption coefficient is greatest at the lower-frequency end of the range, the over-all effect is to increase the rate at which the ultraviolet radiation is extinguished in the nebula. An approximate treatment of the equations of the problem shows that on the inner boundary of the nebula there is a concentration of radiation near the series limit. This "Lyman limit excess" gradually falls to a very small value near the outer boundary of the nebula, where the hydrogen is nearly all in the neutral condition. A more exact treatment of this problem, possibly by the methods being developed by Chandrasekhar, might be of great value in interpreting the observed nebular images. We must recall that in actual practice the ultraviolet radiation field is depleted not by one type of atoms but by several types. Probably neutral and ionized helium play important roles in this connection.

We shall now illustrate how the densities of the neutral atoms of hydrogen and singly ionized helium may be evaluated from the sizes of the nebular images. As an example, let us consider NGC 7009. If we adopt Berman's distance estimate,[21] the main portion of the nebula has a radius of 1.25×10^{17} cm and a shell of thickness 3.7×10^{16} cm. The absorption coefficient of hydrogen at the limit of the Lyman series is

$$a_\nu = \frac{2^5 \pi^2 \epsilon^6 R Z^4}{3\sqrt{3}\, c h^3} \frac{1}{\nu^3} = 5.0 \times 10^{-18} \left(\frac{\nu_1}{\nu}\right)^3$$

per atom. At an optical depth of about 3, extinction will be fairly complete; the portions of the nebula beyond this point will not contribute much to the luminosity of the nebular images. If N is the average number of neutral hydrogen atoms per unit volume, then

$$3 \backsim 5 \times 10^{-18} \times 3.7 \times 10^{16} \times N\,,$$

or $N \sim 20$. Similarly, the thickness of the ionized helium shell is of the order of 2.5×10^{16} cm, $a_\nu = 1.25 \times 10^{-18}$, and $N \sim 100$ for an optical depth of 3. An independent calculation of the upper limit of the *He* II abundance gave for this nebula $N = 800$. Our elementary calculation has taken no account of the re-emission of radiation near the series limit, nor of the fact that the optical depth of 3 at the series limit corresponds to a somewhat smaller optical depth near by. Our calculation of N thus gives us a lower limit to the value of this quantity, but it does not seem likely that our estimates are in error by more than a factor of 2 or 3 from this cause.

In any case, these calculations justify the assumptions we have made to the effect that the hydrogen in the luminous portions of a planetary nebula is nearly all ionized. Hence the masses of the luminous portions calculated on the basis of this assumption seem justified. An examination of slitless spectrograms of a number of planetaries reveals

[21] *Lick Obs. Bull.*, 18, 57, 1937.

that the outer edges of the hydrogen images are often fairly sharp. This shows that toward the boundary of the luminous portion the ionizing radiation becomes depleted over a short distance in the manner indicated by Strömgren[22] and that the gas becomes neutral. As the number of neutral atoms increases, the opacity to the ionizing radiation rises in proportion, and the nebula shows a sharp boundary. With a hydrogen density of about 6×10^3 atoms/cm^3 we see that an optical depth of about 3 can be attained in 10^{14} cm if the hydrogen is neutral. Thus, within a distance amounting to only a few per cent of the nebular radius, hydrogen can change from a condition of high ionization to an almost neutral condition.

The stratification effects for other atoms provide an interesting study. Slitless spectrograms obtained thirty years ago by Wright[23] showed that in many objects of complicated structure the main body of the nebula is surrounded by a faint extended envelope or shell shining with the λ 3727 radiation of [*O* II]. Presumably, radiation escaping from the inner shells has ionized oxygen in these outer regions. Since the ionization potential of neutral oxygen is almost the same as that of hydrogen, we might expect oxygen to be ionized in the regions where hydrogen is ionized and neutral where hydrogen is neutral. The absorption curve for oxygen, however, reaches a maximum at about λ 625 and then falls slowly to half its maximum value at λ 310; and it is still 30 per cent of its maximum at λ 170, which corresponds to an excitation potential of 74 volts. On the other hand, the absorptivity of hydrogen falls off as ν^{-3}, and therefore hydrogen is relatively transparent to radiations in the far ultraviolet while oxyen still absorbs energy appreciably. Thus it is not entirely surprising that oxygen is ionized at greater distances from the nucleus than is hydrogen.

Another point that should be stressed in discussing stratification in the planetary nebulae is the lack of radial symmetry often apparent in their structures. A glance at slitless spectra of NGC 7009, for example, shows that the λ 3727 radiation is concentrated at the ends of the spindle, whereas the λ 4686 radiation is strongest near the broader portions of the spindle. Any detailed discussion of stratification effects must take these irregularities into account. In the ansae of NGC 7009 the degree of excitation is much less than in the main body of the nebula, although even here much *O* III remains. In fact, if we suppose $N_\epsilon \sim 10^3$ and $T_\epsilon \sim 8000°$ K for the ansae, we obtain $N_{O\,\text{II}}/N_{O\,\text{III}} \sim 2$, which is nearly the same value as that for the main body of the planetary.

Line radiation is likely to play an important role in the ionization of atoms in the nebular shell; so we shall develop here the fundamental formulae for the problem. Detailed applications must wait until we have definite knowledge of line radiation intensities in the nebular shell. If an atom is excited from a ground 2p configuration to an excited nd configuration, the frequency of the absorbed radiation may be expressed in the form

$$\nu = RZ^2\left(\frac{1}{n_0^{*2}} - \frac{1}{n^2}\right),$$

where we choose n_0 in such a way as to reproduce the term value of the ground level. We suppose that the upper levels may be represented as hydrogenic; i.e., $T = RZ^2/n^2$. Transitions to the continuum may be represented by

$$\nu = RZ^2\left(\frac{1}{n_0^{*2}} - \frac{1}{\kappa^2}\right) = R\left(\frac{\chi}{\chi_H} + \frac{Z^2}{\kappa^2}\right),$$

where we have replaced n by $i\kappa$ and where χ is the ionization potential of hydrogen. If we let $\chi' = \chi/\chi_H$, the frequency of the bound-free transitions will be

$$\nu = R\chi'\left(1 + \frac{Z^2}{\chi'\kappa^2}\right).$$

[22] *Ap. J.*, **87**, 526, 1939.

[23] *Pub. Lick. Obs.*, **13**, 193, 1918.

Since the relation between the absorption coefficient α_ν and the oscillator strength f is

$$\int \alpha_\nu d\nu = \frac{\pi\epsilon^2}{mc} f ,$$

we may formally express α_ν as $(\pi\epsilon^2/mc)(df/d\nu)$. Just to the redward side of a series limit there will be Δn lines of mean oscillator strength f per unit frequency interval, whereas just to the violet side the f-value per unit frequency interval will be $f\Delta\kappa$, and the f-value corresponding to the ν-interval $d\nu$ will be $df = fd\kappa$, while

$$\alpha_\nu = \frac{\pi\epsilon^2}{mc}\frac{df}{d\nu} = \frac{\pi\epsilon^2}{mc}\frac{df}{d\kappa}\frac{d\kappa}{d\nu} = \frac{\pi\epsilon^2}{mc} f \frac{d\kappa}{d\nu} .$$

Now

$$d\nu = -\frac{2RZ^2}{\kappa^3} d\kappa ,$$

so that

$$\alpha_\nu = \frac{\pi\epsilon^2}{mc} f \frac{\kappa^3}{2RZ^2} ,$$

where

$$f = \frac{1}{(2j'+1)}\frac{8\pi^2 m\nu}{3h}\frac{a^2}{4l^2-1} S\rho^2$$

can be calculated if the radial quantum integral ρ is known.

If the atom is exposed to intense radiation over an interval $\Delta\nu$, the amount of energy absorbed will be

$$\alpha_\nu I_\nu \Delta\nu = \frac{\pi\epsilon^2}{mc}\frac{f\kappa^3}{2RZ^2} I_\nu\Delta\nu ,$$

and the amount of energy absorbed by N atoms exposed to a series of monochromatic sources will be

$$\frac{\pi\epsilon^2}{mc}\frac{1}{2RZ^2}\Sigma f\kappa^3 I_\nu\Delta\nu .$$

The order of magnitude of $\Delta\nu$ as suggested by equation (XVI-12) is about 1×10^{11}. In the case of *O* II an elementary calculation shows that an increase of energy in such a line as λ 303 of *He* II by a factor of 10^4 above the value to be expected from only the geometrical dilution may exert a controlling effect upon the ionization. Thus, photoelectric ionization by line absorption may very well be an important factor in controlling the ionization of the shell.

It should be clear from the foregoing that a theoretical calculation of the ionization in a gaseous nebula becomes a virtually hopeless task. The chief stumbling block is the unknown radiation field within the nebula; but, even if this were predictable, the interlocking effects of different radiations and stratification would render the problem one of forbidding complexity.

If we know the temperature of the central star and the electron density, it should be possible to take ions of the same atom in successive stages of ionization and work out empirical dilution factors. We have done this for the *O* III/*O* II ratio. The calculated W, of course, depends very critically on the assumed stellar temperature and varies, in the cases considered, from 10^{-13} to 10^{-16}. Upon using the *Ne* III/*Ne* V ratios we found that the same central star temperature gave very different dilution factors, and the attempt to calculate ionization from the theoretical formulae was abandoned.

V. ESTIMATES OF TOTAL ABUNDANCES

The final step in the calculation of abundances is to devise some ways of guessing the contribution from the unobserved ions. We have seen that there is no sound theoretical procedure for calculating these contributions, and we shall have to proceed in some purely empirical fashion. We feel reasonably confident that the abundances of various ions deduced from their forbidden lines and given in Table 5 are of the right order of magnitude, but the final step in determining the composition seems necessarily uncertain.

Our method of taking into account the variation of ionization among ions of different ionization potential is as follows: Let A_1 and A_2 denote the abundances of two ions of the same element in successive stages of ionization, and let χ_1 and χ_2 denote their ionization potentials. Then we shall suppose that A_2/A_1 gives us the ratio of the abundances of two ions of any kind with ionization potentials χ_1 and χ_2; i.e., we get a segment of the curve relating abundance with ionization potential. Other pairs of ions give other segments; and when these segments are fitted together by a vertical displacement on the relative-abundance–χ diagram, we get a curve which will tell us how we may expect the ions of a given element to be distributed among their various stages of ionization.

For example, the relative abundances of *O* II and *O* III suggest the relative degree of ionization of atoms of 13.5 and 34.9 volts, ionization potential, the relative numbers of *Ne* III and *Ne* V ions, the relative degrees of ionization of ions of 40.5 and 97 volts, etc. Thus, for NGC 7027 we get the accompanying table relating ionization and ionization

Ionization Potential (Volts)	Relative Ionization	Ionization Potential (Volts)	Relative Ionization
0 (neutral)	0.01	50	0.68
10	0.023	60	.52
20	0.13	70	.42
30	0.81	80	.37
40	1.00	100	0.28

potential. Neutral atoms fall at zero; singly ionized atoms, at the first ionization potential; etc. Thus the relative abundance of *Ne* II is taken as 0.18 (ionization potential of *Ne* I is 21.5 volts); that of *Ne* III is 1.0 (I.P. = 40 volts); that of *Ne* IV is 0.48 (I.P. of *Ne* III is 63 volts); etc. From the relative ionization graphs for each nebula we can calculate the distribution of ions of each element among the various stages of ionization, and therefore the fraction contributed by the observed ions to the total.

Unfortunately, there is a considerable scatter in the points which serve to determine the graphs. In the first place, the abundances derived from approximate target areas may be in error by a factor of 2 or 3 or even more. In the second place, the assumption that the ionization varies smoothly with ionization potential, as assumed in our graphs, cannot be correct if the central star has strong bright lines in its ultraviolet spectrum. Hence the relative proportions of successive ions of a given element may be very different from what we would expect from the graph.

The resultant abundances are listed in Table 6. In some cases, where the discrepancy is particularly bad, we have tabulated the results from two different ions of the same atom. The second of the difficulties listed above may be the dominating trouble here. In most other cases, both ions of a given element give reasonably concordant values. The figures in parentheses are those obtained from a previous calculation, in which it was presumed that the target area for the collisional excitation of the metastable levels varied as Z^2 for the more highly stripped ions. In general, the older calculations give the same order of magnitude as the newer ones, suggesting that effects of selection, ionization, etc., introduce a larger uncertainty than ignorance of the correct target areas.

The scatter in the relative abundances of Table 6 probably arises from the unavoidable approximations and effects of selection more than from variations in chemical composition. For carbon we have had to rely on recombination lines entirely and have given greater weight to *C* III than to *C* IV because of the greater uncertainty in the *b*-factor for the latter. The results for nitrogen are based primarily on the *N* II ion, and in many cases the ratio of total to observed ions is very large. The spread of the computed oxygen abundances, which have been calculated from the intensities of the green nebular lines and good target areas, is less than the spread of the computed abundances of other elements. The relative ionization curves indicate that for oxygen the *O* III ions contribute a large

TABLE 6*

THE COMPOSITION OF THE PLANETARY NEBULAE

Nebula	*He*	*C*	*N*	*O*	*F*	*Ne*	*S*	*Cl*	*A*
NGC 7662...	135	1.06	0.018	0.33	0.00003	0.0016	0.006	0.0087	0.003
	......	......	......	......	....	0.0078	0.033	....	0.0001
	(154)	(0.9)	(0.03)	(0.31)	(0.00009)	(0.02)	(0.02)	(0.0025)	(0.0005)
NGC 6741...	237	......	1.9	0.63	....	0.05	0.29	0.009	0.008
	......	......	......	......	....	......	......	....	0.003
NGC 7027...	240	0.73	1.34	0.42	0.0012	0.02	0.27	0.013	0.0023
	......	......	......	......	....	......	......	0.044	....
	(260)	(0.81)	(1.48)	(0.7)	(0.00008)	(0.04)	(0.21)	(0.006)	(0.0009)
NGC 2440...	280	0.82	0.92	0.18	....	0.015	0.058	0.0021	0.0014
NGC 7009...	52	0.34	0.018	0.16	....	0.002	0.010	0.014	0.0012
	......	......	......	......	....	......	......	0.001	....
	(56)	(0.3)	(0.02)	(0.25)	....	(0.002)	(0.007)	(0.0012)	(0.00035)
NGC 6572...	100	0.18	0.08	0.12	....	0.022	0.016	0.0004	0.0015
	(103)	(0.21)	(0.09)	(0.28)	....	(0.064)	(0.016)	(0.0003)	(0.0011)
IC 418...	19	0.12	0.025	0.19	....	......	0.004	0.0003	....
	(25)	(0.20)	(0.095)	......	....	......	(0.009)	(0.0002)	....

* This table gives the number of atoms of a given kind per thousand hydrogen atoms. In some cases the abundances calculated on the basis of two ions of the same atom disagreed significantly; so we give both results. The figures in parentheses refer to results of an earlier calculation, in which the abundances of the various ions were calculated on the basis of different assumptions about the target area Ω for collisional excitation of the metastable levels.

share of the total in most planetaries. Hence the oxygen abundances ought to be fairly well established as of the order of from 10^{-3} to 10^{-4} that of hydrogen. The data for fluorine are very scanty, and the corrections for ionization are very large and uncertain. Neon turns out to be less abundant than we might have anticipated from the work of Unsöld on τ Scorpii. Argon, the next in line of the noble gases, is of the order of ten times less abundant than neon, while sulphur seems comparable in amount. The forbidden lines of chlorine and argon are both very weak.

When we compare the different nebulae, we notice a considerable scatter in the relative contributions of the various elements. For example, in NGC 7027 the abundances of nearly all elements seem unusually high with respect to hydrogen, the Balmer continuum of which seems unusually weak for such a bright nebula. It may be that a greater proportion of atoms of a given element are represented by observable ions than we anticipated; and until this matter is further investigated, we cannot distinguish between this and the possibility that hydrogen is actually less abundant there.

We have attempted no abundance estimates for the metals, since the corrections for ionization would be so uncertain as to render the results quite meaningless. The target areas are unknown even though we may have transition probabilities.

The "average composition" of a planetary nebula, essentially obtained by taking a geometric mean of the abundances in Table 6, may be compared with the results obtained for the solar atmosphere by Menzel and Goldberg[24] and for the atmosphere of τ Scorpii by Unsöld. (See Table 7.)

The uncertainties in all methods are very great; but the abundances of carbon, nitrogen, and oxygen, with respect to hydrogen, in τ Scorpii seem, nevertheless, very similar to the results for the sun or the gaseous nebulae. For helium, nitrogen, oxygen, and sulphur the agreement between the solar and nebular abundances is surprisingly good. The discrepancy of a factor of 10 for carbon is easily understandable because our estimates are ultimately based on a single recombination line of *C* II. The discrepancy for neon

TABLE 7

COMPARISON OF THE AVERAGE COMPOSITION OF A PLANETARY NEBULA WITH THE COMPOSITIONS OF THE SUN AND τ SCORPII

Element	Relative Numbers of Atoms		τ Scorpii
	Planetary Nebula	Sun	
Hydrogen	1000.0	1000.0	1000.0
Helium	100.0	222.0	175.0
Carbon	0.6	0.04	0.17
Nitrogen	0.2	0.12	0.3
Oxygen	0.25	0.37	1.0
Fluorine	0.0001	....	....
Neon	0.01	....	1.1
Sulphur	0.036	0.037	....
Chlorine	0.002	....	....
Argon	0.0015	....	....

probably arises from other causes. It may be that τ Scorpii is richer in neon content than the average star or that conditions are particularly suitable for the excitation of neon lines.

Thus the conclusion of Bowen and Wyse that the chemical composition of the gaseous nebulae is essentially the same as that of the sun appears to be substantiated by an independent and more elaborate theoretical treatment of the same problem. We have attempted to use the splendid observational material obtained by Wyse to deduce the abundances of a number of ions which are not observable in the sun.

It would seem that further progress along these lines cannot be made until improved observational and theoretical material is available. The identifications of additional ions, especially those having higher ionization potential than does *Ne* V, would be very helpful in estimating the distribution of the various ions among different stages of ionization. Target areas for various ions ought to be computed; such calculations are in progress at present for *O* II.

This paper completes the series of articles entitled "Physical Processes in Gaseous Nebulae."

[24] *Atoms, Stars, and Nebulae* Philadelphia: Blakiston, 1942.

PART 3. INTERPRETATION OF NEBULAR SPECTRA

Reprinted from *The Astrophysical Journal,* Vol. 105, No. 1, January, 1947.

THE ELECTRON VELOCITY DISTRIBUTION IN GASEOUS NEBULAE AND STELLAR ENVELOPES

DAVID BOHM AND LAWRENCE H. ALLER
University of California, Berkeley, California, and
Kirkwood Observatory, Indiana University
Received October 18, 1945; revised October 15, 1946

ABSTRACT

A general method is given for solving Boltzmann's equation for the steady-state velocity distribution of electron-ion gases, such as those present in gaseous nebulae or stellar atmospheres. In a typical planetary nebula it is found that the velocity distribution is very close to Maxwellian. Because the average lifetime of an electron in the assembly is about 10 years and because its energy is reshuffled nearly every second by electrostatic encounters, the deviations from the equilibrium Maxwellian distribution must be very small. The non-Maxwellian distributions obtained by Hagihara are merely first approximations to an expansion of a Maxwellian distribution in terms of the wrong temperature.

Section I gives a general summary of the results and the physical picture behind them. Section II lists the formulae for the probabilities of all processes involved; in Section III Boltzmann's equation is set down, and Hagihara's work is discussed. Section IV gives a general method for solving Boltzmann's equation when electrostatic interaction is the dominating term.

I. INTRODUCTION

In a gaseous nebula, electrons arising from photoemission are responsible for the excitation of forbidden lines, such as those of *O* III. In the process of excitation, the electronic energy is reduced, and, finally, the electrons are recaptured. The steady state of electron velocity distribution is determined by the condition of balance of all processes tending to change it. Since the assembly is clearly not in thermodynamic equilibrium, the velocity distribution may deviate from the Maxwellian. Nevertheless, it has usually been assumed that the deviations were small, although it has never been clearly proved that this is true. In fact, Hagihara[1] has suggested that the distribution need not be even approximately Maxwellian. If this is true, a great deal of existing work must be re-examined. For example, since the estimation of abundances of elements from line intensities requires a knowledge of the velocity distribution, particularly of the high-energy portion usually associated with the "tail" of the Maxwellian curve, a careful investigation of this point is required. Not only is a detailed study of the mechanisms involved in producing a steady state needed for an understanding of the microscopic processes, but it also must lie at the basis of any treatment of macroscopic problems, such as the diffusion of concentrations of matter.

Hagihara has maintained not only that the distribution may deviate widely from the Maxwellian but also that the precise extent of this deviation is not completely determined by specifying the distribution of the intensity of the incident radiation and the density of matter. In fact, he states that, for any temperature, he can obtain a distribution (generally non-Maxwellian) which is a solution of the equations involved. It would certainly seem that the distribution would be defined by the physical conditions mentioned above. Hence, if it is really ambiguous, as maintained by Hagihara, there is something about the physical processes involved that we do not understand.

To clarify some of these points, we have solved Boltzmann's equation by a method especially designed to treat the high-velocity portion of the distribution with the neces-

[1] *Tokyo Astr. Bull.*, Nos. 525, 542, 571, 1941; see also a series of papers entitled "Electron Velocity Distribution in a Planetary Nebula," *Jap. Jour. Astr. and Geophys.*, **17**, 199, 417, 1940; **18**, 89, 149, 1941; **19**, 9, 75, 1941; **20**, 1, 37, 59, 67, 1943.

sary accuracy. Our results show that the distribution must actually be very close to Maxwellian and that the non-Maxwellian distributions considered by Hagihara are really the first approximations to Maxwellian distributions of another temperature. Thus the apparent ambiguity in the electron temperature turns out to have a purely formal character.

A brief physical picture of how the steady state is actually brought about may be helpful here. Neutral atoms, principally hydrogen, are ionized photoelectrically by ultraviolet radiation arising ultimately from the central star. Electrons are emitted with a kinetic energy equal to the difference between the photon energy and the ionization energy. The number emitted with a given energy is determined by the frequency distribution of the radiation and the absorption coefficient as a function of the energy. Now consider what happens to an electron after it is emitted. It can undergo elastic scattering by other electrons, ions, or neutrals; inelastic scattering by ions, giving rise to excitation of metastable ions; or collisions of the second kind with metastable ions, emission of radiation in free-free transitions, or recapture with the emission of radiation. A steady state results from the balance of all these processes. Fortunately, they do not all occur with the same frequency. In fact, elastic scattering by other electrons and ions is enormously more probable than any of the others. Estimates made in Section II show that the average electron exchanges its energy with other electrons by elastic collisions about once a second if the electron density has the typical value of about $10^4/\text{cm}^3$. The next most frequent process is the excitation of *O* III ions to a metastable level. Even if an electron has the optimum energy for excitation, this process happens only once in 4 months. If it does occur, however, a great deal of energy is absorbed, so that the temperature of the electron gas is lowered. The average lifetime of an electron before it is recaptured is about 10 years, while free-free emissions would require 40 years to dissipate the average electronic energy at a typical electron temperature of 10,000° K.

If electrons remain in the gas 10 years, suffer inelastic collisions once in 4 months, and exchange energy with one another every second, the distribution must certainly approach the equilibrium Maxwellian form. Since the high-energy portion of the distribution, which is responsible for the excitation of the forbidden lines, must result from the boiling-up of low-energy electrons, all traces of the initial stellar energy distribution will have been wiped out, except in so far as it has contributed to the total energy supply in the thermodynamic sense. The calculations of Menzel and others who have used a Maxwellian distribution are thus justified. In Sections II, III, and IV we solve the problem mathematically, in order to present a general method useful in treating stellar envelopes and interstellar space and also to justify the physically evident conclusion that the distribution must be very close to the Maxwellian.

Under what circumstances, if any, would we expect appreciable deviations from a Maxwellian distribution? In general, deviations become appreciable when the probability of inelastic scattering or recombination begins to be comparable with that of elastic scattering. The high-energy ranges of the distribution are more likely to be affected than the low-energy ranges.[2] If, as in typical nebulae, inelastic collisions occur every 10^8 seconds and a redistribution of energy by electrostatic forces takes place every second, the deviations must be negligible. Even in a nebula consisting entirely of doubly ionized oxygen atoms and electrons, where we might expect the inelastic collisional processes to be about ten thousand times more frequent, the velocity distribution would scarcely deviate from the Maxwellian. In a pure hydrogen nebula of so high a temperature that most of the Balmer-line excitation took place by collisions rather than by recombination, the velocity distribution would still be nearly Maxwellian. At the very low densities of interstellar space, where the collisional reshuffling of electrons proceeds very slowly, it

[2] It may be shown that the distribution does not deviate appreciably from the Maxwellian for ratios of E/kT below a critical value, of the order of the square root of the ratio of the probability of elastic encounters to that of the most likely type of inelastic encounter.

is still doubtful that detectable deviations exist, since the radiative processes would occur also at a very slow pace. In general, we are tempted to conclude that deviations of electronic velocities from a Maxwellian distribution are likely to be very small in all stellar and interstellar media accessible to observation.

II. CROSS-SECTIONS FOR THE VARIOUS PHYSICAL PROCESSES

We shall now collect the expressions for the cross-sections of the various processes that occur in a gaseous nebular shell exposed to dilute high-temperature radiation. Many of the formulae have been given elsewhere,[3] but we shall need them here in slightly different forms.

A. PHOTO-IONIZATION

The number of transitions from the ground level (essentially all ionizations take place from the lowest level) arising from the absorption of radiation in a frequency interval ν to $\nu + d\nu$ is given by

$$F_{1\kappa} d\nu = 4\pi N_1 I_\nu \alpha(\nu) \frac{d\nu}{h\nu}, \tag{1}$$

where N_1 is the number of atoms in the ground level and I_ν is the intensity of radiation, and by

$$W \frac{2h\nu^3}{c^2} \frac{1}{e^{h\nu/kT_1} - 1},$$

where W represents the effect of the dilution of the radiation by geometric and other factors and may be a function of ν. For convenience, we may choose T_1 as the color temperature of the central star in the ultraviolet. The absorption coefficient is

$$\alpha(\nu) = \frac{\pi \epsilon^2}{mc} f_{1\kappa}, \tag{2}$$

where $f_{1\kappa}$ is the Ladenburg f, or oscillator strength, for the transition involved. Here $h\nu$ is the energy absorbed in the transition. Normally, formulae for f are given for discrete transitions (n–n' in hydrogen). To develop the proper expressions for the transitions involving the continuum, Menzel and Pekeris[4] have introduced the quantum number, κ, by replacing n by $i\kappa$ where κ is a real and continuous variable. Then frequencies of transitions from a lower level n' in hydrogen to the continuum are given by the Balmer formula,

$$\nu = RZ^2 \left(\frac{1}{n'^2} + \frac{1}{\kappa^2} \right). \tag{3}$$

The velocity of a free electron V is related to κ by $\frac{1}{2} mV^2 = hRZ^2/\kappa^2$, so that

$$mV\,dV = h\,d\nu = -\frac{2hRZ^2}{\kappa^3} d\kappa. \tag{4}$$

For convenience we may express the number of neutrals in any level n in terms of the number of ions and electrons, the equilibrium value given by the Saha equation and the

[3] See the series entitled "Physical Processes in Gaseous Nebulae": *Ap. J.* (I) **85**, 330, 1937; (II) **86**, 70, 1937; (III) **88**, 52, 1938; (*IV*) **86**, 313, 1938; (V) **88**, 423, 1938; (X) **92**, 408, 1940; (XI) **93**, 178, 1941; (XII) **93**, 195, 1941; (XIII) **93**, 230, 1941; (XVI) **94**, 30, 1941; (XVII) **94**, 436, 1941; (XVIII) **102**, 239, 1945. The papers of the series are referred to in the text as Paper I, Paper II, etc. Equations are referred to by paper and number. Thus (I-4) means eq. 4 of Paper I.

[4] *M.N.*, **96**, 78, 1935.

Boltzmann law, and a correction factor which takes into account the deviation from thermodynamic equilibrium, according to (I-4),

$$N_n = b_n N_i N_\epsilon \frac{h^3}{(2\pi m k T_\epsilon)^{3/2}} e^{hRZ^2/n^2kT} \frac{\varpi_n}{2}, \tag{5}$$

where N_n is the number of atoms in level n in each cm,[3] N_1 is the number of ions per cm,[3] N_ϵ is the number of electrons per cm^3, ϖ_n is the statistical weight of level n, and b_n is the correction factor. By combining the above formulae, we obtain an expression for the number of electrons entering the range $d\nu$ at ν per cm^3 per second,

$$F_{1\kappa} d\nu = N_i N_\epsilon \frac{KZ^4}{T_\epsilon^{3/2}} b_1 \frac{W e^{h\nu_1/kT_\epsilon}}{e^{h\nu/kT_1}1} \frac{g\,d\nu}{\nu}, \tag{6}$$

where K is a numerical constant equal to 3.202×10^{-6}; g is the Gaunt correction factor, which may be tabulated as a function of the frequency ν; and b_1 is a correction factor for the ground level. An expression for b_1 has been given in (V-2), viz.,

$$b_1 = \frac{G_{T_\epsilon}}{e^{X_1} \int_{y_1}^{\infty} W_{1\kappa} \frac{dy}{y(e^y - 1)}}, \tag{7}$$

where

$$X_1 = \frac{hRZ^2}{kT_\epsilon} = \frac{h\nu_1}{kT_\epsilon} = \frac{1.5703 \times 10^5}{T_\epsilon}, \qquad y = \frac{h\nu}{kT_1},$$

and G_{T_ϵ} has been given graphically in Paper XVI. It is of the order of 0.1 for $T_\epsilon \sim 10^{4\circ}$ K. If W is constant with frequency, we have

$$F_{1\kappa} d\nu = N_i N_\epsilon \frac{3.202 \times 10^{-6}}{T^{3/2}} \frac{G_{T_\epsilon} \frac{d\nu}{\nu}}{(e^y - 1) \int_{y_1}^{\infty} \frac{dy}{y(e^y - 1)}}. \tag{8}$$

It is clear that N_i, N_ϵ, and T_ϵ will adjust themselves so that the total number of ionizations will equal the total number of recombinations. In fact, this is one of the relations used in Paper V to determine T_ϵ.

B. RECOMBINATION

The target area for the capture of an electron on a level n with an energy corresponding to κ is

$$\sigma_{\kappa n} = \frac{2^4}{3\sqrt{3}} \frac{\epsilon^2 h}{m^2 c^3} \frac{1}{\left(\frac{1}{n'^2} + \frac{1}{\kappa^2}\right)} \frac{\kappa^2}{n'^3} g_{II}, \tag{9}$$

where g_{II} is a correction factor. The total number of captures to a single ion is obtained simply by summing over n and multiplying by the velocity V. If we write the energy in terms of electron volts as a convenient unit, we obtain, for the total number of captures on all levels of an ion,

$$\frac{2^4}{3\sqrt{3}} \frac{\epsilon^2 h}{m^2 c^3} \sqrt{\frac{\mathfrak{V}_\epsilon}{150m}} \frac{\mathfrak{V}_i}{\mathfrak{V}} \Sigma \frac{1}{\left(\frac{1}{n'^2} + \frac{\mathfrak{V}}{\mathfrak{V}_i}\right)} \frac{g_{II}}{n'^3} = 1.24 \times 10^{-14} \frac{\mathfrak{V}_i}{\sqrt{\mathfrak{V}}} \Sigma \frac{1}{\left(\frac{1}{n'^2} + \frac{\mathfrak{V}}{\mathfrak{V}_i}\right)} \frac{1}{n'^3},$$

where $\mathfrak{V}_i$ here denotes the ionization potential of hydrogen. If, however, we use c.g.s. units which are more convenient in some cases, we get, for the total number of captures per cm³/sec,

$$Vf(V)\,N_\epsilon^2\Sigma\sigma_{\kappa n}=2.08\times10^{-22}N_\epsilon^2 f(V)\,V\Sigma\frac{1}{\left(\frac{1}{n'^2}+\frac{\mathfrak{V}}{\mathfrak{V}_i}\right)}\frac{1}{n'^3}\frac{E_i}{E}. \qquad (10)$$

Thus the cross-section varies slowly with the energy. To estimate its magnitude, let us consider, for example, an electron of energy 1.35 volts, for which $\mathfrak{V}_i/\mathfrak{V}$ is equal to 10. The sum then is about 1.7, and we get $\sigma = 10 \times 1.7 \times 2.1 \times 10^{-22} = 3.57 \times 10^{-21}/\text{cm}^2$. The rate of recombination, $N_iV\sigma$, for $N_i \sim 10^4$ and $V = 6.9 \times 10^7$ cm/sec, will be simply 2.5×10^{-9} sec. This result means that a single electron, of energy 1.35 volts, will take 4×10^8 sec, or about 13 years, to be captured. A slower electron, of, say, 0.5 volt of energy, will be recaptured in the order of 8 years. We shall adopt 10 years as the order of magnitude of the lifetime of an electron in the nebula.

C. INELASTIC COLLISIONS

The target area for the collisional excitation of the ^1D metastable level of *O* III by an electron moving with velocity V is[5]

$$\sigma=\frac{1}{\varpi_A}\frac{h}{4\pi m^2}\frac{\Omega(AB)}{V^2}=\frac{4.17}{2J_A+1}\frac{\Omega(AB)}{V^2}, \qquad (11)$$

since ϖ_A, the statistical weight of the ground term of *O* III, is 9. If we express the energy in volts, we get

$$\sigma_{AB}=\frac{2.6\times10^{-15}}{\mathfrak{V}}\text{ cm}^2. \qquad (12)$$

In order to produce an inelastic collision with an *O* III ion, an electron must have at least 2.48 volts of energy, so that $\sigma = 0$ when $\mathfrak{V}$ is less than $\mathfrak{V}_0 = 2.48$ volts. An electron with slightly greater energy than $\mathfrak{V}_0$ has the maximum probability of suffering an inelastic collision. Let us estimate the average time taken for such an electron to produce an excitation. The probability of excitation per electron is $N_A\sigma V$, where N_A is the number of *O* III ions per cm³. In most bright nebulae, N_A is of the order of unity. Now $V \sim 9.35 \times 10^7$, and $\sigma = 1.105 \times 10^{-15}$; hence $N\sigma V \sim 9.7 \times 10^{-8}$. The mean time required for an excitation will be 1.03×10^7 seconds, or about 4 months, and this is at the most favorable energy for such a process.

The superelastic collisons are even less important, since they take place only with excited ions which are present in lower than equilibrium concentrations. The target areas for the two inverse processes are related by

$$\varpi_A\sigma_{AB}=\varpi_B\sigma_{BA} \qquad (13)$$

in virtue of the principle of detailed balancing in thermodynamic equilibrium.

To find the number of superelastic collisions, we must know the number of excited ions, a quantity determined by the balance between the production by collisional excitation and destruction by radiation and superelastic collisions. To a first approximation, we may assume a Maxwellian distribution to calculate this balance; it will soon be evident that this assumption in this step introduces a very small error. The results of Paper XIII enable us to work out the relevant expressions as we want them. Under typical conditions the ratio of superelastic to total de-excitations will be about 0.1.

[5] See (X-26).

D. FREE-FREE TRANSITIONS

The target area for a free-free transition from a state κ to a state lying between κ' and $\kappa' + d\kappa'$ is given by

$$\sigma_{\kappa\kappa'} d\kappa' = \frac{2^4}{3\sqrt{3}} \frac{\epsilon^2 h}{m^2 c^3} \frac{1}{\left(\frac{1}{\kappa^2} - \frac{1}{\kappa'^2}\right)} \frac{\kappa^2}{\kappa'^3} g\, d\kappa' . \tag{14}$$ [6]

Since $E = hR/\kappa^2$ and since $Z = 1$ for hydrogen, the loss in energy is $\Delta E = hR(1/\kappa^2 - 1/\kappa'^2)$. To find the mean rate of energy loss, we multiply the probability of collision by the energy lost per collision,

$$N_i \sigma_{\kappa\kappa'} V d\kappa' dE = \frac{2^4}{3\sqrt{3}} \frac{\epsilon^2 h^2 R}{m^2 c^3} N_i V \frac{\kappa^2}{\kappa'^3} g\, d\kappa' . \tag{15}$$

To obtain the total energy loss, we integrate over all values of κ' larger than κ. Thus

$$\frac{d\bar{E}}{dt} = \frac{2^4}{3\sqrt{3}} \frac{\epsilon^2 h^2 R}{m^2 c^3} \tfrac{1}{2} V g N_i . \tag{16}$$

This is the mean rate of energy loss. To find the mean time required to dissipate an energy equal to the initial energy, divide by E and take the reciprocal,

$$\frac{F}{\frac{d\bar{E}}{dt}} = \frac{\frac{1}{2} m V^2}{2.10 \times 10^{-22} n R \frac{1}{2} V g N_i} . \tag{17}$$

Expressing the energy, E, in volts, we get

$$\frac{E}{\frac{d\bar{E}}{dt}} = 1.2 \times 10^9 \sqrt{\mathfrak{V}} \text{ seconds} .$$

For electron energies in the neighborhood of 1 volt, this interval amounts to about 40 years. Hence, at prevailing electron temperatures in gaseous nebulae, free-free emissions dissipate energy very slowly.

E. COULOMBIC ENCOUNTERS

The coulombic scattering, which arises from the electrostatic interactions between charged particles, may be estimated with the aid of the Rutherford expression[7] for the cross-section for scattering of a particle of energy, E, through an angle, θ, thus:

$$\sigma = \frac{\epsilon^4}{16 E^2 \sin^4 \frac{\theta}{2}} . \tag{18}$$

The total number of particles scattered into angles between θ and $\theta + d\theta$ is

$$dn = 2\pi N_\epsilon V \frac{\epsilon^4}{16 E^2} \frac{\sin \theta d\theta}{\sin^4 \frac{\theta}{2}} . \tag{19}$$

[6] If we write Menzel and Pekeris' formula (1.53) for a continuous state of width $d\kappa$ at κ' instead of for a discrete lower level n', we get our equation (14).

[7] This is the Rutherford scattering formula. A discussion along the lines in which we are interested may be found in Kennard, *Kinetic Theory of Gases* (New York: McGraw-Hill Book Co., 1938).

Since the cross-section is so large for small angle deflections, these will play the dominant role, and we may replace $\sin\theta$ by θ for the vast majority of collisions. Then

$$dn = \frac{2\pi N_\epsilon v\epsilon^4}{E^4}\frac{d\theta}{\theta^3}. \tag{20}$$

Because an ion is so very much heavier than an electron, we may neglect energy transfers to it. To estimate the mean energy transfer in electron-electron encounters, we must note that the Rutherford formula applies only in the frame of reference in which the center of gravity is at rest. Since the angle of scattering in the fixed co-ordinate system is one-half that in the center-of-mass system[7] and since the relative energy is one-half that of the incident electron, the above formula is unchanged if θ is allowed to refer to the actual angle of scattering in the fixed system and E to the incident energy. For the time being we shall assume that the scattering particle is moving at a negligible velocity. In Section IV the effect of its finite velocity will be considered. To a first approximation the energy gained by the scattering electron is given by

$$E = E\sin^2\theta \sim E\theta^2 . \tag{21}$$

The mean rate of energy transfer then is

$$\Delta E dn = \frac{2\pi N_\epsilon v\epsilon^4}{E}\frac{d\theta}{\theta}. \tag{22}$$

We obtain the total energy transfer by integrating from some appropriate minimum value of θ to the maximum value of θ, which is π, thus:

$$\frac{dE}{at} = \frac{2\pi N_\epsilon v\epsilon^4}{E} ln\left(\frac{\pi}{\theta_{\min}}\right). \tag{23}$$

The scattering cross-section increases without bound as the angle of deflection approaches zero. Hence, if the formula is to be physically significant, there must be some process or processes which serve to fix a minimum angle for which the scattering expression can be applied.

The minimum angle is determined by the maximum distance for which the electrostatic field of the electron is not seriously modified by the presence of all the other electrons in the gas. In fact, the more distant encounters must be regarded as many-body collisions. At any instant many distant electrons are pulling on any given electron, some in one way, some in another. The average force tends to cancel out; but, because the density fluctuates, there are small residual energy transfers, the effects of which should, in a completely rigorous treatment, be added to the effects of the close encounters. The same problem arises in the determination of the time of relaxation in stellar systems. Spitzer[8] and Chandrasekhar[9] suggest that the assumption of a single encounter, leading to the Rutherford cross-section, breaks down at a collision distance of about two or three times the mean distance between stars. At greater distances they believe that the complex many-body interactions will effectively cause the mean force to be much less than the coulomb force. In fact, the contribution of more distant encounters is so small that it can be ignored by choosing an appropriate maximum value of b, the cutoff distance, beyond which the scattering is negligible. Because the result depends only on the logarithm of this value, it will not depend critically on the cutoff, which may be taken as b_0, the mean distance between stars.

[8] *M.N.*, **100**, 387, 1940.

[9] *Principles of Stellar Dynamics* (University of Chicago Press, 1942), see esp. pp. 56 ff.

The plasma problem differs from the stellar problem in two respects; first, there are repulsive, as well as attractive, forces, and, second, there is plenty of time for equilibrium to be set up. On the other hand, Chandrasekhar and Spitzer have shown that the time of relaxation of the stellar system is much greater than the probable age of the universe. In the plasma there is a tendency to establish an equilibrium state in which every charge is shielded by charges of the opposite sign. This shielding tendency is resisted by the random motion, which tends to mix charges regardless of sign. As a result of the balance of these two tendencies, there is a critical distance at which shielding, on the average, becomes effective. An exact treatment of this problem is complicated, and it will suffice to establish an upper limit to the cutoff distance (i.e., a lower limit to the scattering angle.)

To calculate this distance we may use the methods of Debye. The potential around any charge is determined by Poisson's equation:

$$\nabla^2\phi = 4\pi\epsilon\,(N_\epsilon - N_i)\,. \tag{24}$$

The electronic and ionic densities, N_ϵ and N_i, are determined by the Boltzmann equations

$$N_\epsilon = N_0 e^{\epsilon\phi/kT}\,, \qquad N_i = N_0 e^{-\epsilon\phi/kT}\,.$$

If $\epsilon\phi/kT$ is not too large, the first term of the expansion for the exponentials is adequate. We have

$$\nabla^2\phi = \frac{8\pi\epsilon^2 N_0}{kT}\,\phi\,.$$

The solution corresponding to a unit charge at short distance and finite potential at infinity is

$$\phi = \frac{\epsilon e^{-r/h}}{r}\,.$$

where

$$h^2 = \frac{kT}{8\pi N_0 \epsilon^2}\,. \tag{25}$$

Now h is the distance at which the potential begins to deviate appreciably from the coulombic; it is therefore the distance at which shielding becomes appreciable. Below this distance the potential is roughly coulombic, and the Rutherford cross-section applies; while above this value the potential soon becomes negligible, so that one may take h as the cutoff. This should certainly give at least an upper estimate.

When the value of h is computed for our problem, we find that it is much larger than the mean distance between electrons. Because this quantity appears in the logarithm, the replacement of b_0 by h changes the result by a factor of 2 or less; hence, in practice, the question of cutoff is not critical and certainly does not affect the qualitative validity of our results. At worst, we can say that the true value for the cutoff lies somewhere between b_0 and h.

The relation between the maximum encounter distance, b, and the angle $\theta_{\min}$ is given by the expression

$$\theta_{\min} = 2 \arctan \frac{\epsilon^2}{2Eb} \sim \frac{\epsilon^2}{Eb}\,; \tag{26}$$

and it makes very little difference whether or not we employ for b the mean distance between electrons of the Debye encounter limit, h. At an electron density of $10^4/\mathrm{cm}^3$ and a temperature of 10,000° K, h will be 5 cm, while the mean interelectronic distance, b_0, will be of the order of 0.05 cm. Under these conditions the log $(\pi/\theta_{\min})$ factor in equation (23) will amount to 16.5 or 12, depending on our choice of b; hence the two results for

dE/dt will differ by a factor of less than 2. For a numerical example, let us choose the log factor to be 15. To find the mean time required for an electron to dissipate its energy, we divide the mean rate of energy loss given in equation (23) by E and take the reciprocal, thus:

$$T \sim \frac{E^2}{30\pi N_\epsilon v \epsilon^4}. \tag{27}$$

In an example where $E = 3.3$ volts and $N_\epsilon \sim 10^4$, the corresponding velocity, v, will be 1.07×10^8 cm/sec, and the time will be of the order of 4 seconds. That is, a 3.3-volt electron loses most of its energy to the other electrons in only 4 seconds. Since most electrons have even less energy than this, they may redistribute what they have in an even shorter time. Therefore, 1 second is the order of the time involved in shuffling the energy among the electrons.

Thus it is clear that interelectronic collisions are the main processes acting to determine the distribution among the velocities and that the other physical processes, such as recombination, etc., amount to only small perturbations. An electron liberated within the nebula loses half its initial energy within a few seconds and thereafter obtains an energy primarily determined by the velocities of all the other electrons. Since the distribution cannot differ much from the equilibrium form that it would have if all other processes were absent, it must be very close to the Maxwellian. A method for the calculation of the slight deviations will be considered in the next section.

Spitzer[8] and, more recently, Chandrasekhar[9] have treated the problem of coulombic-type encounters for the corresponding stellar case in great detail. Chandrasekhar has derived general expressions for inverse-square law deflections, valid for all angles of encounter. In the present treatment we have confined our attention to small angles, so that our expressions are somewhat simpler than those of Chandrasekhar. For the detailed theory of encounters, the reader is referred to the excellent treatment given in ***Principles of Stellar Dynamics***.

III. BOLTZMANN'S EQUATION FOR THE ASSEMBLY

A. SOLUTION WITH ELASTIC COLLISIONS ONLY

In order to calculate the deviations from Maxwell's formula, we must solve the Boltzmann integrodifferential equation,[10] which determines the distribution. Let $N_\epsilon = N_i$ denote the number of electrons per cm^3; $f_2(u, v, w)$ denotes the distribution function specifying the fraction of all electrons in the range between u and $u + du$, v and $v + dv$, and w and $w + dw$. We normalize f_2 so that its integral over all velocity space is unity, viz.,

$$\iiint_{-\infty}^{+\infty} f_2(u, v, w)\, du\, dv\, dw = 1. \tag{28}$$

If the distribution function is isotropic in velocity space, then f_2 is a function only of the absolute value of the velocity V, and the normalization reduces to

$$4\pi \int_{-\infty}^{+\infty} f_2(V)\, V^2 dV = 1. \tag{29}$$

Our $f_2(V)$ is related to the $f(v)$ used by Menzel through the relation

$$f(v) = 4\pi V^2 f_2(V) \tag{30}$$

[10] See, e.g., Kennard, *op. cit.*, p. 89.

(notice that we use a V to denote the total speed and v for the y-component). The total number of electrons in any specified interval of velocity space is given by the formula

$$dN = N_\epsilon f_2(V)\, du\, dv\, dw\,. \tag{31}$$

Boltzmann's equation simply expresses the rate of change of the distribution function arising from the various processes, such as collision, excitation, recombination, etc., which are occurring in the assembly. The equation is

$$\left.\begin{aligned}\frac{\partial(N_\epsilon f_2)}{\partial t} = -(V\cdot\nabla)(N_\epsilon f_2) - (\dot{V}\cdot\nabla_V)(N_\epsilon f_2) + \left[\frac{\partial}{\partial t}(N_\epsilon f_2)\right]_{\text{coll}} \\ + \left[\frac{\partial}{\partial t}(N_\epsilon f_2)\right]_p,\end{aligned}\right\} \tag{32}$$

wherein $(\dot{V}\cdot\nabla_V)$ signifies the operator

$$\dot{u}\frac{\partial}{\partial u} + \dot{v}\frac{\partial}{\partial v} + \dot{w}\frac{\partial}{\partial w},$$

$$\left[\frac{\partial}{\partial t}(N_\epsilon f_2)\right]_{\text{coll}}$$

signifies the rate of change of $(N_\epsilon f_2)$ arising from encounters, while

$$\left[\frac{\partial}{\partial t}(N_\epsilon f_2)\right]$$

signifies the rate of change of $N_\epsilon f_2$ arising from all other processes, such as recombination, inelastic collisions, free-free emissions, etc., discussed in Section II.

The first two terms represent the effects of hydrodynamical flow in phase space, as a result of which the density and velocity distribution are changing simply because of the motions and macroscopic accelerations undergone by the particles involved. If the distribution is independent of the spatial co-ordinates and if there are no macroscopic body forces, these two terms vanish. For the present we shall assume that these simple conditions prevail in a nebula, at least over moderate distances. At some time it may be advisable to consider spatial diffusions and the effects of radiation pressure over large distances, which will appear as a body force. These forces could introduce but small corrections to the main collisional terms.

Because elastic collisions have been shown to be far more frequent than any other event, the distribution will differ only slightly from what it would be if they were the only processes. It is therefore worth while, first, to consider the distribution arising from elastic collisions alone. It will be shown presently that the only solutions to Boltzmann's equation for this case must be Maxwellian. Thus we may write

$$f_2(V) = \left(\frac{m}{2\pi kT}\right)^{3/2} e^{-mv^2/2kT}\,. \tag{33}$$

Later, when the effect of other processes is considered, we shall find that the distribution differs slightly from the Maxwellian; but we can write

$$f_2(V) = b(V)\left(\frac{m}{2\pi kT}\right)^{3/2} e^{-mv^2/2kT}\,, \tag{34}$$

where $b(V)$ is a correction factor close to unity.

Hagihara states that there is an approximate solution to Boltzmann's equation for the isotropic case with elastic collisions that is more general than the Maxwellian distribution, viz.,

$$f_2(V) = \left(\frac{m}{2\pi kT}\right)^{3/2} [c + dV^2]\, e^{-mv^2/2kT}\,. \tag{35}$$

For any given temperature, T, and for any assigned spectral energy distribution and matter density, the condition for a radiative steady state and the condition for constant populations in each level determined c and d. But the temperature itself apparently is not determined by any physical process, and the result is ambiguous. To prove that the V^2 term may be present, Hagihara[11] expands the most general solution in a series of Hermite polynomials

$$f_2(u, v, w) = \left(\frac{m}{kT}\right)^{1/2} \sum_{ijk} f_{ijk}\varphi_{ijk}\,, \tag{36}$$

with

$$\varphi_{000} = \frac{1}{(2\pi)^{3/2}}\, e^{-\frac{u^2+v^2+w^2}{2kT/m}}\,, \qquad \varphi_{ijk} = (-1)^{i+j+k} H_{ijk}\varphi_{000}\,,$$

and

$$H_i = H_i\left(u\sqrt{\frac{m}{kT}}\right), \qquad H_j = H_j\left(v\sqrt{\frac{m}{kT}}\right), \qquad H_k = H_k\left(w\sqrt{\frac{m}{kT}}\right),$$

$$H_0(X) = 1\,, \qquad H_1(X) = X \qquad \text{and} \qquad H_2(X) = X^2 - 1\,.$$

The solution with $H_0(X)$ refers to the Maxwellian distribution, while that with $H_1(X)$ refers to mass motion. Since we are considering the case in which there is no mass motion, the H_1 terms must vanish. The only isotropic term of second order is

$$H_{200} + H_{020} + H_{002} = V^2 - 3\,, \qquad \text{where} \qquad V^2 = u^2 + v^2 + w^2\,.$$

The total number of particles is found by integrating f_2 over all velocities. Because of the orthogonality properties of the Hermite polynomials, the only contribution to this number is made by H_0. As for the total energy, it is readily shown that only H_0 and $H_{200} + H_{020} + H_{002}$ contribute. Since neither the number of particles nor the total energy can be changed by elastic collisions and since the number and energy together determine the coefficients of these two terms, the coefficients of H_{000} and $H_{200} + H_{020} + H_{002}$ are not changed by collisions. Thus, to the second order the $V^2 - 3$ term is also a steady-state solution of Boltzmann's equation. A direct substitution in the precise form of the equation, however, shows that this latter term is not a solution to higher orders. Yet, if we wish to know the tail of the distribution function, the higher-order terms are essential, and therefore the approximate solution given by Hagihara cannot be applied unless the very high-order polynomials are also considered. The greater the velocities in which we are interested, the further the expansion must be carried.

It is possible, however, to go further and to show that, for the case of elastic collisions, all steady-state solutions to Boltzmann's equation must be Maxwellian. Boltzmann's H theorem[12] serves to establish this result. The method employed is to show that the function, $H = \Sigma f_2 ln f_2$, which is summed over all parts of phase space, must have a negative derivative unless the distribution is Maxwellian. Since a steady state requires that all functions of the distribution be constant, it can exist only for the Maxwellian case. In the light of this result, as well as that of the previous considerations, showing that the V^2 term is a bad approximation for the high-energy range, Hagihara's approximate solution

[11] *Tokyo Astr. Bull.*, No. 525, eq. (2).

[12] Kennard, *op. cit.*, p. 53.

must be thought of simply as the first term of a Maxwellian distribution expanded about another temperature, thus

$$e^{-mV^2/2kT'} = e^{-mV^2/2kT}\left(1 - \left[\frac{m}{2kT} - \frac{m}{2kT'}\right]V^2 + \ldots\right).$$

When the distribution is not in equilibrium, there is no unique temperature; this is the essential content of Hagihara's result. Yet, from the facts discussed above, we may define a best temperature which makes the correction factor as close as possible to unity. It is clear that the best Maxwellian distribution is that which, by itself and without a correction factor, represents correctly the total number of particles and the total energy. This means that the best temperature is the one in which the V^2 term is absent. We shall see that Boltzmann's equation requires a boundary condition specifying a value of $b(0)$ and the coefficient of the V^2 term in the expansion of $b(V)$. By choosing $b(0) = 1$ and the coefficient of the V^2 term zero, we satisfy our boundary conditions with the best possible temperature. Thus the temperature is made unique, and the ambiguity mentioned above is shown to be merely the consequence of the choice of an inappropriate temperature.

Since a Maxwellian distribution of any temperature is the most general solution of Boltzmann's equation when only elastic collisions are considered, it is clear that the temperature must be determined by processes other than elastic collisions. In the nebular case the distribution is very close to an equilibrium one, and to a first approximation we may assume a Maxwellian distribution, and then determine the temperature and the density by setting up the condition for a balance of the emission and recapture of electrons and for the balance of energy. Menzel and his collaborators have done this in Papers V and XVI of the nebular series.

B. GENERAL FORMULATION OF BOLTZMANN'S EQUATION

Let us now turn to the problem of the detailed formulation of Boltzmann's equation. We shall find it convenient to re-express all quantities as functions of the energy. Let the fraction of particles between E and $E + dE$ be denoted by $f_3(E)dE$. Since $E = \frac{1}{2}mV^2$ and $dE = mVdV$, we have

$$f_3(E)\,dE = f_3(E)\,mV\,dV = f(V)\,dV \quad \text{or} \quad f_3(E) = \frac{f(v)}{mV}. \tag{37}$$

The Maxwellian distribution in energy may be written as

$$f_3(E) = \frac{2E^{1/2}}{\sqrt{\pi}\,(kT)^{3/2}}\,e^{-E/kT}. \tag{38}$$

We shall also find it convenient to write the deviation from the Maxwellian form with the aid of the factor $b(E) = 1 + \delta(E)$ where $\delta(E)$ is small. Furthermore, since the best temperature is chosen with $b(0) = 1$ and the coefficient of E in the expansion of $b(E) = 0$, we also have $b'(0) = 0$ and

$$f_3(E) = \frac{2}{\sqrt{\pi}\,(kT)^{3/2}}[1 + \delta(E)]E^{1/2}e^{-E/kT}. \tag{39}$$

We shall now collect the expressions for the noncoulombic contributions to $(\partial N_\epsilon f_2/\partial t)$.

a) Photo-ionization.—The number of electrons emitted per second in the energy range dE is given by equation (8), with ν replaced by $E/h + E_a/h$, where E_a is the ionization energy of hydrogen, or 13.54 volts,

$$F_{1E}dE = \frac{3.202\times10^{-6}}{T_\epsilon^{3/2}}N_\epsilon^2 hG_{T\epsilon}\frac{dE}{E+E_a}\frac{1}{(e^y-1)}\int_{y_1}^{\infty}\frac{dy}{y(e^y-1)}. \tag{40}$$

b) Recombinations.—The number of electrons possessing energy between E_1 and $E_1 + dE_1$ that are recaptured on all levels is found by multiplying the probability of recapture of electron and ion by the number of electrons and ions concerned. The number of electrons is $N_\epsilon f_3(E_1)dE_1$, and the number of ions is $N_i = N_\epsilon$. We get

$$\left[\frac{\partial (N_\epsilon f_3)}{\partial t}\right]_{\text{recomb}} = 1.24 \times 10^{-14} N_\epsilon^2 f_3(E_1) \frac{\mathfrak{V}_i}{\sqrt{\mathfrak{V}_1}} \Sigma \frac{1}{\left(\frac{1}{n'^2} + \frac{\mathfrak{V}}{\mathfrak{V}_i}\right) n_1^3} \tag{41}$$

if we express E in volts, $\mathfrak{V}$, in this formula. In absolute units the result is

$$\left[\frac{\partial (N_\epsilon f_3)}{\partial t}\right]_{\text{recomb}} = 2.08 \times 10^{-22} N_\epsilon^2 f_3(E_1)\, V\, \Sigma \frac{1}{n_1^3\left(\frac{1}{n'^2} + \frac{E}{E_a}\right)} \frac{E_a}{E}. \tag{42}$$

c) Inelastic collisions.—Inelastic collisions cause electrons to be transferred from a given energy range to a lower one. Provided $E_1 > E_0$, the rate at which electrons leave a given energy range is $dE_1 \sigma(E_1) V(E_1) N_A N_\epsilon f_3(E_1)$, and the rate at which they are caused to enter from an energy exceeding E_1 by the excitation energy, E_0, is $f_3(E_1 + E_0)\sigma(E_1 + E_0) N_A N_\epsilon V(E_1 + E_0)$. In a typical planetary nebula, where the green O III lines are the strongest, N_A denotes the number of O III ions in the ground ^3P term, and N_B the number in the ^{1}D level, while E_0 denotes the excitation energy. The total change is

$$\left.\begin{aligned}\left[\frac{\partial (N_\epsilon f_3)}{\partial t}\right]_{\text{inel}} = N_A N_\epsilon [f_3(E+E_0)\, V(E_1+E_0)\, \sigma(E_1+E_0) - f_3(E_1)\, V(E_1) \\ \times \sigma(E_1)\, Q(E_1)],\end{aligned}\right\} \tag{43}$$

where $Q(E_1) = 0$ when $E < E_0$ and $Q(E_1) = 1$ when $E > E_0$. By similar reasoning, we obtain for the superelastic collisions

$$\begin{aligned}\left[\frac{\partial (N_\epsilon f_3)}{\partial t}\right]_{\text{superel}} = N_B N_\epsilon [f_3(E_1 - E_0)\, \sigma(E_1 - E_0)\, V(E_1 - E_0)\, Q(E_1 - E_0) \\ - f_3(E_1)\, \sigma(E_1)\, V(E_1)].\end{aligned}$$

If E_1 is less than E_0, there can be no gains from superelastic collisions.

d) Free-free emissions.—Free-free emissions arise from electron-ion encounters and will cause electrons to leave a given energy range and be transferred to a lower one, differing by the energy of the photon emitted. The rate of loss is calculated from equation (14) thus:

$$N^2 f_3(E_1)\, V(E_1) \int_0^\infty \frac{2^4}{3\sqrt{3}} \frac{\epsilon^2 h}{m c^3} \frac{1}{\left(\frac{1}{\kappa^2} - \frac{1}{\kappa'^2}\right)} \frac{\kappa^2}{\kappa'^3}\, g\, d\kappa'. \tag{44}$$

The rate of gain of electrons will be

$$N_\epsilon^2 \frac{2^4 \epsilon^2 h}{3\sqrt{3} m c^3} \int_0^\infty f_3\left[E + hR\left(\frac{1}{\kappa^2} - \frac{1}{\kappa'^2}\right)\right] \frac{V\left[E_1 + hR\left(\frac{1}{\kappa^2} - \frac{1}{\kappa'^2}\right)\right]}{\left(\frac{1}{\kappa^2} - \frac{1}{\kappa'^2}\right)} \frac{\kappa^2}{\kappa'^3}\, g\, d\kappa', \tag{45}$$

and the net rate of change is just the difference between the latter and the former expressions.

e) Coulombic terms.—Let us now consider the contribution of the coulombic terms. The rate of change of the distribution resulting from scattering is determined from the coulomb cross-section (cf. Sec. II). The problem, however, is complicated by the fact that both of the electrons involved in the collision are moving. Therefore, we must sum up the contributions to the change of distribution resulting from electrons of all different velocities. When an electron collides with another electron which is moving, it may lose or gain energy, depending on the relative velocities and the angle of collision. It is from those collisions that produce a gain of energy that the high-velocity portion of the Maxwellian distribution arises.

For the present we shall denote the fact that the collision cross-section depends on the energies of both colliding particles by writing it $\sigma = \sigma(E_1, E_2, \Delta E)$, where E_1 is the electron under consideration, E_2 is the electron with which it collides, and ΔE is the energy given to the first electron as a result of collision. The precise form of σ will be derived in Section IV; the number of collisions per second by electrons of energy E with electrons of energy E_2 and energy exchange, ΔE, then is

$$d(\Delta E) \int_0^\infty (dE_2)\, v_r N_\epsilon^2 f_3(E_1)\, f_3(E_2)\, \sigma(E_1, E_2, \Delta E)\,, \tag{46}$$

where v_r is the relative velocity of the two electrons. To find the rate at which the distribution function is changed by collision, let us consider electrons in a given energy range from E_1 to $E_1 + dE_1$. Electrons of energy $(E_1 - \Delta E)$ are scattered into this range by collisions with an exchange of ΔE in the positive sense, as defined above. On the other hand, collisions also cause electrons to leave this range. The net change in the distribution is found by summing over all ΔE's. Although ΔE cannot be greater than E_2, the cross-section decreases so rapidly with E that we may make our limits $-\infty$ to $+\infty$. We obtain

$$\left.\begin{aligned}\left[\frac{\partial(N_\epsilon f_3)}{\partial t}\right]_{\text{coll}} &= N_\epsilon^2 \int_0^\infty dE_2 \int_{-\infty}^{+\infty} d(\Delta E)\, [\, v_r(E_1 - \Delta E, E_2)\, \sigma(E_1 - \Delta E, E_2, \Delta E) \\ &\quad \times f_3(E_1 - \Delta E)\, f_3(E_2) - v_r(E_1, E_2)\, \sigma(E_1, E_2, \Delta E)\, f_3(E_1)\, f_3(E_2)]\,.\end{aligned}\right\} \tag{47}$$

We now have all the terms that enter into Boltzmann's equation.

IV. THE SOLUTION OF BOLTZMANN'S EQUATION

We shall begin by considering the coulomb terms only, which, as we have seen, represent the main processes involved. The term $[\partial N_\epsilon f_3/\partial t]_{\text{coll}}$, in its complete form, is too complicated to allow a solution. It is possible, however, to obtain an approximation which takes advantage of the fact that coulomb scattering involves mainly very small angles and, therefore, very small changes of energy. This is made possible by the fact that the coulomb cross-section, $\sigma(E_1, E_2, \Delta E)$, becomes very large as ΔE goes to zero. Because σ varies continuously as a function of E_1, we can expand $\sigma(E_1 - \Delta E, E_2, \Delta E)$ as a series of powers of ΔE about E_1. This means that, in so far as σ depends on ΔE through E_1, we can expand it, but, in so far as it depends on ΔE directly, we do not expand it. For convenience we shall write

$$v_r \sigma = \Omega(E_1, E_2, \Delta E)\,. \tag{48}$$

Then $[\partial N_\epsilon f_3/\partial t)_{\text{coll}}$ can be expanded up to the second power in ΔE as follows:

$$\left.\begin{aligned}\left[\frac{\partial(N_\epsilon f_3)}{\partial t}\right]_{\text{coll}} &= N_\epsilon^2 \int_0^\infty dE_2 \int_{-\infty}^{+\infty} d\Delta E \Big[\, (-\Delta E) \frac{\partial}{\partial E}\, \Omega(E_1, E_2, \Delta E)\, f_3(E_1) \\ &\quad + \frac{(\Delta E)^2}{2} \frac{\partial^2}{\partial E_1^2}\, \Omega(E_1, E_2, \Delta E)\, f_3(E_1)\Big]\, f_3(E_2)\,.\end{aligned}\right\} \tag{49}$$

Because the explicit dependence of Ω on ΔE is such that Ω is large only for very small ΔE, the higher terms in the expansion integrate out to a negligible value. This expansion results in an equation very similar to the Fokker-Planck equation.[13] The difference is that the terms are to be calculated directly from the known collision cross-sections and not semi-empirically, as is usually done. This procedure is possible only because the cross-section for this case is both known and simple.

If we now introduce equation (39) for $f_3(E)$ and neglect all terms involving $\delta(E)$ twice as a factor (as they are of the second order), we obtain

$$\left[\frac{\partial (N_\epsilon f_3)}{\partial t}\right]_{\text{coll}} = (I) + (II) + (III),$$

where

$$\left.\begin{aligned}
(I) &= \frac{2}{\sqrt{\pi}\,(kT)^{3/2}} N_\epsilon^2 \int_0^\infty dE_2 \int_{-\infty}^{+\infty} d(\Delta E) \Big[(-\Delta E) \frac{\partial}{\partial E_1} \Big(\frac{\Omega E^{1/2} e^{-E/kT}}{(kT)^{3/2}} \Big) \\
&\quad + \frac{(\Delta E)^2}{2} \frac{\partial^2}{\partial E^2} \Omega E^{1/2} e^{-E_1/kT} \Big] [e^{-E_2/kT} E_2^{1/2}] \frac{2}{\sqrt{\pi}\,(kT)^{3/2}}; \\
(II) &= \frac{2N_\epsilon^2}{\sqrt{\pi}\,(kT)^{3/2}} \int_0^\infty dE_2 \int_{-\infty}^{+\infty} d\Delta E \Big[-\Delta E \frac{\partial}{\partial E} \Omega E_1^{1/2} e^{-E_1/kT} \delta(E_1) \\
&\quad + \frac{(\Delta E)^2}{2} \frac{\partial^2}{\partial E_1^2} \Omega E_1^{1/2} e^{-E_1/kT} \delta(E_1) \Big] \Big[e^{-E_2/kT} E_2^{1/2} \frac{2}{\sqrt{\pi}\,(kT)^{3/2}} \Big]; \\
(III) &= \frac{2N_\epsilon^2}{\sqrt{\pi}\,(kT)^{3/2}} \int_0^\infty dE_2 \int_{-\infty}^{+\infty} d\Delta E \Big[-(\Delta E) \frac{\partial}{\partial E_1} (\Omega E_1^{1/2} e^{-E/kT}) \\
&\quad + \frac{(\Delta E)^2}{2} \frac{\partial^2}{\partial E_1^2} (\Omega E_1^{1/2} e^{-E_1/kT}) \Big] [e^{-E_2/kT} E_2^{1/2} \delta(E_2)] \frac{2}{\sqrt{\pi}\,(kT)^{3/2}}.
\end{aligned}\right\} \quad (50)$$

Term (I) would be present even if we had a Maxwellian distribution with $\delta(E) = 0$. Therefore, since a Maxwellian distribution is known to be in equilibrium for scattering alone, term (I) must be zero. In the course of the calculation we shall find it convenient to prove this. Term (II) represents the change in the deviation of the distribution from the Maxwellian, brought about by collision of the non-Maxwellian portion with the much larger Maxwellian component of the distribution which is the principal term. Term (III) represents the change in the Maxwellian component brought about by energy exchange with the small non-Maxwellian component. Terms involving $\delta(E)$ twice as factors represent the changes brought about by the small non-Maxwellian component with itself and are therefore negligible.

The first step in solving the problem is to carry out the integrations over ΔE, which are the same in all three terms. Inspection of these terms shows that they contain as factors the following two quantities:

$$A = \int_{-\infty}^{+\infty} d(\Delta E)\, \Delta E \Omega (E_1, E_2, \Delta E); \qquad B = \int_{-\infty}^{+\infty} d(\Delta E)\, (\Delta E)^2 \Omega (E_1, E_2, \Delta E). \quad (51)$$

A is proportional to the mean rate at which an electron of energy E_1 gains energy from electrons of energy E_2, while B is proportional to the dispersion of this rate about its mean. Terms containing A represent the change in the distribution function arising from the mean rate of energy transfer among the individual electrons, while terms containing

[13] Chandrasekhar, "Stochastic Problems in Physics and Astronomy," *Rev. Mod. Phys.* **15**, 1, 1943.

B represent the modifications introduced by the fact that the rate of energy transfer is not equal to its average but fluctuates in a more or less random way. The question of when it is adequate to represent all effects of a set of collision processes by their mean effects and by the mean deviation from the mean is essentially the same as in the treatment of the Fokker-Planck equation.[13] The mathematical justification for this procedure in our case is that, because the collision cross-section is such a rapidly decreasing function of ΔE, powers of ΔE higher than the second make a negligible contribution.

In order to evaluate A and B, we must first obtain an expression for the energy transfer ΔE as a function of the angle of scattering and of the angle between the motion of the two particles. Let us consider two colliding electrons, labeled "1" and "2," respectively. The velocity of the first particle is represented by V_1 and that of the second by V_2. Let us choose the z-axis in the direction of motion of the first particle, and let the angle between V_1 and V_2 be represented by the polar angle ω. We shall take the azimuthal angle as zero. To find the average value of ΔE, we must integrate over all possible angles of scattering, weighted with the scattering cross-section;[14] the result is

$$A(\omega) = -\frac{4\pi\epsilon^4 S(V_1^2 - V_2^2}{m(V_1^2 + V_2^2 - 2V_1V_2\cos\omega)^{3/2}}, \tag{52}$$

where (cf. eq. [27])

$$S = ln\left[\left(\frac{\pi}{\theta}\right)_{\min}\right].$$

Because S is such a slowly varying function of V, it is practically constant and is shown in Section II; its value is about 10 or 20 for the densities and temperatures of typical nebulae. Further,

$$B(\omega) = \frac{4\pi S\epsilon^4 V_1^2 V_2^2 \sin^2\omega}{(V_1^2 + V_2^2 - 2V_1V_2\cos\omega)^{3/2}}. \tag{53}$$

Inspection of A shows that, if $V_1 > V_2$, the average energy transfer is negative, while if $V_1 < V_2$, it is positive. This result simply means that a particle tends to lose energy on collisions with less energetic particles and to gain energy on collision with more energetic ones.

Since the distribution is isotropic, we must now average over ω, the angle between the motion of the two colliding particles, weighting according to the element of solid angle. Since all azimuthal angles are equally probable, we need merely weight over polar angles ω, with the factor $\frac{1}{2}\sin\omega d\omega$ (which is normalized when integrated over ω from 0 to π). Although these integrals are continuous functions of V_1 and V_2, their analytical form undergoes a change at $V_1 = V_2$. Thus it is necessary to consider the two possibilities: the first in which $V_1 > V_2$ and the second in which $V_2 > V_1$. We get[14]

$$\left.\begin{aligned} &(I)\quad (V_1 > V_2)\quad A = -\frac{4\pi\epsilon^4 S}{mV_1}; \quad B = \frac{8\pi S\epsilon^4 V_2^2}{3V_1}, \\ &(II)\quad (V_2 > V_1)\quad A = \frac{4\pi\epsilon^4 S}{mV_2}; \quad B = \frac{8\pi S\epsilon^4 V_1^2}{3V_2}. \end{aligned}\right\} \tag{54}$$

This result means that the average rate of energy loss to a slower particle is inversely proportional to the velocity of the faster particle only. On the other hand, in collisions with a faster particle, a slower particle gains energy at a rate inversely proportional to the

[14] For discussion of a closely analogous stellar problem see Chandrasekhar, *Ap. J.*, 93, 326, 1941.

velocity of the faster particle only. Obviously, these two results are necessary for each other's consistency. We may now re-write $[\partial N_\epsilon f_3/\partial t]_{\text{coll}}$ in terms of A and B, viz.,

$$\left.\begin{aligned} \left[\frac{\partial}{\partial t}(N_\epsilon f_3)\right]_{\text{coll}} &= \frac{4N_\epsilon^2}{\pi (kT)^3}\int_0^\infty dE_2 E_2^{1/2} e^{-E_2/kT}\Big[-\frac{\partial}{\partial E_1} A(E_1, E_2) E_1^{1/2} e^{-E_1/kT} \\ &\quad + \frac{1}{2}\frac{\partial^2}{\partial E_1^2} B(E_1, E_2) E_1^{1/2} e^{-E_1/kT}\Big] \frac{4N_\epsilon^2}{\pi (kT)^3}\int_0^\infty dE_2 E_2^{1/2} e^{-E_2/kT}\Big[-\frac{\partial}{\partial E_1} \\ &\quad \times A E_1^{1/2} e^{-E_1/kT}\,\delta(E_1) + \frac{1}{2}\frac{\partial^2}{\partial E_1^2} B E_1^{1/2} e^{-E_1/kT}\,\delta(E_1)\Big] \frac{4N_\epsilon^2}{\pi (kT)^3} \\ &\quad \times \int_0^\infty dE_2 E_2^{1/2} e^{-E_2/kT}\,\delta(E_2)\Big[-\frac{\partial}{\partial E} A E_1^{1/2} e^{-E_1/kT} + \frac{1}{2}\frac{\partial^2}{\partial E_1^2} B E_1^{1/2} e^{-E_1/kT}\Big]. \end{aligned}\right\} \tag{55}$$

The next step is to carry out integrations over E_2. For integrals (I) and (II) this operation is possible, since all dependence on E_1 is explicit in terms of known functions. In integral (III), however, E_2 occurs through the unknown $\delta(E_2)$. This raises a serious difficulty in obtaining a complete solution in a straightforward way. We shall see presently, however, that a systematic approximation procedure can be adopted, in which the value of $\delta(E_2)$ to be inserted in integral (III) is always to be obtained from the next lower approximation. Meanwhile, we shall direct our attention to integrals (I) and (II), both of which require the evaluation of exactly the same integrals; these integrals have been evaluated by Chandrasekhar and others. We find

$$\left.\begin{aligned} C(E_1) &= \frac{4N_\epsilon^2}{\pi (kT)^3}\int_0^\infty dE_2 E^{1/2} e^{-E_2/kT} A(E_1, E_2) = \frac{2N_\epsilon^2 8\pi S\epsilon^4}{\pi (kT)^{3/2}(2\pi m kT)^{1/2}} \\ &\quad \times\left[2e^{-E_1/kT} - \frac{W\left(\sqrt{\frac{E_1}{kT}}\right)}{\sqrt{\frac{E_1}{kT}}}\right] \end{aligned}\right\} \tag{56}$$

and

$$\left.\begin{aligned} D(E_1) &= \frac{4N_\epsilon^2}{\pi (kT)^3}\int_0^\infty dE_2 E_2^{1/2} e^{-E_2/kT} B(E_1, E_2) = \frac{2N_\epsilon^2 8\pi S\epsilon^4 (2kT)}{\pi (kT)^3 (2\pi m kT)} \\ &\quad \times\left[\frac{W\left(\sqrt{\frac{E_1}{kT}}\right)}{\sqrt{\frac{E_1}{kT}}} - e^{-E_1/kT}\right], \end{aligned}\right\} \tag{57}$$

where

$$W(t) = \int_0^t e^{-x^2} dx.$$

In order to show that a Maxwellian distribution obtained by setting $\delta(E) = 0$ is a steady state for the case of coulomb collisions alone, we must show that term (I) is zero. We can do this most conveniently by showing that its integral is zero, or that

$$-C(E_1) E_1^{1/2} e^{-E_1/kT} + \frac{1}{2}\frac{\partial}{\partial E_1} D(E_1) E_1 e^{-E_1/kT} = 0. \tag{58}$$

Direct substitution of the formulae for C and D into the above expression shows that it is actually zero. This provides a check on the calculation. With the aid of equation (58) we may now simplify term (II) in Boltzmann's equation thus:

$$\left[\frac{\partial}{\partial t}(N_\epsilon f_3)\right]_{\text{coll}} - \text{term}(III) = \frac{1}{2}\frac{\partial}{\partial E_1} D(E_1)\, e^{-E_1/kT} \frac{E_1^{1/2}\,\partial\,\delta(E_1)}{\partial E_1}. \quad (58a)$$

Thus terms (I) and (II) together reduce to simply the expression above.

AN APPROXIMATION METHOD FOR THE SOLUTION OF BOLTZMANN'S EQUATION

Boltzmann's equation may be written thus for a steady state:

$$\left[\frac{\partial(N_\epsilon f_3)}{\partial t}\right] = \left[\frac{\partial(N_\epsilon f_3)}{\partial t}\right]_{\text{coll}} + \left[\frac{\partial(N_\epsilon f_3)}{\partial t}\right]_p. \quad (59)$$

Because $(\partial N_\epsilon f_3/\partial t)_{\text{coll}}$ is so large, we start with a zeroth approximation, in which $(\partial N_\epsilon f_3/\partial t)_p$ is neglected altogether. The solution for the zeroth approximation is simply a Maxwellian distribution with the temperature and particle density chosen to satisfy the conditions that the total number of electrons emitted is equal to those recaptured and that the total energy absorbed is equal to that going into the various dissipative processes such as excitation, radiation, etc. This is essentially the procedure used by most investigators in the past.

The next step is to solve for the higher approximations by an iterative process. We insert the zeroth-order Maxwellian approximation for $f_3(E)$ into $(\partial N_\epsilon f_3/\partial t)_p$ and then obtain an integrodifferential equation for the correction factor. The justification for this procedure is that $(\partial N_\epsilon f_3/\partial t)_p$ is already multiplied by relatively small coefficients, which are of the first order, so that, to the first order, it may be evaluated by replacing $f_3(E)$ by its zeroth-order approximation wherever it occurs. On the other hand, $(\partial N_\epsilon f_3/\partial t)_{\text{coll}}$ is so large that we cannot do this. Once we obtain a first-order solution, this procedure may be repeated to obtain higher-order approximations if desired. In general, this method is good only when the main part of the distribution is not too far from Maxwellian. For the nebular case this criterion is usually satisfied fairly well. For the first approximation our equation takes the form

$$\left.\begin{aligned} &\frac{1}{2}\frac{\partial}{\partial E_1} D(E_1)\, e^{-E_1/kT} E_1^{1/2} \frac{\partial\,\delta(E_1)}{\partial E_1} + \frac{4N_\epsilon^2}{\pi(kT)^3}\int_0^\infty dE_2 E_2^{1/2} e^{-E_2/kT}\,\delta(E_2) \\ &\qquad \times\left[-\frac{\partial}{\partial E_1} A E_1 e^{-E_1/kT} + \frac{1}{2}\frac{\partial^2}{\partial E_1^2} B E_1^{1/2} e^{-E_1/kT}\right] = -\left[\frac{\partial(N_\epsilon f_3)}{\partial t}\right]_{p_0}, \end{aligned}\right\} \quad (60)$$

where $[\partial(N_\epsilon f_3)\partial t]_{p_0}$ signifies $[\partial(N_\epsilon f_3)\partial t]_p$ evaluated with the assumption of a suitable Maxwellian distribution.

In this form the equation is still difficult to solve, because the unknown variable occurs under the integral sign. If desired, it could be solved as a series of powers of E_1, but, in general, this is not necessary. The reason is that, if the distribution is close to the Maxwellian, the integrand can be shown to be very small. Remembering that our boundary conditions on $\delta(E)$ are that $\delta(0) = 0$, $\delta'(0) = 0$, we see from the differential equation that $\delta(E)$ can become appreciable only at rather large values of E, since it cannot begin to grow until its derivative has grown. The rate of change of its derivative, however, is small because it involves the ratio of probabilities of inelastic to coulomb encounters. But when E_2 is large, the factor $e^{-E_2/kT}$ makes the integrand negligible. Thus the integral itself is much smaller than term (II), and we may regard it as a second-order term, neglecting it altogether in the first approximation.

Because the distribution is so close to the Maxwellian that it is not worth while to find its precise departure therefrom, we shall be content to find an upper limit to this deviation. The processes of photoelectric ionization, excitation of the O III lines, and recombination are most rapid; hence it suffices to consider these alone.

We shall find it convenient to transform to the dimensionless variable

$$\frac{E}{kT} = \lambda . \tag{61}$$

Let us define a new distribution function as

$$dn = f_3(E)\, dE = f_4(\lambda)\, d\lambda = kT f_3(E)\, d\lambda ; \tag{62}$$

thus

$$f_3(E) = \frac{f_4(\lambda)}{kT} . \tag{63}$$

With f_4, the normalized Maxwellian distribution is

$$f_4 = \frac{2}{\sqrt{\pi}} \lambda^{1/2} e^{-\lambda} , \tag{64}$$

and the complete distribution is

$$f_4 = \frac{2}{\sqrt{\pi}} \lambda^{1/2} e^{-\lambda} [1 + \delta(\lambda)] . \tag{65}$$

In evaluating the noncollisional terms, let us first consider the one due to the photoionization (cf. eq. [40]). We want to find the rate of emission per unit of λ. Now equation (40) gives

$$F_{1E}\, dE = F_{1E}\, kT\, d\lambda . \tag{66}$$

Thus

$$\left[\frac{\partial (N_\epsilon f_4)}{\partial t}\right]_{\text{emis}} = \frac{3.202 \times 10^{-6}\, N_\epsilon^2 G_{T\epsilon}}{T_\epsilon^{3/2}} \frac{1}{\lambda + \lambda_a} \frac{1}{e^y - 1} \int_{y_1}^{\infty} \frac{dy}{y(e^y - 1)} . \tag{67}$$

The term due to recombination is found from equation (9) by multiplyiug the rate of recombination for a single electron by the electron density in λ-space for the approximate Maxwellian distribution, which we are using in the noncoulomb terms, as follows:

$$\left[\frac{\partial (N_\epsilon f_4)}{\partial t}\right]_{0\ \text{recomb}} = 2.097 \times 10^{-22} \frac{2}{\sqrt{\pi}} V(\lambda) \frac{\lambda_a}{\lambda} N_\epsilon^2 e^{-\lambda} \lambda^{1/2} \Sigma \frac{1}{\left(\frac{1}{n'^2} + \frac{\lambda_a}{\lambda}\right)} \frac{g}{n^3} . \tag{68}$$

Since $V = \sqrt{2kT\lambda/m}$, we get

$$\left[\frac{\partial (N_\epsilon f_4)}{\partial t}\right]_{0\ \text{recomb}} = \frac{2}{\sqrt{\pi}} \times 2.097 \times 10^{-22} \sqrt{\frac{2kT}{m}}\, \lambda_a N_\epsilon^2 e^{-\lambda} \Sigma \frac{1}{\left(\frac{1}{n'^2} + \frac{\lambda_a}{\lambda}\right)} \frac{g}{n^3} . \tag{69}$$

For excitation terms we refer to Section III, equations (42) and (11), where $(\partial N_\epsilon f_4/\partial t)_{\text{inel}}$ is given by

$$9.28 N_A N_\epsilon \frac{f_4(\lambda + \lambda_a)}{v(\lambda + \lambda_a)} - \frac{f_4(\lambda)}{V(\lambda)} Q(\lambda) , \tag{70}$$

where $Q(\lambda) = 0$, when $\lambda < \lambda_0$; and $Q(\lambda) = 1$ when $\lambda > \lambda_0$. With

$$f \sim \frac{2}{\sqrt{\pi}} e^{-\lambda} \lambda^{1/2}$$

we get

$$\left[\frac{\partial (N_\epsilon f_4)}{\partial t}\right]_{\text{inel}} = \frac{9.28}{\sqrt{\frac{2\pi kT}{m}}} \times 2N_A N_\epsilon e^{-\lambda}\left[e^{-\lambda_0} - Q(\lambda)\right]. \tag{71}$$

Finally, the coulomb collisions must be considered. The term we have worked out is $(\partial N_\epsilon f_3/\partial t)_{\text{coll}}$. What we want is

$$\left[\frac{\partial}{\partial t}(N_\epsilon f_4)\right]_{\text{coll}} = kT\left[\frac{\partial}{\partial t}(N_\epsilon f_3)\right]_{\text{coll}} = \frac{(kT)}{2}\frac{\partial}{\partial E_1} D(E_1)\, e^{-E/kT}E_1^{1/2}\frac{\partial \delta}{\partial E}. \tag{72}$$

If we replace E_1 by kT and D by its value given in equation (57) and use equation (58*a*), we get

$$\left[\frac{\partial}{\partial t}(N_\epsilon f_4)\right]_{\text{coll}} = \frac{16\pi S\epsilon^4}{(kT)^{3/2}\sqrt{\pi}\sqrt{2\pi m}}\frac{\partial}{\partial\lambda}\left[e^{-\lambda}\lambda^{1/2}\frac{W(\sqrt{\lambda})}{\sqrt{\lambda}} - e^{-\lambda}\frac{\partial\delta}{\partial\lambda}\right]. \tag{73}$$

Finally, equating $-(\partial N_\epsilon f_4/\partial t)_{\text{coll}}$ and $(\partial N_\epsilon f_4/\partial t)_{0/p}$, we obtain

$$\left.\begin{aligned}
-\frac{\partial}{\partial\lambda}\left(e^{-\lambda}\lambda^{1/2}\left[\frac{W(\sqrt{\lambda})}{\sqrt{\lambda}} - e^{-\lambda}\right]\right)\frac{\partial\delta}{\partial\lambda} &= \frac{3.202\times 10^{-6}\sqrt{2m}\,(kT)^{3/2}G_{T_\epsilon}}{T_\epsilon^{3/2}16S\epsilon^4(\lambda+\lambda_a)(e^{y}-1)} \\
&\times\int_{y_1}^{\infty}\frac{dy}{y(e^{y}-1)} - \frac{(kT)^2\times 2.1\times 10^{-22}\lambda_a e^{-\lambda}}{4\sqrt{\pi}S\epsilon^4}\Sigma\frac{1}{\left(\frac{1}{n'^2}+\frac{\lambda_a}{\lambda}\right)}\frac{g}{n'^3} \\
&+ 9.28\frac{N_A}{N}\frac{kTm}{8\sqrt{\pi}S\epsilon^4}e^{-\lambda}\left[e^{-\lambda_0} - Q(\lambda)\right].
\end{aligned}\right\} \tag{74}$$

In a typical example, with $T_\epsilon = 10{,}000°$ K, $N_A \sim 1$, $N_\epsilon \sim 10^4$, $y_1 \sim 2$, $\lambda + \lambda_a \sim 14.5$, and $G_{T_\epsilon} \sim 0.1$, the value of the integral on the right turns out to be about 0.04, and

$$\left.\begin{aligned}
-\frac{\partial}{\partial\lambda}\left(e^{-\lambda}\lambda^{1/2}\left[\frac{W(\sqrt{\lambda})}{\sqrt{\lambda}} - e^{-\lambda}\right]\right)\frac{\partial\delta}{\partial\lambda} &= 4\times 10^{-10} - 1.4\times 10^{-9}e^{-\lambda} \\
&+ 7.6\times 10^{-8}e^{-\lambda}\left[e^{-\lambda} - Q(\lambda)\right].
\end{aligned}\right\} \tag{75}$$

Since λ varies between 0 and 5 over the range of interest and since the coefficient $(e^{-\lambda}\lambda^{1/2}[W(\sqrt{\lambda})/\sqrt{\lambda} - e^{-\lambda}])$ is not much less than 0.01 over most of the range of interest, we see immediately that $\delta(\lambda)$ cannot possibly become large. Some difficulties arise from a zero of the quantity $\lambda^{1/2}[W(\sqrt{\lambda})/\sqrt{\lambda} - e^{-\lambda}]$ at the origin which leads to an infinity in the value of $\partial\delta/\partial\lambda$ there. This infinity is spurious, however, because the approximation breaks down when the derivative of δ gets large. Because the coefficients on the right-hand side are so small, the region in which the derivative is large is confined to a small region close to the origin. Thus the equation is in error in this very small region where it predicts large values of δ'. We may neglect this region altogether, however, because it is so narrow that it cannot possibly affect the high-energy tail. Essentially, this singularity arises from the infinite cross-section for recombination at low energies. This cross-section, it will be recalled, integrates out to a finite value in all cases.

From the foregoing discussion it is clear that the inelastic terms could be at least a million times larger than they are now, before there was any doubt that the distribution was Maxwellian.

We are grateful to Dr. S. Chandrasekhar and Dr. Lyman Spitzer, who read the manuscript and made a number of valuable suggestions.

Reprinted from *The Astrophysical Journal,* Vol. 109, No. 1, January, 1949.

MULTIPLET INTENSITIES FOR THE NEBULAR LINES $^4S-^2D$ OF *O* II

L. H. ALLER,* C. W. UFFORD, AND J. H. VAN VLECK
Kirkwood Observatory, Indiana University, Bloomington, Indiana; University of Pennsylvania Philadelphia, Pennsylvania; and Harvard University, Cambridge, Massachusetts
Received July 3, 1948

ABSTRACT

The transitions $1s^22s^22p^3\ ^4S_{3/2}-^2D_{5/2}$ and $1s^22s^22p^3\ ^4S_{3/2}-^2D_{3/2}$ in *O* II are "allowed" only for electric quadrupole and magnetic dipole radiation and only if, in addition, there are deviations from Russell-Saunders coupling. The disagreement between previous theoretical computations of the relative strengths of these transitions and astrophysical observations is shown to be caused by the omission of spin-spin interaction, which is now included. The interesting feature of the calculation is that it is essential to consider not only concomitantly magnetic dipole and electric quadrupole radiation but also simultaneously the first-order effect of spin-spin coupling and the second order of spin-orbit interaction. The radial integral, η, which determines the influence of the spin-spin coupling on the transition probability was calculated by numerical integration with wave functions derived from the self-consistent field of Hartree and Black. The coupling of one spin with the orbit of another does not significantly alter the intensity ratio but does contribute to the multiplet splittings $^2D_{5/2}-^2D_{3/2}$ and $^2P_{3/2}-^2P_{1/2}$, which vanish in the conventional first approximations, since the shell is half-complete. This contribution involves the integral η as a proportionality factor and is hence examined as a control on our estimates of η.

For comparison with the theoretical calculations, the ratio of the intensities for $^4S_{3/2}-^2D_{5/2}$ and $^4S_{3/2}-^2D_{3/2}$ was measured photometrically in a number of representative planetary nebulae. The average measured value of the ratio was 0.49, to be compared with a theoretical value 0.58. The agreement is within the error of calculation.

I. INTRODUCTION

Among the strongest and most frequently observed of the forbidden lines found in the gaseous nebulae are the λ 3726.16 and λ 3728.91 lines of *O* II corresponding to the $^4S_{3/2}-^2D_{3/2}$ and $^4S_{3/2}-^2D_{5/2}$ transitions in the ground p^3 configuration. These lines are prominent in low-excitation planetary nebulae, e.g., IC 418 or NGC 40; in the diffuse nebulae, e.g., Orion or Messier 8; and in the faintly luminous patches of the Milky Way, which have been studied by Struve and his colleagues. Mayall[1] found this [*O* II] radiation which appears upon low-dispersion plates as a single unresolved line at λ 3727, in many external galaxies; and Humason[2] has studied the frequency of its appearance as a function of nebular type. The spiral nebulae show it most frequently. The bright fuzzy patches in Messier 33, the Spiral in Triangulum, show it to be stronger[3] than the well-known green "nebulium" lines of [*O* III], which are the lines usually dominant in planetary nebulae. The widespread appearance of this radiation would suggest that a knowledge of the physical parameters characterizing its production (e.g., transition probabilities) might be useful in deciphering the character of the nebulous regions in which it appears in our own and other galaxies.

Like other forbidden lines, the λ 3727 pair is produced by the collisional excitation of atoms to the upper level of the transition involved, followed by a cascade to the ground term with the emission of electric quadrupole and magnetic dipole radiation. In regions where the electron density is so high that the collisional de-excitations of the upper level are much more frequent than are radiative de-excitations, the population of the upper level follows Boltzmann's formula. Under these conditions, which may be approached in some gaseous nebulae, measures of the intensity of the lines, together with a knowledge of the electron temperature and the transition probabilities, would give us the number

* Research Associate of the W. J. McDonald Observatory of the University of Texas.

1 *Lick Obs. Bull.*, **19**, 33, 1939.

2 *Pub. A.S.P.*, **59**, 180, 1947.

3 L. H. Aller, *Ap. J.*, **95**, 54, 1942.

of ionized oxygen atoms. In regions of lower density we must know the target areas for collisional excitation as well. We shall be concerned here with the calculation of the transition probabilities for the λ 3727 pair, with a view to their subsequent astrophysical applications.

The lowest configuration of *O* II is $1s^2 2s^2 2p^3$ with terms $^4S_{3/2}$, $^2D_{5/2,\,3/2}$, $^2P_{1/2,\,3/2}$ in this order from the ground state up. The spread of the configuration from ^4S to ^2P is roughly 40,000 cm^{-1}, and the multiplet separations of the ^2D and ^2P levels are, respectively, -21.0 and -1.5 cm^{-1}. Therefore, the configuration is very close to pure Russell-Saunders. The strengths of the various transitions within the configuration, all of which are rigorously forbidden for electric dipole radiation, have been calculated by Pasternack[4] and by Shortley, Aller, Baker, and Menzel,[5] henceforth abbreviated to "SABM." The lines joining ^2D and ^2P are allowed for electric quadrupole and magnetic radiation even in Russell-Saunders coupling. The transition $^4S-{}^2P$ is forbidden for such radiations in the ideal Russell-Saunders limit but is allowed as soon as one includes the first-order effect of the spin-orbit interaction. The λ 3727 pair of lines $^4S_{3/2}-{}^2D_{5/2}$, $^4S_{3/2}-{}^2D_{3/2}$, with which we are concerned, are more highly forbidden than the rest. Pasternack and SABM derive nonvanishing intensities for them by including the second-order effect of the spin-orbit coupling. However, their results for the intensities of this particular pair of lines, unlike those for the others, do not agree well with observation. The ratio r of the intensities is

$$r=\frac{I\,(^4S_{3/2}-{}^2D_{5/2})}{I\,(^4S_{3/2}-{}^2D_{3/2})}\,. \tag{1}$$

SABM calculate that this ratio is 0.033 for pure magnetic dipole, 2.33 for electric quadrupole, and 1.64 for both types simultaneously—the actual case. The observed value is $r = 0.49$. Pasternack obtained $r = 1.9$, in slightly worse disagreement with experiment than is the value 1.64 of SABM; Pasternack's estimate is probably the less accurate of the two, as he calculated the quadrupole moment (s_q in equation [4]) by an approximation based on screening constants rather than by means of Hartree wave functions. As Aller and Menzel[6] have emphasized, there is, apparently with either calculation, a very serious and baffling divergence between the results of theory and those of experiment.

The main thesis of the present paper is that the discrepancy is removed when one includes all the magnetic interaction between the various electrons. SABM computed the deviations from Russell-Saunders coupling by means of the conventional spin-orbit potential,

$$H_0=-\frac{eh^2}{8\pi^2 m^2 c^2}\left(\frac{1}{r}\frac{dV}{dr}\right)\Sigma_i \boldsymbol{l}_i\cdot \boldsymbol{s}_i\,, \tag{2}$$

which couples the spin with its own orbit. Here V is the Hartree central potential, and $-dV/dr$ is thus equal to $(Z-\sigma(r))e/r^2$, where $-\sigma(r)e$ is the mean electronic charge, other than that of one $2p$ electron, interior to a radius r. Pasternack and SABM both omitted the magnetic interaction of the spins with each other, and also part of the interaction of the spin of one electron with the orbit of another. The explicit formula for the portion of the Hamiltonian which they omitted is[7]

$$\begin{aligned} H_1=\Sigma_{j\neq i}\frac{e^2 h}{2\pi m^2 c^2 r_{ij}^3}&[\boldsymbol{r}_{ij}\times(\boldsymbol{p}_j-\tfrac{1}{2}\boldsymbol{p}_i)\cdot \boldsymbol{s}_i] \\ &+\Sigma_{j>i}\frac{e^2h^2}{4\pi^2 m^2 c^2 r_{ij}^3}\times[\boldsymbol{s}_i\cdot \boldsymbol{s}_j-3\,r_{ij}^{-2}(\boldsymbol{s}_i\cdot \boldsymbol{r}_{ij})(\boldsymbol{s}_j\cdot \boldsymbol{r}_{ij})]\,, \end{aligned} \tag{3}$$

[4] *Ap. J.*, **92**, 129, 1940. [5] *Ap. J.*, **92**, 178, 1940. [6] *Ap. J.*, **102**, 246, 1945.

[7] W. Heisenberg, *Zs. f. Phys.*, **39**, 499, 1926; or Condon and Shortley, *The Theory of Atomic Spectra* (Cambridge: At the University Press, 1935), p. 211. The factor $\frac{1}{2}$ in the term of eq. (3) proportional to $\boldsymbol{p}_i$ arises from the Thomas correction. In principle, we should add to eq. (3) a term $(e^2h^2/4\pi^2m^2c^2r^3)\sigma_{2p}(r)\Sigma_i$

where $r_{ij} = r_i - r_j$. Here the first sum comes from the interaction of one spin with the orbit of another, while the second sum is the spin-spin interaction. The momentum operator $(h/2\pi i)$ grad $_j$ . . . of an electron is denoted by p_j, and its spin, measured in multiples of $h/2\pi$, by s_j. Thus the components of s_j are proportional to the usual Pauli matrices.

The contribution of the Hamiltonian (3) to the energy has been computed by a variety of authors,[8] though not explicitly in *O* II. In the present paper we shall be concerned with the effect of (3) on the magnetic dipole and electric quadrupole radiation. For this purpose it is, of course, necessary to calculate how the interactions (2) and (3) perturb the wave functions.

When the effect of (3) is included, we find that the intensity ratio (1) of the lines in question becomes 0.58, in more satisfactory agreement with the observational value 0.49. It may seem startling to the reader that the result is so significantly altered by the addition of (3) to (2), since usually (3) is a subordinate correction. The reason why the interaction (3) is abnormally important in our problem is that the spin-spin portion (second sum) of (3) affects the intensity of the lines in question in the first order, whereas the ordinary magnetic coupling (2) does so only in the second order.

In Section IV of this paper we shall examine the multiplet splittings of the $^2P_{1/2,\,3/2}$ and $^2D_{3/2,\,5/2}$ states. The question of multiplet widths is, of course, superficially a quite different one from that of intensities, but the same radial integrals are involved in both quantities. It is for this reason that it is instructive also to study the multiplet splittings. The first sum of equation (3), i.e., the portion which is caused by the interaction of one spin with the orbit of another electron, has no appreciable effect on the intensity ratio. On the other hand, only this part of (3) is of importance in connection with the multiplet intervals. Since the shell is only half-complete, the conventional spin-orbit interaction (2) does not contribute to the multiplet splitting except in the second and higher orders. This fact explains why the multiplet intervals are so small for the states in question.

We use Hartree wave functions[9] throughout. Eigenfunctions inclusive of exchange in the Fock approximation have been calculated by Hartree, Hartree, and Swirles;[10] but it was not thought that the small difference introduced by exchange would be significant for our purposes. The final results depend on the values of three integrals, viz.,

$$s_q = (\tfrac{2}{5})\, e \int_0^\infty r^2 R_{2p}^2(r)\, dr\,, \tag{4}$$

$$\zeta = \frac{e h^2}{8\pi^2 m^2 c^2} \int_0^\infty r^{-1} \frac{-dV}{dr} R_{2p}^2(r)\, dr\,, \tag{5}$$

$$\eta = \frac{e^2 h^2}{16\pi^2 m^2 c^2} \int_0^\infty \int_0^{r_2} r_2^{-3} R_{2p}^2(r_1) R_{2p}^2(r_2)\, dr_1 dr_2\,. \tag{6}$$

$l_i \cdot s_i$, to allow for the fact that the interaction of a spin with the orbital magnetic moment of another electron has already been included in eq. (2) to the extent that the field from the other electron can be regarded as central, and, of course, it is erroneous to count the same effect twice. Here $\sigma_{2p}(r)$ denotes the fraction of the charge of a $2p$ electron, which is, on the average, interior to r. In our particular case of the configuration $2p^3$, however, this correction to eq. (3) is unnecessary to the degree of approximation under consideration, for we are interested only in a matrix element of (3) when the corresponding matrix element of (2), and hence the redundancy correction, are zero. In general, the term of eq. (3) proportional to $-\frac{1}{2}p_i$ is equivalent to the corresponding screening portion of eq. (2), provided that the charge cloud of the jth electron is centrosymmetric, which, of course, it is not for a $2p$ electron. Because of an erroneous factor $\frac{1}{2}$, this equivalence is not brought out in the discussion on p. 212 of Condon and Shortley, as the factor which they give there as $Zr_1^{-3} - \frac{1}{2}r_{12}^{-3}$ should read $\frac{1}{2}Zr_1^{-3} - \frac{1}{2}r_{12}^{-3}$ (or, more accurately, $\frac{1}{2}Zr_1^{-3} - \frac{1}{2}\sigma(r_1)r_1^{-3}$, when Gauss's theorem is applied to the relevant portion of eq. [1], p. 211, with the assumption of central symmetry).

[8] G. Breit, *Phys. Rev.*, **36**, 383, 1930; **39**, 636, 1932; H. Wolfe, *Phys. Rev.*, **41**, 443, 1932; E. David, *Zs. f. Phys.*, **91**, 289, 1934; G. Araki, *Proc. Math. Phys. Soc. Japan*, **19**, 128, 1937; D. R. Inglis, *Phys. Rev.*, **61**, 297, 1942.

[9] D. R. Hartree and M. M. Black, *Proc. R. Soc. London, A*, **139**, 311, 1933.

[10] *Phil. Trans. R. Soc. London, A*, **238**, 229, 1939–40.

Here $R_{2p}(r)/r$ is the radial part of the $2p$ wave function, supposed normalized in the fashion

$$\int_0^\infty R_{2p}^2(r)\,dr = 1\,.$$

The expression (4) gives the second moment involved in the quadrupole strength. The radial integral in (5) is the usual one involved in the conventional spin-orbit interaction (2). The matrix elements of (3), on the other hand, are all proportional to the double integral (5). The way the various integrals are involved is best exhibited by giving the structure of the final formula for the transition probability, which turns out to be

$$A = [\alpha\, s_q^2 \zeta^2 + \rho \zeta^4 + \gamma \zeta^2 \eta + \delta \eta^2]\,. \tag{7}$$

Here α, β, γ, δ are constants which are, of course, different for the $^4S_{3/2}-{}^2D_{3/2}$ and $^4S_{3/2}-{}^2D_{5/2}$ transitions and which will be computed later in the paper in terms of known parameters, such as the $^4S-{}^2D$ and $^4S-{}^2P$ frequency intervals. Both SABM and Pasternack had only the part of equation (7) not involving η.

The integral ζ is best evaluated by extrapolation from the multiplet intervals in adjacent configurations, such as $2p^2$, $2p^4$, where the first-order effect of the spin-orbit interaction does not disappear because of half-completion of the shell. This extrapolation[11] gives $\zeta = 174\ \text{cm}^{-1}$. Condon[12] has calculated s_q to be $0.70ea_H^2$ ($a_H = he/4\pi mc$) by means of Hartree wave functions. In Section IV we shall obtain the value $1.33\ \text{cm}^{-1}$ for η.

II. THE MATRIX OF THE MAGNETIC INTERACTION

The matrix elements of the conventional spin-orbit interaction (2) have been calculated elsewhere in the literature, and so we shall not give details of how they are obtained.

In order to calculate the multiplet splittings, we shall need, in addition, the diagonal elements of the remainder (3) of the Hamiltonian. Also, in connection with the intensity problem, we shall require the matrix elements of (3) of the form $^4S_{3/2}-{}^2D_{3/2}$, since the intensities of the lines under consideration are very sensitive to this element. The only nonvanishing elements of the conventional spin-orbit interaction (2) are off-diagonal ones of the type $^4S-{}^2P$ and $^2P-{}^2D$. This interaction, when present, is much greater than the interelectronic magnetic coupling represented by the expression (3); in fact, the integral ζ is over one hundred times as large as η. Hence for our purposes it is pointless to calculate the matrix elements of (3) of the form $^4S-{}^2P$ or $^2P-{}^2D$, since they would be of much smaller order than the corresponding elements of (2). In other words, in our later secular determinant (15), entries involving η are carried only when nonconcurrent with ζ.

The Hamiltonian (3) involves the co-ordinates of two electrons. The matrix elements for the states in Russell-Saunders coupling in the LSJ system will be linear combinations of these two electron matrix elements. Hence, by expansion of the wave functions, the matrix elements for our three-electron system can be expressed in terms of those for a two-electron system, for which formulae have been given by Marvin.[13] The reduction of the three- to the two-electron problem is simplified by using the Slater method of diagonal sums,[14] together with certain deductions from group theory concerning the structure of the matrix elements. Considerations based on group theory or explicit calculation show that the only nonvanishing element of the second or spin-spin part of (3)

[11] Cf., e.g., H. A. Robinson and G. H. Shortley, *Phys. Rev.*, **52**, 713, 1937.

[12] *Ap. J.*, **79**, 219, 1939.

[13] *Phys. Rev.*, **71**, 102, 1947. The results of the present paper were obtained before Marvin's paper appeared and agree with those from his formulae.

[14] *Phys. Rev.*, **34**, 1293, 1929. For a good explanation of the general method which we use for calculating spectral terms by means of Slater determinants, see chaps. vi–viii of Condon and Shortley, *op. cit.*

is of the structure $^4S_{3/2}-{}^2D_{3/2}$. Furthermore, the diagonal elements of the first member of (3) are of the form

$$(LSJ|H_1|LSJ) = C[J(J+1) - L(L+1) - S(S+1)], \tag{8}$$

where C is independent of J but depends on L and S. This structure is a group-theoretical consequence[15] of the fact that $r_{ij}^{-3}[(\boldsymbol{r}_i - \boldsymbol{r}_j) \times (\boldsymbol{p}_j - \frac{1}{2}\boldsymbol{p}_i)]$ transforms under a rotation of the orbital frame like the vector $\boldsymbol{L}$, and $\boldsymbol{s}_i$ under a rotation of the spin frame like the vector $\boldsymbol{S}$.

For $M = \frac{5}{2}$ there is the single Slater determinant, symbolized in the notation of Condon and Shortley by

$$^2D_{5/2} = |1^+1^-0^+|.$$

By following their *Theory of Atomic Spectra* (p. 171), one can express the corresponding diagonal matrix elements as the sum of twelve two-electron integrals of the typical form

$$\tfrac{1}{2}\int (1^+1^-)\, H_1\, (1^+1^-).$$

These twelve integrals can be calculated by means of Marvin's formula (21). When they are summed algebraically with the proper sign, one finds

$$(^2D_{5/2}|H_1|^2D_{5/2}) = -\frac{37\eta}{5}, \tag{9}$$

with η as in equation (6). From (8), it follows immediately that

$$(^2D_{3/2}|H_1|^2D_{3/2}) = \frac{111\eta}{10}. \tag{10}$$

Furthermore, we have

$$(^4S_{3/2}|H_1|^4S_{3/2}) = 0, \tag{11}$$

since equation (8) vanishes when $J = L$ and $S = 0$.

For $M = \frac{3}{2}$ the Slater determinants in the zero-order scheme are

$$F = |1^+0^+-1^+|, \quad G = |1^+1^-0^-|, \quad H = |1^+1^--1^+|, \quad I = |1^+0^+0^-|.$$

The diagonal elements, found in the same way as above, are

$$\int F^*H_1F = 0, \quad \int G^*H_1G = \frac{37\eta}{5}, \quad \int H^*H_1H = -\frac{8\eta}{5}, \quad \int I^*H_1I = -\frac{23\eta}{5}.$$

Hence the invariance of the diagonal sum gives

$$(^2D_{5/2}|H_1|^2D_{5/2}) + (^2D_{3/2}|H_1|^2D_{3/2}) + (^2P_{3/2}|H_1|^2P_{3/2}) + (^4S_{3/2}|H_1|^4S_{3/2}) = \frac{6\eta}{5}.$$

Using equations (9), (10), and (11), we thus have

$$(^2P_{3/2}|H_1|^2P_{3/2}) = -\frac{5\eta}{2}, \tag{12}$$

and, by equation (8),

$$(^2P_{1/2}|H_1|^2P_{1/2}) = 5\eta. \tag{13}$$

The matrix element connecting $^4S_{3/2}$ and $^2D_{3/2}$ comes entirely from spin-spin interaction. It is calculated by using the Slater determinantal wave functions and noting that the linear combinations appropriate to the LSJ system are

$$^4S_{3/2} = F, \qquad ^2D_{3/2} = \frac{2\sqrt{2}G - H + I}{\sqrt{10}}.$$

[15] H. A. Kramers, *Kon. Akad. v. Wetensch. Amsterdam*, **34**, 965, 1931.

The matrix elements can be expressed in terms of those for the two-electron system by the same procedure as before. The relevant formula of Marvin is his equation (22) rather than (21), since now we are dealing with spin-spin coupling rather than with interaction of the orbit of one electron with the spin of another electron. The final result is

$$({}^4S_{3/2}|H_1|{}^2D_{3/2}) = -\frac{6\eta}{\sqrt{5}}. \tag{14}$$

Before termination of the present section, there are two remarks that seem worth making about the calculation of the matrix elements for a two-electron system. One of these observations is that, in the calculation of the diagonal elements, one can omit entirely the application of the momentum operator $(h/2\pi i)$ grad . . . , to the radial factors in the wave functions. Such an application can give only zero, inasmuch as the radial factors are real; and so a nonvanishing result would give an imaginary diagonal element, impossible for a self-adjoint operator. This remark is based on the assumption that the wave function can be expressed as a simple Slater determinant. If one allows for polarization effects not included in the present paper, the decomposition into sums of products of one-electron functions, each separable into polar co-ordinates, is no longer possible, and the argument ceases to apply.

Our other remark is that the proportionality of the matrix elements given in equations (9)–(14) to the integral η may appear to the reader only an approximation contingent on the retention of terms of lowest nonvanishing order in the expansion of the Hamiltonian in the ratio of the smaller to the larger radius for a pair of electrons. Actually, the proportionality is rigorous for the $2p^3$ configuration or any other configuration of type $2p^x$ (neglecting polarization or, in other words, configuration interaction). This point is not mentioned by Marvin and is proved by the following reasoning not involving orbital co-ordinates. If k is an arbitrary vector not involving orbital co-ordinates, the function $(r_{ij}\cdot k)/r_{ij}^3$, regarded as a function either of the variables x_i, y_i, z_i, or x_j, y_j, z_j, is a solution of Laplace's equation, and so its expansion takes the form

$$\sum_{l=0}^{\infty}\sum_{m=-l}^{l} a_{lm}T_{lm}(\theta_<\phi_<)T_{l+1,-m}(\theta_>,\phi_>)\frac{r_<^l}{r_>^{l+2}}.$$

Consider, first, the case in which the momentum or gradient operator involved in the first part of equation (3) works on the inner electron. Then when we integrate over the co-ordinates of the outer electron, in $\int\Psi^*H_1\Psi$, the factor of $\Psi^*\Psi$ involving the outer electron will transform like the representation $D_1 \times D_1$ of the rotation group and so contains no representation higher than $l = 2$. Hence the orthogonality properties of different irreducible representations of the rotation group show that the integral over the outer ($>$) electron vanishes when the above expansion of the first line of equation (3) is carried beyond $l = 1$. Furthermore, the integral vanishes when $l = 0$, as the factor $T_{l+1,-m}(>)$ is then odd, whereas the factor from $\Psi^*\Psi$ is even. Thus only the term with $l = 1$ persists. This is, of course, of the order of $r_>^{-3}$ corresponding to equation (6). To show that a similar result also obtains when the grad operator works on the outer electron, we note that the integrand now has as a factor Ψ^* grad Ψ, whose nonvanishing component transforms like the representation $l = 1$ of the rotation group, since grad Ψ is effectively an invariant if $\Psi = qf(r)$ $(q = x, y, z)$ and if only the application of the gradient operator to the q factor is relevant. Hence, because of the orthogonality, only the term $l = 0$ in the expansion makes a contribution, again of the order (6), since r^{-2} grad $\sim 1/r^3$. (This reasoning does not preclude the off-diagonal elements of the first part of [3] having terms of higher order than η in $1/r$; but such elements are of no importance for our problem, since we are not interested in elements of [3] of the type $[{}^2P|{}^4S]$ or $[{}^2P|{}^2D]$.) The expansion of the second or spin-spin part of equation (3) is of the form

$$\frac{\Sigma b_{lm}T_{lm}(<)T_{l+2,-m}(>)\,r_<^l}{r_>^{l+3}},$$

with coefficients b_{lm} involving the spin vectors. An analogous argument shows that integration over the outer electron gives a vanishing result except for the term $l = 0$ of the expansion.

III. PERTURBATION THEORY

The secular determinant may now be constructed from the results of the preceding section, along with the formulae for the matrix elements of the conventional spin-orbit interaction given on page 268 of Condon and Shortley. For $M = \frac{3}{2}$, the result is

$$\begin{vmatrix} E_0({}^2\mathrm{D}) + \frac{111}{10}\eta - E & \frac{1}{2}\zeta\sqrt{5} & -\frac{6\eta}{\sqrt{5}} \\ \frac{1}{2}\zeta\sqrt{5} & E_0({}^2\mathrm{P}) - \frac{5\eta}{2} - E & \zeta \\ -\frac{6\eta}{\sqrt{5}} & \zeta & E_0({}^4\mathrm{S}) - E \end{vmatrix} \tag{15}$$

Here $E_0({}^2\mathrm{D})$, for instance, denotes the energy of the ${}^2\mathrm{D}$ state in the absence of any spin-orbit or magnetic coupling. The perturbation calculation is simple, since the system is nondegenerate. The second-order energies are as follows:

$$\begin{aligned} E({}^2\mathrm{D}_{3/2}) &= E_0({}^2\mathrm{D}) + \frac{111}{10}\eta - \frac{5\zeta^2}{4(PD)}, \\ E({}^2\mathrm{P}_{3/2}) &= E_0({}^2\mathrm{P}) - \frac{5\eta}{2} + \frac{5\zeta^2}{4(PD)} + \frac{\zeta^2}{(PS)}, \\ E({}^4\mathrm{S}_{3/2}) &= E_0({}^4\mathrm{S}) - \frac{\zeta^2}{(PS)}, \\ E({}^2\mathrm{D}_{5/2}) &= E_0({}^2\mathrm{D}) - \frac{37\eta}{5}, \\ E({}^2\mathrm{P}_{1/2}) &= E_0({}^2\mathrm{P}) + 5\eta, \end{aligned} \tag{16}$$

where $(PD) = E_0({}^2\mathrm{P}) - E_0({}^2\mathrm{D})$, etc. Here terms in η^2 have been neglected, since numerically η is of the order of magnitude of $\zeta^2/(PD)$. We have also included the formulae for the energies of the other states. It is not necessary to construct a secular determinant for them, since only one level has $J = \frac{5}{2}$ or $J = \frac{1}{2}$. The empirical values of the frequency denominators needed in connection with (16) are

$$(PD) = 13{,}642\ \mathrm{cm}^{-1}, \qquad (DS) = 26{,}828\ \mathrm{cm}^{-1}, \qquad (PS) = 40{,}470\ \mathrm{cm}^{-1}.$$

We prefer to use these values rather than to try to calculate them by means of the Slater F parameters, for the actual intervals do not have ratios in accord with the conventional first-order expressions for the electrostatic energy which does not include polarization. According to the formula of Slater, we should have

$$(PD) = 6F_2, \qquad (DS) = 9F_2,$$

where F_2 is a constant.

The perturbed wave functions corresponding to the approximation (16) for the ${}^2\mathrm{D}_{3/2}$ and ${}^4\mathrm{S}_{3/2}$ states are the following combinations of the unperturbed wave functions:

$$\begin{aligned} \Psi({}^2\mathrm{D}_{3/2}) &= a\Psi^0({}^2\mathrm{P}_{3/2}) + b\Psi^0({}^4\mathrm{S}_{3/2}) + c\Psi^0({}^2\mathrm{D}_{3/2}), \\ \Psi({}^4\mathrm{S}_{3/2}) &= a'\Psi^0({}^2\mathrm{P}_{3/2}) + b'\Psi^0({}^4\mathrm{S}_{3/2}) + c'\Psi^0({}^2\mathrm{D}_{3/2}). \end{aligned} \tag{17}$$

$$\begin{aligned} &a = -\frac{\zeta\sqrt{5}}{2(PD)}, \quad a' = -\frac{\zeta}{(PS)}, \quad b = -\frac{6\eta}{(DS)\sqrt{5}} - \frac{\zeta^2\sqrt{5}}{2(PD)(DS)}, \\ &b' = c = 1, \quad c' = \frac{6\eta}{(DS)\sqrt{5}} + \frac{\zeta^2\sqrt{5}}{2(PS)(DS)}. \end{aligned} \tag{18}$$

Our notation for the various coefficients is the same as that of SABM, but their formulae differ from (18) in not including any of the terms proportional to η.

The line strengths can now be found from equations (11) and (12) of SABM, provided that we use the revised values (18) of the coefficients. The strengths for magnetic-dipole radiation are thus

$$S_m\,({}^2D_{5/2} - {}^4S_{3/2}) = \frac{12c'^2}{5}\left(\frac{he}{4\pi mc}\right)^2, \quad S_m\,({}^2D_{3/2} - {}^4S_{3/2}) = \tfrac{4}{15}(10aa' + 15bb' + 6cc')^2\left(\frac{he}{4\pi mc}\right)^2. \tag{19}$$

The corresponding electric-quadrupole strengths are

$$S_q\,({}^2D_{5/2} - {}^4S_{3/2}) = 14a'^2 s_q^2, \quad S_q\,({}^2D_{3/2} - {}^4S_{3/2}) = (ac' + ca')^2 6 s_q^2, \tag{20}$$

with s_q defined as in equation (4). Here and elsewhere terms involving powers of ζ and η higher than those given have been omitted.

The dipolar and quadrupolar transition probabilities are connected with the corresponding line strengths by the relations[16]

$$A_m = \frac{1}{2J+1}\frac{64\pi^4\nu^3}{3hc^3}S_m, \quad A_q = \frac{1}{2J+1}\frac{32\pi^6\nu^5}{5hc^5}S_q, \tag{21}$$

where $2J + 1$ is the statistical weight of the initial level. The contributions of the magnetic-dipole and electric-quadrupole radiation to the transition probability are additive. A similar additivity does not obtain for the line strengths themselves, since they are, rather artificially, defined in such a way as to make the proportionality factors different in the two relations of equations (21).

Equations (18), (19), (20), and (21) show that, when the dipolar and quadrupolar contributions to the transition probability are summed, the resultant depends on the parameters in the fashion (7). The quadrupole radiation is caused, to our approximation, only by the nondiagonal elements of the conventional spin-orbit interaction. On the other hand, the magnetic-dipole radiation depends on this interaction and the usually omitted spin-spin interaction both separately and in combination.

IV. EVALUATION OF THE RADIAL INTEGRAL η

Before a numerical estimate of the relative intensities can be made, it is necessary to know the value of the parameter η. This can be calculated by the evaluation of the integral (6) numerically with the Hartree and Black wave functions. This procedure gives $\eta = 1.33\ \text{cm}^{-1}$.

As a test of the validity of the Hartree and Black wave functions for calculating η, the radial integral ζ of the conventional spin-orbit interaction was calculated numerically. The value of $-dV/dr$ to be used in equation (5), then, is

$$-\frac{dV}{dr} = er^{-2}\left\{8 - 2\int_0^r [R_{1s}(r)^2 + R_{2s}(r)^2 + R_{2p}(r)^2]\,dr\right\}. \tag{22}$$

When equation (22) is employed, the value of ζ turns out to be 198 cm^{-1}, which is larger than the presumably correct value 174 cm^{-1} obtained from the isoelectronic sequence. The lack of perfect agreement is not disquieting, for evaluation of ζ requires very accurate wave functions near the nucleus. The inner regions are weighted less strongly in the computation of η than in that of ζ, though still more heavily than in the calculation of electrostatic binding energies. The discrepancy between the empirical value of ζ and that calculated by the Hartree and Black wave functions suggests that possibly η should be reduced by a factor of 174/198 to the value 1.17 cm^{-1}, though this is probably an overcorrection.

[16] Cf. Condon and Shortley, *op. cit.*, pp. 98 and 100, eqs. (3) and (8).

If the value 1.17 cm^{-1} were used instead of 1.33, the intensity ratio defined in equation (1) would be increased by 10 per cent, and the agreement with experiment would be somewhat less satisfactory. If the value 198 rather than 174 cm^{-1} were employed for ζ, this ratio would be increased by 3.5 per cent. The relative transition probabilities are thus more sensitive to changes of η than to those of ζ.

V. MULTIPLET SPLITTINGS OF THE ^{2}P AND ^{2}D TERMS

Once the integrals η and ζ are known, the multiplet intervals in the ^{2}D and ^{2}P states can be calculated and compared with experiment. If we take $\eta = 1.33\ cm^{-1}$ and $\zeta = 174\ cm^{-1}$, the values of the splittings obtained from equation (16) are

$$\Delta^2D = E(^2D_{5/2}) - E(^2D_{3/2}) = -21.8\ cm^{-1}\ ,$$

$$\Delta^2P = E(^2P_{3/2}) - E(^2P_{1/2}) = -16.4\ cm^{-1}\ ;$$

whereas the experimental values are[17]

$$\Delta^2D = -21.0\ , \qquad \Delta^2P = -1.5\ cm^{-1}\ .$$

The agreement between theory and experiment for Δ^2D is striking, and appears to confirm our evaluation of η. On the other hand, no reasonable choice of η will possibly remove the discrepancy for Δ^2P; the value $\eta = 0.57\ cm^{-1}$ would be required. The disagreement is probably caused by the neglect of polarization or, in other words, configuration interaction. The calculated values are probably quite sensitive to polarization corrections, as they may make the conventional spin-orbit coupling no longer quite disappear from the diagonal, even though the shell is half-complete. It is puzzling why these corrections are apparently so much larger for Δ^2P than for Δ^2D.

VI. THE TRANSITION PROBABILITIES

With the values of s_q, ζ, and η available, it is now possible to find numerical values for the strengths (19) and (20) and hence for the transition probabilities A. With $\zeta = 174\ cm^{-1}$ and $\eta = 1.33\ cm^{-1}$, we find that[18]

$$A\ (^2D_{5/2} \rightarrow {^4S_{3/2}}) = (4.85 + .02 + 0.17 + 0.37) \times 10^{-5} = 5.41 \times 10^{-5}\ sec^{-1}\ , \quad (23)$$

$$A\ (^2D_{3/2} \rightarrow {^4S_{3/2}}) = (3.12 + 1.19 + 4.82 + 4.88) \times 10^{-5} = 14.01 \times 10^{-5}\ sec^{-1}\ . \quad (24)$$

Because of the statistical weight factor, the intensity ratio (1) is $\frac{3}{2}$ times the quotient of equation (23) by equation (24). The various terms in the sums show the breakdown in terms of the structure (7). By comparison of equations (23) and (24) with equation (7) it is seen that the spin-spin interaction contributes appreciably to the total transition probability both by itself in a term of the form η^2 and by combination with the conventional spin-orbit interaction in a term of the type $\eta\zeta^2$. In fact, the last two terms are the major part of the expression (24). On the other hand, the quadrupole member constitutes the bulk of (23).

With our choice of constants, the theoretical value of the ratio (1) of the intensities of the two components of the ^{4}S$-^2$D multiplet is $r = 0.58$. The observed values to be compared with this will be discussed in the next section.

[17] The values of the intervals which we quote are from *Atomic Energy Levels*, by Charlotte E. Moore, **1**, sec. 1, 48 (Circular 467 of the National Bureau of Standards). These values supersede the less accurate ones given in the older tables of Bacher and Goudsmit. We are much indebted to Mrs. Sitterly for informing us of the re-evaluation of the levels in advance of publication.

[18] Note that we use arrows in connection with transition probabilities but not in connection with line strengths, since the former, unlike the latter, are unsymmetric as regards interchange of initial and final indices.

VII. THE OBSERVED INTENSITY RATIO OF THE $^4S_{3/2}-^2D_{5/2}$ AND $^4S_{3/2}-^2D_{3/2}$ LINES OF *O* II

For comparison with the theoretical calculations, one of us (L. H. A.) decided to measure photometrically, in a number of representative planetary nebulae, the intensity ratio of the two multiplet components $^4S_{3/2}-^2D_{5/2}$ and $^4S_{3/2}-^2D_{3/2}$, which have wave lengths, respectively, 3729 and 3726 A. For ten such nebulae, observations were secured with the 82-inch reflector of the McDonald Observatory and the Cassegrain spectrograph equipped with quartz prisms and 500-mm camera, which gives a dispersion of 33 A/mm at λ 3727. A spot sensitometer and wedge-slit exposures with a lamp provided the photometric calibration. The data for NGC 7027 are taken from Mount Wilson

TABLE 1

INTENSITY RATIO λ 3729/λ 3726 IN THE PLANETARY NEBULAE

Nebula	α	δ	Plate No.	Date	No. of Points	r
NGC 40	$0^h\ 7^m.6$	+71° 32′	CQ 4619	Sept. 1, 1945	4	0.85
			CQ 4625	Sept. 2, 1945	3	.82
NGC 2392	7 23.3	+21 7	CQ 5222	Apr. 10, 1947	1	.46:
IC 4593	16 7.0	+12 20	CQ 4593	Aug. 25, 1945	1	.67
NGC 6210	16 40.3	+23 59	CQ 4599	Aug. 26, 1945	2	.48
			CQ 4628	Sept. 3, 1945	1	.44
NGC 6543	17 58.6	+66 38	CQ 4600	Aug. 26, 1945	1	.46
			CQ 4637	Sept. 8, 1945	2	.52
NGC 6572	18 7.2	+ 6 50	CQ 4589	Aug. 24, 1945	1	.45
BD+30° 3639	19 30.8	+30 18	CQ 4642	Sept. 9, 1945	1	.46
NGC 6826	19 42.1	+50 17	CQ 4590	Aug. 24, 1945	1	.48:
IC 4997	20 15.6	+16 25	CQ 4639	Sept. 8, 1945	1	.36:
			CQ 4641	Sept. 9, 1945	1	.40:
NGC 7009	20 58.7	−11 46	CQ 4607	Aug. 31, 1945	2	.55
			CQ 4617	Sept. 1, 1945	2	.50
			CQ 4623	Sept. 2, 1945	4	.48
NGC 7027	21 3.3	+41 50	E 1730	July 31, 1946	2	.48
			E 1747	Aug. 23, 1946	1	.42
			E 1748	Aug. 24, 1946	2	.47
			E 1752	Sept. 4, 1946	2	0.50
Mean						0.49±0.01

spectrograms obtained by R. Minkowski in connection with a detailed spectrophotometric investigation of this object by Minkowski and Aller.

Table 1 gives the NGC number of the nebula, the position for epoch 1900, the plate number, the date of the observation, the number of points in the nebula at which the λ 3729/λ 3726 intensity ratio *r* was measured, and, finally, the intensity ratio itself.

With but three exceptions, this ratio comes out near 0.5. The ratio appears to be lower for IC 4997 and higher for IC 4593 and especially NGC 40. The discrepant ratio for NGC 40 appears to be real. Among the nebulae studied, this object has the lowest surface brightness and probably also the lowest density throughout most of its volume. It shows a pronounced filamentary structure, and much of the λ 3727 radiation may be contributed by regions of low density.

Our theoretical intensity ratio presupposes that the $^2D_{3/2}$ and $^2D_{5/2}$ levels are populated according to their statistical weights, i.e., in the ratio of 4 to 6. That is, the electron density is assumed sufficiently large for the relative populations of the excited levels to be determined by collisional processes, i.e., collisional de-excitations of the ^{2}D term are assumed to be much more frequent than are radiative de-excitations.

Recent work[19] suggests that the target areas for collisional excitation of the ^{2}D term is so small that, in many gaseous nebulae, ions leave the metastable levels by radiation rather than by superelastic collisions. The exact theoretical value of the ratio, therefore, depends on the electron density, the electron temperature, and the cross-sections for collisional transitions between $^2D_{3/2}$, $^2D_{5/2}$, $^4S_{3/2}$, $^2P_{3/2}$, and $^2P_{1/2}$. The ratio, r, will be larger, the smaller the electron density, while with increasing electron density the intensity ratio should approach the theoretical value. The large value of the ratio for NGC 40, which presumably has the lowest density, and the small value for IC 4997, which is known to be the most dense of these nebulae, are in qualitative agreement with expectations.

Pending a complete treatment of the collisional cross-sections, we compare our theoretical value, 0.58, with the straight mean value 0.49 given by the observations, although it seems likely that the value 0.38 found from IC 4997 may turn out to be the value to be compared with the theoretical calculations.

In many instances the fluctuations in the ratio from nebula to nebula are probably real and arise from actual differences in electron density. Density fluctuations within a given nebula further complicate the problem.

VIII. CONCLUSION

The previous discrepancy between the observed and the calculated relative intensities for the 3729 and 3726 lines of *O* II is explained. The disagreement by about 20 per cent with the revised calculations is not disquieting, inasmuch as the Hartree and Black wave functions used to compute η and s_q may not be perfect and especially inasmuch as the calculation neglects polarization or configuration interaction.

The writers wish to thank Professor D. H. Menzel for his stimulating interest in this paper.

[19] L. H. Aller, *Pub. A.S.P.* (in press).

Reprinted from *The Astrophysical Journal,* Vol. 113, No. 1, January, 1951.

SPECTROPHOTOMETRY OF REPRESENTATIVE PLANETARY NEBULAE

LAWRENCE H. ALLER*
Observatory, University of Michigan

Received June 17, 1950

ABSTRACT

This investigation completes a spectrophotometric study of planetary nebulae initiated some years ago at the Lick Observatory. Special attention is paid to nebulae observed in the λλ 3700–6500 A region by Wyse, to objects with interesting structural features, and to certain stellar nebulae. The observed Balmer decrement agrees with the theoretical predictions by Menzel and Baker for their model *B*, in which all Lyman radiation is assumed degraded into Lyman-α, which alone escapes from the nebula. With the low dispersion employed, most of the central stars show continuous spectra with little evidence of absorption or emission lines. The nuclei of NGC 7026, 6751, and 6905, which show broad Wolf-Rayet emission lines of oxygen, helium, and carbon, are discussed in detail, and relative line intensities are derived.

I. INTRODUCTION

A basic requisite for the understanding of the planetary nebulae is a knowledge of the intensities of the emission lines that comprise their spectra. The brightness of the lines of a given ion differ from nebula to nebula or even within the same nebula in such a way as to reveal important clues to the prevalent conditions of excitation and ionization.

The meaning of the intensity of a nebular emission line merits a moment's attention. First, we would like to have the total intensity of the nebular line—i.e., the total amount of monochromatic energy received from the whole nebula per second by a unit area just outside the earth's atmosphere perpendicular to the line drawn to the nebula. This information should be supplemented by measurements of the distribution of energy throughout the nebular images—the results may be expressed very well by means of isophotic contours such as those published by Berman.[1] Although the total intensity and the isophotic contour give a complete description of the observational data, the emission per unit volume is the item of greatest theoretical interest. An estimation of this quantity requires some knowledge of the spatial structure of the nebula, as well as of its distance. In general, both of these are poorly known.

A number of observers have published measurements of nebular line intensities. Berman observed with the Crossley reflector and quartz spectrograph of the Lick Observatory. Plaskett[2] obtained slit spectra of a number of the brighter objects at the Dominion Astrophysical Observatory. More recently, Bowen and Wyse[3] used the Lick 36-inch refractor and spectrographs equipped with the image slicer. They estimated the intensities of the lines with the aid of scale plates. T. L. Page[4] secured excellent photometric material for a large number of nebulae. He employed the Cassegrain spectrograph of the 82-inch McDonald reflector and a wide slit. A report on the continuous spectra of the planetaries has been published, but an account of the line intensities has not yet appeared.

Some years ago at the Lick Observatory the writer measured the relative total intensities of the nebular lines (and, when possible, the Balmer continuum) in eleven planetary nebulae.[5] In addition to this list, a few objects observed by Stoy were also discussed. The

* The reductions described in this paper were carried out primarily while the writer was at Indiana University.

[1] *Lick Obs. Bull.*, **15**, 86, 1930.

[2] *Pub. Dom. Ap. Obs. Victoria*, **4**, 187, 1928.

[3] *Lick Obs. Bull.*, **19**, 1, 1939.

[4] *Ap. J.*, **96**, 78, 1942.

[5] *Ap. J.*, **93**, 236, 1941.

present investigation comprises a completion of this program. In the present instance, however, the relative total intensities are given for only the stellar planetaries. For the other objects the intensities measured at a particular point (or sometimes two or more points) in the nebula are tabulated, although in a number of instances the total intensities also have been measured.

Section II discusses the selection of the nebulae observed; Section III describes the techniques of observation and reduction; Section IV gives the results, mostly in the form of tables; and Section V treats of the Balmer decrement. A final portion of the paper is devoted to some remarks on the nuclei of the nebulae.

II. SELECTION OF OBJECTS

The observational program was planned in conjunction with quantum-mechanical calculations of the target areas for the collisional excitation of certain of the stronger forbidden lines. Since the transition probabilities are known, it should be possible to utilize such target areas in connection with the observed intensities to get information on the state of ionization and excitation, the electron temperature, and ultimately the composition of the radiating gases.

Particular emphasis was placed on the nebulae studied by Wyse[6]—NGC 2165, 2440, 6741, and IC 5217, which had not been observed in the earlier program. The idea was to supplement the extensive line-intensity measurements in the visual and photographic regions with data for the ultraviolet. Second, there has been included a number of objects of moderate surface brightness that show interesting structural details, e.g., NGC 1535, 2392, 3242, 4361, 6720, 6751, 6778, 6818, 6905, and 7026. Finally, observations were secured of certain stellar nebulae of interest because of their high densities, central stars, and the problem of the Balmer decrement.

In all, 170 spectrograms of 45 planetary nebulae and their nuclei were secured in the 1943–1945 program. Of these objects, 27 have been previously observed by Page and 33 by Wright,[7] and a few have been studied by Minkowski[8] and by Stoy.

For many nebular problems wide-slit spectrograms of the type obtained by Page are probably more useful than slitless plates. Sky fog is greatly cut down, and there is no interference from overlapping stellar spectra. Slitless plates of the larger nebulae are useful only for the stronger, well-separated lines and are capable of giving us no reliable data on the weaker lines or those that fall close to one another. On the other hand, slitless spectrograms often yield more complete information about the structural features of the planetaries and permit the determination of the integrated intensities of the nebular images. The latter data are needed, for example, for the determination of the temperatures of the central stars. A comparison of the line intensities from Page's plates with those obtained in the present series will be of considerable interest, however.

The structural features of a number of planetary nebulae—in particular NGC 1535, 2022, 2392, 2440, 3242, 6309, 6445, 6720, 6751, 6778, 6905, 7026—have been studied on the Crossley slitless spectrograms, and some results have already been published elsewhere.[9] Since these reductions were completed, however, it has become evident that an adequate treatment of the spatial structures of planetary nebulae requires plates taken with a much larger scale, e.g., the slitless spectrograms secured by Olin Wilson at the coudé focus of the 100-inch reflector.[10] Accordingly, reference is made only to the Crossley results for a few of the fainter objects, which cannot be observed at the coudé focus.

[6] *Ap. J.*, **95**, 356, 1942.

[7] *Pub. Lick Obs.*, **13**, 193, 1918.

[8] *Ap. J.*, **95**, 243, 1942.

[9] Cf. the author's *Astrophysics* (New York: Ronald Press Co.), 1954, chap. 5

[10] Cf., for example, the discussion of NGC 6572 and NGC 7662 by Olin Wilson and L. H. Aller, *Pub. A.A.S.*, **55**, 70, 1950.

Limitations imposed by uncertainties in the physical theory.—Recent discussions by Bates, Massey, Fundaminsky, and Leech[11] have shown that the predicted cross-sections for the collisional excitation of the forbidden lines are grossly in error. At the present time it is possible to calculate only a lower limit to the abundances of the ions responsible for the forbidden lines—with no assurance that the real values lie close to this lower limit. For the same reason, electron temperatures derived from a comparison of the 4363 and green nebular lines are open to question, and a precise specification of this parameter is not possible. It is possible, however, that the relative cross-sections are less affected than the absolute values, and for the present we may retain the electron temperatures heretofore assigned. The general character of the excitation in the nebulae, as well as the energy distribution in the continuous spectrum as observed by Page, strongly indicates that T_ϵ must lie in the neighborhood of 10,000° K for most nebulae.

In view of these uncertainties, estimates of ionic densities and the level of excitation and ionization from a study of the emission of [*O* II], [*O* III], [*Ne* V], [*Ne* III], etc., would be insecure. Accordingly, no such discussion is attempted here.

III. OBSERVATIONS AND REDUCTIONS

Photometric procedures.—The observational technique is similar to that employed in the 1938–1939 series.[5] All the plates were secured with the quartz slitless spectrograph at the Crossley reflector of the Lick Observatory. Laboratory standardizations were secured with a spot sensitometer and were supplemented with exposures of the comparison star taken with diaphragms. In the earlier series, π^1 Cygni was employed as the comparison star; in the present investigation the stars observed by Barbier and Chalonge have been utilized.[12] Their list contains objects well distributed over the sky, carefully observed with painstaking spectrophotometric techniques.

Tracings of the plates were made with the Moll microphotometer at the Lick Observatory and with the Beals-type microphotometer at the Yerkes Observatory. Each plate was traced parallel to the dispersion. One trace usually sufficed for the smaller objects; extended objects frequently required several traces through different parts of the nebula. Tracings perpendicular to the dispersion and, in some instances, along the lines of symmetry of the nebula have also been run, in order to study the structural features of certain images in the larger objects.

The construction of deflection log *I*-curves involved both the sensitometer spots and the widened spectra of the comparison stars. Observations of the comparison star consisted of a series of equal exposures, between each of which the effective aperture of the telescope was varied in a known fashion by means of a diaphragm with circular holes. In the region of low densities the *D* log *I*-curves usually coincided; the variations with wave length appeared in the slopes of the straight-line portions. The plates secured on a single night's observations were developed together, and a single set of *D* log *I*-curves used was derived. Since development conditions were standardized and plates of a single emulsion number were used, the reduction-curves were very similar from one set of observations to the next.

Vignetting.—In the slitless spectrograph, vignetting is dependent upon wave length. Hence it is necessary that the nebula and comparison star be photographed in the same part of the field. Since the adjustment is made visually and the spectrum of the star appears continuous, slight errors are sometimes made. The position of the bright-line nebula can be adjusted accurately. The plateholder can be slid perpendicular to the direction of dispersion, so that several spectra of the comparison star or of the comparison star and

[11] *Trans. R. Soc. London, A* (in press). For a discussion of possible effects of the revision of the cross-sections on ionic densities, etc., derived from nebular line intensities see L. H. Aller, *Ap. J.*, **111**, 609, 1950.

[12] *Ann. d'ap.*, **4**, 30, 1941.

the nebula can be photographed upon the same plate. Thus the error in the centering of the comparison star can be found. The effect of vignetting was measured as a function of difference in centering upon a series of the plates taken especially for the purpose. In all instances the error in the observations from that cause would appear to be small, amounting to about $\Delta \log I = 0.01$ in the limit. After August, 1944, the corrections have been determined and applied to each night's observations of the comparison stars whenever necessary.

Effect of background continuum.—As previously mentioned, in extended objects the intensity is measured at some particular point or points—usually the brightest portions. In stellar objects the total area under the reduced intensity-curve of the monochromatic image has been measured. Usually, the intensities so derived are in good agreement with those found simply from the heights of the profile on the tracings. In each instance the contribution of the background continuum of the star or nebula must be estimated and subtracted from the measured intensity. This correction is especially serious for the weaker lines and for the Balmer continuum. With respect to the latter, Page's procedure of extrapolating the observed continuum longward of the Balmer limit to shorter wave lengths has been followed. As in the 1938–1939 observations, the tabulated intensity of the 20 A interval of the Balmer continuum is corrected for this underlying continuum. In general, the slitless plates are unsatisfactory for the determination of the energy distribution in the continuum. Slit spectra are required for this purpose.

Type of plates.—Until September, 1944, Eastman 103*a*-O plates were employed. Thereafter, Eastman II*a*-O plates were used. The gain in quality more than offsets the loss in speed. For photometric purposes the plates are less satisfactory than the standard Eastman 33's or Eastman 40's, in that they often show larger random errors.

Sources of error.—Among the sources of error we may mention, first of all, the limitations imposed by the nonuniformity of the photographic emulsion. Second, the calibration-curves are likely to be in error, particularly in the far ultraviolet. Laboratory and comparison-star data may be utilized in the region from about λ 3600 to λ 5000, but farther in the ultraviolet no satisfactory standards are available, and a mean curve has been employed throughout the ultraviolet. The exposure times on the comparison stars and laboratory standards are shorter than those on the nebulae, so that the deflection log I-curves are appropriate to a shorter time interval than the nebular exposures. To correct for the atmospheric extinction, use has been made of Popper's determination of the absorption as a function of wave length. These mean absorption coefficients are reliable for the summer and fall months, but at other times the fluctuations, particularly in the ultraviolet, may become serious. A troublesome feature of photometry on long-exposure slitless spectrograms is the effect of sky fog. At a given point in the spectrum of a nebula, sky light of many different colors is incident. One has to subtract this sky fog from the monochromatic nebular image, and a precise reduction is not possible. Finally, there are practical difficulties in estimating the position of the background continuum and in correcting for the effects of overlapping images.

Precision of the measures.—Perhaps some idea of the precision of the observations may be gained from the internal agreement of the intensities in a given nebula, measured upon different plates and nights. The computed probable error of the intensity of a line depends upon its position in the spectrum and upon its strength. In the region λλ 3700–5000 the probable error of a single measurement of a line stronger than 1.0 (on the scale $H\beta = 10$) is about 10 per cent. Weaker lines show somewhat larger probable errors, partly because of greater accidental errors and partly because of the difficulty in locating the position of the continuum. Ultraviolet lines, such as λ 3341 or λ 3428, also show larger probable errors, of the order of 14–17 per cent. These are the accidental errors; systematic effects are not included.

A detailed comparison with Wyse's intensities for the four nebulae common to our lists shows reasonably good agreement. The fainter lines are systematically measured slightly weaker in the present set of observations. The differences, however, are much smaller than in the earlier series, in which total intensities were measured and the fainter lines (referred to results obtained with the slit spectrograms) seemed systematically too weak. Perhaps sky fog is responsible for some of these troubles.

IV. THE OBSERVED LINE INTENSITIES

Table 1 gives the designation of the nebula, the α and δ for 1945, the plate number, the exposure time, and, finally, the secant of the zenith distance at midexposure. Tables 2, 3, and 4 give the relative intensities of the principal nebular lines on the scale $H\beta = 10$. Table 2 gives the data for low-excitation objects; Table 3 the results for certain high-excitation nebulae (i.e., objects that show the ultraviolet lines of [*Ne* v] and the Bowen fluorescent mechanism); while Table 4 gives the results for a number of nebulae of low surface brightness or extended surface. For these objects, only the intensity of the strongest lines could be measured. Frequently, the overlapping of images prevented any reliable estimate of the intensities for N1 and N2.

The intensity of the Balmer continuum is expressed in terms of 20 A intervals at the Balmer limit, also on the scale $H\beta = 10$. The tabulated value is corrected for the underlying continuum.

The high-excitation nebulae NGC 2022 and 2392 have recently been the subjects of a detailed spectrophotometric study by Minkowski and the writer. We shall discuss the Lick observations in conjunction with the Mount Wilson data at a later date.

Binuclear nebulae.—NGC 6778 resembles NGC 2440 or 7026, except that its surface brightness is much lower. The *He* II image is small and concentrated perhaps in the nucleus, and the other lines show a binuclear formation. There is no evidence of the spectrum of a central star.

Table 5 gives the intensity distribution along the line joining the two lobes for the [*O* III] $\lambda\lambda$ 4959–5007, $H(H\beta, H\gamma, H\delta)$, and λ 3727 images of NGC 6778, and the corresponding data (plus that for the [*Ne* III] image in NGC 7026). I have averaged the intensities for the two lobes, since the nebulae possess (very nearly) bilateral symmetry. In NGC 7026, $I(H\beta)$ is taken as 10 at the brightest portion of the condensation—3″ from the center, whereas $I(H\beta)$ is chosen as 10 at 5″ from the center of NGC 6778. I have also calculated the distribution of the emission per unit volume on the assumption that the emitting lobes are roughly spherically symmetrical. It now appears that such a hypothesis is inadequate in even the first approximation. In NGC 7026 the λ 4686 emission is strongly concentrated toward the center, as Wright pointed out many years ago, but I have been unable to find marked differences between the other images. On the other hand, the table shows that the maximum of the image of λ 3727 falls some 2″ or 3″ farther from the center than that of the [*O* III] image. Similar studies have also been carried out for NGC 2440.

Ring nebula.—The high-excitation lines of the Bowen fluorescent mechanism and λ 3203 of *He* II appear in NGC 6905 and 6720, for example, although they are not much above the level of sky fog on the plates. Slitless spectra are not well suited for NGC 6720; I have, however, measured the intensities of the ultraviolet lines with respect to $H\beta$, with the following results:

λ	3133	3203	3341	3445
I	1.8	1.4	0.7	1.0

If the assumption is made that, to a first approximation, the emitting material in the shell in NGC 6720 is distributed in a spherically symmetrical fashion, one may deduce the emission per unit volume, $E(\rho)$, from the intensity distribution across the monochro-

TABLE 1

OBSERVATIONS OF PLANETARY NEBULAE

Nebula	α	δ	Plate	Date	Exposure (Minutes)	Secant Zenith Distance
Anon.........	0h25m3	+55°36′	416	Oct. 14, 1944	120	1.07
			423	Oct. 15, 1944	30	1.06
IC 351.......	3 44	+34 53	417	Oct. 14, 1944	90	1.10
			424	Oct. 15, 1944	20	1.22
II 2003.......	3 52.8	+33 43	418	Oct. 14, 1944	93	1.01
			425	Oct. 15, 1944	75	1.10
			426	Oct. 15, 1944	14	1.03
NGC 1535....	4 11.7	−12 53	454	Nov. 18, 1944	120	1.58
J 320.........	5 2.5	+10 39	472	Mar. 9, 1945	60	1.41
			473	Mar. 9, 1945	21	1.65
NGC 2022....	5 36.6	+ 9 2	450	Oct. 17, 1944	232	1.21
II 2149.......	5 43.5	+46 7	448	Oct. 17, 1944	5, 11	1.41
			449	Oct. 17, 1944	22	1.31
			474	Mar. 9, 1945	5, 12	1.45
			475	Mar. 9, 1945	30	1.58
NGC 2165....	6 19.2	−12 57	427	Oct. 15, 1944	5	2.58
			428	Oct. 15, 1944	20	2.35
			429	Oct. 15, 1944	85	1.89
			464	Nov. 19, 1944	150	1.54
J 900.........	6 22.8	+17 50	462	Nov. 19, 1944	100	1.30
			463	Nov. 19, 1944	20	1.32
			468	Jan. 12, 1945	172	1.08
			469	Jan. 12, 1945	20	1.24
NGC 2392....	7 26.0	+21 2	324	Mar. 19, 1944	120	1.34
			325	Mar. 19, 1944	20	2.44
			430	Oct. 15, 1944	87	1.10
			465	Nov. 19, 1944	90	....
			470	Jan. 13, 1945	150	1.36
			471	Jan. 13, 1945	20	1.93
NGC 2440....	7 39.5	−18 5	419	Oct. 14, 1944	20	2.46
			420	Oct. 14, 1944	77	2.06
			439	Oct. 16, 1944	5	2.92
			440	Oct. 16, 1944	10	2.79
			441	Oct. 16, 1944	19	2.58
			442	Oct. 16, 1944	90	2.02
			455	Nov. 18, 1944	180	1.82
NGC 3242....	10 22.1	−18 21	313	Mar. 18, 1944	70	1.84
			314	Mar. 18, 1944	60	2.40
			326	Mar. 19, 1944	60	2.10
			327	Mar. 19, 1944	15	2.31
			333	May 20, 1944	10	2.33
			334	May 20, 1944	60	2.54
			456	Nov. 18, 1944	55	1.90
			482	Mar. 11, 1945	30	1.83
NGC 4361....	12 21.7	−18 28	477	Mar. 9, 1945	195	1.93
			483	Mar. 10, 1945	170	1.79
II 3568.......	12 31	+82 53	488	May 8, 1945	50	1.62
			489	May 8, 1945	10	1.66
II 4593.......	16 9.2	+12 13	336	May 20, 1944	17	1.13
			337	May 20, 1944	45	1.11
			345	May 21, 1944	10	1.29
			346	May 21, 1944	18	1.25
			347	May 21, 1944	43	1.18
			478	Mar. 9, 1945	15	1.10
			479	Mar. 9, 1945	20	1.11
II 4634.......	16 58.3	−21 43	340	May 20, 1944	18	2.10
			349	May 21, 1944	18	2.00
			350	May 21, 1944	60	1.97
			491	May 8, 1945	55	2.54

TABLE 1—*Continued*

Nebula	α	δ	Plate	Date	Exposure (Minutes)	Secant Zenith Distance
NGC 6309....	$17^{h}10^{m}.9$	−12°51′	351	May 21, 1944	87	1.57
			352	May 21, 1944	20	1.93
			358	June 22, 1944	120	1.62
			484	Mar. 11, 1945	17	1.91
			485	Mar. 11, 1945	105	1.64
NGC 6445....	17 46	−20 1	506	June 10, 1945	140	1.85
NGC 6567....	18 10.6	−19 5	353	June 18, 1944	47	1.80
			354	June 18, 1944	24	1.79
			492	May 8, 1945	50	2.37
Anon.........	18 15	+10 6	497	June 9, 1945	65	1.46
			515	June 15, 1945	60	1.34
II 4732.......	18 30	−22 42	399	July 22, 1944	120	2.00
			516	June 15, 1945	60	2.13
			517	June 15, 1945	15	2.28
Anon.........	18 47.3	+20 45	498	June 9, 1945	110	1.12
			502	June 10, 1945	20	2.00
			503	June 10, 1945	63	1.62
NGC 6720....	18 52	+32 57	288	Sept. 18, 1943	....	1.13
			289	Sept. 18, 1943	....	1.25
			509	June 14, 1945	....	1.20
NGC 6741....	19 0	− 0 31	291	Sept. 18, 1943	120	1.36
			292	Sept. 18, 1943	30	1.65
			304	Sept. 21, 1943	105	1.44
			431	Oct. 16, 1944	90	1.61
			432	Oct. 16, 1944	20	2.05
NGC 6751....	19 2.9	− 6 3.1	486	May 7, 1945	115	1.47
			518	June 16, 1945	160	1.39
II 4846.......	19 13	− 9 10	355	June 18, 1944	62	1.66
			355	June 18, 1944	20	1.66
			499	June 9, 1945	60	1.66
NGC 6778....	19 15.5	− 1 42	495	June 8, 1945	110	1.29
			526	July 1, 1945	130	1.38
NGC 6790....	19 20	+ 1 23	381	July 16, 1944	90	1.85
			511	June 14, 1945	60	1.29
			512	June 14, 1945	10	1.35
NGC 6803....	19 29	+ 9 57	385	July 17, 1944	30	1.13
			500	June 9, 1945	60	1.14
			501	June 9, 1945	6	1.17
NGC 6807....	19 32	+ 5 21	400	July 22, 1944	40	1.25
			507	June 10, 1945	15	1.20
Anon.........	19 36.5	+15 48	533	July 2, 1945	90	1.64
			534	July 2, 1945	8	1.30
NGC 6818....	19 40.0	−14 18	297	Sept. 20, 1943	115	1.68
			298	Sept. 20, 1943	30	2.08
			375	July 15, 1944	59	1.81
			379	July 16, 1944	180	1.62
			523	June 16, 1945	60	2.22
NGC 6833....	19 48	+48 49	389	July 18, 1944	83	1.10
			410	Aug. 20, 1944	113	1.03
NGC 6884....	20 9	+46 19	361	June 24, 1944	40	1.61
			362	June 24, 1944	20	1.46
			363	June 24, 1944	119	1.21
NGC 6886....	20 11	+19 50	293	Sept. 19, 1943	60	1.30
			294	Sept. 19, 1943	20	1.39
			299	Sept. 20, 1943	105	1.47
			386	July 17, 1944	170	1.08
			387	July 17, 1944	60	1.33
			443	Oct. 17, 1944	147	1.11
NGC 6891....	20 12	+12 33	401	Aug. 22, 1944	80	1.38
			402	Aug. 22, 1944	25	1.62
			433	Oct. 16, 1944	30	1.47
			434	Oct. 16, 1944	70	1.58

TABLE 1—*Continued*

Nebula	α	δ	Plate	Date	Exposure (Minutes)	Secant Zenith Distance
NGC 6905....	20^h^20^m^	+19°56′	493	May 8, 1945	138	1.31
			524	June 16, 1945	115	1.07
NGC 7026....	21 5	+47 46	295	Sept. 19, 1943	65	1.37
			365	June 24, 1944	115	1.04
			366	June 24, 1944	25	1.02
			406	Aug. 19, 1944	190	1.03
			422	Oct. 15, 1944	120	1.04
IC 5117......	21 30	+44 20	390	Aug. 19, 1944	90	1.06
			391	Aug. 19, 1944	30	1.02
			412	Aug. 20, 1944	90	1.37
			413	Aug. 20, 1944	15	1.62
			460	Nov. 19, 1944	120	1.26
Anon.........	21 31	+39 22	368	June 26, 1944	195	1.32
			392	July 18, 1944	70	1.02
			397	July 22, 1944	20	1.10
			411	Aug. 20, 1944	188	1.01
			435	Oct. 16, 1944	60	1.26
II 5217.......	22 22	+50 42	296	Sept. 17, 1943	45	1.74
			370	June 25, 1944	70	1.07
			371	June 25, 1944	10	1.04
			407	Aug. 19, 1944	120	1.07
			408	Aug. 19, 1944	40	1.25
			452	Nov. 18, 1944	120	1.12
Anon.........	22 29.2	+47 28	528	July 1, 1945	110	1.10
			535	July 2, 1945	90	1.35
			536	July 2, 1945	20	1.21
NGC 7354....	22 38.2	+61 0	510	June 14, 1945	120	1.47
Anon.........	23 24.1	+57 54	538	July 2, 1945	75	1.16
			539	July 2, 1945	20	1.11

TABLE 2

LOWER-EXCITATION NEBULAE

λ	Ion	Anon 0^h25^m	J 320	II 2149	II 3568	II 4593	II 4634	Anon 19^h36^m
Bac 20 A....	....	....	....	....	0.88	0.73	0.70	....
3703.9......	*H* I	....	....	....	....	....	....	....
3712.0......	*H* I	....	....	....	....	....	....	....
3727.........	[*O* II]	26	2.1	11	0.95	5.9	3.1	3.7
3750.2......	*H* I	....	....	0.73	....	....	....	....
3770.6......	*H* I	....	....	0.9	....	....	....	0.36
3797.9......	*H* I	0.33	0.52	0.92	....	....	....	0.53
3835.4......	*H* I	1.20	0.75	1.2	0.62	....	....	0.51
3868.7......	[*Ne* III]	11	11	2.4	6.6	3.4	6.0	8.9
3889.1......	*H* I	3.1	2.6	1.5	2.2	1.9	2.0	1.10
3967.4......	[*Ne* III]	5.5	5.5	2.3	3.0	2.3	4.0	3.3
4026.2......	*He* I	....	....	....	0.56	....	....	0.46
4069+76....	[*S* II]	0.78	....	....	0.39	....	....	1.1
4101.8......	*H*δ	3.3	3.0	3.0	2.5	2.6	2.3	2.0
4340.5......	*H*γ	4.7	5.0	5.2	4.2	5.0	3.9	3.4
4363.2......	[*O* III]	1.6	1.8	0.60	0.88	....	0.52	1.0
4471.5......	*He* I	2.1	0.8	0.76	0.43	0.64	0.46	0.65
4634+40....	*N* III	....	....	....	....	....	....	0.26
4685.8......	*He* II	2.6	....	....	....	....	....	0.52
4711.4......	[*A* IV]	....	....	....	....	....	....	0.58
4740.3......	[*A* IV]	....	....	....	....	....	....	0.77
4861.3......	*H* I	10.0	10.0	10.0	10.0	10.0	10.0	10
4959.5......	[*O* III]	35	34	14	50	18	28	52
5007.6......	[*O* III]	88	89	41	140	57	78	19.7
C...........	....	....	0.27	0	0.27	0.28	0.0	0.36

TABLE 2—*Continued*

λ	Ion	NGC 6567	Anon 18^h15^m	II 4732	Anon 18^h47^m	II 4846	NGC 6790	NGC 6803	NGC 6807
Bac 20 A	...	...	0.92	...	1.30	0.58	0.60	2.4	...
3703.9	*H* I	...	0.22	...	0.19	...	...	...	...
3712.0	*H* I	...	0.26	...	0.25	...	0.21	...	...
3727	[*O* II]	2.3	16	1.6	7.8	4.0	1.3	4.0	2.9
3750.2	*H* I	...	0.47	...	0.37	0.49	0.15	0.33	0.38
3770.6	*H* I	...	0.47	...	0.41	0.41	0.24	0.47	0.42
3797.9	*H* I	...	0.50	...	0.54	0.75	0.31	0.91	0.56
3835.4	*H* I	0.98	0.81	0.91	0.77	0.88	0.40	1.4	0.72
3868.7	[*Ne* III]	5.6	...	10	1.07	8.7	8.6	9.6	8.4
3889.1	*H* I	2.1	1.1	1.6	1.4	2.1	1.0	2.1	1.4
3967.4	[*Ne* III]	2.7	1.6	3.4	2.0	4.2	3.2	4.8	3.8
4026.2	*He* I	0.53	0.20	...	0.22	0.43	0.16	0.30	0.4
4069+76	[*S* II]	0.57	0.51	...	0.19	0.47	0.14	0.56	0.42
4101.8	*H*δ	2.50	2.6	2.3	2.5	2.7	1.8	2.3	2.7
4340.5	*H*γ	4.40	5.1	4.3	4.6	4.9	4.1	4.6	4.9
4363.2	[*O* III]	0.57	...	2.4	0.22	1.4	1.4	0.98	1.1
4471.5	*He* I	0.60	0.27	0.91	0.48	0.94	0.56	0.96	0.65
4634+40	*N* III	...	...	...	0.11	...	0.21	0.32	...
4685.8	*He* II	...	...	...	...	...	...	0.78	...
4711.4	[*A* IV]	...	...	...	0.12	...	0.21	0.62	...
4740.3	[*A* IV]	...	...	...	...	...	0.29	0.71	...
4861.3	*H* I	10.0	10.0	10	10	10	10	10	10
4959.5	[*O* III]	32	0.48	47	13	36	56	38	40
5007.6	[*O* III]	79	1.61	...	38	97	147	107	...
C	...	0.06	0.14	0.14	0.17	0.04	0.68	0.23	0.17

TABLE 2—*Continued*

λ	Ion	NGC 6833	NGC 6884	NGC 6891	NGC 7026		II 5117	Anon 22^h29^m	Anon 23^h24^m
					Center	Lobes			
Bac 20 A	...	0.54	...	1.34	0.51	...	...	1.63	...
3712.0	*H* I	...	...	...	...	...	0.17	0.39	...
3727	[*O* II]	1.1	1.1	1.9	3.1	4.7	2.0	4.0	1.0
3734	...	...	...	...	...	...	...	0.55	0.48
3750.2	*H* I	0.17	0.11	...	...	...	0.22	0.54	0.40
3770.6	*H* I	0.26	0.12	...	...	...	0.35	0.62	0.41
3797.9	*H* I	0.41	0.20	0.22	...	...	0.38	0.90	0.49
3835.4	*H* I	0.54	0.42	0.52	...	...	0.51	1.0	0.60
3868.7	[*Ne* III]	6.4	6.0	7.3	6.6	7.4	9.3	7.5	4.5
3889.1	*H* I	1.3	0.78	2.3	1.9	2.1	1.2	2.5	1.1
3967.4	[*Ne* III]	2.9	2.6	3.5	3.1	3.6	4.1	4.1	2.8
4026.2	*He* I	0.18	0.07	0.64	...		0.30	0.61	...
4069+76	[*S* II]	0.22	0.06	...	0.58	0.41	0.60	0.23	...
4101.8	*H*δ	2.0	0.98	2.5	2.2	2.2	2.3	3.5	1.7
4340.5	*H*γ	3.4	2.23	5.10	3.7	4.4	4.2	5.2	3.6
4363.2	[*O* III]	1.1	0.52	...	0.18	...	1.9	0.83	2.2
4471.5	*He* I	0.38	0.40	0.66	0.44	0.56	0.63	1.22	0.74
4634+40	*N* III	...	0.26	0.25	0.60	0.46	0.54	0.36	...
4685.8	*He* II	...	1.24	0.20	2.8	1.3	1.38	...	...
4711.4	[*A* IV]	...	0.42	...	0.27	0.38	0.50	0.30	...
4740.3	[*A* IV]	...	0.49	...	...	...	0.75		
4861.3	*H* I	10	10	10	10.0	10	10	10	10
4959.5	[*O* III]	22	60	30	42	40	61	20	13
5007.6	[*O* III]	54	170	83	96	84	160	...	31
C	...	0.46	0.84	0.66	0.37	...	0.46	0.0	0.33

matic image, $I(x)$. At a given point x_0 from the center of the projected nebular image, the intensity $I(x_0)$ in the cross-section is given by the integral equation

$$I(x_0) = 2\int_{x_0}^{\infty} \frac{E(\rho)\,\rho\,d\rho}{\sqrt{\rho^2 - x_0^2}}.$$

Solutions of this equation have been given by Von Zeipel, Plummer, and others.[13] Convenient tables for the numerical solution are given by Wallenquist.[14] Chandrasekhar and Münch[15] have shown that the accuracy attainable in the solution for $E(\rho)$ cannot be high. Probably only the general character of the emission function can be derived.

Measurements of the intensity distribution along the major axis of the images of λλ 4686, 3869, 3727, and 4340 are published elsewhere.[9] Whereas the cross-sections of the H and He II images may be interpreted in terms of a spherically symmetrical emis-

TABLE 3

HIGH-EXCITATION NEBULAE

λ	Ion	IC 351	IC 2003 (a)	IC 2003 (b)	NGC 1535 Bright Ring	NGC 2165	J 900
3133	*O* III	10	8.5	...	...	8.8	3.1
3178	*He* I	...	...	...	...	0.81	...
3203	*He* II	3.4	2.7	...	2.3	1.9	1.4
3299	*O* III	...	...	...	...	0.74	...
3312	*O* III	1.7	0.84	...	...	0.97	0.57
3346	*Ne* V+*O* III	2.1	1.35	1.56	0.8	2.2	1.05
3425	*Ne* V+*O* III	...	0.70	0.71	...	6.0	1.8
3444	*O* III	3.0	2.8	3.0	0.95	2.2	1.7
3650(20 A)	*H*	...	...	...	(1.9)	0.7	0.25
3727	[*O* II]	0.87	4.2	5.0	0.80	3.13	5.2
3750.2	*H*	...	0.5	...	...	0.56	0.55
3770.6	*H*	...	...	...	...	0.40	0.42
3797.9	*H*	0.62	0.48	...	...	0.48	0.50
3819	*He* I	...	...			...	...
3835.4	*H*	0.61	0.81	...	...	0.53	0.74
3868.7	[*Ne* III]	9.0	9.0	7.5	7.4	8.7	7.4
3889.1	*H*	1.7	2.0	2.0	2.1	1.8	1.6
3967.4	[*Ne* III]+*H*	3.8	5.0	4.2	3.0	4.3	3.4
4026.2	*He*	...	0.28	...	...	0.18	0.33
4069+76	[*S* II]	...	0.25	...	...	0.20	0.30
4101.8	*H*	2.3	3.0	3.0	2.1	2.65	2.1
4200	*He* II	...	...	...	...	0.12	...
4340.5	*H*	5.8	5.0	6.2	4.9	4.6	4.2
4363.2	[*O* III]	1.6	1.2	...	1.9	2.1	1.3
4471.5	*He* I	2.8	0.37	...	0.60	0.34	0.40
4542	*He* II	0.32	0.43	...	...	0.18	0.27
4634–40	*N* III	0.37	0.34	...	...	0.47	0.32
4685.8	*He* II	6.0	5.6	7.5	2.4	6.0	4.7
4711.4	[*A* IV]	0.85	0.49	...	...	0.72	...
4740.3	[*A* IV]	0.77	0.44	...	...	0.65	0.40
4861.3	*H*	10.0	10	10.0	10	10	10
4959.5	[*O* III]	44	40	30	...	53	39
5007.6	[*O* III]	176	136	98	...	128	126
C	...	0.12	0.0		...	0.0	0.17

[13] See, e.g., Smart, *Stellar Dynamics* (Cambridge: At the University Press, 1938), pp. 297–304.

[14] *Upsala Medd.*, No. 65, p. 42, 1936.

[15] *Ap. J.*, **111**, 144, 1950, esp. secs. 2 and 3.

TABLE 3—*Continued*

λ	Ion	NGC 2440	NGC 3242 Bright Ring	NGC 6309	NGC 6741	NGC 6886	Anon 21^h31^m	II 5217
3133	*O* III	4.4	7.0	....	....	....	3.5	
3178	*He* I	....	....	....	....	....	....	....
3203	*He* II	2.0	2.0	....	1.3	....	3.0	....
3299	*O* III	....	....	....	....	....	....	....
3312	*O* III	0.73	1.1	....	....	0.07	....	....
3346	*Ne* V+*O* III	6.7	1.6	....	1.9	2.0	5.6	....
3425	*Ne* V+*O* III	16	0.55	4.7	4.1	4.4	18.0	....
3444	*O* III	1.6	3.3	....	1.7	0.9	1.0	0.6
3650(20 A)	*H*	0.20	0.80	....	....	....	0.48	0.75
3727	[*O* II]	8.3	1.7	4.7	8.2	9.3	4.5	2.0
3750.2	*H*	....	....	....	....	0.19	0.25	0.29
3770.6	*H*	....	....	....	....	0.17	0.37	0.32
3797.9	*H*	....	....	....	....	0.20	0.47	0.52
3819	*He* I	....	....	....	....	....	....	....
3835.4	*H*	....	....	....	0.74	0.60	0.69	0.60
3868.7	[*Ne* III]	8.3	11.7	9.5	8.6	9.7	6.2	9.8
3889.1	*H*	....	....	....	1.2	1.1	1.7	2.0
3967.4	[*Ne* III]+*H*	3.7	5.0	4.5	3.2	3.3	3.4	4.2
4026.2	*He*	....	....	....	0.62	0.12	0.26	0.20
4069+76	[*S* II]	....	....	....	0.60	0.44	0.28	0.16
4101.8	*H*	2.6	3.2	2.6	3.0	1.7	2.25	2.8
4200	*He* II	....	....	....	0.36	....	....	....
4340.5	*H*	5.2	6.3	4.2	3.7	4.0	4.6	5.5
4363.2	[*O* III]	2.9	....	....	1.3	1.5	2.0	1.3
4471.5	*He* I	0.27	0.52	....	0.50	0.22	0.53	0.62
4542	*He* II	0.30	....	....	....	....	0.82	....
4634–40	*N* III	1.0	0.66	....	0.62	0.42	0.20	0.34
4685.8	*He* II	7.6	5.1	7.7	3.4	4.0	9.0	1.30
4711.4	[*A* IV]	1.6	....	....	0.81	0.4	1.0	0.70
4740.3	[*A* IV]		....	....	1.32	0.4	0.5	0.74
4861.3	*H*	10	10	10	10.0	10.0	10	10
4959.5	[*O* III]	75	59	40	58	65	29	55
5007.6	[*O* III]	....	135	101	141	160	75	150
C	....	....	....	....	0.16	0.49	0.21	....

sion per unit volume, the [*Ne* III] and [*O* II] images clearly show that the emission departs far from spherical symmetry. In fact, negative values of $E(\rho)$ are required for the region just inside the edge of the ring, in order to represent the observed data. Therefore, departures from spherical symmetry are severe. In view of the filamentary character of this object this result is hardly surprising.

Comments on other nebulae are given in the notes to Table 4.

The structural features of a number of planetary nebulae are being studied in collaboration with Olin C. Wilson, who has obtained slitless spectra with the 100-inch coudé. The present measures may be of value in relating the intensity measures in one monochromatic image with those in another.

V. THE BALMER DECREMENT

Because of the uncertainties in the collisional cross-sections, we are unable to derive ionic abundances and electron temperatures from the nebular line intensities, nor does it seem worth while to calculate electron densities, since the nebular surface brightnesses and distances are so poorly known. Accordingly, our discussion is restricted here to a single problem—the Balmer decrement.

The most detailed observational discussion of the Balmer decrement is that by Ber-

TABLE 4

Nebulae with Extended Surfaces

λ	Ion	NGC 4361*	NGC 6445†	NGC 6571‡	NGC 6778§	NGC 6818 West Side	NGC 6818 East Side	NGC 6905‖	NGC 7354#
3133	*O* III	...	...	...	...	...	...	2.2	...
3203	*He* II	...	...	...	...	...	...	2.8	...
3346	[*Ne* V]	5.4	...	...	...	2.6	4.8	...	...
3428	[*Ne* V]	...	...	...	...	6.7	8.6	...	...
3444	*O* III	20	...	...	...	...	3.2	...	...
3650(20 A)	*H*	...	...	1.2	0.19	1.0	...	...	...
3727	[*O* II]	...	20	17	5.8	4.7	6.8	0.87	...
3868	[*Ne* III]	5	12	7.5	4.6	11	14	5.2	5.5
3889	*H*	...	...	...	2.2	...	2.0	...	...
3967.4	[*Ne* III]	3.7	...	3.8	2.2	4.5	4.5	2.2	2.6
4101.8	$H\delta$	4.2	...	3.2	2.8	3.0	2.6	1.6	2.8
4340.5	$H\gamma$	9.7	4.1	5.4	4.3	5.9	6.6	2.4	3.6
4363.2	[*O* III]	...	...	...	1.3	...	2.1	...	...
4471.5	*He* I	...	...	...	1.1	...	...	...	...
4634–40	*N* III	...	...	...	0.8	...	...	...	...
4686	*He* II	25	5.6	3.2	1.62	8.2	9.3	3.9	6.2
4740	[*A* IV]	...	...	...	...	...	1.3	...	...
4861	$H\beta$ }		10	10	10	10	10	10	10
4959	[*O* III] }	60	{110 }	160 }	24	...	170 }	78 }	140 }
5007	[*O* III] }		{ }	}	65	...	}	}	}

* High excitation. Overlapping of irregular monochromatic images prevents accurate measurements.

† The monochromatic images resemble one another closely. The appearance of NGC 6445 recalls a coral atoll and inclosed lagoon.

‡ Tabulated values are means of two sides of nebulae.

§ Both lobes have the same excitation. Mean values are tabulated.

‖ Mean of measures in three parts of nebula.

\# The monocromatic images resemble one another closely.

TABLE 5

Structure of NGC 7026

Distance	*O* III	*O* II	*Ne* III	Hydrogen	Distance	*O* III	*O* II	*Ne* III	Hydrogen
x	...	...	...	...	4	81	3.62	4.66	6.51
0	71	2.32	3.31	4.25	5	53.5	2.29	2.33	3.86
1″	70	2.6	4.41	5.85	6	23.8	1.33	0.98	2.3
2	77	3.51	6.62	8.60	7	9.15	0.64	0.417	1.0
3	82	4.72	7.35	10.0	8	4.88	0.29	0.07	0.13

Structure of NGC 6778

Distance	*O* II	*O* III	Hydrogen	Distance	*O* II	*O* III	Hydrogen
0	3.7	57.5	8.2	8	3.86	29.4	4.7
1″	3.8	58.5	8.5	9	3.11	20	3.1
2	3.9	62.7	9.2	10	2.2	12	1.8
3	4.0	64.5	9.8	11	1.3	6.3	1.0
4	4.3	63	10.0	12	0.81	3.0	0.36
5	4.8	60	10.0	13	0.49	1.15	...
6	4.87	53	9.0	14	0.26	...	...
7	4.48	42	7.1				

man,[16] who calibrated Wright's intensity estimates with the aid of photometric measurements of his own. He divided the nebulae into three classes, depending on the ratio of N2 to $H\beta$. In order to correct for space absorption, he assumed the interstellar material to be uniformly distributed in galactic longitude and to fall off uniformly with height above the galactic plane. The distance of each nebula had been derived from studies of galactic rotation.[17] With the aid of the distances, line intensities, and galactic latitudes, he made a least-squares solution for the coefficient of absorption per kiloparsec, the exponent n in the assumed law of selective absorption λ^{-n}, and the Balmer decrement for each group. We summarize his results for the Balmer decrement in Table 6. The difficulty with Berman's procedure is that the absorbing material is so unevenly distributed throughout the galaxy that the basic assumptions are not even approximately fulfilled.

TABLE 6

THE BALMER DECREMENT

(After Berman)

λ	I	II	III	λ	I	II	III
4861......	10.0	10.0	10.0	3969.....	1.25	1.37	1.79
4340......	4.45	4.80	5.00	3889.....	1.15	1.00	1.24
4101......	2.13	2.4	2.63	3835.....	0.77	0.52	0.89

In an attempt to overcome some of these difficulties, the nebulae have been divided into three classes: S, stellar low-excitation objects; L, low-excitation objects with extended surfaces; and H, high-excitation objects (those showing ultraviolet *He* II [*Ne* V], or *O* III lines). Smoothed Balmer decrements are derived for each of these objects. If the true decrements are the same for each group, the smoothed intensities should differ from one another only because of differential space absorption. In particular, the nebulae with the slowest decrements should be the least affected by space absorption. Accordingly, the Balmer decrements for each group have been reduced on the assumption that the one with the slowest decrement is essentially unaffected. In this way a set of tentative, observed Balmer decrements may be found.

With observations of this precision, it suffices to take the absorption proportional to $1/\lambda$. Hence one may write

$$\log \frac{I_n}{I_4} = \log \left(\frac{I_n}{I_4}\right)_0 - C\left[\frac{1}{\lambda_n} - \frac{1}{0.4861}\right],$$

where I_n/I_4 is the intensity of the nth line referred to $H\beta$ as 1.0; the subscript "0" refers to the value that this ratio would have if corrected for space absorption; and C is a constant to be determined for each nebula, whose value, computed by least squares, is listed in Tables 2 and 3 for several nebulae. The wave lengths, λ_n, are assumed expressed in microns. With the aid of these C's, it is possible to correct the intensities of the nebular lines for the effects of space absorption.

The mean Balmer decrements, calculated for groups S, L, and H, are compared with the theoretical calculations by Menzel and his colleagues.[18] Models A and B assume a nebula in which the central star radiates no energy in the Lyman lines. Model A is transparent in the lines, so that all the Lyman radiation produced in the nebula by cyclic processes escapes. In model B, all Lyman radiation is assumed degraded into Lyman-α,

[16] *M.N.*, **96**, 890, 1936.

[17] L. Berman, *Lick Obs. Bull.*, **18**, 57, 1937.

[18] See J. G. Baker and D. N. Menzel, *Ap. J.*, **88**, 52, 1938; Baker, Menzel, and Aller, **88**, 423, 1938; **89**, 587, 1939.

which alone escapes from the nebula. The Balmer decrement is rather insensitive to the electron temperature; I have chosen a value of 10,000° as representative of typical planetary nebulae. Model *C* refers to a thin nebula (similar to *A*), whose central star radiates like a black body in the Lyman lines. The stellar radiation steepens the theoretical decrement markedly; it would have little influence in a thick nebula.

The Balmer decrement is quite similar for the three groups of nebulae, and there is no clear-cut indication that the high-excitation objects have a systematically different decrement than do the low-excitation objects. The decrements I have found are less steep than those published by Berman but are in substantial agreement with the calculations for theoretical model *B*. This result comes as no surprise, since both models *A* and *C* correspond to thin shells, which should be much fainter than the objects studied in the present survey. It is, of course, possible that slight variations will be found in the Balmer decrements for objects of comparable density and excitation; but, until such differences have been established, it would appear that space absorption remains the chief cause of the variation in the observed decrements, an opinion expressed by Berman some years ago.

VI. THE SPECTRA OF THE CENTRAL STARS

The spectra of the central stars of the planetary nebulae are difficult to observe. Absorption lines originating in the nucleus are often smothered under strong nebular emission, and nuclear emission lines can be observed only when they do not coincide in position with nebular lines.[16]

Stars with continuous spectra.—Most of the central stars showed continuous spectra. Thus Anon 0^h25^m, NGC 6807, II 4732, and probably also NGC 6309 and 6567 showed weak continua probably of stellar origin, with no trace of any stellar emission lines. The well-exposed spectrum of the central star of Anon 18^h15^m showed no trace of any emission lines. A number of central stars, those of IC 351, Anon 18^h47^m, NGC 6790, NGC 6803, Anon $19^h36^m.5$, NGC 6884, and NGC 5217 show λ 4686 *He* II, λλ 4712–4740 [*A* IV], and the λ 4640 group. Probably the λ 4640 lines, as well as the others, originate in the inner portion of the nebula rather than in the star. The central stars of Anon 22^h29^m and 23^h24^m seem to show the λ 4640 group, but not λ 4686. These lines may

TABLE 7

THE BALMER DECREMENT: COMPARISON BETWEEN THEORY AND OBSERVATION

λ	*n*	Observed			Theoretical		
		Low Excitation		High Excitation			
		Stellar *S*	Nonstellar *L*	*H*	*A*	*B*	*C*
4861	4	....	....	....	....	....	....
4340	5	5.25	5.66	5.65	5.76	5.1	3.5
4101	6	2.97	3.31	3.13	3.74	3.1	1.6
3969	7	1.93	2.20	1.99	2.55	2.06	0.89
3889	8	1.44	1.50	1.32	1.82	1.43	0.55
3835	9	1.00	1.03	0.93	1.36	1.05	0.36
3797	10	0.74	0.85	0.66	1.05	0.79	0.25
3770	11	0.57	0.56	0.49	0.81	0.59	....
3750	12	0.45	0.46	0.36	0.65	0.46	....
3734	13	0.42	....	....	0.52	0.37	....
3722	14	0.38	....	....	0.42	0.30	....
3712	15	0.30	....	....	0.35	0.25	....
3707	....	0.26	....	....	....	....	....

originate in the star itself. On the other hand, J 320 shows λ 4686 but no λ 4640; presumably the emission here is of nebular origin.

The intensity distribution in the continuous spectra of some of the brighter stars has been measured. Table 8 gives the results for NGC 1535, NGC 6891, and II 4593. No correction has been made for space absorption. Preferably, such studies should be made with carefully guided wide-slit spectrograms (to eliminate sky fog) and with somewhat higher dispersion than I have employed.

TABLE 8

INTENSITY DISTRIBUTION IN THE CONTINUOUS SPECTRA OF CENTRAL STARS OF NGC 1535, NGC 6891, AND II 4593

λ	LOG (I_λ/I_{4800})			λ	LOG (I_λ/I_{4800})		
	IC 4593	NGC 6891	NGC 1535		IC 4593	NGC 6891	NGC 1535
3300......	0.67	....	0.68	4100....	0.21	0.26	0.15
3400......	.59	0.74	.59	4200....	.19	.21	.11
3500......	.53	.67	.51	4300....	.16	.16	.09
3600......	.47	.58	.43	4400....	.13	.12	.07
3700......	.41	.49	.36	4500....	.11	.09	.06
3800......	.35	.42	.30	4600....	.08	.06	.04
3900......	.30	.36	.24	4700....	.05	.03	.02
4000......	0.25	0.31	0.19	4800....	0.00	0.00	0.00

TABLE 9

THE SPECTRA OF THE NUCLEI OF NGC 6751 AND NGC 6905

NGC 6751		NGC 6905		IDENTIFICATION
λ	*I*	λ	*I*	
3203.........	0.36	3203(prob.neb.)	5	3203 *He* II
3276.........	0.25	....	....	3275 *O* V
3338–3357....	0.63	....	....	3360 *N* III
~3380.......	0.39	....	....	3381 *O* III+*O* IV
3397–3429....	1.3	....	....	3411 *O* IV
~3430.......	0.49	3420–3437	2.9	3429–3445 *O* III
3469–3489....	0.55	....	....	3483 *N* IV
3696–3722....	1.2	3672–3710	2	3708–3750 *O* III, *O* IV, *O* V
3808–3837....	0.42	3794–3854	22	3811–3834 *O* VI
3947–3987....	0.33	3920–3982	2.3	3965 *O* III
4082–4125....	0.55	4096–4141	....	4070 *O* III 4100 *N* III, *He* II 4123 *O* V
~4200.......	0.19	....	....	4200 *N* III, *He* II
	0.18	....	....	....
4427–4470....	0.26	4410–4470	0.76	4441 *C* IV
4492–4572....	0.13 0.17	4490–4565	1.2	4512–4540 *N* III? 4542 *He* II 4515 *C* III
4616–4714....	10.0	4640–4702	10	4650 *C* III, *C* IV 4686 *He* II

The central stars of NGC 7026, 6751, and 6905 show broad emission lines. In the nucleus of NGC 7026 no lines are found in addition to those near λ 3800 and λ 4650, although there is some suggestion of weak diffuse features in the ultraviolet. The spectrum of the central star of NGC 6751 is of the carbon-sequence, Wolf-Rayet type, with high-excitation lines of carbon, helium, and oxygen in various stages of ionization. Nitrogen, represented by the λ 3483 *N* III group, may be present also. The spectrum of the nucleus of NGC 6905 is similar, although there is no suggestion of any nitrogen. Presumably it and the nucleus of NGC 7026 are typical stars of the carbon sequence, in which the stronger features are the *C* IV λ 4650 and *O* VI λ 3811 groups. Table 9 gives the intensities for the nuclear emission lines of NGC 6751 and 6905. Measures are difficult because of the overlapping nebular spectrum and the faint diffuse character of the stellar lines.

I want to thank Director C. D. Shane and express my gratitude to the late Dr. J. H. Moore, who made it possible for me to obtain these plates at the Lick Observatory. I am grateful to Dr. Otto Struve, then director of the Yerkes Observatory, who allowed me the use of their microphotometer with which many of the tracings were made. Thanks are due to the Graduate School of Indiana University and to the Draper Fund of the National Academy for grants to aid the reduction of the observations. Miss Carolyn Mooshy assisted with the measurement and reduction of the tracings.

Reprinted from *The Astrophysical Journal*, Vol. 118, No. 3, November, 1953.

THERMAL EFFECTS OF COLLISIONAL LINE EXCITATION IN GASEOUS NEBULAE

LAWRENCE H. ALLER
University of Michigan Observatory
Received May 18, 1953; revised July 9, 1953

ABSTRACT

The cooling effects of ions of oxygen, neon, nitrogen, and other light elements may be estimated from the observed intensities of all the relevant forbidden lines. The electron temperatures in nebulae such as IC 418, with relatively weak forbidden lines (as compared with the Balmer lines), must be substantially greater than in nebulae with relatively strong forbidden lines. Thus the electron temperatures in high-excitation planetaries are not necessarily higher than those in low-excitation planetaries but may even be relatively low. The electron temperature cannot rise substantially above 25,000° or 30,000° K in a planetary nebula because of the dissipation of energy in the excitation of the hydrogen lines, unless the hydrogen is substantially all ionized.

Many years ago, Bowen[1] pointed out that the forbidden lines in the gaseous nebulae were produced by the collisional excitation of metastable levels in ions of oxygen, nitrogen, neon, etc., in various stages of ionization. The atoms returned to the ground level with the emission of forbidden lines. The emitted quantum simply escaped from the nebula without further absorption. This process removes kinetic energy from the gas and serves to lower the electron temperature below the values it would have for a pure hydrogen nebula in radiative equilibrium with a hot central star.

Some years ago[2] the electron temperatures of gaseous nebulae were estimated from the relative intensities of the green nebular lines and λ 4363 of doubly ionized oxygen with the aid of target areas for collisional excitation calculated by Hebb and Menzel.[3] The resulting temperatures came out mostly in the neighborhood of 10,000° K. Adopting these temperatures as valid, it was shown that the observed intensities of the forbidden lines would account for a lowering of the temperature from the earlier predicted values[4] obtained for a pure hydrogen nebula (e.g., of the order of 34,000° K for a central star of temperature 40,000° K) to a value near 10,000° K.[5] T. L. Page's[6] measures of the intensity distribution in the Balmer continuum favored an electron temperature near 10,000° K. This latter estimate, however, supposes that the background continuum which must be subtracted is constant in intensity with wave length.

More recently, doubts have been cast on the reality of an electron temperature of 10,000° K. The theory from which the target areas of collisional excitation were computed was shown to be inexact by Bates, Fundaminsky, Leech, and Massey.[7] They concluded that the target areas were systematically too large; furthermore, even the relative cross-sections may be in error. Menzel has asserted that the target areas computed by Hebb and himself are not necessarily too large and that they do not violate the conservation theorem of Bohr, Peierls, and Placzek. He has not, however, demonstrated that the other objections raised against this calculation of target areas can be disre-

[1] *Ap. J.*, **67**, 1, 1928.

[2] D. H. Menzel, L. H. Aller, and M. H. Hebb, *Ap. J.*, **93**, 240, 1941.

[3] *Ap. J.*, **92**, 408, 1940.

[4] J. G. Baker, D. H. Menzel, and L. H. Aller, *Ap. J.*, **88**, 422, 1938.

[5] D. H. Menzel and L. H. Aller, *Ap. J.*, **94**, 30, 1941.

[6] *Ap. J.*, **96**, 78, 1942.

[7] *Phil. Trans. R. Soc., London*, **243**, 93, 1950.

garded. Furthermore, Minkowski[8] finds that the observed value of the Balmer discontinuity in IC 418, interpreted with the aid of Spitzer and Greenstein's[9] calculations on the continuum produced by double-photon emissions, cannot be reconciled with an electron temperature as low as 10,000° K.

Therefore, it becomes important to re-examine the hypothesis that the collisional excitation of the nebular lines will necessarily lower the electron temperature to about 10,000° K. The temperature of the electron gas is controlled by the condition that in each unit volume the total amount of energy brought into the continuum by photo-ionizations is equal to that dissipated in free-free transitions and in collisional excitations of forbidden lines. We suppose, for the moment, that the electron temperature will not be so high that collisional excitation of the hydrogen lines will be important. The effect of possible hydrogen-line excitation will be discussed later. The equation of conservation of energy for the continuum may be written, in the symbolic notation of Menzel, Baker, and the writer, in the following form:[10]

$$\sum_{1}^{\infty} \int_{\nu_n}^{\infty} F_{\kappa n} h\nu_{\kappa n} d\nu + \sum_{2}^{\infty} \int_{\nu_n}^{\infty} F_{\kappa n} h\nu_{n1} d\nu + \int_{0}^{\infty}\int_{0}^{\infty} F_{\kappa\kappa'} h\nu_{\kappa\kappa'} d\nu_{\kappa} d\nu_{\kappa'} + \Sigma E_i = \int_{\nu_1}^{\infty} F_{1\kappa} h\nu_{1\kappa} d\nu \,, \tag{1}$$

where the left-hand terms represent, respectively, the energy liberated by the free electrons upon recapture and subsequent cascade to the lower levels, the free-free emissions, and the energy dissipated in the excitation of forbidden lines of various ions. The summation is carried out over the forbidden emissions of all different ions. The right-hand side of the equation represents the energy brought into the continuum by photo-ionizations.

The quantities E_i are related to the intensities of the forbidden lines as measured by various observers. Neglecting, for the moment, the fact that the different forbidden-line emissions have different spatial distributions in the nebula, we may set

$$E_i = E(H\beta)\frac{I_i}{I(H\beta)}, \tag{2}$$

where the ratio of the forbidden-line intensity I_i to the intensity of $H\beta$ has been measured by a number of observers.[11] Now $E(H\beta)$, the emission per unit volume in $H\beta$, is given by[12]

$$E(H\beta) = b_4 N_i N_\epsilon \frac{2hRKZ^6}{T_\epsilon^{3/2}} \frac{g e^{X_4}}{512}, \tag{3}$$

where

$$X_4 = \frac{hRZ^2}{4^2 kT_\epsilon} = \frac{9800}{T_\epsilon}, \tag{4}$$

[8] *Pub. A.S.P.*, **65**, 161, 1953. I am grateful to Dr. Minkowski for communicating this result to me in advance of publication.

[9] *Ap. J.*, **114**, 407, 1951.

[10] *Ap. J.*, **88**, 313, 1938, eq. (9).

[11] See, e.g., Louis Berman, *Lick Obs. Bull.*, **15**, 86, 1930; *M.N.*, **96**, 893, 1936; A. B. Wyse, *Ap. J.*, **95**, 356, 1942; and L. H. Aller, *Ap. J.*, **93**, 236, 1941; **113**, 125, 1951. M. L. White (*Ap. J.*, **115**, 71, 1952) has measured the relative intensities of the [*N* II] lines and *Hα* in a number of nebulae.

[12] D. H. Menzel, *Ap. J.*, **85**, 330, 1937; see esp. eqs. (4), (9), (14), (25), and (30).

$K = 3.260 \times 10^{-6}$, and the Gaunt factor, g, is equal to 0.822. Here b_4 is a measure of the deviation of the population of the fourth level from the values appropriate to thermodynamic equilibrium at T_ϵ. The actual numerical value of b_4 depends on the electron temperature. We may make an estimate of it from the calculations of the Balmer decrement by Baker and Menzel.[13] Observation shows that their "case B" appears to be followed for most planetary nebulae, so we shall take b_4 from their tables as a function of temperature. If explicit expressions[12] for the $F_{n\kappa}$'s, $F_{\kappa n}$'s, etc., are now substituted in equation (1), use is made of equations (2) and (3), and the common factor $N_i N_\epsilon 2hRZ^6 T_\epsilon^{-3/2}$ is canceled from both sides of the equation, we obtain

$$b_1 e^{X_1}\left(\frac{k}{2hRZ^2}\right) T_1 \int_{y_1}^{\infty} W_{1\kappa} \frac{dy}{e^y - 1} = \left(\frac{k}{2hRZ^2}\right)^2 T_\epsilon^2 + \left(\frac{k}{2hRZ^2}\right) aT_\epsilon + \tfrac{1}{2}G_{T_\epsilon} - \frac{1}{2}\sum_1^\infty \frac{S_n}{n^5} + \sum_i E(H\beta)\frac{I_i}{I(H\beta)}. \tag{5}$$

The factor b_1 is eliminated with the aid of the equation[14]

$$b_1 e^{X_1} \int_{y_1}^{\infty} W_{1\kappa} \frac{dy}{y(e^y - 1)} = \sum_1^\infty \frac{S_n}{n^3} = G_{T_\epsilon}. \tag{6}$$

Here $W_{1\kappa}$ is the dilution factor of the radiation, while G_{T_ϵ} may be calculated as a function of the electron temperature. We then obtain[15]

$$\frac{(k/2hRZ^2)T_1 \int_{y_1}^{\infty} [W_{1\kappa} dy/(e^y - 1)]}{\int_{y_1}^{\infty} W_{1\kappa}[dy/y(e^y - 1)]} = \frac{(k/2hRZ^2)^2 T_\epsilon^2 + a(k/2hRZ^2)T_\epsilon + \frac{1}{2}G_{T_\epsilon} - \frac{1}{2}\sum_1^\infty \frac{S_n}{n^5}}{G_{T_\epsilon}} + 1.61 \times 10^{-3} \frac{b_4(T_\epsilon)}{G_{T_\epsilon}} e^{X_4} \sum_i \frac{I_i}{I(H\beta)}. \tag{7}$$

The left-hand side of the equation depends only on the temperature T_1 of the central star and the dilution factor $W_{1\kappa}$, which varies so slowly with frequency that it may be canceled out in most instances of practical interest. The right-hand side of the equation depends only on T_ϵ. The last term differs from that given previously[4] by Menzel and the writer, in that observed quantities are introduced directly. Equation (7) may be written symbolically as [14]

$$f_1(T_1) = f_2(T_\epsilon) + C(T_\epsilon). \tag{8}$$

The correction term $C(T_\epsilon)$ may be written as

$$C(T_\epsilon) = A(T_\epsilon)\Sigma \frac{I_i}{I(H\beta)}, \tag{9}$$

[13] *Ap. J.*, **86**, 70, 1937; **88**, 52, 1938.

[14] J. G. Baker, D. H. Menzel, and L. H. Aller, *Ap. J.*, **88**, 423, 1938.

[15] Compare with *ibid.*, eq. (5).

where the coefficient $A(T_\epsilon)$ may be tabulated as shown in Table 1. Notice that the function $A(T_\epsilon)$ varies slowly with the temperature.

For a given value of T_1, T_ϵ may be estimated by calculating f_1 as a function of T and calculating f_2 as a function of T, and reading from the resultant tabulation the value of T_ϵ that makes both sides of the equation equal. For purposes of the present illustration, however, I shall assume a series of values of T_ϵ, calculate T_1 with the aid of the

TABLE 1

VALUES OF $A(T_\epsilon)$

T_ϵ	$A(T_\epsilon)$	T_ϵ	$A(T_\epsilon)$	T_ϵ	$A(T_\epsilon)$
8000	5.54×10^{-3}	14,000	5.00×10^{-3}	20,000	4.68×10^{-3}
10,000	5.33	16,000	4.86	25,000	4.49
12,000	5.14	18,000	4.76	30,000	4.36

TABLE 2

THEORETICAL RELATION BETWEEN ELECTRON TEMPERATURES AND CENTRAL-STAR TEMPERATURES FOR PLANETARY NEBULAE

	IC 418	NGC 2149	IC 4593	NGC 6210	NGC 6543	NGC 6826	NGC 7662
Assumed electron temperature (° K):							
8000	17,000	22,000	26,000	36,000	26,000	31,000	62,000
10,000	19,000	24,000	28,000	38,000	28,000	33,000	65,000
12,000	22,000	26,000	30,000	40,000	30,000	35,000	65,000
14,000	24,000	28,000	32,000	42,000	32,000	37,000	70,000
16,000	26,000	30,000	34,000	44,000	35,000	40,000	72,000
18,000	28,000	33,000	36,000	46,000	38,000	42,000	74,000
20,000	31,000	36,000	39,000	48,000	40,000	44,000	77,000
25,000	37,000	41,000	44,000	56,000	46,000	52,000	86,000
30,000	46,000	52,000	55,000	68,000	57,000	62,000	98,000
$\Sigma I_i/I(H\beta)$	3.5	7.0	8.7	13.6	9.1	11.5	24
T_1	33,200	32,500	33,400	32,900	35,000*	34,600	55,000*
Suggested T_ϵ	22,000	18,000	16,000	8000	17,000	11,000	....

* Temperatures obtained by Zanstra method.

observed quantity $\Sigma[I_i/I(H\beta)]$ and equation (7), and compare the resulting series of values of T_1 with various estimates of the central-star temperature.

The body of Table 2 gives the central-star temperatures for a number of planetary nebulae as evaluated from equation (7) for a series of values of the electron temperature T_ϵ. The row labeled $\Sigma[I_i/I(H\beta)]$ gives the sum of the intensities of the forbidden lines in terms of that of $H\beta$. The values of the central-star temperature, T_1, are taken from their spectral classes,[16] except that the entries marked with an asterisk (*) refer to Zanstra temperatures.[17] The row labeled "Suggested T_ϵ" gives a value of T_ϵ consistent with the theoretical T_1-T_ϵ relationship.

Notice that there is a considerable range in the suggested electron temperatures, from

[16] *Ap. J.*, **108**, 462, 1948; temperatures for the nuclei of IC 418 and NGC 2149 were obtained by O. C. Wilson and the writer, *Ap. J.* (in press).

[17] See L. Berman, *Lick Obs., Bull.*, **18**, 73, 1937.

8000° to 25,000° K. The nebulae with relatively weak forbidden lines would tend to have the highest electron temperatures, and there would not necessarily be any close connection between the central-star temperature and the electron temperature. The state of ionization in the nebula which determines the number and kinds of ions present would play a big role in fixing T_ϵ. Clearly the preceding calculation does not permit the determination of the electron temperature. All it shows is that the observed intensities of the forbidden lines are consistent with electron temperature between 8000° and 25,000° K for accepted values of the central-star temperatures. In particular, there is no difficulty in reconciling the cooling mechanism with an electron temperature in IC 418 of the order of 22,000° K or even higher. We cannot give more precise estimates because we do not know the ultraviolet energy distribution in the spectra of the central stars. Calculations by A. Underhill,[18] by Jean McDonald,[19] by William Liller,[20] by James Milligan and the writer, and by others indicate that for stars in this temperature range, the energy distribution beyond the Lyman limit may correspond to a color temperature very different from the effective temperature or the Zanstra temperature for the same star. Qualitatively, the effect is that the flux, F_ν, is cut down far below that of a black body of the same effective temperature but that the blocking of the outgoing radiation is greatest near the Lyman limit and falls off to higher frequencies. Hence, F_ν may fall off more slowly with increasing ν than it would for a black body of the same effective temperature, so that a greater proportion of high-energy electrons would be photo-ejected. The effect is as though the central star had a much higher temperature.

Preliminary calculations show that the theoretical T_1-T_ϵ relation is extremely sensitive to the exact energy distribution in the far ultraviolet. For example, the *H/He* ratio enters in a fundamental way. Comparison of model atmospheres of pure hydrogen with atmospheres in which an *H/He* ratio of 5 is introduced show that the effect of the *He* I absorption limit at λ 504 is very marked. The depletion of high-frequency quanta appreciably lowers the theoretical electron temperature.

Furthermore, the energy distribution in the Lyman continuum is modified in an optically thick nebula. Some calculations carried out long ago by Baker, Menzel, and the writer[21] indicated that the theoretical electron temperature would tend to increase near the outer boundary of the ionized zone. The radiation field throughout the main luminous body of the planetary, however, would correspond to that appropriate to small optical depths, in accordance with Strömgren's theory of the ionized hydrogen zone.[22] At these small optical depths the older calculations indicated that W could be canceled on the left-hand side of equation (7) without much loss in accuracy. At the present time a more elaborate study would not appear to be worth while.

The electron temperatures of many nebulae implied by these considerations may be compared with those obtained with the aid of new target areas recently calculated by M. J. Seaton.[23] In Table 3, I list the adopted intensities of $N_1 + N_2$ and λ 4363 and of T_ϵ as computed from Seaton's cross-sections and Garstang's A-values.[24] The nebular-line intensities are approximately corrected for space absorption by assuming that $I(H\gamma)/I(H\beta) = 0.55$, as implied by measures of the Balmer decrement.[25] For IC 418, the intensity of λ 4363 with respect to $H\gamma$ is taken from the work of Wyse. The intensities

[18] *Pub. Copenhagen Obs.*, No. 151, 1950; *Pub. Dom. Ap. Obs. Victoria*, **8**, 357, 1951.

[19] Unpublished thesis, University of Michigan, 1952.

[20] Unpublished thesis, University of Michigan, 1953.

[21] *Ap. J.*, **90**, 271, 601, 1939.

[22] *Ap. J.*, **87**, 526, 1939.

[23] *Proc. R. Soc., London, A*, **218**, 400, 1953. I am indebted to Dr. Seaton for communicating his results to me in advance of publication.

[24] *M.N.*, **111**, 115, 1951.

[25] See L. H. Aller, *Ap. J.*, **113**, 138, 1951.

of the green nebular lines are taken from photometric measures made in 1938–1939[26] and 1943–1945[25] at the Lick Observatory. For IC 4593, NGC 6210, NGC 6543, and NGC 6826, the intensity of λ 4363 with respect to $H\gamma$ was measured on slit spectrograms secured at the McDonald Observatory.[27] The electron temperature is very sensitive to small errors in the $(N_1 + N_2)/4363$ ratio unless this quantity is very large. Unfortunately, the great strength of the green nebular lines in most planetaries renders very difficult the accurate measurement of this ratio by photographic photometry. Photoelectric measures of the $(N_1 + N_2)/H\beta$ ratio in a number of planetaries are planned by William Liller and the writer as a check on the photographic work.

A comparison of the last rows of Tables 2 and 3 is of some interest. Usually the theoretical temperature is higher than the "nebular-line" temperature, the one covering a range from 8000° K to 22,000° K, the other a range from 6700° K to about 20,000° K. Some of the differences might be eliminated by an adjustment of T_1 or perhaps by improved photometric measures, or by a more accurate theoretical treatment of the ultraviolet radiation field in the nebula.

Andrillat[28] has published electron temperatures based on the older cross-sections for a

TABLE 3

ELECTRON TEMPERATURES OBTAINED FROM NEBULAR LINE INTENSITIES

	IC 418	NGC 2149	IC 4593	NGC 6210	NGC 6543	NGC 6826	NGC 7662
$I(N_1+N_2)$	15.7	55	75	116	76	100	198
$I(4363)$	0.33	0.60	0.12	0.8	0.32	0.55	2.7
T_ϵ	19,500	14,000	7600	12,000	10,000	10,700	15,700

number of planetaries. Seaton points out that these temperatures should be greatly increased on the basis of the new cross-sections. If the nebula has pronounced density concentrations, the $I(4363)/I(N_1 + N_2)$ ratio will be increased, as appears to be true for IC 4997.

Unless hydrogen is almost completely ionized, it would appear unlikely that the electron temperatures can be much higher than 25,000° K or 30,000° K. Otherwise, collisional excitation of the hydrogen lines would be important, and the Balmer decrement should be markedly steepened. This effect has been considered by S. Miyamoto[29] and more recently by J. W. Chamberlain.[30] The fact that the observed Balmer decrement appears to agree with the purely radiative theory of the hydrogen-line formation suggests that collisional excitations of the Balmer lines cannot be significant in most planetaries and therefore that the electron temperature cannot be too high.

As an example, we may consider a nebula such as IC 418, for which we may adopt the following tentative values: $T_1 = 30{,}000°$ K, $R = 3R_\odot$, $N_\epsilon \sim 15 \times 10^3$ electrons/cm^3, radius of shell $= 1.6 \times 10^{17}$ cm. If $T_\epsilon = 20{,}000°$ K, we find with the aid of Chamberlain's tables that the Balmer decrement will not be affected by an observable amount. On the other hand, for an electron temperature as high as 30,000° K, deviations from the purely radiative Balmer decrement can become important. At a temperature well

[26] *Ibid.*, **93**, 236, 1941.

[27] *Ap. J.*, **108**, 462, 1948. The plates were obtained primarily for the planetary nuclei but may be used for nebular lines also.

[28] *C.R.*, **231**, 1432, 1940; **234**, 62, 1952. The intensities actually measured are not given in these brief reports.

[29] *Mem. Coll. Sci., Kyoto Imperial U.*, **A21**, No. 26, 173, 1938.

[30] *Ap. J.*, **117**, 387, 1953.

below that at which the Balmer decrement would be observably altered from the radiative model, the cooling effect of collisional excitation of the hydrogen levels could become important. Suppose that only 2 per cent of the atoms in the third level of hydrogen were excited there by electron collisions. Then about 2 per cent of the intensity of $H\alpha$ is to be attributed to collisional excitations. On the other hand, about eight times as much energy is radiated in Lyman β as in $H\alpha$, so that the amount of energy actually dissipated in the excitation of $H\alpha$ is closer to 18 per cent of the intensity of $H\alpha$ than it is to 2 per cent. As soon as collisionally excited atoms become numerous, the b-factors expressing the deviation from thermodynamic equilibrium increase, and the decrement approaches that of "case C" considered by Menzel, Baker, and the writer,[31] as Chamberlain has pointed out. Detailed consideration of the influence of the hydrogen-line excitation on the conservation-of-energy equation will be reserved for a later paper.

Within the luminous shells of certain planetaries, where the emission per unit volume in the hydrogen and forbidden lines may become very small, the electron temperature

TABLE 4

THEORETICAL ELECTRON TEMPERATURE DISTRIBUTION IN IC 418

Distance from Center (Seconds of Arc)	$\Sigma E_i/E(H\beta)$	T_ϵ (° K)	Distance from Center (Seconds of Arc)	$\Sigma E_i/E(H\beta)$	T_ϵ (° K)
0.0	5.6	21,000	4.82	7.0	19,000
0.96	5.7	21,000	5.78	3.8	23,000
1.93	5.6	21,000	6.75	4.3	22,000
2.90	5.8	21,000	7.72	5.9	20,000
3.86	7.3	18,000			

may approach the theoretical values given in Paper V of the Harvard nebular series.[31] Such high electron temperatures would require so complete an ionization of hydrogen that the collisional excitation of the higher levels would be unimportant.

One important consequence of a relation of the type of equation (7) is that, since the relative emissions in the forbidden and in $H\beta$ (per unit volume) vary from point to point in the nebula, the electron temperature will also vary. We may illustrate the nature of the effect by an application to IC 418, where O. C. Wilson and the writer attempted to analyze the structure of the nebula to determine the relative emission of the various characteristic radiations per unit volume.[32] If the intensity distribution across the monochromatic nebular images is analyzed on the assumption that the nebula is spherically symmetrical, it is found that hydrogen is concentrated in the ring but that there is also a rise in density toward the center, as is shown also by the [*O* III] and [*Ne* III] images. On the other hand, the [*O* II] and [*N* II] images indicate essentially hollow shells. The published hydrogen and O^{++} distributions are substantiated, at least qualitatively, by more recent observations.

Table 4 gives the emission in the forbidden lines expressed in terms of the emission in $H\beta$ as a function of distance in seconds of arc from the central star, and finally the theoretical electron temperature, T_ϵ. In the central portions of the nebula, T_ϵ is controlled mainly by the excitation of [*O* III], whereas O^+ becomes important only in the outer portions. The temperature is predicted to be highest in the bright ring, where the relative emission in the forbidden lines is lowest. We neglect the modification of the radiation field by the nebula itself, but this influence of the continuous absorption is presumed to become important only near the outer edge of the luminous zone.

[31] *Ap. J.*, **88**, 422, 1938; **89**, 587, 1939.

[32] *Ap. J.*, **114**, 421, 1951.

The refinements required by the nongray character of the radiation of the central star and by the excitation of the hydrogen lines will be presented in a later paper, in which application will be made to specific models of nebulae that represent the observed objects as closely as possible.

I am indebted to Dr. R. Minkowski for calling my attention to the observational objections to the lower electron temperatures previously assigned. Dr. M. J. Seaton kindly informed me of his results on the target areas for the collisional excitation of the forbidden lines in advance of publication. Dr. D. H. Menzel and Dr. David Layzer have made interesting suggestions and comments concerning the theoretical target areas.

Reprinted from *The Astrophysical Journal,* Vol. 120, No. 1, July, 1954.

PHOTOELECTRIC SPECTROPHOTOMETRY OF PLANETARY NEBULAE

William Liller and Lawrence H. Aller
The Observatory, University of Michigan, Ann Arbor, Michigan
Received January 22, 1954

ABSTRACT

Photoelectric spectrophotometric measures of the N_1, N_2, and $H\beta$ line intensities in twelve bright planetary nebulae have been made with the aid of the 10° objective prism on the Curtis Schmidt telescope. The absolute surface brightnesses are in good accord with results obtained by the filter photoelectric method. The mean N_1/N_2 intensity ratio is 3.03, while the $(N_1 + N_2)/H\beta$ intensity ratios are in excellent agreement with those found previously by MacRae and Stock and substantiate the photographic photometric measures made some years ago by one of us. The electron densities range from 10^3 to about 2×10^4 electrons/cm^3; they are of the same order of magnitude as those found some years ago by Menzel and one of us. The electron temperatures fall between 8000° and 20,000° K and show no strong correlation with the Zanstra temperature of the central star. This lack of correlation is in accordance with recent theoretical expectations.

I. INTRODUCTION

In the last thirty years the relative intensities of the principal monochromatic radiations of the planetary nebulae have been measured by a number of investigators. Measures or estimates of line intensities have been published by H. H. Plaskett (1929), Louis Berman (1930), Bowen and Wyse (1939; Wyse, 1942), R. Minkowski (1942), M. L. White (1952), and others, including one of the present writers (Aller 1941, 1951). Berman and one of us employed the slitless spectrographs on the Crossley reflector at the Lick Observatory and have given the relative total brightnesses of the monochromatic images. Others, using slit spectrographs, measured the relative intensities referred to some fixed point in the nebula or averaged in some rough fashion over the entire nebula.

These data, combined with estimates of the distances and surface brightnesses of the planetaries, have been utilized in conjunction with theories of physical processes to evaluate the densities, temperatures, and chemical compositions of the nebulae. With the Hebb and Menzel target areas for the collisional excitation of the metastable levels in *O* III, these observations gave electron temperatures in the neighborhood of 8000–10,000° K (Menzel *et al.* 1941). The electron densities appeared to lie in the neighborhood of 10^3–10^4 electrons/cm^3.

More recently Henri Andrillat (1950, 1952) has published, for a number of planetaries, electron temperatures that are in sharp disagreement with those found by others. He has not published the observational data used in his calculations of electron temperatures with the aid of the Hebb and Menzel target areas. His results require intensity ratios of the green nebular lines to 4363 markedly different from those previously obtained. The ratio of the green nebular line intensity to that of λ 4363, $I(N_1 + N_2)/I(\lambda\, 4363)$, can be measured most conveniently by comparing 4363 with $H\gamma$ and $N_1 + N_2$ with $H\beta$. The $I(4363)/I(H\gamma)$ ratio can be measured reasonably well by photographic photometry; the $I(N_1 + N_2)/I(H\beta)$ ratio is much more difficult to establish, because the green nebular lines are often ten to twenty times as strong as $H\beta$ and fall in a spectral region where the sensitivity of the photographic emulsion changes rapidly with wave length.

Hence a check on the photographic spectrophotometry is very much to be desired. The photoelectric cell can reach the stronger lines easily, but the photometry of the fainter lines would be much more difficult. At least for the present, the spectrophotometry of the fainter lines would appear to require the use of the photographic plate. Accordingly, we decided to measure the intensity ratios of λ 5007(N_1), λ 4959(N_2), and $H\beta$ in as many nebulae as could be reached with our equipment in the late-summer observing season. Section II of this paper describes the observational procedures employed; Section III gives the methods of reduction, the results, and the comparison with those of other observers; Section IV treats of the intensity ratio of the green nebular lines; while in Section V we discuss the electron densities and temperatures obtained for the planetary nebulae. Photoelectric measures of the $I(N_1+N_2)/I(H\beta)$ ratio have been made previously by Donald MacRae and Jurgens Stock (1954). Their results are in excellent agreement with our own, as will be seen in Section III of this paper.

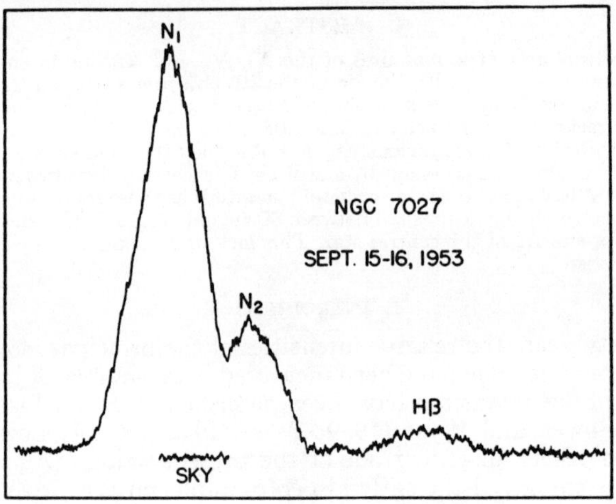

FIG. 1.—Photoelectric tracing of the spectrum of a planetary nebula. NGC 7027 represents a typical high-excitation object.

II. THE OBSERVATIONAL PROCEDURE

The 4° and 6° objective prisms of the Curtis Schmidt telescope, when used in their 10° combination, give a dispersion of 160 A/mm at $H\beta$. Because of the relatively short focal length of the telescope, the pair of [*O* III] images in the green will be completely separated only for planetaries with diameters less than 29″; the oxygen image at λ 4959 will overlap $H\beta$ only if the nebula is more than 62″ in diameter. Thus it is possible to measure photoelectrically the integrated intensities of the N_1, N_2, and $H\beta$ images of the majority of planetary nebulae with this equipment.

On the nights of September 13, 14, 15, and 16, 1953, we observed a total of thirteen planetaries by two different methods. The first consisted of securing spectral traces of all the nebulae by orienting the dispersion in the east-west direction and by driving the telescope in hour angle at a speed slightly different from the sidereal rate. Then the monochromatic nebular images slowly passed over a slit 0.17 mm wide, corresponding to 27 A at $H\beta$. Frequency-control equipment, designed by staff members of the McMath-Hulbert Observatory primarily for the widening of photographic spectra, provided the telescope motion in hour angle. This method is similar to that employed by MacRae and Stock (1954).

In addition, we measured the spectra of all but two of the planetary nebulae by taking individual deflections of the monochromatic nebular images and of the nebular and central star continua to the violet of $H\beta$ and to the red of the λ 5007 image of [*O* III]. Because the faintness of the nebulae often made it difficult to center the images accurately in the diaphragm, we made no attempt to measure the N_1 and N_2 images separately in this fashion. The electronic equipment used in this investigation is identical with that described in a separate paper by one of us (Liller 1954). It was necessary to alter the optical design of the photometer, however, to accommodate the $f/3.5$ beam of the Curtis Schmidt.

Tracings of the spectra of α Lyrae and α Cygni secured at different zenith distances enable the determination of the atmospheric extinction and of the spectral response of the photometer plus the optical system of the telescope. In these reductions we have adopted the energy distributions published by R. C. Williams (1939) and by D. Barbier and D. Chalonge (1940).

Reproduced in Figure 1 is the spectral trace of the high-excitation planetary, NGC 7027. It is a rather small object; hence the separation of the N_1 and N_2 images is clear. The Ring Nebula in Lyra, NGC 6720, was the only nebula for which the diameter (70″) was too large to permit the measurement of the relative intensities of the green [O III] pair.

III. THE REDUCTION OF THE OBSERVATIONS

To convert the raw deflections of the N_1, N_2, and $H\beta$ lines into relative intensities, we took into account the effects of the spectral response of the photometer, the atmospheric

TABLE 1

OBSERVED DATA ON PLANETARY NEBULAE

Nebula (1)	α(1950) (2)	δ(1950) (3)	$I(N_1+N_2)/I(H\beta)$ (4)	n (5)	$I(N_1)/I(N_2)$ (6)	A'' (7)	log (S_1+S_2) (8)
NGC 6210	16ʰ42ᵐ5	+23°9	15.1	2	2.95	4″	+0.55
NGC 6543	17 58.6	+66.6	8.86	2	3.04	10	− .05
NGC 6572	18 09.6	+ 6.8	15.6	2	3.11	7	+ .38
NGC 6803	19 29	+10.0	....	....	....	2.5	− .16
NGC 6818	19 40.0	−14.3	11.6	1	3.00	9	− .35
NGC 6826	19 43.4	+50.4	9.32	3	3.18	13	− .57
NGC 7009	21 01.4	−11.6	16.1	3	2.72	9	+ .22
NGC 7027	21 05.2	+42.0	20.2	4	3.00	5	+ .36
NGC 7662	23 23.5	+42.3	18.2	4	3.02	8	+ .15
NGC 1535	4 11.7	−12.9	15.4	4	3.00	9	− .45
IC 2149	5 43.5	+46.1	5.4	3	2.90	6	− .68
IC 418	5 25.2	−12.7	1.78	2	3.40	6	−0.18

extinction, and the contribution made to the deflections by the continuous spectra of the nebulae and the central stars. As mentioned above, the first two corrections are easily derived from the tracings of Vega and Deneb. The intensity of the stellar and nebular continua required separate measures when the diaphragms were used. This background continuum can be measured directly from the slit tracings.

The diaphragm deflections and slit tracings gave $I(N_1 + N_2)/I(H\beta)$ ratios that were in good agreement. We regarded the slit tracings as more reliable than the diaphragm deflections and assigned them greater weight. The final mean values are listed in column 4 of Table 1, and column 5 lists the number of observations secured on each object. The average probable error of a single result is ± 2.3 per cent. The $H\beta$ image of NGC 6803 was too faint to measure; about all that can be said is that it is several times fainter than N_2.

In Figure 2 we compare the values of $I(N_1 + N_2)/I(H\beta)$ derived in this investigation with the values of this ratio found photoelectrically by MacRae and Stock (1954) (*circles*) and photographically by one of the present writers (Aller 1941, 1951) (*crosses*).

The two photoelectric investigations are in close agreement, reflecting the accuracy of the photoelectric measures. Also in good agreement, but showing a somewhat larger scatter, are the photographic results. There does not appear to be any systematic error in the latter results, only the random scatter that would naturally arise from the uncertainties in photographic photometry. If we assume that the photoelectric values are the true ones, we find that the mean error of a single photographic result is ± 18 per cent. NGC 1535 and 6818 show the largest errors. Both photographic determinations were based on one observation, and if these two nebulae are omitted, the mean error of a single photographic result drops to ± 12 per cent.

From the slit tracings we also determined the quantity $I(N_1)/I(N_2)$. For no planetary were the N_1 and N_2 lines completely separated, and it was necessary to construct "profiles" of at least one of the two lines, since it could not be safely assumed that the nebulae were spherically symmetric. Each profile was constructed by assuming the N_1 and N_2 images to be identical in shape and intensity distribution and by making the redward wing of the N_2 line geometrically similar to the redward wing of the N_1 line.

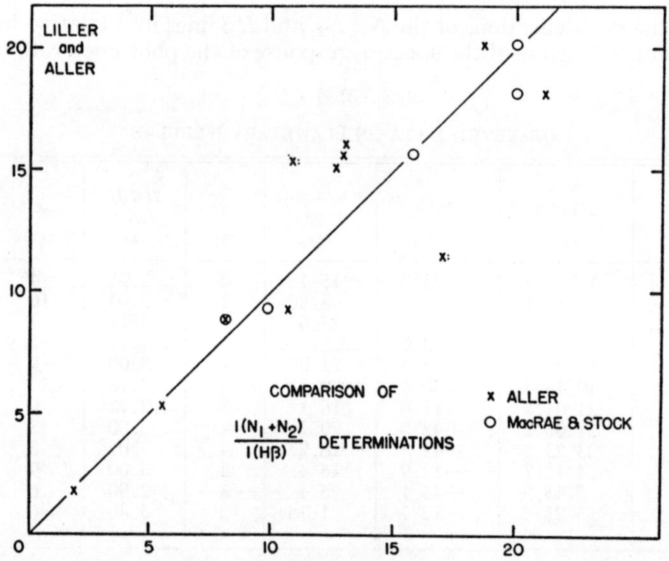

FIG. 2.—A comparison of the values of $I(N_1 + N_2)/I(H\beta)$ found by several observers. The open circles show the photoelectric results of MacRae and Stock; the crosses indicate the photographic determinations of one of us. Colons following two of the crosses mean that the photographic value is based on a single observation.

A subtraction of the area underneath the N_2 profile from the total area underneath the $N_1 + N_2$ tracing then gave the size of the N_1 deflection. As a further check, the spectra of a number of the nebulae were measured by constructing profiles in the above manner for both the [*O* III] lines.

The $I(N_1)/I(N_2)$ ratios are listed in column 6 of Table 1. NGC 1535 was observed twice; the spectra of the remaining planetaries were traced only once with a slit. As a result, it is difficult to tell whether the variation in this ratio from nebula to nebula is real or whether it is due to observational error. The largest source of error seems to arise from the uncertainty in drawing in the background continuum, and we are inclined to believe that the fluctuations are due to these errors.

Finally, by comparing the nebular tracings with the spectral tracings of Vega and Deneb, we can derive monochromatic surface brightnesses for the nebulae. Ambarzumian (1933) has described a method of deriving these surface brightnesses by comparing the intensities of the nebulae with the spectrum of a solar-type star. Therefore, we should like to find the relative intensities of the sun and Vega at the wave lengths of the green nebular lines. If we take the photovisual magnitude of Vega to be $+0.03$, as given by Johnson and Morgan (1953) and that of the sun to be -26.87, the mean of the values published by Kuiper (1938) and by De Vaucouleurs (1949), then the difference in the photovisual magnitudes (26.90) should correspond closely to the ratio of intensities of the two objects at the effective wave length of the International Photovisual System. The mean of the effective wave lengths is 5433 A, according to Seares and Joyner (1943). If we assume that both Vega and the sun radiate as black bodies over the limited wave-length range from 4860 to 5440, we should be able to calculate the ratio of intensities at the mean wave length of the N_1 and N_2 lines, λ 4995. Here we took the color temperatures of the two objects in this wave-length region to be 16,500° and 5700°. The resulting difference in magnitude at λ 4995 is 26.70, which is equivalent to an intensity ratio of 4.79×10^{10}. Hence we are able to calculate what the deflection would be from 1 A of the solar spectrum at the mean wave length of the green forbidden oxygen lines.

In calculating the mean surface brightnesses, we treated each nebula as an evenly illuminated disk of radius A. The values of the radii are those given by Curtis (1918) and are listed in column 7 of Table 1. The last column of this table lists the resulting surface fluxes of the planetary nebulae in the light of the green [*O* III] radiations. The units are the logarithms of the *fluxes* expressed in ergs/cm^2/sec. Because of the uncertainties arising from the use of the effective wave lengths of the International Photovisual System, these values should not be expected to be of as high precision as those given elsewhere by one of us. However, these results give us an excellent independent check of the surface brightnesses derived photoelectrically with the use of narrow-band interference filters. The two sets of determinations have already been compared in the other paper, and we feel that the surface fluxes derived here should be given half the weight of the other values.

IV. THE INTENSITY RATIO OF THE GREEN NEBULAR LINES

The smaller planetaries should give the best determination of the N_1/N_2 ratio, but we have found no dependence of the value upon the size of the nebula. The mean ratio from 12 planetaries is 3.03 ± 0.11 (mean error). The chief error seems to arise from the uncertainty in drawing the background continuum.

Half a century ago, Wilsing and Scheiner (1902) found this ratio to be 2.5; many years later, H. H. Plaskett published a ratio of 2.73 (1929), while L. Berman found 2.63 (1930). Minkowski found a ratio of 2.99 ± 0.15 for the Orion nebula (1934).

Theoretical calculations of the $I(^1D_2-{}^3P_2)/I(^1D_2-{}^3P_1)$ ratio by Pasternack (1940) and independently by Shortley and his associates (1941) were based on the theory of intermediate coupling. They yielded an intensity ratio of 2.83. This theory, which assumes that the perturbation from the LS-coupling situation arises from the interaction of an electron's spin with its own orbit, gives results that contradict the observed intensity ratios in the $^4S-{}^2D$ transition of [*O* II]. The discrepancy was removed by later investigators, who took into account the effects of (spin) — (spin) interactions and of (spin of one electron) — (orbit of another) interactions (Aller *et al.* 1949). R. H. Garstang (1951) has applied this improved theory to the calculation of the energy levels and transition probabilities in the p^2 and p^4 configurations of *C* I, *N* II, *O* III, *F* IV, and *Ne* V and of *O* I, *F* II, and *Ne* III, respectively. He predicts an intensity ratio of 2.93. Thus our observational data favor Garstang's result, but the scatter is still comparable with the differences between the two theoretical predictions.

V. THE ELECTRON DENSITIES AND TEMPERATURES IN THE PLANETARY NEBULAE

Let S_{Bac} denote the flux in ergs/cm²/sec radiated by the nebula in a 20 A interval of the Balmer continuum near its limit (3630–3650 A). If the nebula can be approximated as a shell of thickness d and outer radius A, we define

$$D \equiv 3d\left[1 - \frac{d}{A} + \frac{d^2}{3A^2}\right]. \tag{1}$$

Further, if $N_i = N_\epsilon$, the electron density may be calculated from (Menzel and Aller 1941)

$$N_\epsilon = 1.68 \times 10^{10} S_{Bac}^{1/2} D^{-1/2} T_\epsilon^{3/4}. \tag{2}$$

If observations of S_{Bac} are not available or are not of sufficient accuracy, an estimate of N_ϵ may be found from $H\beta$. The flux in $H\beta$ in ergs/cm²/sec is related to the electron density by

$$S_\beta = 7.37 \times 10^{-20} \frac{N_i N_\epsilon}{T_\epsilon^{3/2}} b_4(T_\epsilon) D e^{X_n}. \tag{3}$$

Here $b_4(T_\epsilon)$ is a factor that measures the departures of the population in the fourth level from the value appropriate to thermodynamic equilibrium at the temperature T_ϵ (Menzel 1937, eq. [4]). Extensive measures of the Balmer lines in a large number of planetaries (Aller 1951) indicate that when corrections are made for the effects of interstellar absorption, Menzel and Baker's "model B" (1938) fits the observed Balmer decrement. In this model the central star radiates no energy in the Lyman lines, and each ultraviolet Lyman quantum absorbed in the nebula is ultimately degraded into a quantum of Lyman-α. The agreement between theory and observation justifies our use of the theoretical b_n's given by Menzel and Baker.

The electron temperature may be computed from the relative intensities of the 4363 and green nebular lines of [O III], provided that the target areas for collisional excitation are known. Earlier determinations of the electron temperature were based on Hebb and Menzel's calculations (1940). More recently, M. J. Seaton (1953) has calculated target areas for O^{++} and other ions by an improved theory. Using his results, we may derive rather easily the electron temperature in terms of the observed ratio of $I(N_1 + N_2)/I(4363)$.

The intensity ratio $I(\lambda\, 4363)/I(N_1 + N_2)$ is related to the populations of the $^1S_0$ and $^1D_2$ levels by

$$\frac{I(4363)}{I(N_1 + N_2)} = \frac{N(^1S_0)\, A(^1S, {}^1D)\, h\nu(^1S, {}^1D)}{N(^1D_2)\, A(^1D, {}^3P)\, h\nu(^1D, {}^3P)}, \tag{4}$$

where A denotes the Einstein coefficient of spontaneous emission and ν the frequency.

The ratio $N(^1S_0)/N(^1D_2)$ depends on the electron density and temperature, the target-area parameters, Ω_{PS}, Ω_{PD}, and Ω_{DS}, and the Einstein A coefficients. Menzel, Aller, and Hebb (1941) give the necessary relationship as equation (9) of their paper. It is necessary to correct the last term in the denominator of their equation for a typographical error by replacing A_{PD} by Ω_{PD}, however. If we use

$$\Omega_{PS} = 0.195, \quad \Omega_{PD} = 1.73, \quad \Omega_{DS} = 0.61 \text{ (Seaton 1953)},$$

$$\left.\begin{aligned} A(^1S, {}^1D) &= 1.6 & A(^1D, {}^3P_2) &= 0.021 \\ A(^1S, {}^3P) &= 0.23 & A(^1D, {}^3P_1) &= 0.0071 \end{aligned}\right\} \text{(Garstang 1951)},$$

we obtain the following expression:

$$\frac{I(4363)}{I(N_1+N_2)} = 13.1\left\{\frac{1.012+2300\sqrt{T_\epsilon}/N_\epsilon}{1.01+2.70\times 10^5\sqrt{T_\epsilon}/N_\epsilon}\right\}10^{-14,300/T_\epsilon} \qquad (5)$$

$$\sim 0.113 \times 10^{-14,300/T_\epsilon}$$

for such low densities as prevail in planetary nebulae.

The basic data needed for the calculations of the electron densities and temperatures are given in Table 2. For each nebula are given the average angular radius, A'', and the relative thickness of the shell, d/A, according to the data of Curtis (1918), the distance in parsecs from the work of Berman (1937), the quantity D, the logarithm of the mean flux in ergs/cm²/sec for the green nebular lines corresponding to the assumed angular size of the nebula, and in column 7 the ratio of the intensities of the green nebular lines and $H\beta$. All the data of column 6 are obtained from the photoelectric work, but for those nebulae not included in Table 1 we have used the photographic determinations of

TABLE 2*

BASIC DATA ON PLANETARY NEBULAE

Nebula (1)	A'' (2)	d/A (3)	r (Parsecs) (4)	$D\times 10^{-17}$ (5)	log (S_1+S_2) (6)	log $[I(N_1+N_2)/I(H\beta)]$ (7)	log (S_{Bac}/S_β) (8)	$I(4363)$ $[H\beta=10]$ (9)	T_ϵ (° K) (10)
NGC 1535....	9″	0.3:	1720	1.52	−0.45	1.19	−0.80 (1)	1.75	14,300
							−0.72 (2)		
IC 2149....	6	1.0	2250	2.02	−0.68	0.74	−1.04 (3)	0.60 (2)	14,200
IC 418....	6	0.2	1800	0.79	−0.18	0.25	−1.14 (1)	0.33	18,300
							−0.82 (4)		
NGC 2392....	23	0.2	860	1.44	−1.31	1.16	−1.62: (3)	3.0	19,600
NGC 3242....	10	0.2	1640	1.20	+0.10	1.29	−1.10 (2)	2.1 (3)	17,700
IC 4593....	5	1.0	2000	1.50	−0.44	0.88	−1.14 (2)	0.12 (5)	7,700
NGC 6210....	4	1.0	1720	1.03	+0.56	1.18	−0.75 (4)	0.8 (5)	10,800
NGC 6543....	10	0.2	1080	0.79	0.00	0.95	−1.07 (1)	0.32 (5)	9,500
NGC 6572....	7	1.0	1230	1.29	+0.34	1.19	−1.17 (1)	1.5	13,500
NGC 6803....	2.5	1.0	2640	0.98	−0.16	1.16	−0.6: (2)	1.07 (2)	12,800
NGC 6818....	9	0.3	2380	2.10	−0.35	1.06	−1.00 (2)	1.6	15,700
NGC 6826....	13	1.0	1050	2.04	−0.55	0.97	−1.12 (1)	0.55 (5)	11,200
NGC 7009....	9	0.3	930	0.82	+0.15	1.21	−1.06 (1)	1.35	13,900
NGC 7027....	5	1.0	2130	1.59	+0.40	1.30	−1.57 (1)	3.15	16,800
NGC 7662....	8	0.2	1200	0.70	+0.08	1.26	−1.21 (1)	2.25	15,000

* The numbers in parentheses in cols. 8 and 9 refer to the source of the numerical data as follows:

1. D. H. Menzel and L. H. Aller, *Ap. J.*, **93**, 195, 1941.
2. L. H. Aller, *Ap. J.*, **113**, 125, 1951.
3. Heretofore unpublished values obtained at the Lick Observatory, 1943–1945, by one of us.
4. Slit-spectra observations obtained at the McDonald Observatory in 1945 by one of us.
5. L. H. Aller, *Ap. J.*, **118**, 547, 1953.

The $I(4363)$ values that are not followed by a number in parentheses were adopted from all data available to us, including isophotic contour measures of 4363 and 4340 on Mount Wilson data, and Minkowski and Aller's spectrophotometry of selected planetaries (*Ap. J.*, in press).

$I(N_1+N_2)/I(H\beta)$ made by one of us (Aller 1941, 1951). Column 8 gives log (S_{Bac}/S_β), the ratio of the flux in the 20 A interval of the Balmer continuum to the flux in $H\beta$, as determined by photographic photometry. Detailed references are given to the sources of the data. Most of the ratios are based on plates obtained with the quartz slitless spectrograph attached to the Crossley reflector at the Lick Observatory, although slit spectrograms obtained at the McDonald Observatory have also been used. The procedure then was to compare the intensity in the Balmer continuum with near-by Balmer lines whose relative intensities are assumed to follow Menzel and Baker's model B. Since the position of the underlying continuum cannot be fixed with very great accuracy, either from slit or slitless plates, it is very difficult to determine S_{Bac}. Accordingly, we have estimated N_ϵ both from S_{Bac} and from the photoelectric determinations of S_β. The chief uncertainty in the calculation of N_ϵ then arises from the error in the distance of the planetary. Column 9 gives the intensity of λ 4363 [*O* III], referred to $H\beta$ as 10. These values are taken from slitless spectrograms obtained at the Lick Observatory and from slit spectrograms secured at the McDonald Observatory by one of us. Column 10 gives the value of the electron temperatures as computed with the aid of Seaton's cross-sections.

With the aid of these electron temperatures, S_{Bac}'s, S_β's, and assumed nebular distances, we now compute N_ϵ by equations (2) and (3). The results are given in Table 3.

TABLE 3

CALCULATION OF ELECTRON DENSITIES IN PLANETARY NEBULAE

NEBULA (1)	T_ϵ (° K) (2)	LOG b_4 (3)	LOG N_ϵ $H\beta$ (4)	LOG N_ϵ *Bac* (5)	LOG N_ϵ Adopted (6)	*Ap. J.*, **102**, 239, 1945 log N_ϵ (7)	*Ap. J.*, **102**, 239, 1945 T_ϵ (° K) (8)
NGC 1535	14,300	−0.55	3.40	3.55	3.44	3.98	11,600
IC 2149	14,200	−0.56	3.45	3.45	3.45	....	....
IC 418	18,300	−0.43	4.20	4.19	4.20	4.04	9,700
NGC 2392	19,600	−0.40	3.05	2.82	3.04	....	....
NGC 3242	17,700	−0.44	3.72	3.72	3.72	3.60	9,500
IC 4593	7,700	−1.02	3.46	3.33	3.46	3.63	7,000
NGC 6210	10,800	−0.72	3.93	4.05	3.94	....	....
NGC 6543	9,500	−0.82	3.80	3.74	3.80	4.07	6,500
NGC 6572	13,500	−0.58	4.01	3.75	4.01	4.02	9,000
NGC 6803	12,500	−0.62	3.63	3.62	3.62	....	....
NGC 6818	15,700	−0.50	3.46	3.50	3.47	3.50	9,400
NGC 6826	11,200	−0.70	3.35	3.30	3.34	3.65	8,200
NGC 7009	13,900	−0.57	3.81	3.80	3.80	4.31	8,900
NGC 7027	16,800	−0.46	3.69	3.46	3.69	3.77	9,600
NGC 7662	15,000	−0.52	3.81	3.75	3.80	4.12	10,000

Column 2 again gives the electron temperature for each nebula listed in column 1. Column 3 gives the corresponding log b_4. The values of log N_ϵ found from S_β with the aid of equation (3) are given in column 4, while column 5 gives the values of log N_ϵ deduced with the aid of log S_{Bac} by means of equation (2). Since the S_{Bac} values are subject to rather large uncertainties, the finally adopted values of log N_ϵ are based primarily upon the $H\beta$ measurements. For comparison we give for these same nebulae the T_ϵ and log N_ϵ values derived from the older data.

The new electron temperatures are all higher than those previously assigned, partly because of the new cross-sections adopted and partly because no attempt had been made to correct the older observations of $I(4363)/I(N_1 + N_2)$ for space absorption. The new electron densities are of the same order of magnitude as those published in 1945. The effect of the lowering of the adopted surface brightness is compensated for by the increase in T_ϵ.

A comparison of the log N_ϵ values obtained from $H\beta$ and from the Balmer continuum (see Fig. 3) shows no systematic differences between the two sets of results. We conclude that the $H_{Bac}/H\beta$ ratio is in satisfactory agreement with the theory of the Balmer decrement as given by Menzel and Baker. This opinion differs from that of T. L. Page (1942), who found that the intensity of the Balmer continuum as compared with that of $H\delta$ was less than the value predicted by the theory of a pure recombination spectrum.

In order to compare Andrillat's electron temperatures with our results, we must compute T_ϵ from our data with the aid of the Hebb-Menzel target areas. The results are given in Table 4. If Seaton's cross-sections are used, Andrillat's electron temperatures would be much higher. Since the $I(4363)/I(H\beta)$ ratio can be measured photographically more easily than the $I(N_1 + N_2)/I(H\beta)$ ratio, one suspects that the discrepancy may lie in the latter ratio.

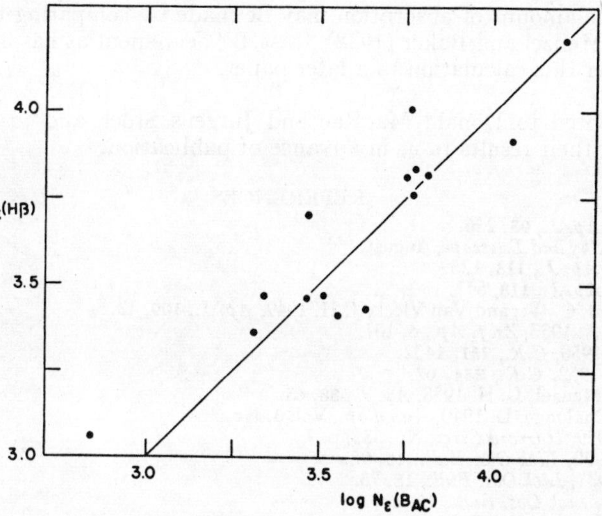

FIG. 3.—A comparison of the logarithms of the electron densities derived from the intensity measurements of $H\beta$ and of a 20 A interval of the Balmer continuum.

TABLE 4

ELECTRON TEMPERATURES DERIVED FROM HEBB AND MENZEL CROSS-SECTIONS

Nebula	T_ϵ (Present Data) (° K)	T_ϵ (Andrillat)	Nebula	T_ϵ (Present Data) (° K)	T_ϵ (Andrillat)
NGC 6543..........	7,850	9,500	NGC 7009..........	10,600	15,000
NGC 6572..........	10,000	15,000	NGC 7027..........	12,300	21,500
NGC 6803..........	9,800	16,500	NGC 7662..........	11,200	25,500

We find no strong correlation between the degree of excitation of the nebula and its electron temperature, although high-excitation nebulae tend, on the average, to have higher electron temperatures than do low-excitation nebulae. For example, IC 418, one of the low-excitation planetaries, almost certainly has a higher electron temperature than does NGC 7009. This result is in accordance with theoretical expectations (Aller 1953).

The mean values of S_β or $S_{N_1+N_2}$ employed in this paper may be replaced by more accurate values when use is made of the isophotic contours of the corresponding monochromatic images of these nebulae. Many years ago Berman derived such contours (1930) for a number of planetaries. Recently, Olin C. Wilson has secured slitless spectrograms with the 100-inch coudé spectrograph, and R. Minkowski has obtained direct photographs with the 200-inch reflector that permit much more satisfactory isophotic contours to be derived. Many of these plates now are being analyzed with the isophotometer at the University of Michigan. Thus the intensity of a single line can ultimately be given in terms of an isophotic contour diagram, each closed curve of which corresponds to a known intensity expressed in ergs/cm^2/sec/unit solid angle. Finally, the surface brightnesses we have tabulated will also have to be corrected for space absorption. For some nebulae, such corrections may turn out to be large. In NGC 7027, for example, interstellar absorption appears to cut down the surface brightness by a factor of about 10! Estimates of the amount of absorption may be made by comparing the Balmer decrement with the Menzel and Baker (1938) "case B" decrement as has been done by Aller (1951). We defer this calculation to a later paper.

We are indebted to Donald MacRae and Jurgens Stock and to M. J. Seaton for communicating their results to us in advance of publication.

REFERENCES

Aller, L. H. 1941, *Ap. J.*, **93**, 236.
Aller, L. H. 1942, *Sky and Telescope*, August.
Aller, L. H. 1951, *Ap. J.*, **113**, 125.
Aller, L. H. 1953, *Ap. J.*, **118**, 547.
Aller, L. H.; Ufford, C. W.; and Van Vleck, J. H. 1949, *Ap. J.*, **109**, 42.
Ambarzumian, V. A. 1933, *Zs. f. Ap.*, **6**, 107.
Andrillat, Henri. 1950, *C.R.*, **231**, 1432.
Andrillat, Henri. 1952, *C.R.*, **234**, 62.
Baker, J. G., and Menzel, D. H. 1938, *Ap. J.*, **88**, 63.
Barbier, D., and Chalonge, D. 1940, *Ann d'ap.*, Vol. **3**, No. 2.
Berman, Louis. 1929, *Harvard Circ.*, No. 335, p. 7.
Berman, Louis. 1930, *Lick Obs. Bull.*, **15**, 98.
Berman, Louis, 1937, *Lick Obs. Bull.*, **18**, 73.
Bowen, I. S. 1939, *Lick Obs. Bull.*, **19**, 1.
Curtis, H. D. 1918, *Lick Obs. Pub.*, **13**, 55.
Garstang, R. 1951, *M.N.*, **111**, 115.
Hebb, M. H., and Menzel, D. H. 1940, *Ap. J.*, **92**, 408.
Johnson, H. L., and Morgan, W. W. 1953, *Ap. J.*, **117**, 313.
Kuiper, G. P. 1938, *Ap. J.*, **88**, 429.
Liller, William. 1954, *Ap. J.* (in press).
MacRae, D., and Stock, J. 1954, *Nature*, **173**, 589.
Menzel, D. H. 1937, *Ap. J.*, **85**, 332.
Menzel, D. H., and Aller, L. H. 1941, *Ap. J.*, **93**, 197; see especially eq. (10), which should be corrected for the error in K, whose numerical value is equal to 3.26×10^{-6}.
Menzel, D. H.; Aller, L. H.; and Hebb, M. H. 1941, *Ap. J.*, **93**, 240.
Minkowski, R. 1930, *Lick Obs. Bull.*, **15**, 98.
Minkowski, R. 1934, *Zs. f. Ap.*, **9**, 202.
Minkowski, R. 1942, *Ap. J.*, **95**, 243.
Page, T. L. 1942, *Ap. J.*, **96**, 78.
Pasternack, S. 1940, *Ap. J.*, **92**, 129.
Plaskett, H. H. 1929, *Harvard Circ.*, No. 335, p. 7.
Seares, Frederick, and Joyner, Mary. 1943, *Ap. J.*, **98**, 312.
Seaton, M. J. 1953, *Proc. R. Soc. London, A*, **218**, 400.
Shortley, G. H.; Aller, L. H.; Baker, J. G.; and Menzel, D. H. 1941, *Ap. J.*, **93**, 178.
Vaucouleurs, G. de. 1949, *Ann. d'ap.*, **12**, 287.
White, M. L. 1952, *Ap. J.*, **115**, 71.
Williams, R. C. 1939, *Pub. Obs. U. Michigan*, **7**, 93.
Wilsing, J., and Scheiner, J. 1902, *Ap. J.*, **16**, 234.
Wyse, A. B. 1942, *Ap. J.*, **95**, 356.

Reprinted from *The Astrophysical Journal*, Vol. 120, No. 3, November, 1954.

THE COMPOSITION OF THE PLANETARY NEBULA NGC 7027

LAWRENCE H. ALLER
University of Michigan Observatory
Received April 21, 1954

ABSTRACT

New spectrophotometric data on the bright, high-excitation planetary NGC 7027 are discussed with the aid of improved transition probabilities and Seaton's improved target areas for forbidden-line excitation. The ionic abundances are computed, and the probable filamentary structure of the nebula is discussed. The electron temperature is estimated to be near 15,000° K, and the density to be near 7×10^4 ions/cm^3 in the filaments. The chief difficulty in the chemical analysis lies in allowing for the distribution of atoms of a given element among different stages of ionization. The abundances are comparable with those found in the stars, a result in harmony with the conclusion reached by Bowen and Wyse in 1939.

I. INTRODUCTION

It has long been recognized that the unusual character of the spectra of the gaseous nebulae arose from the nature of the excitation conditions prevailing in these objects. Hence any attempt to ascertain their quantitative chemical composition requires careful attention to the excitation mechanism responsible for each line utilized. The determination of the composition of a gaseous nebula from the spectral-line intensities involves essentially the following steps:

1. The relative intensities of the relevant nebular lines are measured spectrophotometrically.
2. The transition probabilities and target areas for collisional excitation are utilized to deduce the electron temperature T_ϵ and density N_ϵ of the electron gas from the intensities of the forbidden lines of selected ions. The electron density can also be found from the intensities of the Balmer continuum and the Balmer lines, if the latter can be measured in flux units, i.e., ergs per square centimeter per second, and if the distance of the nebula is known. Some assumption about the uniformity of density within the nebula is necessary.
3. Once values of N_ϵ and T_ϵ are adopted for the emitting regions of the nebula, the relative line intensities may be used to determine the relative abundances of the corresponding ions.
4. The last step is to estimate what fraction of the atoms of a given element exist in the observed stages of ionization. This calculation requires that the ionization equilibrium existing in the nebula be ascertained, and it turns out to be perhaps the most difficult and uncertain step in the whole procedure.

The necessary quantitative data on the line intensities are available for only a very few nebulae. The relative intensities of the stronger lines have been measured in a large number of planetaries, e.g., Berman (1930), Minkowski (1942), and Aller (1941, 1951). The intensities of the weaker lines in a few planetaries were estimated by Bowen and Wyse (1939) and by Wyse (1942). Wyse made eye-estimates with the aid of a scale plate, but his line intensities appear to have been remarkably accurate, considering the roughness of the methods employed. Recently, spectrophotometric measurements have been made for a few bright planetaries, including NGC 7027 (Bowen, Minkowski, and Aller 1955), NGC 7662, and NGC 2392 (Minkowski and Aller 1955); the observational difficulties inherent in the measurement of the intensities of lines embracing more than a thousand fold range in brightness will be discussed in these communications. Fortunately the bright end of the intensity scale, specifically the brightness ratio of the green nebular lines $N_1 + N_2$ to $H\beta$, has been checked photoelectrically (MacRae and Stock 1954; Liller and Aller 1954). The intensities of the weakest lines are inevitably affected by large accidental errors. Accurate Einstein A-values for the hydrogen lines are known from the work of Menzel and Pekeris (1935).

In so far as the recombination lines of helium are concerned, the A-values are known with probably adequate accuracy (Goldberg 1939; Huang 1948). Sufficiently good A-values have also been obtained for the forbidden lines involving equivalent p-electrons (Pasternack 1940; Shortley, Aller, Baker, and Menzel 1941; Ufford, Van Vleck, and Aller 1949; Ufford and Gilmour 1950; Garstang 1951, 1952; Naqvi 1951; Osterbrock 1951).

The forbidden lines represent downward transitions from metastable levels belonging to the ground configurations of various ions. Unfortunately, the target areas for the collisional excitation of these metastable levels are difficult to compute. Calculations for *O* I were carried out by Yamanouchi, Inui, and Amemiya (1940); for *O* III by Hebb and Menzel (1940); for *O* II by Aller (1950), and for *N* II by White and Aller (1949). Bates, Fundaminsky, Massey, and Leech (1950) showed that the theory upon which these calculations were based was inexact and that the computed target areas were probably much too high. Layzer (1951) proposed a new approximate method for calculating target areas for collisional excitation, but no numerical computations have been published. Seaton (1953) has also given an improved theory and has carried out actual calculations of target areas for a number of ions of astrophysical interest. It is not certain that even these target areas are sufficiently accurate for a definitive study of nebular composition.

Finally, the state of ionization in the planetary nebulae cannot be represented by the approximation of an assembly of atoms illuminated by dilute, high-temperature, black-body radiation. The apportionment of atoms among various ionization stages must be estimated empirically (Bowen and Wyse 1939; Wyse 1942; Aller and Menzel 1945; Menzel 1951; Aller 1953*a*). This step seems to be particularly uncertain for objects like NGC 7027, which show a huge range of excitation.

Page (1936) discussed the abundances of some of the elements represented by the strongest forbidden lines in the planetaries, specifically, *N*, *O*, *A*, *Ne*, and *S*. He used the ionization formula for a black body of 50,000° K and a dilution factor obtained from an assumed ratio of O^+ to O^{++} of 5×10^{-3} derived from Berman's data. In view of the approximations involved, little emphasis could be placed on the resulting differences between stellar and nebular abundances. The first definitive attack on the problem was that made by Bowen and Wyse (1939) and Wyse (1942). They utilized their observations of both bright and faint lines and were able to establish the presence of a number of ions that had not been observed previously in the spectra of the planetaries. They concluded that the spectra of the nebulae could be explained on the assumption that these objects had the same composition as the sun. Wyse's observations were re-discussed by Aller and Menzel (1945) from the point of view of the theory developed in the Harvard Observatory series entitled "Physical Processes in Gaseous Nebulae" (1937–1945). The necessary cross-sections for the collisional excitation of the metastable levels were estimated from those obtained for *O* III by Hebb and Menzel (1940). The nebulae were regarded as homogeneous shells or spheres in which the electron temperature and density were constant. Later, Menzel (1951) made an attempt to correct the abundance estimates for the effects of the filamentary structure of the nebulae. He assumed that in all planetaries the ratio of volume occupied by the gaseous filaments to the total volume occupied by the gross structure of the nebula was the same as in the giant planetary in Aquarius, NGC 7293. In this object Menzel estimated that about one three-hundredth of the volume of the nebula was occupied by matter, the rest of the space being essentially empty. An examination of the best direct photographs and slitless spectrograms convinces me that one cannot assume that *all* planetary nebulae possess the same fine structure. For example, NGC 3587, the "Owl Nebula," shows no evidence of any small-scale structure at all, even though this object is probably comparable in size and density with NGC 7293 (Minkowski and Aller 1954)!

Menzel may have overcorrected for the influence of filamentary structure. Furthermore, the target areas employed in these studies appear to have been much too large. Recently semiempirical target areas were employed in a reanalysis of Wyse's data for NGC 7027 (Aller 1953*a*), but the results still could not be considered satisfactory. The appearance of Seaton's improved cross-sections (1953) and the new photometric data make a re-examination of the problem worth while.

II. THE OBSERVATIONAL DATA

Of the planetaries observed in our spectrophotometric study, I chose NGC 7027 as the first object for a detailed analysis because of its exceptionally rich bright-line spectrum. The line intensities were measured at two points in NGC 7027, denoted as *a* and *b*. My quantitative discussion is based on point *a*, which is the brightest patch in the nebula.

The observed Balmer decrement in NGC 7027 is very steep, indicating either unusual conditions of excitation or a pronounced space absorption. The latter interpretation seems the more likely. A discussion of the Balmer decrement in a large number of

TABLE 1

BALMER DECREMENT IN NGC 7027 AND ITS CORRECTION FOR SPACE ABSORPTION

λ (A) (1)	n (2)	I_{th} (3)	I_{obs} (4)	I_c (5)	λ (A) (1)	n (2)	I_{th} (3)	I_{obs} (4)	I_c (5)
6563	3	25.5	65	25.7	3835	9	0.99	0.46	0.91
4861	4	10.0	10	10.0	3797	10	.74	.33	.66
4340	5	5.05	3.26	4.6	3770	11	.57	.33	.67
4101	6	3.05	1.63	2.70	3750	12	.46	.27	.56
3969*	7	1.99	.....	.....	3734	13	0.35	0.23	0.48
3889†	8	1.37	0.88	1.66					

* Blend with [*Ne* III], which is much stronger than *H*7 and not separated from it with our dispersion.
† Blend with *He* I λ 3888.65, which is probably weaker than *H*8.

planetary nebulae (Aller 1951) shows that the observations, after correction for space absorption, agree most closely with Menzel and Baker's (1938) case "B," i.e., an optically thick nebula in which all the Lyman-line quanta are broken down into Lyman-α and the usual Balmer quanta, etc.

Thus, if we compare the observed Balmer decrement with the theoretical Balmer decrement for case B and $T_\epsilon = 14{,}000°$ K, and employ the wave-length law of extinction suggested by Whitford (1948), we find the results given in Table 1. Column 3 gives the theoretical Balmer decrement for case B as interpolated from the data of Menzel and Baker (1938). This theoretical decrement is not sensitive to the electron temperature. Column 4 gives the observed Balmer decrement, and column 5 the observed Balmer decrement after correction for space absorption. The required differential correction for space absorption amounts to 2.75 mag. between λ 4220 and λ 10,300! That is, Whitford's k_λ's are to be multiplied by 2.75, and the observed relative intensities by antilog 1.10 k_λ, to correct them for selective space absorption. The amount of the correction can be stated in other terms by noting that the total extinction at $H\beta$ amounts to a factor of 11.

A comparison of the corrected and theoretical values of the Balmer line intensities shows that the higher members ($n \geq 12$) appear to have been measured systematically too strong. Probably the position of the underlying continuum was incorrectly placed. Other lines of the series show deviations of predominantly the opposite sense. Perhaps

I have underestimated the absorption correction. In any event, it appears safe to say that NGC 7027 is about ten times fainter than it would be, could we but observe it without the interference of the intervening interstellar medium.

III. ESTIMATION OF IONIC ABUNDANCES FROM THEIR FORBIDDEN LINES

Let us consider, first of all, the analysis of the forbidden-line spectrum of NGC 7027. Let the subscript 1 denote the lowest term of the ground p^n configuration of a given ion. The middle term is denoted by 2, and the highest term by 3. For example, in a p^2 configuration the ^3P term is denoted as "1," the $^1D_2$ term as "2," and the $^1S_0$ term as "3." We denote the statistical weights of the three terms by ϖ_1, ϖ_2, and ϖ_3.

Atoms will enter term 2 from term 1 by inelastic collisions with electrons and from term 3 by superelastic collisions and by auroral transitions. They escape by superelastic collisions and by radiative transitions of the nebular type. Atoms enter term 3 by inelastic collisions from terms 1 and 2 and escape by superelastic collisions and radiative transitions of the auroral and transauroral types. The corresponding equations of statistical equilibrium (see Menzel, Aller, and Hebb 1941, eqs. [7] and [8]) are

$$N_1 \frac{\Omega_{12}}{\varpi_1} C \exp\left(-\frac{\chi_2}{kT_\epsilon}\right)+\frac{N_3 C\Omega_{23}}{\varpi_3}+N_3 A_{23} = N_2 \frac{\Omega_{23}}{\varpi_2} C \exp\left(-\frac{\chi_{23}}{kT_\epsilon}\right)+\frac{N_2 C\Omega_{21}}{\varpi_2}+N_2 A_{21} \quad (1)$$

and

$$N_1 \frac{\Omega_{13}}{\varpi_1} C \exp\left(-\frac{\chi_3}{kT_\epsilon}\right)+N_2 \frac{\Omega_{23}}{\varpi_2} C \exp\left(-\frac{\chi_{23}}{kT_\epsilon}\right)=\frac{N_3 C\,(\Omega_{23}+\Omega_{13})}{\varpi_3}+N_3\,(A_{23}+A_{13})\,, \quad (2)$$

where Ω is the target-area parameter employed by Hebb and Menzel (1940) and by Seaton (1953), while.

$$C \equiv \frac{8.54\times 10^{-6}}{h_\epsilon}; \qquad h_\epsilon = \frac{\sqrt{T_\epsilon}}{N_\epsilon}; \quad (3)$$

and

$$\chi_{23} = \chi_3 - \chi_2 \quad (4)$$

denotes the difference in excitation potential between the second and third levels. From equations (1) and (2) we find the relative populations of the second and third levels to be given by:

$$\frac{N_3}{N_2}=\frac{\varpi_3}{\varpi_2}\exp\left[-\frac{\chi_{23}}{kT_\epsilon}\right]\left\{1+\frac{\Omega_{23}\exp\left(-\chi_{23}/kT_\epsilon\right)}{\Omega_{12}\left(1+\Omega_{23}/\Omega_{13}\right)}+\frac{A_{21}\varpi_2}{C\Omega_{12}\left(1+\Omega_{23}/\Omega_{13}\right)}\right\}$$
$$\times\left\{1+\frac{(A_{32}+A_{31})\,\varpi_3}{\Omega_{13}C\left(1+\Omega_{23}/\Omega_{13}\right)}+\frac{(\Omega_{23}+A_{32}\varpi_3/C)\exp\left(-\chi_{23}/kT_\epsilon\right)}{\Omega_{12}\left(1+\Omega_{23}/\Omega_{13}\right)}\right\}^{-1}, \quad (5)$$

which may be written as

$$\frac{N_3}{N_2}=\frac{b_3}{b_2}\frac{\varpi_3}{\varpi_2}\exp\left(-\frac{\chi_{23}}{kT_\epsilon}\right), \quad (6)$$

to exhibit the factor b_3/b_2 that measures the deviation from thermodynamic equilibrium. As the density increases, $b_3/b_2 \to 1$. Similarly, the ratio N_2/N_1 may be written as

$$\frac{N_2}{N_1}=\frac{\varpi_2}{\varpi_1}\exp\left(-\frac{\chi_2}{kT_\epsilon}\right)\frac{\{\Omega_{12}+\Omega_{13}d\exp\left(-\chi_{23}/kT_\epsilon\right)\}}{\{\Omega_{12}+A_{21}\varpi_2/C+\Omega_{23}\,(1-d)\exp\left(-\chi_{23}/kT_\epsilon\right)\}}, \quad (7)$$

where

$$d=\frac{\Omega_{23}+A_{23}\varpi_3/C}{\Omega_{13}+\Omega_{23}+(A_{32}+A_{31})\,\varpi_3/C}. \quad (8)$$

In analogy with equation (6), we may now write

$$\frac{N_2}{N_1}=\frac{b_2}{b_1}\frac{\varpi_2}{\varpi_1}\exp\left(-\frac{\chi_2}{kT_\epsilon}\right). \tag{9}$$

Finally,

$$\frac{N_3}{N_1}=\frac{b_3}{b_1}\frac{\varpi_3}{\varpi_1}\exp\left(-\frac{\chi_3}{kT_\epsilon}\right). \tag{10}$$

Here

$$\frac{b_3}{b_1}=\frac{\{\Omega_{13}+\Omega_{12}\,[p+\exp\left(-\chi_{23}/kT_\epsilon\right)]^{-1}\}}{\{(\Omega_{23}+\Omega_{13})+(A_{32}+A_{31})\,\varpi_3/C-(\Omega_{23}+A_{23}\varpi_3/C)\,[1+p\,\exp(\chi_{23}/kT_\epsilon)]^{-1}\}}, \tag{11}$$

where p is defined by

$$p\equiv\frac{\Omega_{12}}{\Omega_{23}}+\frac{A_{21}\varpi_2}{C\Omega_{23}}. \tag{12}$$

The intensity ratio I_a/I_n of the auroral to nebular lines of [O III] have been employed to estimate T_ϵ (Menzel, Aller, and Hebb 1941). Now

$$\frac{I_a}{I_n}=\frac{N_3A_{3,2}\nu_{3,2}}{N_2A_{2,1}\nu_{2,1}}=\frac{b_3}{b_2}\frac{\varpi_3}{\varpi_2}\exp\left(-\frac{\chi_{23}}{kT_\epsilon}\right)\frac{A_{3,2}\nu_{3,2}}{A_{2,1}\nu_{2,1}}. \tag{13}$$

Equations (11), (12), and (13) may be combined to yield an expression exhibiting the dependence of I_a/I_n upon N_ϵ and T_ϵ with the A's and Ω's entering as parameters. At the temperatures and densities prevailing in the gaseous nebulae, an approximate form of equation (13) may be obtained that is sufficiently good for most calculations. Menzel, Aller, and Hebb adopted values of N_ϵ from the surface brightnesses of the nebulae in $H\beta$ or the nebular continuum (Menzel and Aller 1941), Pasternack's A-values, the Hebb-Menzel target areas, and computed T_ϵ from the observed $I(4363)/I(N_1+N_2)$ ratios (Aller 1941).

Some years later the cross-sections for the collisional excitation of the auroral and nebular [N II] lines were determined by the same methods as had been employed for [O III]. If one assumes that the [N II] and [O III] lines arise in the *same* strata, simultaneous transcendental equations of the form (13) involving N_ϵ and T_ϵ must hold (Aller and White 1949). Thus the electron temperature and density in the radiating elements can be found directly. The electron temperatures that were thus obtained did not differ much from those previously found, but the electron densities in the radiating strata turned out to be higher by factors ranging from 1 to 7. The detailed results were not published, since it was shown soon thereafter that the theoretical basis for the target-area calculations was unsound. Very recently, Seaton (1954) has employed similar considerations but with his improved target areas for collisional excitation.

If we employ Seaton's (1953) target areas with A-values from the work of Garstang (1951), equation (13) becomes, for [O III],

$$\frac{I(4363)}{I(N_1+N_2)}=13.1\times10^{-14{,}300/T_\epsilon}\left\{\frac{1.01+2300h_\epsilon}{1.01+270{,}000h_\epsilon}\right\}\sim0.113\times10^{-14{,}300/T_\epsilon} \tag{14}$$

for low densities, while for the [N II] lines, equation (13) becomes

$$\frac{I(5755)}{I(6560)}=61.5\times10^{-10{,}820/T_\epsilon}\left\{\frac{1.01+320h_\epsilon}{1+194{,}000h_\epsilon}\right\}. \tag{15}$$

The observed intensity ratios on the left-hand side of equations (14) and (15) are 0.0155 and 0.056, respectively. The simultaneous solution of these equations then gives $N_\epsilon\sim7.7\times10^4$ electrons/cm^3 and $T_\epsilon\backsim14{,}500^\circ$ K. On the other hand, if we employ only the

[O III] data together with the photoelectrically measured surface brightness (Liller and Aller 1954) corrected for space absorption and the distance derived by Berman (1937), N_ϵ turns out to be only 1.7×10^4 electrons/cm³. The electron temperature is then 15,600° K. If the target areas are trustworthy, these results imply that there must be a considerable concentration of the material in filaments, although it is also possible that the [N II] and [O III] radiation originates predominantly in *different* regions within the nebula. In the calculations carried out in the remainder of this paper we have taken $N_\epsilon = 7.7 \times 10^4$, $T_\epsilon = 14{,}500°$ K, but it must be noted that the adopted temperature is somewhat higher than that implied by the magnitude of the observed Balmer discontinuity, which suggests $T_\epsilon \leq 10{,}000°$ K.

The emission per unit volume in nebular or auroral-type transitions by ions of type i will be

$$E_n = N_2 A_{2,1} h\nu_{2,1} = \eta_2 N_i \frac{\varpi_2}{\varpi_1} h\nu_{2,1} A_{2,1} \exp\left(-\frac{\chi_2}{kT_\epsilon}\right),$$

$$E_a = N_3 A_{3,1} h\nu_{3,1} = \eta_3 N_i \frac{\varpi_3}{\varpi_1} h\nu_{3,1} A_{3,1} \exp\left(-\frac{\chi_3}{kT_\epsilon}\right), \quad (16)$$

where $\eta_2 = b_2/b_1$ and $\eta_3 = b_3/b_1$. The emission in the green nebular lines of [O III] can be written as

$$E_O = N_2^0 A_0 h\nu_0 = \eta_2^0 N_1^0 \frac{\varpi_2}{\varpi_1} h\nu_0 A_0 \exp\left(-\frac{\chi_2^0}{kT_\epsilon}\right). \quad (17)$$

If $I_{n,i}$ is the total intensity of the nebular transitions of the ion of type i and I_0 is the intensity of the two green nebular lines, the relative numbers of ions of type i and O^{++} ions will be

$$\frac{N_i}{N_0} = \frac{I_{ni}}{I_0} \left\{ \frac{5\lambda \varpi_1 A_0 \eta_2^0}{9\lambda_0 \varpi_2 A \eta_2} \right\} 10^{(\chi - \chi_0)\theta}, \quad (18)$$

where $\theta = 5040/T_\epsilon$. The appropriate formulae for the η's may be developed from equations (7) and (11). Thus we obtain the following expressions which are valid for nebular temperatures and densities.

Nebular lines of [N II]:

$$\eta_2 = \frac{\{2.39 + 0.217 \times 10^{-2.15\theta}\}}{\{2.39 + 2340 h_\epsilon\}} \sim (1 + 960 h_\epsilon)^{-1}. \quad (19)$$

Nebular lines of [O III]:

$$\eta_2 = \frac{\{1.73 + 0.171 \times 10^{-2.84\theta}\}}{\{1.73 + 16400 h_\epsilon + 0.085 \times 10^{-2.84\theta}\}} \sim [1 + 9380 h_\epsilon]^{-1}. \quad (20)$$

Auroral/nebular-line ratio in [O II] requires

$$\frac{\eta_3}{\eta_2} = \frac{\{1 + 0.136 \times 10^{-1.69\theta} + 6.38 h_\epsilon\}}{1 + 85200 h_\epsilon + [0.136 + 10770 h_\epsilon]\, 10^{-1.69\theta}}. \quad (21)$$

Nebular lines of [O II]:

$$\eta_2 = \frac{1.44 + 0.167 \times 10^{-1.69\theta}}{1.44 + 90.2 h_\epsilon + 0.444 \times 10^{-1.69\theta}}. \quad (22)$$

The equations for the fluorine and neon ions are relatively simple:

$$\eta_2 = [1 + 35{,}600 h_\epsilon]^{-1} \qquad [F\ \text{II}], \quad (23)$$

$$\eta_2 = [1 + 200{,}000 h_\epsilon]^{-1} \qquad [Ne\ \text{III}]\,, \tag{24}$$

$$\eta_3 = [1 + 363{,}000 h_\epsilon]^{-1} \qquad [Ne\ \text{V}]\,. \tag{25}$$

The large target areas for [S II] result in a relatively complicated expression for η_3, viz.,

$$\eta_3 = \frac{0.383 + 2.02\, a_\epsilon \times 10^{1.19\theta}}{13.1 + 656{,}000 h_\epsilon - [12.7 + 475{,}000 h_\epsilon]\, a_\epsilon}, \tag{26}$$

where

$$a_\epsilon = [1 + (0.159 + 98.5 h_\epsilon)\, 10^{1.19\theta}]^{-1}\,,$$

while for the nebular lines we need the expression

$$\eta_2 = \frac{2.02 + 0.279 \times 10^{-1.19\theta}}{2.02 + 1260 h_\epsilon + 3.43 \times 10^{-1.19\theta}}. \tag{27}$$

Equations (14)–(27) embrace the ions for which Seaton has calculated target-area parameters. The spectrum of NGC 7027 includes many lines of ions for which no target areas are available.

Following a suggestion by Seaton, I have estimated Ω's by comparing the Ω-parameters for S II and O II and assuming that the same ratio holds for the other ions. Thus

$$\frac{\Omega_{1,2}(S\ \text{II})}{\Omega_{1,2}(O\ \text{II})} = 1.40\,,$$

so we assume

$$\Omega_{1,2}(S\ \text{III}) = 1.40\,\Omega_{1,2}(O\ \text{III})\,, \qquad \Omega_{1,2}(Cl\ \text{IV}) = 1.40\,\Omega_{1,2}(Ne\ \text{IV})\,, \text{ etc.}$$

Table 2 gives the estimated cross-section parameters for the ions of the $3p^n$ configuration. Obviously, abundances computed from these assumed Ω's are subject to large uncertainties.

TABLE 2

ASSUMED CROSS-SECTION PARAMETERS $\Omega_{i,j}$ FOR IONS OF $3p^n$ CONFIGURATIONS

Ion	$\Omega_{1,2}$	$\Omega_{1,3}$	$\Omega_{2,3}$
		$3p^2$	
S III	2.42	0.341	4.03
Cl IV	1.69	.301	3.82
A V	1.18	.275	3.5
		$3p^3$	
Cl III	1.40	.387	20.5
A IV	0.95	.410	23.2
		$3p^4$	
A III	1.16	.161	1.78
K IV	0.855	.135	1.98
Ca V	0.755	0.20	1.98

Expressions similar to equations (19)–(27) may be derived for the forbidden lines of ions in the $3p^n$ configuration. We shall not give the detailed formulae here, as they are based on rather rough cross-sections for collisional excitation.

We calculate the relative concentrations of the various ions on the assumption that all are found in the same layers. In nebulae where stratification exists or where the actual filamentary structure can be seen, this condition is usually not fulfilled. No marked stratification appears in NGC 7027, and such filamentary structure as exists lies below the resolution of the best photographs. It is worth remembering that in nebulae where the filaments are observed in [N II] and [O III], for example, the [O III] filaments are often much more diffuse and softer in appearance than the [N II] filaments.

Since the aforementioned calculations give the concentration of ions in terms of those of O^{++}, we must estimate the density of the O^{++} ions. It is assumed that all the material of the nebula is concentrated in filaments with empty space in between. The emission per unit volume per second in $H\beta$ is given, e.g., by (Aller 1953*b*)

$$E(H\beta) = 0.228 \times 10^{-18} N_i N_\epsilon b_4 T_\epsilon^{-3/2} \exp\left(\frac{hRZ^2}{16kT_\epsilon}\right), \tag{28}$$

whereas that in the green nebular lines is given by

$$E(N_1 + N_2) = \tfrac{5}{9} N(O^{++}) \eta_2 A_0 h\nu 10^{-12,500/T_\epsilon}. \tag{29}$$

If we now divide equation (29) by equation (28) and put

$$\frac{E(N_1+N_2)}{E(H\beta)} = \frac{I(N_1+N_2)}{I(H\beta)} = 18.6, \quad N_\epsilon = 7.7 \times 10^4, \quad \text{and} \quad T_\epsilon = 14{,}500^\circ K,$$

we find $N(O^{++}) = 14.3$.

Table 3 gives the results obtained from the forbidden-line data. Column 2 gives, for

TABLE 3

ABUNDANCES OF IONS IN NGC 7027 FORBIDDEN-LINE DATA

Ion (1)	Trans. (2)	λ (3)	A_n (4)	χ (5)	b_i/b_1 (6)	I_c (7)	N_i (8)
N II.	N	6565	0.0040	1.88	0.40	124	0.87
	A	5755	1.08	4.03	.0020	7.1	0.91
O II.	N	3727	0.0577	3.31	.87	29	1.91
	A	7325	0.200	5.00	.0065	22	0.96
O III.	N	5000	0.028	2.48	.064	1850	14.3
F IV	N	4040	0.133	3.12	.010	0.9	0.012
Ne III.	N	3895	0.26	3.19	.0032	124	2.8
Ne IV.	A	4724	0.534	7.38	.00165	9.7	1.3
Ne V.	N	3400	0.52	3.70	.00176	236	6.4
S II.	N	6722	0.00107	1.84	.400	4.7	0.027
S II.	TA	4072	0.257	3.03	.00053	12.0	0.058
S III.	A	6311	3.6	3.30	.00056	6.8	0.57
Cl III.	N	5520	0.0042	3.69	.1265	2.5	0.0134
Cl IV.	N	7840	0.277	1.60	.0066	1.26	0.007
Cl IV.	A	5322	4.2	3.92	.00026	0.73	0.16
A III.	N	7360	0.404	1.68	.0032	22	0.18
A IV.	N	4720	0.0135	2.61	035	17	0.12
A IV.	A	7250	1.75	4.31	.00032	2.8	0.16
A V.	N	6800	0.743	1.85	.0017	6.5	0.06
K IV.	N	6101	1.037	1.95	.00090	1.2	0.014
K VI.	N	6034	1.75	2.044	.00048	≦0.2	0.02
Ca V.	N	5000	2.37	2.22	0.00035	1.6	0.02

each ion, the type of transition observed, nebular (N), auroral (A), or transauroral (TA). Column 3 gives λ, the mean wave length of the forbidden-line multiplet, while column 4 gives the corresponding transition probability. The excitation potential of the upper term of the transition in electron volts is listed in column 5. Column 6 gives the quantity $\eta_2 = b_2/b_1$ (for the nebular transitions) or $\eta_3 = b_3/b_1$ (for auroral and transauroral transitions). In column 7 I_c denotes the intensity of the whole forbidden-line multiplet (corrected for space absorption) and for the contribution of lines that may be masked by blending or otherwise hidden. Column 8 gives N_i, the relative number of ions, expressed approximately as numbers of ions per cubic centimeter.

If our assumptions are fulfilled, the concentration of a given ion should be the same whether it is computed from the auroral, transauroral, or nebular transition. This condition is, of course, fulfilled for the O^{++} and N^{+} ions, as these were employed to fix N_ϵ and T_ϵ. It is not fulfilled for the S^{+} or for the O^{+} ions. A comparison of the intensities of the auroral ($\sim\lambda$ 7319) and nebular ($\sim\lambda$ 3727) transitions of O^{+} indicates that the density in the filaments where these radiations are predominantly produced is substantially lower than 7×10^4 ions/cm³. On the other hand, the [*S* II] radiations appear to be produced in filaments of much higher density. Hence our results can be understood in terms of a nebula consisting of numerous filaments, knots, and tenuous regions such that the density may range from perhaps less than 10^4 ions/cm³ to perhaps something like 2×10^5 ions/cm³. It has long been recognized that the [*S* II] lines tend to be strong in objects of high density, particularly in small bright nebulae, whereas the [*O* II] radiation is found to predominate in the outer, less dense regions of many large planetaries. The probable, but not directly observable, filamentary structure of NGC 7027 thus imposes a fundamental limitation on the accuracy with which the abundances of the ions can be found in this object. No credence can be placed in the discrepancies between the abundances derived from the auroral and from the nebular lines of *Cl* IV or *A* IV, since the cross-sections for the collisional excitation of these lines can only be guessed.

IV. THE RECOMBINATION LINES OF HELIUM

The intensities of the permitted lines of *He* I depend essentially on the concentration of He^{+} ions, while the intensities of λ 4686, λ 3202, and the Pickering lines of *He* II depend on the concentration of He^{++}. Let us consider, first, the line emission from ionized helium.

The total radiation emitted in the $n \rightarrow n'$ transition per cubic centimeter per second by a strictly hydrogen-like ion is (Menzel 1937)

$$E_{nn'} = N_i N_\epsilon K Z^4 T_\epsilon^{-3/2} b_n(T_\epsilon)\, g_{nn'}\, 2hRZ^2 (n'n)^{-3} \exp(X_n)\,. \tag{30}$$

The emission in λ 4686 of He^{+} is found by setting $Z = 2$, $n = 4$, and $n' = 3$. Thus

$$E_{4686} = 3.92 \times 10^{-18} N(He^{++})\, N_\epsilon T_\epsilon^{-3/2} b_4 \frac{T_\epsilon}{4} \exp(X_4)\,, \tag{31}$$

since the solution for the b_n's for hydrogen at a temperature T_ϵ is the same as the solution for ionized helium at the same value of

$$X_n = \frac{hRZ^2}{n^2 kT_\epsilon} = \frac{hR}{n^2 k(T_\epsilon/Z^2)}\,. \tag{32}$$

Thus the values of b_n for ionized helium at a temperature of 16,000° K are the same as those for hydrogen with $T_\epsilon = 4000°$ K.

The intensity ratio $I_{4686}/I_{H\beta}$ will be related to the concentration ratio of He^{++} and H^{+} ions by

$$\frac{I_{4686}}{I_{H\beta}} = 17.2 \frac{N(He^{++})\, b_4(He^{+})}{n(H^{+})\, b_4(H)} \exp\left(\frac{3hRZ^2}{16kT_\epsilon}\right). \tag{33}$$

If the observed intensity ratio $I(4686)/I(H\beta) = 0.50$ be employed in this equation with the estimate $b_4(He^{+}) = 0.016$ for $T_\epsilon = 14{,}500°$ K, the ratio of doubly ionized helium atoms to ionized hydrogen atoms is found to be 0.066. This figure is sensitive to T_ϵ. Similar calculations can be made for the Pickering lines and other lines of the *He* II spectrum, but all involve uncertain b_n-factors.

Turning, now, to the lines of neutral helium, we find that the number of helium atoms in level j will be given by the combined Boltzmann and Saha equations as

$$N_j = b_j N(He^{+})\, N_\epsilon \left(\frac{h^2}{2\pi m kT_\epsilon}\right)^{3/2} \frac{\varpi_j}{4} \exp\left(\frac{I_0 - \chi_j}{kT_\epsilon}\right). \tag{34}$$

Here I_0 denotes the ionization potential of the *He* I atom, whereas χ_j and ϖ_j are the excitation potential and statistical weight, respectively, of the upper level of the observed transition. The ratio of the number of atoms in the jth level of *He* I to the number in the nth level of hydrogen will be

$$\frac{N_j}{N_n}=\frac{\varpi_j b_j N(He^+)}{4n^2 b_n N(H^+)} \exp\left(\frac{I_j}{kT_\epsilon}-X_n\right), \tag{35}$$

where

$$I_j = I_0 - \chi_j . \tag{36}$$

If the helium and hydrogen recombination lines originate to the same degree in the same layers,

$$\frac{I_{jj'}}{I_{H\beta}}=\frac{E_{jj'}}{E_{H\beta}}. \tag{37}$$

Hence

$$\frac{N_j}{N_4}=\frac{I_{jj'}A_{H\beta}\lambda_{jj'}}{I_{H\beta}A_{jj'}\lambda_{H\beta}}. \tag{38}$$

Equations (35) and (38) now yield

$$\frac{N(He^+)}{N(H^+)}=\frac{64 b_4 I_{jj'}A_{H\beta}\lambda_{jj'}}{\varpi_j b_j I_{H\beta}A_{jj'}\lambda_{H\beta}} \exp\left(X_n-\frac{I_j}{kT_\epsilon}\right). \tag{39}$$

The necessary A-values have been computed by Goldberg (1938), who has also calculated the b_j-factors for idealized helium atoms exposed to the dilute temperature-radiation field of a planetary nebula. Since he treated all the high singlet terms as though they constituted a single level, one value of b_j is obtained for all. In reality, b_j must certainly approach 1 as j increases without bound.

Table 4 gives the results obtained from the three best observed lines of the triplet

TABLE 4

ABUNDANCE OF He^+ IONS CALCULATED FROM TRIPLET TRANSITIONS

λ	Transition	$A\times10^{-6}$	$I(H\beta=1)$	$N(He^+)/N(H^+)$
5876	$2\,^3P-3\,^3D$	64.7	0.144	0.028
4471	$2\,^3P-4\,^3D$	25.9	.050	.035
4026	$2\,^3P-5\,^3D$	12.8	0.035	0.056

series. I have used $b_j(He\,\text{I}) = 0.50$ for all the high triplet levels. The weighted mean value is taken as 0.035. The increase in the computed values with increasing j is to be expected.

The final *H*/*He* ratio turns out to be about 10, in probably fortuitous agreement with earlier results for the average of several planetaries (Aller and Menzel 1945). Wurm (1953), however, concludes that the ratio must lie between 6 and 2. If Wurm is right, the *H*/*He* ratio may be very different for the planetaries than it is for the stars. He regards the estimation of abundances from the neutral helium lines as unreliable and bases his abundance estimates exclusively on the *He* II λ 4686 line in high-excitation planetaries. The determination of an accurate value of the *H*/*He* ratio is complicated by the filamentary structure of most planetary nebulae and by the fact that the temperature may vary from one point in the nebula to another. The ionized helium lines may actually be formed in regions of higher electron temperature than are the hydrogen lines.

V. DETERMINATION OF THE ABUNDANCES

The determination of the true relative abundances of the elements which are represented by the forbidden lines of ions in one or two stages of ionization is an extremely difficult task. We must devise some means of guessing the contributions of the unobserved ions. Since the usual ionization formula for dilute temperature radiation cannot represent the observed distribution of the ions in nebulae such as NGC 7027, an empirical approach must be used. We have employed the method used in the Harvard nebular series (Aller and Menzel 1945), which consists in deriving an empirical ionization-curve from the ratios of the numbers of ions of a given element in successive stages of ionization. Such a procedure necessarily introduces a number of serious uncertainties. The ionic abundances themselves are uncertain because of the accidental observational errors, inaccuracies in the target areas, and the complications that may arise from a possible concentration of different ions in different filaments. It is also assumed that the degree of ionization depends only on the ionization potential. The influence of possible strong, bright lines in the far ultraviolet is neglected. In an object like NGC 7027, it is conceivable that such lines might build up to considerable intensity. The factors for reducing the observed number of ions to the total number of atoms of a given element become very large for the metals, whereas they are much smaller for neon. Table 5 compares the abundances (by numbers of atoms) of elements observed in NGC 7027 with the abundances found in a standard star (Aller 1953*a*). In this compilation I have taken the stellar *H*/*He* ratio as about 19, however. The agreement for *H*, *He*, *N*, and *O* is good, considering the uncertainties in the analysis. Fluorine was observed in the nebulae by Wyse (1942) but not by Minkowski and me; I have tentatively adopted the Wyse intensity in my estimates. The stellar data for fluorine are likewise unsatisfactory. Neon would appear to be less plentiful, and sulfur more so, than in the stars. It is probable that the effects of the filamentary structure are operative here. The [*S* II] emission is greatly favored in dense filaments. The data for chlorine, argon, potassium, and calcium are based on extrapolated target areas; the derived abundances must be uncertain not only from this cause but even more so from the insecure allowance for distribution among unobserved ionization stages.

TABLE 5

COMPOSITION OF NGC 7027 COMPARED WITH THAT OF A "NORMAL" STAR

(Logarithm of Relative Numbers of Atoms)

Element	NGC 7027	Star	Element	NGC 7027	Star
H	13.04	13.27	*S*	9.12	8.52
He	12.0	12.0	*Cl*	7.93	8.3:
N	9.65	9.49	*A*	8.05	9.0:
O	10.00	10.00	*K*	7.1	6.39
F	6.8:	7.7:	*Ca*	7.2	7.72
Ne	9.38	10.04			

The chief virtue of NGC 7027 for a composition study comes from the large number of lines observed in its spectrum, so that the estimate of the abundance of any given element is often based on ions of several different ionization stages. The chief disadvantage of this object lies in its possible filamentary structure and the circumstance that the ionization cannot be regarded as a simple phenomenon of dilute temperature radiation. A discussion of other nebulae, such as NGC 7662, NGC 6572, and NGC 1535, is limited by the less complete information on the lines. In a later paper we shall attempt an analysis of NGC 7662, taking the structural features into account.

I am indebted to Dr. R. Minkowski for suggestions and comments, and I should like to express my gratitude to Dr. M. J. Seaton, who communicated his target-area parameters to me in advance of publication.

REFERENCES

Aller, L. H. 1941, *Ap. J.*, **93**, 236.
———. 1950, *ibid.*, **111**, 609.
———. 1951, *ibid.*, **113**, 125.
———. 1953*a*, *The Atmospheres of the Sun and Stars* (New York: Ronald Press Co.), pp. 327, 329.
———. 1953*b*, *Ap. J.*, **118**, 547.
Aller, L. H., and Menzel, D. H. 1945, *Ap. J.*, **102**, 239.
Aller, L. H., Ufford, C. W., and Van Vleck, J. H. 1949, *Ap. J.*, **109**, 42.
Aller, L. H., and White, M. L. 1949, *A.J.*, **54**, 181.
Bates, D. R., Fundaminsky, A., Leech, J. W., and Massey, H. S. W. 1950, *Phil. Trans. R. Soc. London, A*, **243**, 117.
Berman, L. 1930, *Lick Obs. Bull.*, **15**, 86.
———. 1937, *ibid.*, **18**, 73.
Bowen, I. S., Minkowski, R., and Aller, L. H. 1955 (in press).
Bowen, I. S., and Wyse, A. B. 1939, *Lick Obs. Bull.*, **19**, 1.
Garstang, R. H. 1951, *M.N.*, **111**, 116.
———. 1952, *Ap. J.*, **115**, 506.
Goldberg, L. 1939, *Ap. J.*, **90**, 414.
———. 1941, *ibid.*, **93**, 244.
Huang, Su-shu. 1948, *Ap. J.*, **108**, 354.
Layzer, D. 1951, *Phys. Rev.*, **84**, 1221.
Liller, W., and Aller, L. H. 1954, *Ap. J.*, **120**, 48.
MacRae, D., and Stock, J. 1954, *Nature*, **173**, 589.
Menzel, D. H. 1937, *Ap. J.*, **85**, 330.
———. 1951, *Trans. I.A.U.*, **7**, 471.
Menzel, D. H., and Aller, L. H. 1941, *Ap. J.*, **93**, 145.
Menzel, D. H., Aller, L. H., and Hebb, M. H. 1941, *Ap. J.*, **93**, 230.
Menzel, D. H., and Baker, J. G. 1938, *Ap. J.*, **88**, 52.
Menzel, D. H., and Hebb, M. L. 1940, *Ap. J.*, **92**, 408.
Menzel, D. H., and Pekeris, C. L. 1935, *M.N.*, **96**, 77.
Minkowski, R. 1942, *Ap. J.*, **95**, 243.
Minkowski, R., and Aller, L. H. 1954, *Ap. J.*, **120**, 261.
———. 1955 (in press).
Naqvi, Ali. 1951, Harvard thesis.
Osterbrock, D. 1951, *Ap. J.*, **114**, 469.
Pasternack, S. 1940, *Ap. J.*, **92**, 129.
Seaton, M. J. 1953, *Proc. R. Soc. London, A*, **218**, 400.
———. 1954, *M.N.* (in press). The target area parameters $\Omega_{2,3}$ were communicated to me in advance of publication.
Shortley, G. H., Aller, L. H., Baker, J. G., and Menzel, D. H. 1941, *Ap. J.*, **93**, 178.
Ufford, C. W., and Gilmour, R. M. 1950, *Ap. J.*, **111**, 580.
Wurm, K. 1953, contribution in *Les Processus nucléaires dans les astres* (Liége University, 1954), p. 387.
Wyse, A. B. 1942, *Ap. J.*, **95**, 356.
Yamanouchi, T., Inui, T., and Amemiya, A. 1940, *Proc. Physico-Math. Soc. of Japan*, Ser. 3, **22**, 848.

Reprinted from *The Astrophysical Journal,* Vol. 122, No. 1, July, 1955.

THE SPECTRUM OF NGC 7027

L. H. Aller, I. S. Bowen, and R. Minkowski
University of Michigan Observatory
Mount Wilson and Palomar Observatories
Carnegie Institution of Washington, California Institute of Technology
Received January 31, 1955

ABSTRACT

A total of 263 lines was observed in the spectrum of NGC 7027 with the spectrographs of the 100- and 200-inch telescopes. Intensity measurements and identifications were made for most of these lines. This nebula shows an exceedingly wide range of ionization, forty ions ranging from *Mg* I to *Fe* VII being observed. The 4F–nG and 5G–nH series of *He* II were followed to $n = 29$ and $n = 40$, respectively, and the somewhat abnormal progression of observed intensities along these series was given a theoretical interpretation.

I. INTRODUCTION

A comprehensive study of the physical state and chemical composition of gaseous nebulae is dependent upon an accurate knowledge of their spectra. The most detailed studies were those of Bowen and Wyse (1939), who investigated NGC 6572, NGC 7027, and NGC 7662, and of Wyse (1942), who extended the investigation to six additional planetaries and the Orion Nebula. These studies, carried out with the aid of the Lick 36-inch refractor and a spectrograph equipped with Bowen's (1938) image slicer, gave the wave length, identifications, and approximate intensities of nebular lines observed between the Balmer limit and λ 6755. The wave lengths, the identifications, and even the reality of many of the fainter lines required confirmation by the more powerful spectrographic equipment that has been developed in the last fifteen years. The intensities published by Wyse represent eye-estimates made with the aid of scale plates; they should be replaced by photometrically measured intensities whenever possible.

The most rewarding of planetary nebulae for a detailed spectroscopic study is NGC 7027. This object shows the richest line spectrum of all gaseous nebulae. Probably every line observable in any planetary is represented here. There is a huge range in excitation from *N* I to *Fe* VII. Furthermore, the nebula has a higher surface brightness than any other planetary nebula. All observable radiation comes from the nebular lines plus continuum. There is no direct contribution from any central star. The latter must have a temperature well in excess of 150,000° K. A number of observers have published reproductions of the spectrum (see, e.g., Wyse 1942 and Aller 1954).

II. OBSERVATIONAL PROCEDURE

The present investigation comprises two sets of data: (*a*) wave lengths and identifications of the nebular lines based on plates taken with the coudé spectrograph of the Hale 200-inch telescope and (*b*) relative line intensities primarily from a photometric study carried out with the 100-inch reflector and Cassegrain spectrograph.

The detailed spectrophotometric procedures will be described in a separate communication (Minkowski and Aller 1955). They consisted in photographing the spectrum of NGC 7027 beside that of a suitable comparison star thrown out of focus on a ground-quartz diffuser placed in front of the slit. Vega and ε Persei served as comparison stars. A laboratory calibration provided the transmissivity of the diffuser. Step-slit spectra of a lamp, obtained with the photometric calibration accessory of the 100-inch coudé, provided the density-intensity relationship. In addition to the Mount Wilson data, which covered the spectral range λλ 3600–9000, observations of the λλ 3200–5000 range were obtained at the McDonald Observatory with the Cassegrain spectrograph equipped with quartz prisms.

The essential information relative to the equipment, emulsions, and dispersions employed at Palomar is given in Table 1. Successive columns give the plate number, the camera used, the order of the grating, the wave-length region, the dispersion in A/mm, the emulsion type, and the exposure times. Overlapping order occurred in some spectral regions.

III. OBSERVATIONAL RESULTS

Table 2 contains the essential results of this investigation. The four columns give the wave length, the ion, the electron transition, and the intensity measured at the brightest part of the nebula.

The plates used for the intensity measures had a much lower dispersion than those employed for the wave lengths. Hence many of the lines whose intensities are measured as blends are resolved on the Palomar plates. Accordingly, we have used tracings of the Palomar plates to allocate the relative intensity contributions in blends. Unfortunately, the earlier Palomar plates were secured before the calibration spectrograph was available; hence they contain no calibration. We can interpolate the relative intensities of the lines in a blend with the aid of other lines on the plate whose intensities had been determined photometrically. The continuum is so very strong that even on the calibrated Palomar plates the intensities of the weaker lines can be measured only very roughly. The footnotes to the table contain remarks on certain individual lines. A colon (:) means that the tabulated line intensity is very uncertain.

TABLE 1

OBSERVATIONS OF NGC 7027 OBTAINED AT PALOMAR

Plate	Camera	Grating Order	Wave-Length Range (A)	Dispersion (A/Mm)	Emulsion	Exposure (Min.)
Pc 151.....	36-in	3–4	3300–5000	6.7–9.0	Baked 103-O	1265
Pe 358.....	8.4-in	2	5700–6700	56	103*a*-E	435
Pd 1039....	18-in	2	5500–8500	27	103*a*-F+IN	1132
Pd 1040....	18-in	2	5400–8100	27	103*a*-F+IN	100
Pd 1042....	18-in	2–3–4	3300–5400	13.5–18–27	Baked 103*a*-J	1192
Pd 1555....	18-in	2	5300–8600	27	103*a*-D+IN	430
Pe 1556....	8.4-in	2	4700–5700	56	103*a*-D	330
Pe 1811....	8.4-in	3	4400–5000	38	Baked II*a*-O	440
Pe 1814....	8.4-in	3	3700–4400	38	Baked II*a*-O	425

The intensities of the far-ultraviolet lines λ 3202 *He* I and λ 3133 are adopted from earlier work by one of us (Aller 1941). Their intensities relative to the nearest lines for which we have photometric data from the present study have been utilized to reduce them to the present system. The intensities of all lines are given on the scale $I(H\beta) = 100$.

A comparison of the present relative intensities with those given by Wyse shows that his eye-estimates were remarkably accurate, considering the crudeness of his calibration. No sizable scale error appears to have existed, but the zero point of his intensity scale in the visual region appears to be shifted with respect to the intensity in the photographic region.

IV. THE PROBLEM OF THE IDENTIFICATIONS

The Palomar plates showed many weak lines that could not be resolved or observed on the Mount Wilson spectra. On the other hand, there were a few lines that were ob-

TABLE 2

SPECTRUM OF NGC 7027

Wave Length	Ion	Transition	Intensity
3132.82	*O* III	$3p^3S-3d^3P_2$	20:
3203.15	*He* II	3^2D-5^2F	7:
3299.58	*O* III	$3s^3P_0-3p^3S$	3.5
3312.35	*O* III	$3s^3P_1-3p^3S$	11.3
3340.78	*O* III	$3s^3P_2-3p^3S$	16.0
3345.82	[*Ne* V]	$^3P_1-^1D$	24:
3386.03	....	....	0.5
3411.84	*O* IV	$3^2P_{3/2}-3^2D_{5/2}$	1.1
3425.84	[*Ne* V]	$^3P_2-^1D$ }	
3428.70	*O* III	$3p^3P_1-3d^3P_2$ }	60:
3444.05	*O* III	$3p^3P_2-3d^3P_2$	30:
3554.4	*He* I	2^3P-10^3D	0.3
3587.05	*He* I	2^3P-9^3D	0.8
3634.20	*He* I	2^3P-8^3D	0.8
3666.00	*H* I	2^2P-27^2D	....
3667.68	*H* I	2^2P-26^2D	....
3669.47	*H* I	2^2P-25^2D	....
3671.48	*H* I	2^2P-24^2D	0.6
3673.70	*H* I	2^2P-23^2D	0.8
3676.31	*H* I	2^2P-22^2D	0.9
3679.39	*H* I	2^2P-21^2D	0.9
3682.81	*H* I	2^2P-20^2D	1.0
3686.88	*H* I	2^2P-19^2D	1.1
3691.53	*H* I	2^2P-18^2D	1.3
3697.21	*H* I	2^2P-17^2D	1.4
3703.81	*H* I	2^2P-16^2D	1.7
3705.04	*He* I	2^3P-7^3D	0.7
3711.95	*H* I	2^2P-15^2D	1.8
3715.27	*He* II	4^2F-29^2G	0.7
3721.81	*H* I	2^2P-14^2D [*S* III] $^3P_1-^1S$	4.3
3725.99	[*O* II]	$^4S-^2D_{3/2}$	9.4
3728.76	[*O* II]	$^4S-^2D_{5/2}$	4.6
3732.97	*He* II	4^2F-26^2G	0.7
3734.34	*H* I	2^2P-13^2D	2.3
3736.70	*O* IV	$3p^4D_{7/2}-3d^4F_{9/2}$	0.9
3740.16	*He* II	4^2F-25^2G	0.6
3744.78	....	....	....
3748.77	*He* II	4^2F-24^2G	0.6
3750.19	*H* I	2^2P-12^2D	2.7
3754.69	*O* III	$3s^3P_1-3p^3D_2$	1.4
3757.24	*O* III	$3s^3P_0-3p^3D_1$	0.8
3759.87	*O* III	$3s^3P_2-3p^3D_3$	3.8
3768.86	*He* II	4^2F-22^2G	0.7
3770.66	*H* I	2^2P-11^2D	3.3
3773.95	*O* III	$3s^3P_1-3p^3D_1$	0.7
3781.74	*He* II	4^2F-21^2G	0.8
3791.29	*O* III	$3s^3P_2-3p^3D_2$	0.7
3797.91	*H* I	2^2P-10^2D	3.5
3813.50	*He* II	4^2F-19^2G	0.7
3819.66	*He* I	2^3P-6^3D	1.1
3833.75	*He* II	4^2F-18^2G	0.6
3835.41	*H* I	2^2P-9^2D	4.6
3839.57	[*Fe* V]	$^5D_3-^3F_3$	....
3851.81	[*Fe* V]	$^5D_3-^3F_2$	....
3856.17	*Si* II	$3p^{2\,2}D_{5/2}-4^2P_{3/2}$	0.3
3857.99	*He* II	4^2F-17^2G	0.8
3862.67	*Si* II	$3p^{2\,2}D_{3/2}-4^2P_{1/2}$	0.3
3868.73	[*Ne* III]	$^3P_2-^1D$	51.
3888.91	*H* I	2^2P-8^2D *He* I 2^3S-3^3P	8.8

TABLE 2—*Continued*

Wave Length	Ion	Transition	Intensity
3891.28	[*Fe* v]	$^5D_4-{}^3F_4$	0.4
3895.61	[*Fe* v]	$^5D_3-{}^3P_2$	0.4
3920.89	*C* II	$3^2P_{3/2}-4^2S$	....
3923.56	*He* II	4^2F-15^2G	0.6
3926.50	*He* I	2^1P-8^1D	0.4
3956.82	....	....	0.2
3964.88	*He* I	2^1S-4^1P	
3967.46	[*Ne* III]	$^3P_1-{}^1D$	24.4
3970.07	*H* I	2^2P-7^2D	
3997.36	[*F* IV]	$^3P_1-{}^1D$	0.2
4003.28	....	....	....
4009.17	*He* I	2^1P-7^1D	0.4
4026.12	*He* I	2^3P-5^3D *He* II 4^2F-13^2G	2.0
4060.26	[*F* IV]	$^3P_2-{}^1D$	0.3
4068.62	[*S* II]	$^4S-{}^2P_{3/2}$	4.5
4070.27	*C* III	$4^3F_4-5^3G_5$ [*Fe* III] $^5D_4-{}^3G_5$	0.4
4072.14	[*Fe* v]	$^5D_2-{}^3P_1$	0.3
4076.33	[*S* II]	$^4S-{}^2P_{1/2}$	2.2
4089.15	*O* II	$3d^4F_{9/2}-4f^4G_{11/2}$	....
4097.37	*N* III	$3^2S-3^2P_{3/2}$	2.0
4100.08	*He* II	4^2F-12^2G	
4101.74	*H* I	2^2P-6^2D	16.3
4103.46	*N* III	$3^2S-3^2P_{1/2}$	
4119.16	*O* II	$3p^4P_{5/2}-3d^4D_{7/2}$	....
4120.80	*He* I	2^3P-5^3S	0.3
4122.60	[*K* v]	$^4S-{}^2D_{5/2}$	0.2
4128.74	....	....	0.3
4143.76	*He* I	2^1P-6^1D	0.5
4146.05	....	....	....
4156.35	*C* III	$3^3D_2-5^3F_3$	0.4
4163.33	[*K* v]	$^4S-{}^2D_{3/2}$	0.3
4180.92	[*Fe* v]	$^5D_1-{}^3P_0$	0.2
4186.98	*C* III	4^1F-5^1G	0.5
4189.84	*O* II	$3p^2F_{7/2}-3d^2G_{9/2}$	....
4199.88	*He* II	4^2F-11^2G	1.5
4227.37	....	....	0.5
4267.13	*C* II	3^2D-4^2F	0.9
4325.74	*C* III	$3s\ ^1P-3p^1D$	....
4338.62	*He* II	4^2F-10^2G	33
4340.44	*H* I	2^2P-5^2D	
4363.19	[*O* III]	$^1D-{}^1S$	21
4379.07	*N* III	4^2F-5^2G	0.5
4387.97	*He* I	2^1P-5^1D	1.7
4437.59	*He* I	2^1P-5^1S	....
4471.50	*He* I	2^3P-4^3D	3.9
4510.79	[*K* IV]	$^1D-{}^1S$	0.7:
4541.60	*He* II	4^2F-9^2G	2.3
4552	*Si* III	$4^3S-4^3P_2$	0.4
4571.24	*Mg* I	$3^1S-3^3P_1$	0.9
4606.51	[*Fe* III]	$^5D_4-{}^3F_3$	0.5
4625.54	[*A* v]	$^1D-{}^1S$	0.5
4631.96	....	....	....
4634.15	*N* III	$3^2P_{1/2}-3^2D_{3/2}$	2.8
4640.62	*N* III	$3^2P_{3/2}-3^2D_{5/2}$	3.6
4641.93	*N* III	$3^2P_{3/2}-3^2D_{3/2}$	0.6
4647.46	*C* III	$3^3S-3^3P_2$	1.4
4649.08	*O* II	$3s^4P_{5/2}-3p^4D_{7/2}$	0.4
4650.27	*C* III	$3^3S-3^3P_1$	0.6
4657.68	....	....	1.6
4658.42	*C* IV	5^2G-6^2H [*Fe* III] $^5D_4-{}^3F_4$	
4685.71	*He* II	3^2D-4^2F	45

TABLE 2—*Continued*

Wave Length	Ion	Transition	Intensity
4701.45	[*Fe* III]	$^5D_3-{}^3F_3$	....
4711.34	[*A* IV]	$^4S-{}^2D_{5/2}$	2.8
4713.17	*He* I	2^3P-4^3S	0.7
4714.26	[*Ne* IV]	$^2D_{5/2}-{}^2P_{3/2}$	0.8
4715.65	[*Ne* IV]	$^2D_{5/2}-{}^2P_{1/2}$	0.5
4724.16	[*Ne* IV]	$^2D_{3/2}-{}^2P_{3/2}$	1.3
4725.60	[*Ne* IV]	$^2D_{3/2}-{}^2P_{1/2}$	1.3
4733.68	[*Fe* III]	$^5D_2-{}^3F_2$	....
4740.20	[*A* IV]	$^4S-{}^2D_{3/2}$	11
4754.9	[*Fe* III]	$^5D_3-{}^3F_4$	0.4
4859.32	*He* II	4^2F-8^2G	100 (both lines)
4861.30	*H* I	2^2P-4^2D	
4893.42	*Fe* VII	$^3F_2-{}^3P_1$	0.6
4921.96	*He* I	2^1P-4^1D	3.1
4930.97	[*O* III]	$^3P_0-{}^1D$	<0.5
4944.01	[*Fe* VII]	$^3F_3-{}^3P_2$	....
4950.48	....	....	....
4958.95	[*O* III]	$^3P_1-{}^1D$	420
4972.12	[*Fe* VI]	$^4F_{5/2}-{}^2G_{7/2}$	....
5006.88	[*O* III]	$^3P_2-{}^1D$	1170
5015.55	*He* I	2^1S-3^1P	....
5041.31	*Si* II	$4^2P_{1/2}-4^2D_{3/2}$	1.6
5047.92	*He* I	2^1P-4^1S	1.4
5056.45	*Si* II	$4^2P_{3/2}-4^2D_{5/2}$	1.5
5090	....	....	0.5
5112	*O* V	3^1S-3^1P	0.5
5131.18	....	....	0.6
5145.78	[*Fe* VI]	$^4F_{7/2}-{}^2G_{7/2}$	0.9
5158.94	[*Fe* VII]	$^3F_3-{}^3P_1$	0.6 (both lines)
5161.67	....	....	
5176.42	[*Fe* VI]	$^4F_{9/2}-{}^2G_{9/2}$	0.9
5191.81	[*A* III]	$^1D-{}^1S$	0.9:
5197.95	[*N* I]	$^4S-{}^2D_{3/2}$	1.2:*
5200.41	[*N* I]	$^4S-{}^2D_{5/2}$	0.3:*
5216	....	....	0.5
5232	....	....	0.4
5262.00	....	....	....
5270.28	[*Fe* III]	$^5D_3-{}^3P_2$	0.3:
5277.76	[*Fe* VII]	$^3F_4-{}^3P_2$	0.3:
5309.17	[*Ca* V]	$^3P_2-{}^1D$	1.7
5323.26	[*Cl* IV]	$^1D-{}^1S$	1.0
5335.22	[*Fe* VI]	$^4F_{3/2}-{}^4P_{1/2}$	1.2
5345.98	....	....	1.2
5411.54	*He* II	4^2F-7^2G	9
5423.87	[*Fe* VI]	$^4F_{5/2}-{}^4P_{3/2}$	0.4:
5426.57	[*Fe* VI]	$^4F_{7/2}-{}^4P_{5/2}$	0.2:
5470.93	....	....	0.4:
5484.79	[*Fe* VI]	$^4F_{5/2}-{}^4P_{1/2}$	0.5:
5517.72	[*Cl* III]	$^4S-{}^2D_{5/2}$	1.0
5537.80	[*Cl* III]	$^4S-{}^2D_{3/2}$	2.7
5577.36	[*O* I]	$^1D-{}^1S$	1.5
5592.17	*O* III	$3s^1P-3p^1P$	1.1
5604	[*K* VI]	$^3P_1-{}^1D$	0.6
5614.65	[*Ca* VII]	$^3P_2-{}^1D$	....
5630.88	[*Fe* VI]	$^4F_{7/2}-{}^4P_{3/2}$	0.7
5677.01	[*Fe* VI]	$^4F_{9/2}-{}^4P_{5/2}$	1.1
5696.06	*C* III	3^1P-3^1D	....
5721.13	[*Fe* VII]	$^3F_2-{}^1D$	1.2

* These lines correspond to the 3726/2729 pair of [*O* II]. Their intensity ratio, which does not agree with the theoretical predictions of Ufford and Gilmour (*Ap. J.*, **111**, 580, 1950) and of Garstang (*Ap. J.*, **115**, 506, 1952), is probably affected by collisional effects, as is the 3727 pair of [*O* II].

TABLE 2—*Continued*

Wave Length	Ion	Transition	Intensity
5726.12			
5754.59	[*N* II]	$^1D-^1S$	12
5776.37	[*Mn* VI]	$^3F_3-^3P_1$	
5786.59	*He* II	5^2G-40^2H	
5790.85	*He* II	5^2G-39^2H	
5794.64	*He* II	5^2G-38^2H	
5801.38	*C* IV	$3^2S-3^2P_{3/2}$	2.2
5806.29	*He* II	5^2G-36^2H	
5812.12	*C* IV	$3^2S-3^2P_{1/2}$	1.1
5820.14	*He* II	5^2G-34^2H	
5828.64	*He* II	5^2G-33^2H	
5836.50	*He* II	5^2G-32^2H	0.3:
5847.14	*He* II	5^2G-31^2H	
5858.33	*He* II	5^2G-30^2H	
3862.34 } 5864.81 }	[*Mn* V]	$^4F_{7/2}-^2G_{7/2}$	
5867.82	*He* II	5^2G-29^2H *Si* II $3p4s\,^4P_{5/2}-3p4p\,^4P_{5/2}$	
5875.63	*He* I	2^3P-3^3D	24
5882.35	*He* II	5^2G-28^2H	
5885.90			
5890.98	[*Mn* V]	$^4F_{9/2}-^2G_{9/2}$	
5894.01	[*Mn* VI]	$^3F_4-^3P_2$	
5896.87	*He* II	5^2G-27^2H	
5913.52	*He* II	5^2G-26^2H	
5932.22	*He* II	5^2G-25^2H	0.8
5953.00	*He* II	5^2G-24^2H	0.9
5977.05	*He* II	5^2G-23^2H	0.7
6004.81	*He* II	5^2G-22^2H	1.2
6037.16	*He* II	5^2G-21^2H	1.1
6074.30	*He* II	5^2G-20^2H	1.1+
6086.88	[*Ca* V]	$^3P_1-^1D$ [*Fe* VII] $^3F_3-^1D$	2.5:
6101.76	[*K* IV]	$^3P_2-^1D$	2.5:
6118.31	*He* II	5^2G-19^2H	1.1+
6151.43			
6166.22	[*Mn* V]	$^4F_{7/2}-^4P_{5/2}$	0.3
6170.73	*He* II	5^2G-18^2H	1.1
6218.64 } 6221.04 }	[*Mn* V]	$^4F_{5/2}-^4P_{1/2}$	
6228.41	[*K* VI]	$^3P_2-^1D$	
6233.80	*He* II	5^2G-17^2H	0.9
6300.31	[*O* I]	$^3P_2-^1D$	20
6312.10	[*S* III]	$^1D-^1S$	16
6347.54	*Si* II	$4^2S-4^2P_{3/2}$	
6363.82	[*O* I]	$^3P_1-^1D$	10
6371.34	*Si* II	$4^2S-4^2P_{1/2}$	0.3:
6393.62	[*Mn* V]	$^4F_{9/2}-^4P_{5/2}$	
6406.53	*He* II	5^2G-15^2H	0.8
6435.11	[*A* V]	$^3P_1-^1D$	5.2
6462.14			
6518.34	[*Mn* VI]	$^3F_2-^1D$	
6527.31	*He* II	5^2G-14^2H	0.8:
6548.09	[*N* II]	$^3P_1-^1D$	80
6560.24	*He* II	4^2F-6^2G	
6562.90	*H* I	2^2P-3^2D	650
6583.36	[*N* II]	$^3P_2-^1D$	240
6598.76	[*Fe* VII]	$^3F_4-^1D$	
6678.16	*He* I	2^1P-3^1D	10
6716.52	[*S* II]	$^4S-^2D_{5/2}$	5.4
6730.74	[*S* II]	$^4S-^2D_{3/2}$	7.5
6890.80	*He* II	5^2G-12^2H	2.1
7005.67	[*A* V]	$^3P_2-^1D$	14

TABLE 2—*Continued*

Wave Length	Ion	Transition	Intensity
7065.31........	*He* I	2^3P-3^3S	20
7135.78........	[*A* III]	$^3P_2-^1D$	56
7170.62........	[*A* IV]	$^2D_{3/2}-^2P_{3/2}$	3:
7177.77........	*He* II	5^2G-11^2H	2:
7237.26........	[*A* IV]	$^2D_{5/2}-^2P_{3/2}$	2.6
7262.77........	[*A* IV]	$^2D_{3/2}-^2P_{1/2}$	3.1
7281.32........	*He* I	2^1P-3^1S	2.8
7319.92........	[*O* II]	$^2D_{5/2}-^2P$ }	77
7330.19........	[*O* II]	$^2D_{3/2}-^2P$ }	
7530.54........	[*Cl* IV]	$^3P_1-^1D$	1.4
7592.79........	*He* II	5^2G-10^2H	1.2†
7726.51........	[*S* I]	$D-^1S$	....
7751.04........	[*A* III]	$^3P_1-^1D$	20
8045.58........	[*Cl* IV]	$^3P_2-^1D$	5.3
8196.61........	....	....	2.0
8236.64........	*He* II	5^2G-9^2H	5.2
8502.5........	*H* I	3^2D-16^2F	0.6
8545.53........	*H* I	3^2D-15^2F	0.8
8598.35........	*H* I	3^2D-14^2F	2.6
8665.0........	*H* I	3^2D-13^2F	2.7
8750.5........	*H* I	3^2D-12^2F	3.1
8862.8........	*H* I	3^2D-11^2F	3.5
9014.9........	*H* I	3^2D-10^2F	5.5
9069........	[*S* III]	$^3P_1-^1D$	40:

† In atmospheric absorption band.

served by Wyse and that seemed to appear on the Mount Wilson spectra but were not found on the Palomar plates. In view of the higher dispersion and greater effective exposure of the latter, we are inclined to regard the lines not confirmed by the Palomar plates as largely spurious and have omitted most of them. We have listed all the lines that appear definitely on one or more Palomar plates and a few of the more uncertain lines that are readily identified with ions whose lines appear in the spectrum. Weak, elusive features that appear on only one plate and for which we have no identification are likely to be spurious. We have not listed these possible lines, although some of them may turn out to be real.

One fact to be emphasized is that the limiting intensity reached differs from one region of the spectrum to another. For example, in the infrared region of the spectrum the plates are less sensitive, while in the ultraviolet the lowered transparency of the corrector plate in the Schmidt camera of the spectrograph cuts down the transmitted intensity. Furthermore, the blaze of the grating renders it more efficient in some spectral regions than in others. The presence of an underlying continuum also influences the detectability of weak lines. If a faint continuum exists, weak lines may be seen that would otherwise be lost below the threshold of the plate.

In general, the identification of lines can be made with some certainty in the cases in which the wave lengths are accurately known from laboratory or other measurements and in which the relative intensities of the lines in a given multiplet can be predicted from theory. These conditions are satisfied for most of the permitted lines and for a substantial part of the forbidden lines (cf. Bowen 1955). For some of the fainter forbidden lines, however, the wave lengths are still uncertain by an angstrom or more, and no information is available at the present time for predicting relative intensities. In these cases identification may be very uncertain; for instance, the assignment of a few lines to [*Ni* II] and [*Ni* VII] suggested by one of us (Aller 1954*a*) is probably to be ruled out.

V. THE DEGREE OF EXCITATION OF THE NEBULA

Lines of the following ions are represented in NGC 7027: *H*, *He* I, *He* II, *C* II, *C* III, *C* IV, *N* I, *N* II, *N* III, *O* I, *O* II, *O* III, *O* IV, *O* V, *F* IV, *Ne* III, *Ne* IV, *Ne* V, *Mg* I, *Si* II, *Si* III, *S* I, *S* II, *S* III, *Cl* III, *Cl* IV, *A* III, *A* IV, *A* V, *K* IV, *K* V, *K* VI, *Ca* V, *Ca* VII, *Mn* V, *Mn* VI, *Fe* III, *Fe* V, *Fe* VI, and *Fe* VII. This list differs from those of Bowen and Wyse and of Wyse in a few particulars. For example, the $^5D-^3F$ multiplet of [*Fe* III] is well represented, as is the $^5D-^3P$ multiplet of [*Fe* V]. There is no evidence for [*F* II], although [*F* IV] is definitely present.

Phosphorus does not appear; its lines fall in an unfavorable region of the spectrum. Among the heavier elements, no definite evidence is found for nickel, which is perhaps not surprising in view of the weakness of the forbidden lines of much more abundant iron.

As has been remarked by previous observers, NGC 7027 shows a wide range in excitation from *Mg* I to *Fe* VII. A careful search has failed to reveal any lines of [*Fe* X]. Since the lines of [*Fe* VII] appear in exposures 1/25 as long as the longest exposures (which do not reveal [*Fe* X]) and since all the [*Fe* X] radiation is concentrated in one line, while that of *Fe* VII is spread over several lines, we may conclude that the abundance of *Fe* X is less than 3 per cent of that of *Fe* VII. Hence this nebula, although as high in excitation as any planetary, is not so high as T Coronae Borealis or RS Ophiuchi in their coronal-line stages or even CI Cygni in its highest excitation phase.

One of the most interesting features of the spectrum of NGC 7027 is the behavior of the Pickering (4^2F-n^2G) or Brackett series of ionized helium and of the Pfund (5^2G-n^2G, etc.) series of this ion. The latter series is followed to $n = 40$, whereas the former (whose alternate members nearly coincide with the Balmer lines) cannot be followed quite so far. On the hypothesis of the "primary mechanism" of excitation, i.e., recombination followed by cascade, the emission in ergs per cubic centimeter per second per unit volume is given by (Menzel 1937)

$$E_{nn'} = N_i N_\epsilon \frac{KZ^4}{T_\epsilon^{3/2}} b_n \frac{g}{n'} \frac{2hRZ^2}{n^3} e^{X_n}, \tag{1}$$

where Ni is the number of doubly ionized helium atoms, $N\epsilon$ is the electron density, $Z = 2$ for ionized helium, K is a constant numerically equal to 3.26×10^{-6}, n' is the quantum number of the lower level of the transition ($n' = 4$ for the Pickering and 5 for the Pfund series), n is the quantum number of the upper level, and R is the Rydberg constant. The Kramers-Gaunt factor, g, has been tabulated by Menzel and Baker (1938). The electron temperature T_ϵ is also involved in the exponent

$$X_n = \frac{hRZ^2}{n^2 kT_\epsilon}. \tag{2}$$

The factor b_n measures the deviation of the population of the nth level from that appropriate to thermodynamic equilibrium. If T_ϵ is 20,000° K, Menzel and Baker's "case B" ($Z = 2$) yields the dependence of b_n on n shown in the accompanying table.

n........	9	10	15	20	25	30
b_n........	0.244	0.271	0.371	0.428	0.464	0.486

We select case B because it applies to an optically thick nebula. If the relative intensities of the Pfund series lines are computed by equation (1) and plotted against n, it is found that the values fall off nearly as $1/n^3$ for the higher members. The observed intensities (which are subject to large accidental errors) fall off roughly in accord with theoretical expectations for the earlier members of the series, but for $n > 15$ the decline in intensity with n is very slow. Allowance for the effects of space absorption only aggravates the discordance between theory and observation.

The trouble probably lies in the values of the b_n's. As n increases without bounds, b_n approaches 1, and it is possible that this limiting value is reached near $n = 30$ or 40 as a consequence of the redistribution of atoms among the higher levels by collisions. In fact, there is no reason why b_n could not exceed unity before the limiting value is reached. Further speculation is unwarranted until the problem of the collisional redistribution is treated quantitatively. The Pfund lines of ionized helium have been found by Wilson (1954) with about the expected intensities in certain O-type stars.

The Pickering lines are observed to $n = 29$ and have a behavior somewhat similar to that of the Pfund series, since the higher members of this series likewise seem to be abnormally strong, falling off in intensity at a rate much slower than $1/n^3$. The fact that this series cannot be followed quite so far as the Pfund series appears to be a consequence of observational selection. The higher members of the Pickering series (Brackett) lines are lost between the Balmer lines.

VI. CORRECTION OF THE OBSERVED INTENSITIES FOR SPACE ABSORPTION

The correction of a stellar magnitude for the effects of space absorption is found from the color excess, but for gaseous nebulae no normal "color" exists. Hence the space absorption must be evaluated from a critical examination of the line intensities themselves. For example, we might compare Paschen and Balmer lines that arise from the same upper level or auroral and transauroral lines that fall in widely separated spectral regions. The difficulty in comparing the Balmer and Paschen lines is that the latter have been measured on plates of very low dispersion and considerable difficulty is found in estimating the position of the continuum. The forbidden lines are difficult to use because of uncertainties in the intensities and transition probabilities.

TABLE 3

CORRECTION $\Delta \log I$ TO BE ADDED TO LOGARITHMS OF LINE INTENSITIES TO ALLOW FOR SPACE ABSORPTION

λ 3200. .	+0.53	λ 4000. .	+0.27	λ 4800. .	+0.02	λ 5400. .	−0.13	λ 6500. .	−0.37
3400. .	+ .45	4200. .	+ .20	4861. .	.00	5600. .	− .17	7000. .	− .47
3600. .	+ .39	4400. .	+ .14	5000. .	− .04	5800. .	− .21	7500. .	− .57
3800. .	+0.33	4600. .	+0.08	5200. .	−0.08	6000. .	−0.26	8000. .	−0.66

Perhaps the most reliable estimate can be made from the observed Balmer decrement. Observations of a large number of planetaries show that their Balmer decrements follow Menzel and Baker's case B unless they are affected by space absorption (Aller 1951). Hence, by comparing the decrement in NGC 7027 with this theoretical decrement and assuming that the law of space absorption is that given by Whitford (1947), it is possible to evaluate the amount of extinction (cf. Aller 1954*b*). In order to bring the observed and theoretical decrements into harmony, a differential correction for space absorption amounting to 2.75 mag. between λ 4220 and λ 10,300 A is required. This correction depends on the lower members of the Balmer series and is not affected by the deviation from case B which may be expected to occur for high members of the *H* series, similar to the deviations found in the *He* II series. The surface brightness of the nebula, as observed in the radiation of the green nebular lines, is cut down by about a factor of 10. Table 3 gives the correction to be applied to the observed line intensities as a function of wave length in order to allow for the effects of selective extinction. These corrections should be applied to the observed line intensities when they are used, for example, for theoretical discussions.

VII. COMPARISON WITH OTHER SPECTROPHOTOMETRIC DATA: FURTHER OBSERVATIONS NEEDED

A comparison of the present observations with those of Aller (1941) shows that the latter appear to be affected by a scale error in the sense that the weaker lines are systematically too weak. Evidently the continuum was drawn too high. The photoelectric observations of Liller and Aller (1954) and of Code and Whitford (1954) indicate that the intensity ratio of the green nebular lines to $H\beta$ was measured too small. Code and Whitford have kindly sent us their photoelectric line intensities for the stronger lines (down to about 11 on our scale) in NGC 7027. Except for N_1 and λ 3727, the agreement is excellent, the average deviation between the measured intensities being less than 10 per cent. In particular, there is no indication of any scale error in the photographic results. The great strength of the green nebular lines makes them difficult subjects for photographic photometry, whereas the photoelectric observation of the λ 3727 intensity can be explained by blending with near-by lines.

No checks are available for the weaker lines, and none appear to be forthcoming in the near future. The intensities of the very weakest lines are subject to large accidental errors, but we feel that the intensities are free from serious scale errors.

With the telescopic and spectrographic equipment at present available, it is not likely that many new lines can be added in the region from λ 3600 to λ 8200. On the other hand, the identifications and intensities can be greatly improved in the region from λ 3600 to the limit of transmission of the earth's atmosphere. It is our intention to make observations in this region as soon as appropriate spectrographic equipment is available.

REFERENCES

Aller, L. H. 1941, *Ap. J.*, **93**, 236.
———. 1951, *ibid.*, **113**, 125.
———. 1954*a*, *Nuclear Transformations, Stellar Interiors, and Nebulae* (New York: Ronald Press Co.), chap. 4, Fig. 7.
———. 1954*b*, *Ap. J.*, **120**, 401. The following misprints should be corrected. The second term in the denominator of eq. (11) should read $(A_{32} + A_{31})\varpi_3/C$, while in eq. (12) Ω_{13} should be replaced by Ω_{23} and A_{23} by A_{21}.
Bowen, I. S. 1939, *Ap. J.*, **88**, 113.
———. 1955, *Ap. J.*, **121**, 306.
Bowen, I. S., and Wyse, A. B. 1939, *Lick Obs. Bull.*, **19**, 1.
Code, A., and Whitford, A. E. 1954, private communication.
Liller, W., and Aller, L. H. 1954, *Ap. J.*, **120**, 48.
MacRae, D., and Stock, J. 1954, *Nature*, **173**, 589.
Menzel, D. H. 1937, *Ap. J.*, **85**, 330.
Menzel, D. H., and Baker, J. G. 1938, *Ap. J.*, **88**, 52.
Minkowski, R., and Aller, L. H. 1955, *Ap. J.* (in press).
Whitford, A. E. 1948, *Ap. J.*, **107**, 102.
Wilson, R. 1953, *M.N.*, **113**, 557.
Wyse, A. B. 1942, *Ap. J.*, **95**, 356.

Reprinted from *The Astrophysical Journal,* Vol. 124, No. 1, July, 1956.

SPECTROPHOTOMETRY OF PLANETARY NEBULAE

R. Minkowski and L. H. Aller
Mount Wilson and Palomar Observatories
Carnegie Institution of Washington, California Institute of Technology
and
University of Michigan Observatory
Received February 13, 1956

ABSTRACT

Measurements of the line and continuous spectra of several bright planetary nebulae by methods of photographic spectrophotometry have been carried out with the Cassegrain spectrographs at the 60- and 100-inch reflectors of the Mount Wilson Observatory. The results of extensive observations are reported for NGC 40, NGC 1535, NGC 2022, NGC 2392, and NGC 7662. Additional material has been obtained for IC 351, IC 1747, J 320, NGC 2165, and NGC 2440. In each object the slit was placed on a selected position in the nebula and the exposures carefully guided. The line intensities, expressed on the scale $I(H\beta) = 100$, are given for lines from the Balmer limit to λ 9000.

I. INTRODUCTION

For a number of astrophysical problems it is desirable to have reliable measurements of emission-line intensities in typical gaseous nebulae. On the one hand, for theoretical studies, it is necessary to know the emission per typical unit volume in the radiation of hydrogen and other permanent gases. On the other hand, nebulae in which line intensities have been accurately measured may serve as photometric standards to which other, and often much fainter, objects can be compared.

Recent investigations (Liller and Aller 1954; MacRae and Stock 1954; Code 1956) have demonstrated the value of photoelectric techniques for the measurement of the intensities of the stronger lines. For the present, at least, the weaker lines must be measured by photographic photometry. It also is possible to select smaller regions in the nebula for detailed analysis. Previous photographic spectrophotometric studies (Berman 1930; Plaskett 1931; Page 1936, 1942; Aller 1941, 1951) have been confined to the stronger lines. Wyse (1942) attempted to measure the weaker lines with the aid of eye-estimates on scale plates. The most interesting planetary nebula from the standpoint of the richness of spectrum and range in excitation is NGC 7027 (Bowen, Minkowski, and Aller 1955). Similar spectrophotometric techniques have been applied to other nebulae.

We selected a number of nebulae ranging from low to high excitation which were well placed for the long exposures required to record moderately faint lines and continua. For obvious reasons, most of them were rather bright. We included the ring objects NGC 1535, NGC 2022, NGC 2392, and NGC 7662 and the somewhat irregular structure, NGC 40. A few additional observations were secured of IC 351, IC 1747, J 320, NGC 2165, and NGC 2440. The faintest object on our list, NGC 2022, was chosen because of its high excitation.

II. THE OBSERVATIONAL TECHNIQUE

The observations discussed in this paper were all secured with Cassegrain spectrographs at the 60- and 100-inch reflectors at the Mount Wilson Observatory. At the 60-inch reflector we used a two-prism spectrograph with the 3-inch Schmidt camera, which gives a dispersion of about 170 A/mm at $H\gamma$. At the 100-inch telescope we used the two-prism spectrograph with the Schmidt camera of 6-inch focal length, which gives a dispersion of 110 A/mm at $H\gamma$. This optical arrangement permitted us to obtain plates in good definition from the near infrared to the Balmer limit. Since both spectrographs were equipped with auxiliary guiding eyepieces, it was possible to keep the slit placed accurately upon a selected portion of the nebula for the entire exposure. In this way, differences of excitation in different parts of the nebula may be compared. No attempt has been made to obtain integrated intensities of the nebular images. Since the lines have an intensity range of over a thousand fold, it is necessary to secure observations ranging in exposure time from a few seconds to many hours.

Table 1 gives the nebulae and comparison stars observed and the spectrographs employed. The JS plate series refers to the 60-inch Cassegrain spectrograph; the E series to the 100-inch Cassegrain spectrograph. For the photographic and visual regions, the 103*a*-F emulsion was used; for the infrared, hypersensitized I-N plates were employed. The sixth and eighth columns give the range of exposure times on nebula and comparison stars for each plate secured.

III. PHOTOMETRIC CALIBRATION

To measure the intensities of the nebular lines and continua, it is necessary to establish the relation between photographic blackening and intensity as well as the effects of plate sensitivity, atmospheric transparency, and telescope reflectivity plus spectrograph transmission. The first step is accomplished by a suitable photometric standardization; the second is carried out most expeditiously by photographic suitable comparison stars.

Because of the small size of the plates used in the spectrographs, it was necessary to impress the photometric standards on a larger piece cut from the same plate and then to develop nebular and calibration plate together. We photographed the calibration standards with the 32-inch camera and the coudé spectrograph, adjusting the exposure times to be comparable with those at the telescope. Thus calibration exposures ranging from a minute to many hours had to be secured. The change in slope of the calibration-curve with exposure time is very small. In a single calibration exposure it is difficult to obtain the proper densities for *all* relevant wave lengths. Fortunately, the development procedures were kept the same for all plates.

To compare the standard stars, of known energy distribution, with the nebulae, which are several magnitudes fainter, we adopted the following device. A small piece of ground quartz was placed over the slit of the spectrograph. We then ran the image of the comparison star out of focus by a prearranged amount and placed a portion of the luminous ring of the out-of-focus image on the slit. Thus we provided a diffuse illumination of the slit with a source of known energy distribution.

The transmission of the quartz diffuser as a function of wave length was calibrated photoelectrically in the laboratory by Edison Pettit, under conditions closely simulating those obtained at the telescope. In addition, the calibration was checked photographically at the telescope. For this purpose, stellar spectra obtained with the diffuser were compared with spectra of the same stars obtained without the diffuser and widened by a rapid trailing back and forth of the focal stellar image on the slit with a frequency adequate to avoid the intermittency effect.

The standard energy-distribution stars which served for comparison stars include α Lyrae, which was observed by Williams (1938), Williams and Hall (1941), Kienle Strassl, and Wempe (1938), and Barbier and Chalonge (1940), as well as ϵ Persei, ζ Cephei, and η Ursae Majoris, which have been observed by Kienle, Strassl, and Wempe by Barbier and Chalonge, or by both groups.

The energy distribution in Vega was measured by photographic photometry by R Williams for $\lambda\lambda$ 3700–9000 and by photoelectric photometry for the near infrared by Williams and Hall. The most frequently used comparison star, ϵ Persei, has been measured by Kienle's group and by Barbier and Chalonge, whose results are in good agreement with one another. Redward of λ 3700 the energy distribution corresponds to a color temperature of 26000° K; shortward of λ 3700 the color temperature is 23000° K. The

TABLE 1

SPECTROPHOTOMETRIC OBSERVATIONS OF PLANETARY NEBULAE

	α	δ	Plate	Emul.	Exposures	Comparison Star	Exposures
NGC 40....	$0^h10^m.2$	+72° 11′	JS 753	I-N	420^m	β Cephei	60^m
			JS 754	103*a*-F	420^m	β Cephei	60^m
			JS 755	103*a*-F	420^m	β Cephei	60^m
			JS 756	103*a*-F	150^m	....	....
			JS 757	103*a*-F	64^m, 64^m*	β Cephei	15^m
			JS 758	103*a*-F	100^m	....	....
			JS 759	103*a*-F	480^m	β Cephei	60^m, 15^m
			JS 761	103*a*-F	60^m, 60^m*	β Cephei	60^m
			JS 762	103*a*-F	70^m	ϵ Persei	60^m
			JS 763	103*a*-F	180^m	β Cephei	60^m, 15^m
			JS 766	103*a*-F	90^m	β Cephei	60^m
						ϵ Persei	60^m
IC 351......	3 44.3	+34 55	JS 775	103*a*-F	91^m	....	....
NGC 1535..	4 11.9	−12 52	JS 769	103*a*-F	225^m	ϵ Persei	60^m
			JS 771	103*a*-F	322^m	ϵ Persei	60^m
			JS 774	103*a*-F	60^m, 20^m, 7^m	ϵ Persei	60^m
J 320.......	5 2.8	+10 39	JS 775	103*a*-F	60^m	....	....
NGC 2022..	5 39.3	+ 9 02	E 2283	103*a*-F	435^m	η Ursa Major	60^m
			JS 769	103*a*-F	110^m	ϵ Persei	60^m
			JS 770	103*a*-F	510^m	ϵ Persei	60^m
			JS 772	103*a*-F	145^m	....	....
IC 2165.....	6 19.4	−12 57	JS 774	103*a*-F	180^m	ϵ Persei	60^m
NGC 2392..	7 26.3	+21 02	E 2236	103*a*-F	120^m	ϵ Persei	10^m
					8^m, 2^m	ϵ Persei	60^m
			E 2237	103*a*-F	60^m, 30^m, 15^m, 4^m	....	....
			E 2238	103*a*-F	180^m, 90^m	ϵ Persei	60^m
			E 2268	103*a*-F	480^m	ϵ Persei	60^m
			E 2270	103*a*-F	360^m	ϵ Persei	60^m
			E 2269	I-N	520^m	ϵ Persei	60^m
			E 2282	I-N	430^m	ϵ Persei	60^m
			E 2284	I-N	120^m	ϵ Persei	60^m
			JS 762	103*a*-F	120^m	ϵ Persei	60^m
			JS 764	103*a*-F	194^m	ϵ Persei	15^m, 60^m
			JS 765	103*a*-F	60^m, 60^m†	ϵ Persei	60^m
			JS 766	103*a*-F	336^m	ϵ Persei	60^m
						β Cephei	60^m
			JS 768	103*a*-F	60^m	ϵ Persei	60^m
			JS 776	103*a*-F	410^m	ϵ Persei	60^m
NGC 2440..	7 39.7	−18 05	JS 771	103*a*-F	116^m	ϵ Persei	60^m
NGC 7027..	21 05.2	+42 01	E 1729	I-N	360^m	α Lyrae	60^m
			E 1730	I-N	390^m	α Lyrae	60^m
			E 1731	I-N	330^m	α Lyrae	60^m
						ϵ Persei	60^m
			E 1746	I-N	67^m, 25^m	α Lyrae	60^m
			E 1747	I-N	10^m, 150^m	ϵ Persei	60^m
NGC 7027..	21 05.2	+42 01	E 1748	I-N	360^m	α Lyrae	60^m
						ϵ Persei	60^m
			E 1751	103*a*-F	120^m	α Lyrae	60^m
					4^m	α Lyrae	10^m, 1^m
			E 1752	103*a*-F	80^s, 25^s, 10^s, 40^m	....	....
			E 1753	103*a*-F	13^m	ϵ Persei	60^m, 10^m
			E 1754	103*a*-F	13^m	α Lyrae	10^m
				,	4^m	α Lyrae	2^m, 20^s
			E 1755	103*a*-F	80^s, 25^s, 10^s, 5^s	....	....
			E 1757	103*a*-F	....	ϵ Persei	10^m, 3^m

* A 2-mag. diaphragm was placed in the optical system for this exposure.

† A 1-mag. diaphragm was placed in the optical system for this exposure.

TABLE 1—*Continued*

	α	δ	Plate	Emul.	Exposures	Comparison Star	Exposures
NGC 7662. .	$23^{h}23^{m}.5$	+42°16′	E 2203	I-N	480^{m}	ε Persei	60^{m}
			E 2205	I-N	480^{m}	ε Persei	60^{m}
			E 2220	I-N	120^{m}, 40^{m}	ε Persei	60^{m}
			E 2202	I-N	420^{m}	ε Persei	60^{m}
			E 2204	I-N	450^{m}	ε Persei	60^{m}
			E 2222	I-N	120^{m}	ε Persei	60^{m}
			E 2224	I-N	60^{m}, 2^{m}, 4^{m}, 8^{m}, 15^{m}	. . . .	. . . .
			E 2225	I-N	60^{m}, 30^{m}, 30^{s}, 1^{m}, 2^{m}	. . . .	. . . .
			E 2226	I-N	6^{m}	ε Persei	10^{m}
					15^{m}, 30^{m}	ε Persei	60^{m}
			JS 760	103a-F	425^{m}	. . . .	. . . .
			JS 765	103a-F	360^{m}	ε Persei	15^{m}
NGC 7662. .	23 23.5	+42 16	JS 767	103a-F	15^{m}, 15^{m}†, 15^{m}*, 61^{m}	ε Persei	15^{m}
			JS 773	103a-F	15^{m}, 15^{m}†, 15^{m}*, 30^{m}	ε Persei	15^{m}

Balmer discontinuity, D, is 0.06. For β Cephei (which served as a comparison star for NGC 40), only the Barbier and Chalonge data are available. The Kienle data for η Ursae Majoris give a less steep slope than do the Barbier and Chalonge data. If the two curves are forced to fit at λ 5000, the deviation is about $\Delta \log I = 0.04$ at λ 3900. In the region λλ 5000–6000 both sets of curves are in good agreement. We have adopted the Barbier and Chalonge results.

In the comparison of the nebular-line intensities with the stellar energy distribution, the confluence of the stellar lines at the Balmer limit provides some difficulty. One may draw the energy-distribution envelope over the stellar lines and, using the data of Barbier and Chalonge, make some allowance for this effect. Nevertheless, the intensities of the higher members of the Balmer series may be given only imperfectly.

The observational procedure consisted in observing the nebula and comparison star at as nearly as possible the same mean zenith distance. Exposures on the nebula ranged from a few minutes to about 8 hours, whereas those on the comparison star varied from a few minutes to about 1 hour. It was manifestly impossibly to give the star and the nebula the same exposure for the observations intended to reach the fainter lines and the nebular continuum.

The ultraviolet spectrum of NGC 7027 was observed at McDonald in 1946, but for the other objects it was possible to secure observations only from the limit of the Balmer series to the infrared. All our spectrograms have been traced parallel to dispersion with the Mount Wilson photographically recording microphotometer.

Great care had to be taken to avoid even a tiny error in alignment of these small-scale plates, which can result in considerable difficulties. If the alignment is not perfect, the intensity of Hβ, for example, is measured at a different point in the nebula than is that of Hα or the high numbers of the Balmer series. The only nebula in which difficulties were encountered was NGC 7662. Here the Balmer decrement measured at the center of the bright ring did not agree with those measured at two different points of the outer fainter ring. The discrepancies were systematic and corresponded to a slight systematic error in the alignment of the spectrograms in the microphotometer, amounting to about 40 μ between Hα and λ 3727. Once the alignment error has been estimated, for each wave length it is possible to find the point on the nebular image where the intensity was actually measured. If a plot of intensity versus distance along the slit is now made for each line and a smooth curve drawn connecting the points, one may determine the small correction needed to reduce all intensities to the same selected points in the nebula. The maximum correction for any line at any point in NGC 7662 was about 20 per cent. For the brightest central strip the corrections are negligible. In NGC 40 and particularly in NGC 2392 condensations appear so conspicuous as easily to permit an accurate alignment of the spectra upon the microphotometer. The measurements of NGC 7027, NGC 1535, and NGC 2022 do not indicate any evidence of alignment troubles.

The reduction of the microphometric tracings gives the fundamental observational datum, the ratio $I_\lambda(\text{neb.})/J_\lambda(*)$, where

$$J_\lambda(*) = AI_\lambda(*)\frac{d\lambda}{d\chi}T_\lambda \exp[-k_\lambda(\sec z_s - \sec z_n)] . \quad (1)$$

Here $I_\lambda(\text{neb.})$ is the intensity of the nebular line, A is a constant independent of wave length for each plate, $I_\lambda(*)$ is the intensity in the continuum of the comparison star at the wave length λ, and $d\lambda/d\chi$ is the dispersion on the tracing; T_λ is the transmission of the quartz diffuser (determined by Pettit); k_λ is the atmospheric extinction coefficient; and z denotes the zenith distance of the star or nebula.

We must now discuss the correction for atmospheric extinction. We attempted to observe the comparison star at the same zenith distance as the nebula; and, for the Mount Wilson observations of NGC 7027, the atmospheric extinction corrections are so small that they have been neglected. The maximum error in comparing lines in the λ 6400 region with lines in the λ 3700 region is about 0.01 in the logarithm.

For the subsequent observations of NGC 40, NGC 2392, NGC 2022, and NGC 7662 the effects of differential atmospheric extinction are small. They are relatively large for the few miscellaneous observations we have secured of NGC 1535 and other objects.

During a long nebular exposure, say about 8 hours, z varies. We have calculated the effects of atmospheric extinction and have found that it is sufficiently accurate to use in equation (1) the mean value of z during an exposure for all the observations that have been secured.

The determination of the appropriate absorption coefficient, k_λ, is a somewhat more difficult matter. For the reduction of the observations of NGC 7027 obtained at the McDonald Observatory in 1946, Popper's (1937) values for Mount Hamiltón, corrected for the altitude of Mount Locke in accordance with the procedures suggested by Fowle (1933), have been used. Unfortunately, no direct measures of the extinction were then available.

The atmospheric extinction at Mount Wilson has been studied by a number of observers since the solar investigations of C. G. Abbot (1908). On a number of nights on which we secured observations, photoelectric measures were carried out by Edison Pettit or by Harold Johnson. These measures indicate a sizable fluctuation in the atmospheric extinction from night to night, so we have deemed it best to use for the entire series of our observations a set of mean coefficients adopted from the six-color photometry of Stebbins and Whitford (1945). Since (except for NGC 1535 and a few other objects) the zenith distances of the star and nebula are about the same, the correction for atmospheric extinction is not serious. On the nights devoted to the long exposures the transparency of the sky seemed constant throughout.

Each exposure yields a series of relative intensities of emission lines, the shorter exposures giving the stronger lines; the longer exposures being necessary to get the weaker lines. The derived intensities are to be put on a common system by combining the results from the various plates. To combine results from exposures of two different lengths, lines common to the two plates have to be used. To be determined is the shift of zero point between the two intensity systems. It now appears that more plates of intermediate exposure would have been desirable.

The selection of the proper slit-width for these objects is a serious problem. The slit-width should be sufficiently wide that the top of the line on the trace represents the actual intensity of the line integrated over its width. The photometric accuracy increases as the slit is widened beyond this minimum width; but at the same time the resolution decreases, and faint lines can no longer be separated from the continuous spectrum. The slit-width used at the 100-inch was 0".68, corresponding to 25 μ on the plate for NGC 7027 with its very rich spectrum. For the other objects the slit-width was 1".3, corresponding to 49 μ on the plate for the 100-inch, and 3".7, corresponding to 42 μ on the plate for the 60-inch. From the agreement obtained between various plates, the internal probable error is about 15 per cent for the weaker lines measured on several plates and about 5 per cent for lines of intermediate intensity measured on stronger plates. For individual lines the actual errors may be substantially greater, particularly for the weakest lines, where errors of the order of 30–50 per cent are possible. Smaller errors might have been secured for the stronger lines by widening the spectrograph slit, but the loss of resolution would have obliterated many of the weaker lines.

IV. RESULTS FOR INDIVIDUAL NEBULAE

The results for NGC 7027 have been described elsewhere (Bowen, Minkowski, and Aller 1955). These intensities refer to the brightest condensation in the north-preceding part of the nebula. Our plates also permitted the determination of the intensities in the fainter north-following part of the nebula. The intensities measured in this fainter part are less certain but show no definite signs of stratification effects in the nebula.

The Ring Nebula in Andromeda NGC 7662 has been more intensively studied than almost any other planetary. It is a bright object, favorably placed for long exposures during the best observing season. Olin Wilson has studied the internal motions with the multislit; isophotic contours have been traced from his slitless spectrograms and from direct photographs secured at Palomar by one of us (Aller 1956). The integrated magnitude has been measured photoelectrically by Liller (1955), while Code has measured the stronger lines with his photoelectric scanner.

The slit of the spectrograph was placed north of the central star, across the brightest portion of the bright ring. We traced the 100-inch spectrograms at five points—the center of the bright ring, the inner portions of the outer ring, and the outer portions of the outer ring. The much smaller scale of the 60-inch plates precluded such a detailed study, and we have measured only the brightest part of the bright ring.

Table 2 gives the intensities measured on the 100-inch spectrograms. The measurements in the outer ring showed some scatter, much of which must certainly be attributed to the extremely mottled structure of the outer parts of the nebula.

Table 3 gives the intensities derived from the 60-inch plates. With the spectrograph employed there, it was possible to introduce diaphragms into the optical path in such a way as to cut down the beam by 1 or 2 mag. Thus, keeping the exposure time at the telescope fixed at some convenient interval (15 minutes was actually used), it was possible to measure the intensities of the stronger lines with respect to those of moderate intensity, without resort to the method of combining intensities otherwise employed. Fewer lines were measured on the 60-inch spectrograms. Nevertheless, a comparison of the results for the 100-inch and 60-inch spectrograms shows a reasonably good over-all agreement.

The intensity measurements in the bright ring should be reasonably reliable—at least, probably as good as can be attained with a narrow slit and photographic photometry without a much more extended program. The scatter of the values in the outer ring renders these measures less reliable. The probable errors are larger, not only because the accidental errors are presumed to be greater, but also because the irregularities in the nebular structure are much more marked. It is impossible to set the analyzing beam in the microphotometer always upon the same point in the nebula—and, since the alignment is not perfect and the nebular knots are small, different parts of the knots will be

TABLE 2

Line Intensities in NGC 7662 Measured at the 100-Inch Reflector

Wave Length	Ident.	−10″.8	−5″.4	Center of Ring	+5″.4	+10″.8
8862.8	H			6.2		
8750.5	H			2.8		
8665.0	H			2.0		
8196.6+8236.4	He II			1.5		
8045.6	[Cl IV]		1.0	2.6	2.0	
7751.0	[A III]		2.7	3.7	1.4	
7592.8	He II			0.8		
7530.5	[Cl IV]			1.5		
7319.9+7330.2	[O II]		1.2	2.4	1.5	
7135.8	[A III]	4.6	5.9	13.8	5.0	1.3
7065.3	He I		3.2	4.9	1.7	
6716+6730	[S II]		2.1	1.9	1.5	
6678.2	He I	2.3	5.4	5.9	3.9	1.6
6562.9	Hα	30	155	390	161	85
6363.8	[O I]				0.7	
6300.3+6312.1	[O I]+[S III]		1.6	2.6	1.8	
6086.9	[Fe VII] / [Ca V]			0.68	1.1	
5875.6	He I	3.3	10.2	16	7.0	5.8
5801.4	C IV			pr		
5537.8	[Cl III]			0.9	0.7	
5517.7	[Cl III]			0.6		
5426.6	[Fe VI]			0.33		
5411.5	He II		1.8	6.0	4.8	
5006.9	[O III]	98	710	1700		340
4959.0	[O III]	36	236	550	200	110
4922.0	He I			1.3		
4861.3	Hβ	10.8	36	100	48.7	26.5
4740.2	[A IV]	1.3	3.3	8.0	5.4	2.0
4724.2+4725.6	[Ne IV]			0.8		
4711.3	[A IV]		3.6	7.7	5.9	2.4
4685.7	He II		6.0	51	36	4.9
4658.4	C IV			0.3		
4647.5+4649.1	C III+O II		0.8	1.8		
4640.6	N III		2.1	{5.3	2.6	
4634.2	N III			2.2}		
4541.6	He II		0.7	2.4	1.8	
4471.5	He I	0.75	2.9	4.3	1.7	1.6
4388.0	He I			0.6		
4379.1	N III			0.4		
4363.2	[O III]	2.2	7.1	18.7	9.8	4.8
4340.4	Hγ	3.5	14.9	44	22.5	11.0
4267.1	C II		0.45	1.15	0.8	
4237				0.14		
4227.4				0.20		
4219				0.21		
4199.9	He II		0.3	1.4	1.1	
4187.0	C III		0.3	1.0	0.4	
4180.9	[Fe V]			0.16		
4143.8	He I			0.33		
4120.8	He I			0.31		
4101.7	Hδ	2.05	8.8	23.2	13.3	6.3
4068.6+4076.3	[S II]		0.75	1.8	1.2	
4026.1	He I		1.6	2.8	1.4	
3967.4+3970.1	[Ne III]+H I	3.0	11.6	33	18.1	9.2
3888.9	H I	2.1	6.8	17	8 4	5.1
3868.7	[Ne III]	6.1	30	89	42	21.9
3858.0	He II			0.62		

TABLE 2—*Continued*

Wave Length	Ident.	−10″8	−5″4	Center of Ring	+5″4	+10″8
3835.4	H I	0.8	3.0	7.7	5.0	2.1
3819.7	He I	....	0.83	1.6	1.0	....
3797.9	H I	0.66	2.2	5.7	3.8	1.4
3791.3	O III	....	....	0.8	....	....
3770.7	H I	0.55	1.8	4.6	2.9	1.3
3759.9	O III	....	1.3	5.8	4.5	1.4
3757.2	O III	....	....	2.2	1.7	....
3750.2	H I	0.7	1.4	3.8	2.6	1.2
3734.3	H I	0.5	1.2	3.4	2.3	1.0
3728.8	[O II]}	2.1	{2.9	3.1	2.0	1.0
3726.0	[O II]}		{4.5	5.6	2.9	1.6
3721.8	H I	....	1.5	3.6	1.8	....
3712.0	H I	....	1.1	2.2	1.7	....
3703.8	H I+He I	....	1.2	2.3	1.6	....
3697.2	H I	....	0.7	1.5	....	....
3691.5	H	....	0.6	1.3	....	....
3686.9	H	....	....	0.9	....	....
3683	H	....	....	0.73	....	....

traced at different wave lengths. These difficulties seemed to be particularly great for NGC 7662.

NGC 2392

As in our observations of NGC 7662, the slit of the spectrograph was placed on the brightest portion of the brighter ring south of the nucleus but also included portions of the fainter outer ring at either side. The orientation difficulties, which turned out to be so troublesome in NGC 7662, were minimized for this object because a bright knot in

TABLE 3

LINE INTENSITIES IN NGC 7662 MEASURED AT THE 60-INCH REFLECTOR

λ	I	λ	I
6716+30	2.28	4143.8	0.6
6678.2	4.05	4101.7	28
6563	404	4068+76	1.6
6300.3+6312.1	3.4	4026.1	2.7
5875.6	20	3967+3970	39
5411.5	8.7	3924	0.4
4959.0	370	3888.9	19.6
4861.3	100:	3868.7	87
4740.2	9.5	3835.4	9.7
4711.3	9.6	3819.7	1.1
4685.7	49	3797.9	6.8
4640.6	5.1	3770.7	5.0
4541.6	2.9	3759.9	4.9
4471.5	4.3	3756	1.9
4388.0	0.4	3750.2	3.6
4363.2	19.7	3734.3	3.0
4340.4	46.5	3726+29	10.7
4267.1	0.7	3721.8	3.1
4200	1.2		

the nebulosity fell on the slit on one side so that the plate could be aligned in the microphotometer with its aid.

We give the intensities measured on the E plates at several positions listed in Table 4 near the center of the bright ring and at points chosen on either side, all expressed on the scale $H\beta = 100$ in the center of the bright ring. Also the intensities derived from the 60-inch plate are listed for comparison in Table 5.

TABLE 4

LINE INTENSITIES IN NGC 2392 E SPECTROGRAPH
100-INCH REFLECTOR*

Wave Length	−15″	−10″	Center of Bright Ring	+10″	+15″
7751.0	...	...	4.3	...	...
7592.8	...	...	...	...	...
7319.9+7330.2	...	...	13.5	...	...
7135.8	...	...	17.3	...	...
6716+30	...	5.8	21.2	3.8	...
6678.2	...	...	3.3	...	...
6650	...	...	...	...	...
6562.9	75	72	380	54	56
6300.3+6312.1	...	...	8.7	...	...
5875.6	...	...	9.9	...	...
5411.5	...	...	5.4	...	...
5006.9	212	227	1100	185	194
4959.0	64	70	326	49	53
4861.3	21	23	100.0	19.6	18
4740.2	...	4.1	9.2	4.4	...
4724.2	...	...	2.0	...	...
4711.3	...	5.0	10.0	4.2	...
4685.7	13.2	15.5	57	15.6	13.4
4658.4	...	...	2.4	...	...
4640.6	...	...	1.1	...	...
4634.2	...	...	pr	...	...
4541.6	...	...	2.7	...	...
4471.5	...	...	2.7	...	...
4363.2	6.0	8.2	26.4	7.0	6.0
4340.4	10.9	11.7	44.1	9.7	9.1
4227.4	...	...	3.6	...	...
4200.0	...	...	1.5	...	...
4143.8	...	...	1.0	...	...
4101.7	7.0	7.7	28	6.8	5.8
4076.3+4068	...	...	1.2	...	...
4069	...	...	4.5	...	...
4026.1	...	...	1.7	...	...
3968.7	6.9	8.9	30.5	6.6	6.0
3888.9	5.0	5.8	17.4	4.0	3.0
3868.7	17.4	17.3	78	12.6	14.6
3835.4	...	2.9	7.7	2.6	...
3797.9	...	...	5.6	...	...
3770.7	...	...	4.8	...	...
3760.0	...	...	1.7	...	...
3750.2	...	...	2.9	...	...
3734.3	...	...	2.1	...	...
3727	16.7	13.0	71	5.7	...
3712.0	...	...	1.5	...	...
3703.8	...	...	1.4	...	...

* The sign on the settings +15″, +10″, etc., is arranged so as to agree with the convention for NGC 7662 in the sky. λ 3770 may be blended. [N II] is strengthened at −0″.15 with respect to Hα. Notice the peculiar distribution of λ 4686 with respect to Hβ! The 4227 line is certainly real; no identification has been found.

The remarkable features of this nebula are many. It certainly is not a simple shell structure—and, indeed, may not be shell-like at all. The outer "ring" consists of a number of knots and condensations. The inner ring is broken and irregular. The excitation level is likewise unusual, in that strong [Ne v] lines are observed simultaneously with a strong λ 3727 [O II]. The spectrum of the central star has been studied in some detail. It seems to correspond to a spectral class of about O9.5 and an excitation temperature in the neighborhood of 30000° K (Wilson and Aller 1954).

At first glance, it seems that the outer portion of the nebulosity may be regarded as an amorphous mass whose intensity gradually falls off with distance from the central star. Upon this amorphous mass are superposed numerous knots and condensations, some of which attain a fair size. The knots show a lower level of excitation than does the main body of the nebula. In 1944 and 1945 a number of slitless spectrograms were obtained with the quartz spectrograph at the Crossley reflector of the Lick Observatory. The photometric procedures and reduction methods are described elsewhere (Aller 1951). Measurements of line intensities in the brightest portion of the bright ring are compared in Table 6. Note the great strength of the [Ne v] lines. In addition, the intensities of the lines in some of the bright knots have been measured. We give in Table 7 the values for one of these knots on the scale Hγ = 50. Hβ was not measurable on the slitless plate used. The [Ne v] lines are missing, and λ 3727 is relatively much stronger than in the main body of the nebula. In a weaker knot the relative intensities of λλ 3727, 3868, 4101, and 4340 were found to be 3.0, 3.4, 1.2, and 1.8, respectively. Here the intensity of the amorphous background has been subtracted so that the values refer to the knots them-

TABLE 5

INTENSITIES OF LINES OF INTERMEDIATE STRENGTH IN NGC 2392 AS MEASURED WITH THE 60-INCH REFLECTOR (BRIGHT RING)

λ	*I*	λ	*I*	λ	*I*
6716+30	19.2	4712	9.2	4101.7	19.8
6678.2	4.4	4658.4	3.5	4069+76	4.1
6300+6312	9.4	4640.6	1.6	4026.1	2.8
5875.6	9.5	4541.6	2.7	3888.9	17.0
5411.5	6.7	4471.5	2.5	3835.4	6.7
4861.3	(100.0)	4363.2	18.3	3797.9	6.3
4740.2	6.9	4227.4	3.9	3770.7	5.4
4724	3.0	4200.0	1.7	3727	55.2

TABLE 6

COMPARISON OF INTENSITIES IN NGC 2392 AS MEASURED WITH THE CROSSLEY REFLECTOR AND AT MOUNT WILSON

λ	Ident.	*I* (Mt. W.)	*I*(L)	λ	Ident.	*I* (Mt. W.)	*I*(L)
3203	He II	....	26	3970	[Ne III]	30	30
3341	[Ne V]	....	36	4101	Hδ	28	23
3428	[Ne V]	....	84	4340	Hγ	44	53
3727	[O II]	71	76	4686	He II	57	60
3868	[Ne III]	78	78	4861	Hβ	100	88

selves. The intensities derived from the slit spectra include both the contribution of the knot and that of the amorphous background.

Since much superior slitless spectrograms are now available, it does not seem worth while to discuss the older material further. Isophotic contours of the nebula provide more definitive data on the knots and condensations (Aller 1956).

Fewer lines have been observed than in NGC 7662 or NGC 7027. The nebula is intrinsically fainter, and for a given exposure a smaller number of lines is recorded.

TABLE 7

INTENSITIES IN AN OUTER KNOT OF NGC 2392

λ	Ion	I	λ	Ion	I
3727	[O II]	168	4101	Hδ	26
3868	[Ne III]	73	4340	Hγ	50
3970	[Ne III] +H	26	4686	He II	30

TABLE 8

INTENSITIES IN NGC 40

λ	Ident.	Weak Knot	Center	Strong Knot
6716+30	[S II]	43.7	54.6	97.0
6562.9	Hα+[N II]	541.0	800.0	1500.0
6300	[O I]	7.8	20.0	18.0
5876	He I	19.8	15.3	21.0
5755	N II+Sky	4.9	29.5	9.8
5007	[O III]	33.3	26.6	32.0
4959	[O III]	15.0	13.1	13.0
4861	Hβ	93.1	100.0	166.0
4340	Hγ	34.6	43.0	73.0
4101	Hδ	18.4	22.5	40.0
3969	H	11.2	11.9	25.0
3889	H	12.5	11.6	24.7
3835	....	....	....	10.5
3797	....	....	....	18.7
3770	....	....	....	13.5
3750	....	....	....	7.9
3727	[O II]	172.0	206.0	341.0

NGC 40

This low-excitation nebula is intrinsically rather faint, and only the strongest lines of he usual prominent elements have been measured. The results obtained are disappointng in view of the effort expended. Because of the far northern declination (+70°), it was possible to observe this nebula only with the 60-inch reflector. The slit was set 4″ north of the central star; tracings were made at three different points in the nebula.

Table 8 gives the measured intensities at the three points, the intensity of Hβ being ake as 100. Some of the strongest and some of the weakest lines were measured on one plate only; these are indicated by an asterisk. The lines of [O II], [N II], [S II], and He I re strong. The higher numbers of the Balmer series were measured on only one plate nd gave somewhat discordant values; λ 3869 is blended with [Ne III], λ 3889 is confused with He I, while Hα is strongly blended with [N II]. M. L. White (1952) estimated the average 6584 [N II]/Hα ratio to be 0.76 in this nebula; hence about half the radiation in the red line is to be attributed to [N II].

The N_1 and N_2 lines seem to be concentrated toward the center of the nebula. Furthermore, the positions of the condensations observed in [O III] do not always agree with the positions of the [O II] knots. Hence conditions of excitation vary from point to point in the extremely complex nebula NGC 40. Direct photographs indicate that this nebula does not have a spherically symmetric structure, although there are some vague hints of a basic cylindrical symmetry.

Even after allowing for these blends, the Balmer decrement seems to be too steep, as though the nebula was affected by interstellar absorption. Observations obtained by one of us in 1938 with a slitless spectrograph (Aller 1941) showed also a steep decrement, although the individual intensity values are not in good agreement. The λ 3727 [O II] line is almost twice as bright as Hβ, according to our present measures. Relative to other

TABLE 9

LINE INTENSITIES IN NGC 2022

λ	Ident.	I(Mt. W.)	I(L)	λ	Ident.	I(Mt. W.)	I(L)
6562.9.....	H I	306.0	....	4101.7.....	Hδ	19.4	15.8
5875.6.....	He I	6.6	....	3967.4.....	[Ne III]+H	21.4	20.6
5411.5.....	He II	16.7	....	3888.9.....	H I	8.3	8.7
5006.9.....	O III	....	550.0	3868.7.....	[Ne III]	36.7	35.6
4959.0.....	O III	189.0	197.0	3727.......	[O II]	....	15.4
4861.3.....	Hβ	100.0	100.0	3444.0.....	O III	....	10.9
4740.2.....	[A IV]	9.5	....	3428.7.....	[Ne V]	....	53.2
4711.......	[A IV]	13.2	....	3345.8.....	[Ne V]	....	26.2
4685.7.....	He II	101.0	86.0	3312.3.....	O III	....	8.0
4541.6.....	He II	3.4	....	3203.1.....	He II	....	44.0
4363.2.....	[O III]	10.5	11.2	3132.8.....	O III	....	45.0
4340.4.....	Hγ	41.7	39.1				

lines, it is stronger than in other planetaries, although the λ 3727/Hβ ratio is only slightly larger than in certain of the knots in NGC 2392 or in the ansae of NGC 7009. The [O III] lines are very weak in this nebula.

Further long exposures on this nebula would appear unprofitable. If an intensive study of a low-excitation nebula is planned, it would be better to observe some object such as BD+30°3639, or Anon 18^h15^m, whose surface brightnesses are very much greater. These objects are small, a circumstance that provides no particular difficulty when they are observed with a large telescope. BD+30°3639 has a complex nuclear spectrum; the central star of Anon 18^h15^m seems to have a perfectly continuous spectrum.

NGC 2022

Although it is an intrinsically somewhat faint object, NGC 2022 was selected for study because of its high excitation. It is a rather regular ring nebula, characterized by a particularly strong λ 4686. Earlier work (Aller 1951) had suggested that it was perhaps the highest-excitation regular-shaped nebula.

Our adopted intensities, I(Mt. W.) (see Table 9), are based on plates secured with both the 60-inch and the 100-inch reflectors. For comparison we give the intensities I(L), measured on a spectrogram secured with the Crossley quartz slitless spectrograph at the Lick Observatory in 1944. The two sets of intensities are in good agreement. Although the Lick plate did not register the fainter lines, it does show that the ultraviolet

[Ne v], [O III], and He II lines are strong. The [O II] λ 3727 radiation is present, and the image in this line is large. Tentative intensities previously published by one of us (Aller 1941) were based on a plate of questionable photometric calibration taken in 1934 by Stoy. The present set of intensities should be much more reliable, even though the intensities of some of the lines in the far ultraviolet may be overestimated.

NGC 1535

This beautiful double-ring planetary has a more nearly regular structure than most other representatives of its class. Even so, large-scale slitless spectrograms reveal irregularities in both rings. In spite of its unfavorable declination, the object was included in our program, and exposures ranging from 7 minutes to 5 hours and 22 minutes were secured with the JS spectrograph at the 60-inch telescope.

TABLE 10

LINE INTENSITIES IN NGC 1535 AS MEASURED WITH THE 60-INCH REFLECTOR

λ	Ident.	Bright Ring	Outer Ring	Previous Values* Bright Ring
6562.9	Hα	361.0	48.0	....
5875.6	He I	20.0	8.6	....
4959.0	[O III]	363.0	....	....
4861.3	Hβ	100.0	27.6	100.0
4740.2	[A IV]	5.2	....	....
4712.0	[A IV]	6.7	....	....
4685.7	He II	19.8	....	24.0
4640.2	N III	4.1	....	....
4541.6	He II	4.3	....	....
4471.5	He I	7.2	....	6.0
4363.2	[O III]	16.4	4.3	19.0
4340.4	Hγ	51.6	13.9	49.0
4101.7	Hδ	28.8	7.6	21.0
3969	[Ne III], Hε	40.0	11.0	30.0
3888.9	Hζ	21.8	6.9	21.0
3868.7	Ne III	89.0	22.0	74.0
3835.4	Hη	7.7	....	....

* *Ap. J.*, **113**, 134, Table 3, 1951.

In spite of the small scale of the resulting spectrograms, we measured the intensities both in the bright inner ring and in the outer fainter ring. They are given in Table 10, with the intensity of Hβ in the brighter ring chosen as 100. The high-excitation lines are missing from the outer ring.

Comparison of the measures for the bright ring with values previously published by one of us shows reasonably good agreement. Ultraviolet spectra show that the lines of O III, λ 3444, λ 3346, and He II λ 3203 are present, although much weaker than in other high-excitation objects considered—in accordance with our expectations. Further exposures on this object seem unprofitable unless much faster optics can be employed.

We have also secured observations of the low-excitation objects IC 1747 and J 320 and the higher-excitation objects NGC 2165, NGC 2440, and IC 351. In NGC 2165 and NGC 2440 only lines of intermediate strength have been measured. Wyse (1942) has given eye-estimates of the intensities of all the lines he could observe, and one of us has published intensities for the stronger lines (Aller 1951). Table 11 contains the results for IC 351 and J 320. They are based on one plate per nebula.

In Table 12 we compare the results for NGC 2165 and NGC 2440 with those obtained some years ago by one of us. The Mount Wilson plates cover only lines of intermediate intensity. Exposures were not obtained for the stronger lines; for the weaker lines a higher dispersion would have been appropriate, in order to minimize the effects of the continuum.

NGC 2165 appears as a small, nonuniform disk, whereas NGC 2440 resembles NGC 7027 in a number of features. In particular, NGC 2440 seems to show a combination of low- and high-excitation features. For example, the low-ionization lines of [O I] λ 5755; [O II] λ 3727, and also λ 5200 are stronger in NGC 2440 than in NGC 1535. On the other hand, He II, λ 5411, and λ 4686, as well as λ 4640 N III, and 4740 [A IV] are stronger

TABLE 11

LINE INTENSITIES IN IC 351 AND J 320

λ	IC 351	J 320	λ	IC 351	J 320	λ	IC 351	J 320
6563	350	280	4686	52	...	4101	24	27
4959	300:	310	4363	15	20	3969	31	46
4861	100	100	4340	49	44	3868	78	93

TABLE 12

INTENSITIES FOR NGC 2165 AND NGC 2440

λ	NGC 2165		NGC 2440	
	I Mt. W.	*I* Lick	*I* Mt. W.	*I* Lick
3797	0.50	0.48	...	...
3835	0.55	0.53	1.01	...
3889	1.41	1.80	2.04	...
3967	...	4.30	4.60	3.70
4026	0.22	0.18	...	...
4076	0.21	0.20	0.30	...
4101	1.95	2.60	...	2.60
4200	0.10	0.12	...	...
4340	...	4.60	...	5.20
4363	1.58	2.10	...	2.90
4471	0.44	0.34	0.42	0.27
4541	0.21	0.18	0.22	0.30
4640	0.35	0.47	0.87	1.00
4712	0.66	0.72	1.60	1.60
4740	0.56	0.65	1.45	...
4861	...	10.00	...	10.00
5192–99	...	...	1.95	...
5411	0.72	...	1.1	...
5755	...	...	2.8	...
5875	1.74	...	2.1	...
6085	0.18	...	...	...
6300	0.74	...	4.8	...
6364	0.30	...	1.6	...
6678	0.64	...	1.2	...
6713+6730	0.93	...	4.27	...

than in NGC 1535. The λ 4640 line may arise from a nuclear star, since it is shorter than any of the other lines on the plate. There is little evidence for any stratification effects; the low- and high-excitation radiations appear to arise in contiguous or well-mixed regions.

V. THE CONTINUOUS SPECTRA OF THE PLANETARIES

In addition to their characteristic bright emission-line spectra, the planetaries also show an underlying continuum. Some of it must originate from the recombination of protons and electrons; some of it may come from double photon emission, a mechanism suggested by Breit and Teller (1940) and favored by Greenstein and Spitzer (1951).

In principle, the determination of the energy distribution in the continuum should be a simple process; one compares the nebula with the standard star. In practice, it is difficult primarily because of the *extreme* faintness of the continuum in all objects investigated except NGC 7027. With practicable exposure times, the measured intensities in the blue region are derived from very low photographic densities where slight errors in the calibration-curves lead to appreciable errors in the intensities. If a wide slit is used, this trouble can be partly overcome, but at the price of a spectrum of lower purity, in which the contributions of weak lines are added to the true continuum. Even at best, the presence of the weaker lines makes it difficult to estimate the position of the continuum in much the same way as the presence of stellar absorption lines makes it hard to find the continuum in an absorption-line star. Because of the very small dispersion in the infrared, it is also difficult to take into account the effects of the atmospheric-absorption bands which may be allowed for only approximately.

If the central star is bright, the scattering and diffraction effects produced by the web supporting the secondary mirror may cause trouble. The measured nebular emission, therefore, may be partly affected by diffracted light from the central star. In NGC 2392, which has low surface brightness and a very bright central star, the results obtained suggest strongly that the measured energy distribution in the continuum is affected by stellar light; the measured intensity increases by a factor of more than 6 from λ 8000 to λ 4000. In the light of present knowledge, this result appears impossible, and it seems better to disregard it until the possible effect of scattered light in this case has been investigated. On the other hand, in NGC 7027 the central star is too faint to be observed, and this difficulty disappears, of course.

The energy distribution in the continuum of the nebula, $I_c(n)$, follows from a direct measure of the ratio $I_c(n)/I_\lambda(s)T$, where $I_\lambda(s)$ denotes the energy distribution in the stellar continuum and T the transmission of the quartz diffuser.

NGC 7027 is the only object for which results of some value could be obtained. This nebula has a relatively strong continuum which is prominent at the red end of even moderately exposed plates. The line spectrum, however, is exceedingly rich, so that the location of the continuum is difficult. Observations covering the visual and infrared regions (λλ 4000–8500) were obtained on 103*a*-F and I-N emulsions at the Mount Wilson Observatory, whereas the McDonald plate covers the interval λλ 3200–5000. In the λλ 4000–4700 region the Texas plate shows a steeper gradient than do the Mount Wilson plates, although the fit is not much poorer than one might expect. Using the best average fit on a $1/\lambda$ plot, the deviations in log I are within ± 0.075. When we consider the difficulty in drawing in the continuum, about a 10 per cent discrepancy over the common range is not surprising.

The measured energy distribution is given in Table 13. Both Vega and α Persei were used as comparison stars. The Balmer continuum is much stronger than the general continuum, and the influence of numerous weak and moderately strong emission lines, which operates so disadvantageously in other parts of the spectrum, should play no role. On the assumption that the Balmer continuum is a pure recombination spectrum and that there is no self-reversal, the electron temperature T_ϵ may be obtained from the emission intensity E_λ at two points λ_1 and λ_2 from

$$\log E_1 - \log E_2 = \frac{0.4343hc}{kT_\epsilon}\left(\frac{1}{\lambda_1} - \frac{1}{\lambda_2}\right). \quad (2)$$

The observed intensities are, of course, to be corrected for interstellar reddening. With the differential correction of 1.43 mag. between λ 3750 and λ 8750 found from the ratio of the Paschen and Balmer lines in the following paper (Aller and Minkowski 1956), the measured intensities in the Balmer continuum vary reasonably well as $1/\lambda$ and lead to $T_\epsilon = 12300 \pm 3600$, in satisfactory agreement with the value 14300 derived from line intensities.

The intensities immediately to the red of the Balmer limit are low and could therefore not be determined with great accuracy. They contradict, however, the result by Page

TABLE 13

ENERGY DISTRIBUTION MEASURED IN THE SPECTRUM OF NGC 7027

λ	LOG *I* McDonald *Qf*/2 8270	LOG *I* Mt. Wilson	λ	LOG *I* Mt. Wilson
3300	1.85	....	4800	1.74
3400	1.91	....	5000	1.77
3500	1.97	....	5200	1.85
3600	2.02	....	5400	1.89
3700	1.16	....	5600	1.92
3800	1.19	....	5800	1.95
3900	1.25	....	6000	1.98
4000	1.34	1.39	6200	2.00
4100	1.40	1.44	6400	2.03
4200	1.45	1.49	6600	2.05
4300	1.53	1.53	6800	2.07
4400	1.59	1.58	7000	2.07
4500	1.65	1.62	7500	2.04
4600	1.70	1.67	8000	1.98
4700	1.74	1.70		

(1942), who found constant intensity in this region. The McDonald results lead to a value of $\Delta \log I$ of about 0.9 for the Balmer discontinuity; the Mount Wilson results above λ 4000 carry some suggestion that the McDonald intensities below λ 4300 are too low and that $\Delta \log I$ may be slightly smaller. The observed value would lead to an electron temperature of only 6600° if the two-quantum emission contributed in the amount computed by Spitzer and Greenstein (1951). An electron temperature as low as this is contradicted by the gradient of the continuous spectrum and by the values of around 14000° found from line intensities. It seems necessary to assume that the two-quantum emission is weak in NGC 7027. Actually, the two-quantum emission may be squelched at high densities. Seaton (1954) has made quantitative estimates of the degree of suppression of two-quantum emission as a function of density. Its weakness in NGC 7027 agrees with the suggestion by Seaton and with the result of Aller (1954), who found for the average electron density the value of 1.7×10^4 cm^{-3} and concluded that the local densities in condensations must be much higher, to explain the [S II] transauroral- and nebular-line intensities. It is impossible to find out exactly how much of the two-quantum emission is present. If this emission were entirely absent, the observed discontinuity would lead to an electron temperature of 20000°. This is to be considered as an upper limit.

Redward from the Balmer discontinuity the intensity increases, as would be expected. This contradicts the result of Page, who found nearly constant intensity. But the shape of the intensity distribution between the Balmer limit and λ 8000 defies any theoretical interpretation. This discordance is most probably due to the difficulty in determining the true level of the continuous spectrum. In view of this difficulty, it would not be justifiable to discuss the discordance, which cannot be considered as having any physical significance.

REFERENCES

Abbott, C. G. 1908, *Smithsonian Ann.*, Vol. **2**.
Aller, L. H. 1941, *Ap. J.*, **93**, 236.
———. 1951, *ibid.*, **113**, 125.
———. 1956, *Gaseous Nebulae* (New York: John Wiley & Sons, Inc.), p. 256.
Aller, L. H., Bowen, I. S., and Minkowski, R. 1955, *Ap. J.*, **122**, 62.
Barbier, D., and Chalonge, D. 1940, *Ann. d'ap.*, **3**, 2.
Berman, L. 1930, *Lick Obs. Bull.*, **15**, 86.
Breit, G., and Teller, E. 1940, *Ap. J.*, **91**, 215.
Code, A. 1956, private communication.
Fowle, F. E. 1933, *Smithsonian Misc. Coll.*, **88**, 611.
Kienle, H., Strassl, H., and Wempe, J. 1938, *Zs. f. Ap.*, **16**, 201.
MacRae, D., and Stock, J. 1954, *Nature*, **173**, 589.
Liller, W. 1955, *Ap. J.*, **122**, 235.
Liller, W., and Aller, L. H. 1954, *Ap. J.*, **120**, 48.
Page, T. L. 1936, *M.N.*, **96**, 627.
———. 1942, *Ap. J.*, **96**, 78.
Plaskett, H. H. 1931, *Pub. Dom. Ap. Obs. Victoria*, **4**, 187.
Popper, D. M. 1937, thesis, University of California.
Seaton, M. J. 1954, *Ann. d'ap.*, **17**, 296.
Spitzer, L., and Greenstein, J. L. 1951, *Ap. J.*, **114**, 407.
Stebbins, J., and Whitford, A. E. 1945, *Ap. J.*, **102**, 273.
White, M. L. 1952, *Ap. J.*, **115**, 71.
Williams, R. 1938, *Pub. Obs. Michigan*, **7**, 93.
———. 1939, *ibid.*, pp. 147 and 159.
Williams, R., and Hall, J. 1942, *Ap. J.*, **95**, 225.
Wilson, O. C., and Aller, L. H. 1954, *Ap. J.*, **119**, 243.
Wyse, A. B. 1942, *Ap. J.*, **95**, 359.

Reprinted from *The Astrophysical Journal,* Vol. 124, No. 1, July, 1956.

THE INTERPRETATION OF THE SPECTRUM OF NGC 7027

L. H. ALLER AND R. MINKOWSKI
University of Michigan Observatory
and
Mount Wilson and Palomar Observatories
Carnegie Institution of Washington, California Institute of Technology
Received February 13, 1956

ABSTRACT

The spectrophotometric data for NGC 7027 are analyzed theoretically with a new value of the space absorption. Ionic concentrations derived from different forbidden-line multiplets of oxygen and sulphur substantiate this newer correction. Revised atomic abundances are obtained, but any substantial improvement in the analysis requires theoretical calculations of transition probabilities and target areas for forbidden-line excitation in the ions of elements such as iron.

I. REVISION OF THE ESTIMATE OF SPACE ABSORPTION

The first theoretical analysis of the newer spectrophotometric data of NGC 7027 (Bowen, Minkowski, and Aller 1955) was based on the assumption that the true Balmer decrement was given by Menzel's and Baker's "case B" (1938). This hypothesis was employed to evaluate the space absorption in accordance with the results suggested by an analysis of the Balmer decrements in some forty-odd planetaries (Aller 1951). A critical examination of the spectrophotometric data for NGC 7027 showed that this postulate of a case B Balmer decrement led to serious difficulties for two reasons.

a) Independently of any excitation mechanism, the ratio of the Paschen line intensities to the Balmer line intensities should be

$$\frac{I\,(Pa_n)}{I\,(Ba_n)}=\frac{A_{n3}\nu_{n3}}{A_{n2}\nu_{n2}}. \qquad (1)$$

The Paschen line intensities corrected on the assumption of a case B Balmer decrement were very much too weak. The observational data are not completely decisive on this point, inasmuch as the underlying continuum could have been drawn too high.

b) The energy distribution in the nebular continuum appeared to be very abnormal.

Accordingly, we have attempted to make a new estimate of the space absorption by taking the Paschen line intensities at their face value. The lines utilized in this study are the transitions originating on the levels n = 10, 11, 12, 13, 15, and 16. The theoretical ratio $I(Pa)/I(Ba)$ is nearly 0.306 for all these lines. The observed values are exhibited in Table 1. The intensities of the weakest lines ($I < 1.0$) are very uncertain. The 9015 line falls in a region where the estimate of the background is very insecure. The effect of space absorption is to cut down the intensities of the lines by a factor of 3.76 over a base line $\lambda_1 = 3750$ to $\lambda_2 = 8750$. All the absorption corrections tabulated by Whitford (1948) must therefore be multiplied by a factor of 1.42 for this object.

II. DETERMINATION OF N_ϵ AND T_ϵ FROM THE [N II], [O III] LINES

We must now correct the observed intensities of the emission lines on the basis of our revised space absorption. The suggested corrections, $\Delta \log I$, are given in Table 2. The new estimates of N_ϵ and T_ϵ from the relative intensities of the auroral and nebular lines of [N II] and [O III] differ only slightly from those obtained previously (Aller 1954). The auroral/nebular line-intensity ratios are 0.0129 and 0.046 for [O III] and [N II], respectively. Then, using equations (14) and (15) of the last-cited reference, we find $N_\epsilon = 6.3 \times 10^4$ and $T_\epsilon = 14000°$ K as compared with 7.7×10^4 and 14500° K, respectively.

III. THE REVISED BALMER DECREMENT IN NGC 7027

If the space-absorption correction of Table 2 is now employed in conjunction with the observed intensities (col. 2) of the Balmer lines, we obtain the corrected intensities listed in the third column of Table 3. These are compared with the theoretical intensities

TABLE 1

COMPARISON OF BALMER AND PASCHEN LINE INTENSITIES

n	$\lambda(Pa)$	I	$\lambda(Ba)$	I	$I(Pa)/I(Ba)$
10........	9015	5.5	3798	3.5	1.57
11........	8863	3.5	3770	3.3	1.06
12........	8750	3.1	3750	2.7	1.15
13........	8665	2.7	3734	2.3	1.18
15........	8545	0.8:	3712	1.8	0.44
16........	8502	0.6:	3704	1.7	0.35
Adopted ratio........	....	....	....	....	1.15

TABLE 2

CORRECTIONS FOR SPACE ABSORPTION FOR LINE INTENSITIES IN NGC 7027

λ	$\Delta \log I$	λ	$\Delta \log I$	λ	$\Delta \log I$
3200..........	+0.260	4861..........	0.000	6500..........	−0.190
3600..........	+ .200	5000..........	− .020	7000..........	− .242
4000..........	+ .130	5500..........	− .076	7500..........	− .292
4500..........	+0.053	6000..........	−0.134	8000..........	−0.330

TABLE 3

COMPARISON OF OBSERVED AND THEORETICAL BALMER DECREMENTS FOR NGC 7027

λ (1)	I(Obs.) (2)	I(Corr.) (3)	$I(B)$ (4)	I(Coll.) (5)	$I(\mathfrak{H}=0.15)$ (6)
Hα 6563.....	650	410	255	532	405
Hβ 4861.....	100	100	100	100	100
Hγ 4340.....	33	40	50.5	32.2	40.7
Hδ 4101.....	16.3	21.3	30.5	15.0	22.1
H8 3889.....	8.8	12.6	13.7	5.4	9.1
H9 3835.....	4.6	6.6	9.9	3.6	6.5
H10 3797.....	3.5	5.2	7.4	2.5	4.8

for the thick nebula in radiative equilibrium (Menzel and Baker, case B), column 4, and with the intensity for pure collisional excitation for an assumed electron temperature of 14000° K (Chamberlain 1953), column 5. The observed intensity, corrected for space absorption, is seen to fall between these two extremes, thus suggesting that collisional excitation may play a role in the production of the Balmer line spectrum in this nebula.

Chamberlain (1953) showed that, if both collisions and radiative processes act to populate the levels, the factors b_n, introduced by Menzel (1937) to measure the deviation from thermodynamic equilibrium, may be written as

$$b_n = (1 - \mathfrak{H})\, b_n^B + \mathfrak{H}\, b_n^C \,, \tag{2}$$

where b_n^B and b_n^C are, respectively, the b_n factors computed for pure radiative excitation and for collisional processes in an optically thick nebula. They are given in Table 3 of Chamberlain's paper. The function $\mathfrak{H}$ is given by

$$\mathfrak{H} = \frac{a_\kappa(T_\epsilon)\, T_\epsilon^{3/2}}{G(T_\epsilon)\, K} \frac{N_1}{N_\epsilon} . \tag{3}$$

Here $a_\kappa(T_\epsilon)$ is the collision-ionization parameter,

$$a_\kappa(T_\epsilon) = \int_{v_0}^{\infty} Q_\kappa(v)\, v f(v)\, dv \,, \tag{4}$$

where Q_κ is the cross-section for ionizing collisions, $f(v)$ is the Maxwellian velocity distribution, and v_0 is the velocity corresponding to the ionization energy. $G(T_\epsilon)$ has been tabulated as a function of the electron temperature (Menzel, Baker, and Aller 1938), $K = 3.26 \times 10^{-6}$, and N_1 is the number of atoms in the ground level.

If both collisional and radiative processes are operative, we may write:

$$\frac{I_{n,2}}{I_{H\beta}} = \frac{(1-\mathfrak{H})}{(1-\mathfrak{H}) + (b_4^C/b_4^B)\,\mathfrak{H}} \left(\frac{I_{n,2}}{I_{H\beta}}\right)_B + \frac{\mathfrak{H}}{\mathfrak{H} + (b_4^B/b_4^C)\,(1-\mathfrak{H})} \left(\frac{I_{n,2}}{I_{H\beta}}\right)_{\text{coll}} . \tag{5}$$

At $T = 14000°$ K, $b_4^C/b_4^B = 6.6$. With $\mathfrak{H} = 0.15$, we predict the Balmer decrement given in column 6 of Table 3. The agreement is satisfactory except for Hζ, which is blended with helium λ 3889; Hϵ has to be excluded because of blending with [Ne III]. Let us adopt $T_\epsilon = 14000°$ K and calculate the ratio N_1/N_ϵ from equation (3). We interpolate $a_\kappa(T_\epsilon) = 8.8 \times 10^{-14}$ from Chamberlain's data and take $G(T_\epsilon) = 0.165$ from the Harvard nebular series. With $\mathfrak{H} = 0.15$, we find $N_1 \sim \frac{1}{2} N_\epsilon$, which would imply that in the filaments where the hydrogen lines were produced the gas was only two-thirds ionized! An alternative possibility is that T_ϵ is higher, so that a_κ is much larger and N_1/N_ϵ is correspondingly smaller. At any rate, NGC 7027 appears to be one nebula in which Miyamoto's (1938) suggestion of some collisional excitation appears to be fulfilled, although collisional excitation does not dominate, as it appears to do in certain radio-frequency sources (Minkowski and Aller 1954).

IV. IONIC CONCENTRATIONS OF OTHER ELEMENTS

We now employ the same methods as given previously (Aller 1954) to calculate the concentrations of the other ions. The line intensities to be employed are different because of the lessened influence of the space absorption, and, because of the new choice of N_ϵ and T_ϵ, the η-factors are slightly different. The largest change in the results is produced by the revision in the space-absorption correction. Table 4 lists for each ion the type of transition, the η-factor, the intensity I_n, and, finally, the ionic concentrations both for the revised or "small" space absorption and for the initially assumed or "large" space absorption.

The nebular and auroral lines of [O II] give roughly similar ionic concentrations, and the nebular and transauroral [S II] lines likewise yield concordant results. It now appears that a temperature of 14000° K and $N_\epsilon = 6.5 \times 10^4$ explains the relative intensities of the principal forbidden lines in NGC 7027. It is possible that different filaments have different temperatures and densities, i.e., that no unique value of T_ϵ and N_ϵ can be assigned to the nebula as a whole. The [S II] radiation may be concentrated in slightly denser filaments, the [O II] radiation in less dense filaments. Quite possibly, the hydrogen radiation and the forbidden-line radiation are concentrated in different filaments.

The results for the heavier ions, for which the cross-sections have been estimated rather than calculated, are less encouraging. There are big discrepancies in the con-

TABLE 4

CONCENTRATIONS OF IONS IN NGC 7027

Ion	Type of Transition	η	I_n	Small Space Absorption (ions/cm³)	Large Space Absorption (ions/cm³)
N II	N	0.36	202	1.1	0.9
	A	.00184	9.3	1.0	0.9
O II	N	.860	21.2	1.1	1.9
	A	.0054	40.5	1.6	1.0
O III	N	.0537	1920	12.7	14.3
F IV	N	.0082	0.68	0.008	0.012
Ne III	N	.0026	95	2.0	2.8
Ne IV	N	.00115	4.1	1.3	2.0
Ne V	N	.00146	144	3.5	6.4
S II	N	.374	7.9	0.035	0.027
S II	TA	.00415	8.8	0.039	0.058
S III	A	.00045	10.6	0.80	0.57
Cl IV	N	.132	3.1	0.011	0.013
Cl IV	N	.0055	3.2	0.02	0.007
Cl IV	A	.000214	0.87	0.17	0.16
A III	N	.00259	40.6	0.28	0.18
A IV	A	.031	14.4	0.08	0.12
A IV	N	.000234	5.2	0.3	0.16
A V	N	0.00145	11.4	0.09	0.06
K IV	N	....	2.2	0.022	0.015
K VI	N	....	1.5	0.018	0.018
Ca V	N	....	1.8	0.023	0.025

centrations of Cl IV and A IV ions, depending on whether they are computed from the nebular or the auroral transitions. The discordances clearly indicate the need for more accurate calculations of target areas for collisional excitation. The ions in particular need of attention are S III, Cl III, Cl IV, A III, A IV, and A V. Eventually, cross-sections for K IV, K VI, and Ca V and other metals also should be calculated. We return to this question in Section V.

An additional estimate of the density in the regions where the [O II] lines appear may be obtained from the circumstance that the $I(3726)/I(3729)$ ratio depends on the density (Ufford, Van Vleck, and Aller 1949; Seaton 1954). If we use the notation and cross-sections suggested by Seaton, we find:

$$r = \frac{I\,(^2D_{5/2} - {}^4S_{3/2})}{I\,(^2D_{3/2} - {}^4S_{3/2})} = 1.5\left\{\frac{1 + 3.26\,\chi}{1.02 + [10.4 + 4.0\exp(-\chi_{23}/kT_\epsilon)]\,\chi}\right\}. \quad (6)$$

Here $\chi = N_\epsilon/100\sqrt{T_\epsilon}$. Our formula differs from that of Seaton, in that we have included the influence of the ^2P term in setting up the equation of statistical equilibrium With a ratio $r = 0.49$ for NGC 7027 and $T_\epsilon = 14000°$ K, N_ϵ then turns out to be 1.7×10^4 electrons/cm^3, a value similar to that estimated from the surface brightness (Aller 1954; Liller and Aller 1954) but much smaller than the N_ϵ obtained from the ratio of auroral to nebular line intensities. The effect of averaging over filaments of divers densities may affect the 3726/3729 intensity ratio differently than it affects the nebular/auroral line ratio. Errors in the observed intensity ratio, as well as in the cross-sections estimated by Seaton, may contribute further to the discrepancy.

V. THE COMPOSITION OF THE NEBULA

We must now examine what effects these revised ionic concentrations will have on the estimated composition of the nebula. The second and fifth columns of Table 5 give

TABLE 5

ABUNDANCES OF ELEMENTS

Atom	Small Absorption			Large Absorption	
	Obs. No.	Ionization Factor	Relative No.	Obs. No.	Relative No.
N	1.0	30:	5000:	0.9	4500:
O	14.0	3.8	10000	15.8	10000
F	0.008	3.2	5	0.012	6
Ne	6.8	1.2	1500	11.2	2500
S	0.83	8.2	1300	0.97	1300
Cl	0.18	3.5	120	0.16	86
A	0.69	1.8	240	0.37	110
K	0.040	2.5	18	0.032	13
Ca	0.023	3.6	16	0.025	15
H	....	....	1.17×10^7	....	1.17×10^7
He	....	....	1×10^6	....	1×10^6

the observed relative members of all the ions of each element for both space absorption assumptions. Thus, nitrogen is represented only by [N II], oxygen by [O II] and [O III], whereas neon is represented by [Ne III], [Ne IV], and [Ne V]. We must estimate the corrections necessary to reduce these observed abundances to the actual relative abundances of the elements involved. These ionization corrections appear to be about the same as those estimated earlier; the correction for the more highly ionized atoms seems very uncertain. The same ionization factor is applied to the second and fifth columns. For some elements the effect of the change in space absorption upon the abundance is pronounced.

It is obvious that the compositions of the planetary nebulae can never be estimated properly until the stage of ionization can be accurately determined. The element most suited to such a study (in NGC 7027) is iron. Were the transition probabilities and target areas for the collisional excitation of the lines of [Fe III], [Fe V], and [Fe VII] available, we could make a much better estimate of the distribution of atoms among various stages of ionization.

The correction factors are disagreeably large for sulphur and nitrogen, which are observed only in the lowest stages of ionization. Our factors may be too large for both these elements, which may be concentrated in filaments, so that their actual abundances may be smaller.

It is extremely difficult to find the concentration of neon. The proportion of Ne IV seems to be three or four times too small as compared with Ne III and Ne V. Putting it another way, the [Ne IV] lines are about three times as weak as they should be. Possibly the target areas for the ^2P terms in Ne IV are too high, or maybe there are two types of filaments—one a "high"-excitation and the other a "low"-excitation filament, in which Ne III and still lower-excitation stages prevail. Our estimate of the abundance of neon is therefore extremely shaky. The chlorine and argon concentrations are based on uncertain cross-sections. They do indicate values more nearly in harmony with those found for the stars. The metals are very poorly determined because of the uncertainty in the ionization correction. Further discussion does not appear warranted at this time.

REFERENCES

Aller, L. H. 1951, *Ap. J.*, **113**, 125.
———. 1954, *ibid.*, **120**, 401.
Bowen, I. S., Minkowski, R., and Aller, L. H. 1955, *Ap. J.*, **122**, 62.
Chamberlain, J. 1953, *Ap. J.*, **117**, 387.
Liller, W., and Aller, L. H. 1954, *Ap. J.*, **120**, 48.
Menzel, D. H. 1937, *Ap. J.*, **117**, 387.
Menzel, D. H., and Baker, J. G. 1938, *Ap. J.*, **88**, 52.
Menzel, D. H., Baker, J. G., and Aller, L. H. 1938, *Ap. J.*, **88**, 422.
Minkowski, R., and Aller, L. H. 1954, *Ap. J.*, **119**, 232. An error in the entry for α_3 (10000° K) requires a correction to the values of γ (O I) and γ(O III) in Table 2 as follows: γ(O I) = 62900, 2100, 330; and γ(O III) = 190000, 6400, and 1300; for T_ϵ = 10000, 15000, and 18000, respectively.
Miyamoto, S. 1938, *Jap. Mem. Coll. Sci.*, **21**, 173.
Seaton, M. J. 1954, *Ann d'ap.*, **17**, 74.
Ufford, C. W., Van Vleck, J. H., and Aller, L. H. 1949, *Ap. J.*, **109**, 42.
Whitford, A. E. 1948, *Ap. J.*, **107**, 102.

Reprinted from *The Astrophysical Journal*, Vol. 125, No. 1, January, 1957.

CHEMICAL COMPOSITIONS OF SELECTED PLANETARY NEBULAE

LAWRENCE H. ALLER
University of Michigan Observatory
Received July 25, 1956

ABSTRACT

By methods similar to those used in a previous study of NGC 7027, analyses are carried out for NGC 1535, 2022, 2165, 6741, IC 5217, NGC 2440, 2392, and 7662 and Anon $+21^h31^m$, $+39°$. The observations employed are primarily those made by Minkowski and the writer at the Mount Wilson Observatory. The effects of space absorption and of filamentary structure are discussed, but the most serious errors appear to be those arising from trying to allow for the distribution of atoms among the various stages of ionization. These difficulties are particularly troublesome for nitrogen, which is observed only in the lowest stages of ionization. The neon/oxygen ratio appears to be smaller than in the stars, whereas the sulfur/oxygen ratio appears to be larger. The hydrogen/helium ratio appears to be about 5. No conclusive evidence is found for actual composition variations from one object to another, although the hydrogen/helium and particularly the nitrogen/oxygen ratios present some puzzling questions. If abundance differences are ruled out, we must postulate in certain objects, such as Anon $+21^h31^m$, $+39°$ and NGC 2440, the simultaneous presence of high- and low-excitation filaments.

I. INTRODUCTION

Since the classical investigations of the spectra and chemical compositions of the gaseous nebulae by Bowen and Wyse (1939) and by Wyse (1942), a number of studies have been published. Menzel and the present writer (1945) rediscussed Wyse's data from the point of view of the theoretical developments in the Harvard nebular series. Subsequently, Menzel (1948) and the present writer (1953) attempted to obtain improved abundances with the aid of the older cross-section data or simple modifications thereof. Seaton (1954) discussed ionic concentrations in several objects, whereas the present writer (1954) made a detailed discussion of new observations of NGC 7027. An improved calculation has been made with the aid of a seemingly better estimate of the space absorption (Minkowski and Aller 1956*b*).

Extensive photometric and spectrophotometric data have become available recently for several planetary nebulae (Liller and Aller 1954; Liller 1955; Minkowski and Aller 1956*a*). The theoretical analysis of these data with the aid of target areas for collisional excitation and the Einstein *A*-values of the principal forbidden lines permit some inferences to be drawn concerning excitation conditions of electron temperatures and filamentation. Several of these objects have been discussed by Seaton (1954), but the availability of improved and more extended observational data make a rediscussion desirable. Let us consider each of these nebulae in turn.

II. NGC 7662

a) Observational Data

The bright Ring Nebula in Andromeda ranks with NGC 7027 as one of the most intensively studied planetaries. Following the pioneer spectrophotometric studies of Plaskett (1928) and of Berman (1930), discussions of the spectrum were given by Page (1936, 1942) and by the present writer (1941). The analysis of the important spectroscopic observations of Bowen and Wyse (1939) and of Wyse (1942) led them to the conclusion that the compositions of the gaseous nebulae could be regarded as essentially the same as those of the stars. The same conclusion was reached by Menzel and the present writer (1943, 1945) from a treatment of the same data by the methods of the Harvard nebular series. They also discussed the data for the Bowen fluorescent mechanism (1942) and concluded that it fitted the observed line intensities quantitatively as well as qualitatively.

The observational data have been greatly improved and extended since 1945. Wilson's earlier measurements of the internal motions (1950) have been supplemented by coudé spectrograms secured with the multislit and imager rotator to obtain measurements of the internal motions at many points in the nebular image. Slitless spectrograms secured by Wilson plus Minkowski's direct photographs obtained at the 200-inch with a narrow-band-pass filter-plate combination have been used to obtain isophotic contours (Aller 1956). A combination of the multislit velocity data with the isophotic contours will permit the construction of a three-dimensional model of this nebula. Reliable measurements of the surface brightness in this object have been secured by photoelectric photometry (Liller and Aller 1954; MacRae and Stock 1954; Liller 1955). Tracings with a photoelectric monochromator have also been secured by Code.

b) Structural Characteristics

Small-scale photographs or direct telescopic inspection show that NGC 7662 consists of a bright inner ring and a fainter outer one. Large-scale photographs and the isophotic contours thereof display an extremely mottled structure in the outer ring and a non-uniform distribution of brightness in the inner ring. These irregularities tend to be smoothed over in the hand-drawn isophotic contours. The λ 4686 He II and the [Ne V] radiations are restricted to the inner ring. The [O II] λ 3727 radiation has about equal surface brightness in the inner and outer rings. Hence a uniform shell model cannot be expected to give very reliable results for NGC 7662, and we must be prepared to accept the influence of pronounced density fluctuations within the nebula.

TABLE 1

COMPARISON OF BALMER AND PASCHEN LINE INTENSITIES IN NGC 7662

n	λ (Paschen)	I_P	λ (Balmer)	I_B	Ratio
11.....	8862.8	6.2	3770	4.6	1.35
12.....	8750.5	2.8	3750	3.8	0.73
13.....	8665.0	2.0	3734	3.4	0.59

A Wallenquist type of analysis shows that in the crudest approximation the nebula can be regarded roughly as a two-shell structure. The emission in Hβ in the inner shell is about eight times greater than in the outer shell.

c) Corrections for Space Absorption; Estimates of Ionic Concentrations in Filaments

In order to estimate the amount of space absorption, we may compare the observed and "standard" (Aller 1951) Balmer decrement. Another method, used here, is to compare the Balmer and Paschen line intensities (Table 1).[1] Unfortunately, the data are less accurate for NGC 7662 than for NGC 7027. The mean ratio I_P/I_B is adopted as 0.89, whereas the theoretical ratio is 0.306. We may correct, then, for extinction with the aid of Whitford's data (1947). The total extinction at Hβ is found to be 1.25 mag.

From the surface brightness of Hβ and of the Balmer continuum uncorrected for space absorption, Liller and the writer (1954) estimated the electron density to be 6.8×10^3

[1] This method postulates that the relative populations of the individual states in a given upper configuration are proportional to their statistical weights. In other words, the population of a state is assumed to depend only on n; the dependence on the azimuthal quantum member l is neglected.

electrons/cm^3. Berman's (1937) value of the parallax was assumed. If the surface brightness be increased by 1.25 mag., the corresponding N_ϵ will be 1.2×10^4 electrons/cm^3.

Another estimate of the electron density may be made from the λ 3726, λ 3729 lines of [O II]. Following the notation of Seaton (1953), where

$$x = \frac{N_\epsilon}{100\sqrt{T_\epsilon}}, \tag{1}$$

we substitute in the expression

$$\frac{I\ (3729)}{I\ (3726)} = r = 1.5\left[\frac{1+3.3x}{1.02+(10.4+4.0\,e^{-\chi_{23}/kT_\epsilon})\,x}\right] \tag{2}$$

the value $r = 0.56$ in the bright ring. Adopting, tentatively, $T_\epsilon = 14300°$ K, we find $N_\epsilon = 7.8 \times 10^3$ electrons/cm^3. In the outer ring, ratios of 0.64 and 0.69 are observed. From these ratios we find $N_\epsilon \sim 4.2 \times 10^3$ electrons/cm^3 and 3.25×10^3 electrons/cm^3 for the outer ring.

If we take the density, 7.8×10^3 electrons/cm^3, as applying to the effects of both the inner and the outer shell integrated along the line of sight, we estimate the outer shell as having a density of about 4.0×10^3 electrons/cm^3, and the inner shell as having an electron density of about 1.1×10^4 electrons/cm^3. The latter value checks with the surface-brightness measurements and assumed parallax. It is to be emphasized, however, that the [O II] radiation is probably concentrated in different regions than are those of the other ions.

Let us tentatively adopt $N_\epsilon = 1.5 \times 10^4$ electrons/cm^3 as representative of the lower limit to the density in the filaments where the [O III] lines are formed and take the intensity of λ 4363 = 2.08 from the Mount Wilson photometry (Minkowski and Aller 1956), while that of $(N_1 + N_2)$ is taken as 180.0, from the photoelectric photometry (Liller and Aller 1954). We employ the expression

$$\frac{I\ (4363)}{I\ (N_1+N_2)} = 13.1 \times 10^{-14300/T_\epsilon}\left\{\frac{1.01+2300\sqrt{T_\epsilon}/N_\epsilon}{1.01+2.7\times 10^5\sqrt{T_\epsilon}/N_\epsilon}\right\} \tag{3}$$

to find $T_\epsilon = 14200°$ K. In objects where both the auroral and the nebular lines of [O III] and of [N II] are present, we may solve for both N_ϵ and T_ϵ, provided that both sets of lines are formed in the same strata. Unfortunately, in NGC 7662 the data for the [N II] lines are less complete. These [N II] lines were not observed by Minkowski and Aller, although Wyse listed the λ 5755 and λ 6564 line intensities as 1.5 and 12, respectively. This auroral/nebular line intensity ratio cannot be reconciled with any reasonable T_ϵ unless the [N II] lines are formed in denser filaments, as is, indeed, possible.

We may also compare the auroral and nebular lines of [O II] with those of [O III]. With an electron temperature T_ϵ of the order of 13000°–14000° K, we find that an electron density of about 7×10^3 electrons/cm^3 will reproduce the observed intensity ratio of the [O II] λ 7330 and λ 3727 lines. This density is in accord with that found from the 3729/3727 line intensity ratio and tends to confirm the assignment of the [O II] lines to the regions of lower density.

A comparison of the intensities of the transauroral and nebular lines of [S II] shows no accord with the previously mentioned densities. The observed intensity ratio cannot be reproduced even with a density as high as 10^5 electrons/cm^3. Possibly the measured intensities are subject to abnormally large errors or the theoretical cross-sections may not be sufficiently accurate. There is no substantial difference in the [S II] line ratios between the inner and the outer ring, suggesting that the [S II] lines always tend to be produced in filaments of relatively high density.

In calculating the occupation numbers for the metastable levels of the various ions in NGC 7662, the selection of an average density based on the surface brightness of a smooth shell of uniform density would appear unrealistic. The [O II] lines certainly appear to originate in the regions of lower density, but the [S II] lines must be produced in regions of much higher densities. Probably a variety of filaments exists, which differ from one another in size, density, and temperature.

Unless one were able to observe the spectrum of a single filament, one would have to accept the composite contributions of a number of condensations differing in size and density. In the spectrophotometric work on NGC 7662 the slit was placed across the nebula, passing through the bright ring. Hence the lines photographed upon the plate represented contributions from elements along the line of sight. The larger filaments give the mottled appearance to the ring, but smaller ones may lie below the limit of resolution of the photographs. The probable range in the densities and temperatures in the condensations and filaments in NGC 7662 severely limits the accuracy of our analysis. As a compromise between the high densities suggested by the [S II] lines and the much lower ones demanded by the [O II] lines, I have adopted, for the "average" ion,

$$N_\epsilon = 7 \times 10^4 \text{ electrons/cm}^3 , \qquad T_\epsilon = 13000°\text{K}$$

for the purposes of calculating the corrections to the Bolzmann formula (cf. Aller 1954, Sec. III). The calculation of these η-factors and the estimation of the ionic concentrations then proceed exactly as in the NGC 7027 discussion.

d) Calculation of Ionic Concentrations

Let N_{1i} denote the number per cubic centimeter of ions of type i in the ground term (essentially the total number of ions of the given kind) and let N_{2i} be the number in the second metastable term, while N_{3i} denotes the number in the third metastable term. Then we define the η-factors as follows:

$$\frac{N_{3i}}{N_{1i}} = \eta_3 \frac{\varpi_3}{\varpi_1} \exp\left(-\frac{\chi_3}{kT_\epsilon}\right), \tag{4}$$

$$\frac{N_{2i}}{N_{1i}} = \eta_2 \frac{\varpi_2}{\varpi_1} \exp\left(-\frac{\chi_2}{kT_\epsilon}\right). \tag{5}$$

Formulae for the η's for which cross-sections have been published by Seaton (1953) have been given previously. To this list we may add the expression for the auroral lines of [Ne IV] which are observed in some nebulae. Let us define

$$A_1 = \left[\left(0.19 + 500 \frac{\sqrt{T_\epsilon}}{N_\epsilon}\right) + 10^{-2.62\theta}\right]^{-1} \tag{6}$$

and

$$B_1 = 3.5 + 3.7 \times 10^5 \frac{\sqrt{T_\epsilon}}{N_\epsilon}, \tag{7}$$

where $\theta = 5040/T_\epsilon$. Then

$$\eta_3 = \frac{0.23 + 0.7\,A_1}{0.23 + B_1 (1 - A_1 \times 10^{-2.62\theta})}. \tag{8}$$

Lines of [S III], [Cl III], [A III], and especially [A IV] are observed in the nebulae, but the concentrations of the corresponding ions may be estimated only with the aid of extrapolated cross-sections (Aller 1954). The corresponding expressions for the η's will not be given here.

The ionic concentrations are computed from the nebular-line intensities (cf. Aller 1954, eq. [18]) as follows:

$$\frac{N_i}{N_0}=\frac{I_{ni}}{I_0}\left[\frac{5}{9}\frac{\lambda}{\lambda_0}\frac{\varpi_1}{\varpi_2}\frac{A_0}{A}\right]\frac{\overset{0}{\eta_2}}{\eta_2}10^{(\chi-\chi_0)\theta}$$
$$=C_\lambda\frac{I_{ni}}{I_0}\frac{\overset{0}{\eta_2}}{\eta_2}10^{(\chi-\chi_0)\theta}. \tag{9}$$

The subscripts "0" refer to the [O III] lines, λ 5007 and λ 4959. Thus η_2^0 is the correction factor for the $^1D_2$ level of O^{++}, $\chi_0 = 2.48$ ev, $A_0 = 0.028$, I_0 is the sum of the intensities of the two green nebular lines, and I_n is the sum of the intensities of the nebular transitions. The statistical weights of levels (2) and (1) are $\tilde{\omega}_2$ and $\tilde{\omega}_1$, respectively. For auroral or transauroral lines, I_n is replaced by I_A or I_{TA}, $\tilde{\omega}_2$ by $\tilde{\omega}_3$, and η_2 by η_3. In each instance A refers to the transition probability of the line in question and χ to the excitation potential of the upper level.

TABLE 2

IONIC CONCENTRATIONS IN NGC 7662 AS DEDUCED FROM FORBIDDEN LINES

Ion (1)	Transition (2)	C_λ (3)	η (4)	$I_n/I_0 \times 10^4$ (5)	N_i (6)
N II........	*N*	9.20	0.390	51.5	0.08
Ne III.......	*N*	0.0832	.00306	690	4.0
Ne IV.......	*A*	0.0183	.00154	3.8	0.4:*
Ne V........	*N*	0.0366	.0017	114	0.8:
S II.........	*N*	7.82	.402	5.7	0.007
S II.........	*TA*	0.033	.0050	10.1	0.012
S III........	*A*	0.049	.00053	8.5	0.19
Cl III.......	*N*	1.63	.126	5.7	0.007
Cl IV.......	*N*	0.159	.0063	10.3	0.014
A III........	*N*	0.102	.0030	48	0.092
A IV........	*N*	0.441	.034	75	0.12
A IV........	*A*	0.0086	0.00029	....	....

* A colon indicates a very uncertain value.

Column 1 of Table 2 lists the ions whose forbidden-line intensities have been measured in NGC 7662. Column 2 gives the type of transition—nebular (*N*), auroral (*A*), or transauroral (*TA*). Column 3 gives the coefficient, C_λ, while Column 4 gives the coefficient, η ($\eta_0 = 0.062$ for O^{++} for the adopted T_ϵ and N_ϵ). Column 5 gives the ratio of the intensity of the forbidden line in question to the green nebular lines. These intensities refer to the whole forbidden-line multiplet as corrected for space absorption. Finally, Column 6 gives the ionic concentration, N_i, of the number of ions per cubic centimeter. The concentration of O^{++} ions is found to be $N(O^{++}) = 18.6$ ions/cm³ with the aid of equations (28) and (29) of the NGC 7027 discussion. Here $N_{\text{hydrogen}} = 7 \times 10^4$ and $T_\epsilon = 13000°$ K.

The hydrogen/helium ratio may be found from the recombination He I and He II lines. A hydrogenic theory may be applied to the He II lines. The Pickering lines are weak, and no reliable intensity ratio is available for λ 3203; hence I have confined my attention to the $I(4686)/I(4861)$ ratio. The ratio 0.54, after correction for space absorption, gives 0.076. I have adopted here Wurm's (1955) extrapolation of the factor $b_4(He^+)$. To the degree that the He II and hydrogen emissions are concentrated in regions of different density and temperature, the results are necessarily uncertain. If the temperature in the zone where the He II emission is primarily produced is substantially higher than has been assumed, the concentration of helium will be greater.

The He I λλ 5876, 4471, and 4026 line intensities were analyzed first with Goldberg's (1941) theory and then with Mathis' (1956) new treatment of the triplet levels. The older theory gives a ratio of He^+/H of 0.038, whereas Mathis' theory, which explicitly allows for the finite lifetime of the 2^3S level, gives 0.107. Adopting the results from the Mathis theory, we find the final hydrogen/helium ratio to be 5.5.

e) Distribution of Atoms among Various Stages of Ionization

The distribution of atoms among various levels of ionization is calculated by comparing the relative numbers of O^+ and O^{++} ions, S^+ and S^{++} ions, He^+ and He^{++} ions, and Ne^{++}, Ne^{+++}, and Ne^{++++} ions. The low-energy shape of the curve seems reasonably well defined, but the high-energy end of the curve is uncertain. I have adopted the distribution function shown in Table 3 to represent the distribution of atoms among various stages of ionization (Aller and Menzel 1945; cf. table, p. 261). The distribution function is narrower than that employed for NGC 7027. That is, there is a smaller spread of atoms among the possible ionization stages, i.e., there are relatively fewer neutral atoms or highly ionized atoms. The high-energy portion of the curve can be improved only with the aid of measurements of the lines of highly ionized atoms for which target areas and transition probabilities are known. The absence of lines of iron in high stages of ionization, [Fe VII],]Fe X], etc., gives a limit to the distribution function. Unfortunately, a meaningful ionization-curve cannot be obtained from theoretical considerations.

TABLE 3

ASSUMED IONIZATION DISTRIBUTION FUNCTION FOR NGC 7662

Ion. Potential (ev)	Distr. Func.	Ion. Potential (ev)	Distr. Func.	Ion. Potential (ev)	Distr. Func.
0.00	0.00	50	0.88	100	0.09
10.0	.08	60	.74	110	.03
20	.38	70	.58	120	0.00
30	.70	80	.41		
40	0.90	90	0.34		

TABLE 4

TOTAL CONCENTRATION OF OXYGEN IN NGC 7662

Ionization stage	II	III	IV	V	VI	S_1	S_2
Ionization potential of next lower stage (ev)	13.6	35.0	54.7	77.1	113.4		
Ionization function	12	82	82	45	0	82	223
N_{ion}		18.6				2.7	

The use of the ionization distribution function for the estimation of elemental abundances is illustrated as follows: Successive rows of Table 4 list, for oxygen, the stage of ionization, the ionization potential of the next lower stage of ionization, the relative concentrations from the assumed distribution function, and, finally, the last row gives the observed numbers of ions. In this instance we give only the number of O III (O^{++}) ions, since the [O II] radiation is produced primarily in different strata than is the [O III] radiation. The sum, S_2, of the third row is 223, while the observed ions contribute only the fraction 82/223 of this amount. Hence the observed ionic concentration has to be multiplied by the factor 2.7 to obtain the total concentration of 50 oxygen atoms/cm^3 in all stages of ionization. Since the adopted hydrogen concentration is 7×10^4 ions/cm^3, the hydrogen/oxygen ratio is about 1400. A broader distribution-curve with a flatter maximum would give a greater oxygen concentration and hence a smaller hydrogen/oxygen ratio.

The final results for the ionic concentrations in NGC 7662 are shown in Table 5. The second column gives the relative numbers of ions summed over all observed stages of ionization. The third column gives the factor by which the observed concentrations have to be multiplied in order to obtain the total concentration, and the fourth column gives the relative numbers of atoms, expressed on the scale oxygen = 10000. Notice the embarrassingly large factor by which the observed nitrogen concentration (as N^+) has to be multiplied in order to get the total nitrogen concentration! For comparison I include also the results published previously for the nebula and based on approximate cross-sections (Aller and Menzel 1945) and the values for NGC 7027 (Minkowski and Aller 1956), but with an improved estimate of the helium abundance.

TABLE 5

COMPOSITION OF NGC 7662

Atom	Relative N_i	Correction Factor	Total Number	Aller-Menzel 1945	NGC 7027
H	See text	1.00	1.4×10^7	3×10^7	1.1×10^7
He	See text	1.00	2.5×10^6	4×10^6	2.1×10^6
N	16	13	200	540	5400
O	3700	2.7	10000	10000	10000
Ne	1070	1.2	1260	240	1400
S	40	5.5	220	(180)	1300
Cl	4.1	2.3	9.4	260	120
A	43	2.1	91	90	240

Comparing NGC 7662 and NGC 7027, we notice the much larger concentrations in the latter of sulfur, chlorine, and especially nitrogen. Before attributing these discrepancies to real differences in composition, we must stress again the much greater range of ionization in NGC 7027. It is probable that the ionization effects are much more pronounced than are the possible differences. That is, our allowance for the unobserved stages of ionization in nitrogen, chlorine, and sulfur may have been inadequate. The [N II] lines in NGC 7662 are remarkably weak, and I am not convinced that the observed differences are due to abundance fluctuations rather than to excitation differences. The N III lines of the Bowen fluorescent mechanism are fairly strong in NGC 7662. Unfortunately, they cannot be used to estimate the concentration of the doubly ionized nitrogen atoms because of the extreme complexity of the radiation field of the N III line. If we assume a much steeper drop of the ionization-curve for low energies or that the higher ionization stages are more fully occupied or both, larger abundances of the elements nitrogen, chlorine, and sulfur will be obtained.

The composition differences found between NGC 7027 and NGC 7662 illustrate the difficulties inherent in this type of analysis. Several nebulae must be analyzed in the hope that a judicious comparison of results obtained from different objects may permit a meaningful "average" abundance to be derived.

III. NGC 2392

a) Line-Intensity Data and Space-Absorption Corrections

The extremely complex nebula NGC 2392 has a high level of excitation. The filamentary structure is obvious in visual examination in the telescope and is strikingly exhibited in the direct photographs secured by Minkowski and the slitless spectrograms by Wilson. The isophotic contours are very intricate (Aller 1956). We would scarcely expect densities derived from average surface brightnesses to be strongly correlated with the density found from the individual condensations. The published photographic spectrophotometric data refer to particular filaments. Since the nebula has a relatively low surface brightness as compared with NGC 7662 or NGC 7027, it has not been possible to reach the fainter lines.

The electron density derived from photoelectric measurements of the surface brightness and Berman's (1937) parallax (Liller and Aller 1954) was 1.1×10^3 electrons/cm^3. The electron temperature derived from the [O III] lines was 19600° K, one of the highest values found for any planetary nebula.

Unfortunately, no lines of the Paschen series have been observed, so that the space absorption must be estimated in some other way. On the assumption that the true Balmer decrement corresponds to Menzel and Baker's (1938) case B, I have estimated the amount of space absorption (cf. Table 6). This procedure gives an upper limit to the space-absorption correction, as collisional excitation tends to steepen the Balmer decrement, but the assumption of a purely radiative decrement appears to hold for most planetary nebulae. Calculations will be carried out with both assumptions, i.e., that the space absorption is as given in Table 6 and that there is no space absorption.

TABLE 6

CORRECTIONS FOR SPACE ABSORPTION TO BE APPLIED TO LINE INTENSITIES IN NGC 2392

λ	Δ log I	λ	Δ log I	λ	Δ log I
7751..........	−0.28	4861.........	0.00	3727.........	+0.120
6563..........	− .17	4340.........	+ .05	3428.........	+ .160
5876..........	− .10	4101.........	+ .08	3203.........	+0.180
5411..........	−0.05	3868.........	+0.10		

The larger correction for space absorption leads to an extinction of about 1 mag. at Hβ; hence the electron density derived from the smooth-shell assumption should be increased to about 1.8×10^3 electrons/cm^3. If we use this density to solve for T_ϵ from the intensity ratio, $\lambda\, 4363/(N_1 + N_2) = 0.021$, we find $T_\epsilon = 19700°$ K. For our present calculations we shall take $T_\epsilon = 20000°$ K throughout.

b) The Ionic Concentrations

Let us now fix our attention on the auroral and nebular lines of [O II] and on the transauroral and nebular lines of [S II]. We start with the electron density $N_\epsilon = 1.8 \times 10^3$ suggested by the surface-brightness measurements and try successively higher densities. Let us consider, first, the data for ionized sulfur (Table 7). Successive columns in the table give the assumed electron density, the correction factors to the Boltzmann formula η_3 and η_2 for S II, and the correction factor η_0 for the [O II] lines. The intensities of the nebular and transauroral lines of [S II] are 13.7 and 6.4 if correction is made for space absorption and 21.2 and 5.7 if no correction is made for space absorption. If the correct density is chosen, the ionic concentration calculated from the nebular and transauroral lines must be the same. If we assume no space absorption, the density of the filament turns out to be in the neighborhood of 1.5×10^4 electrons/cm^3.

The situation with respect to [O II] is shown in Table 8. Successive columns give the electron densities, the correction factors η_2 and η_3, and the computed concentrations based on nebular- and auroral-line intensities 94 and 7.7 (absorption) or 71 and 13.5 (no absorption). If absorption is assumed, the required density is about 2.5×10^3 electrons/cm^3. If no absorption is assumed, the required density becomes about 7×10^3 electrons/cm^3. In order to illustrate the influence of space absorption and the filamentary structure on the ionic concentrations, I have carried out the calculation of abundances

TABLE 7

ESTIMATED CONCENTRATION OF SINGLY IONIZED SULFUR IN NGC 2392

N_ϵ	$\eta_3 \times 10^4$	$\eta_2 \times 10^2$	$\eta_0 \times 10^3$	Transauroral Lines		Nebular Lines	
				Abs.	No Abs.	Abs.	No Abs.
1.8×10^3........	0.127	2.11	1.36	31	28	4.8	7.4
2.5.............	0.204	2.88	1.90	27	24	4.9	7.6
3.4.............	0.322	3.85	2.58	23	21	5.0	7.7
7..............	1.00	7.40	5.30	15	14	5.4	8.3
10..............	1.79	10.0	7.57	12	11	5.7	8.8
20..............	4.90	17.1	15.15	9	8	5.6	10.2

TABLE 8

IONIC CONCENTRATION OF SINGLY IONIZED OXYGEN IN NGC 2392

N_ϵ	η_2	$\eta_3 \times 10^4$	Nebular Lines		Auroral Lines	
			Abs.	No Abs.	Abs.	No Abs.
1.8×10^3.........	0.173	0.39	72	54	88	150
2.5...............	.225	0.63	77	59	75	130
3.4...............	.281	1.00	84	64	64	113
7................	.441	2.9	111	84	45	80
20................	0.675	11.8	206	156	32	43

for three assumptions concerning the density in the filaments, i.e., $N_\epsilon = 1.8 \times 10^3$ electrons/cm^3 (as suggested by the mean surface brightness or by the [O II] lines with absorption), $N_\epsilon = 7 \times 10^3$ electrons/cm^3 (as suggested by the [O II] lines with no absorption), and $N_\epsilon = 2 \times 10^4$ electrons/cm^3 (as indicated by the [S II] lines). The electron temperature is adopted as 20000° K throughout. Calculations have been carried out on two assumptions, with and without space absorption (cf. Table 6). In deriving the final abundances, I have, however, assumed in each instance the same ionization-curve. The columns headed "c" denote the values derived from the intensities corrected for space absorption as indicated by fitting the observed Balmer decrement to Menzel and Baker's case B. The columns headed "u" indicate the results obtained when it is assumed that there is no space absorption. The lower part of Table 9 gives the final abundances.

Notice that the final abundances do not depend very sharply on whether we assume low, medium, or high densities in the filaments, provided that all different types of ions are mixed together and that O^+ ions are not concentrated, for example, in filaments of one density and S^+ ions in filaments of other densities. On the other hand, the results do depend on whether or not we assume space absorption.

The hydrogen/helium ratio depends, of course, on the temperature assumed. If we suppose that both the He II and the He I radiations are produced in the same filaments as are the hydrogen lines, we get a hydrogen/helium ratio of about 7. The hydrogen/

TABLE 9

NGC 2392: IONIC CONCENTRATIONS N_i AND TOTAL ABUNDANCES

ASSUMED ELECTRON DENSITY		$N_\epsilon=1.8\times10^3$		$N_\epsilon=7\times10^3$		$N_\epsilon=2\times10^4$	
		$N_i\times10^{-4}$		$N_i\times10^{-3}$		$N_i\times10^{-3}$	
Observed Ion	Type of Transition	*c*	*u*	*c*	*u*	*c*	*u*
O II	N	62	46	37	28	200	148
	A	75	129	15	27	31	41
O III	N	1200	1200	463	463	1360	1360
Ne III	N	290	225	112	89	330	255
Ne IV	A	85	84	27	27	76	74
Ne V	N	410	285	160	113	465	322
S II	N	4.1	6.2	1.8	2.8	5.5	9.9
	TA	26	23	5.1	4.5	8.8	7.6
S III	A	68	90	26	36.1	72	99
A III	N	14.4	24.5	5.6	97	16.4	28
A IV	N	13.2	12.5	5.1	5.0	15.7	14.9
H		3.2×10^7					
He		4.5×10^6					
O		10000	10000	10000	10000	10000	10000
Ne		1500	1100	1450	1100	1450	1060
S		1900	2500	1900	2300	1700	2300
A		180	240	180	240	195	240

oxygen ratio is near 3×10^3. As in the other gaseous nebulae, the chief uncertainty lies in the apportionment of the various unobserved ions in the different stages of ionization.

One important difference between NGC 2392 and such objects as NGC 7009 and NGC 7662 is that in the latter objects the He II radiation is confined to the inner shell, whereas in NGC 2392 there appears to be no substantial blocking of the far-ultraviolet radiation by the inner regions. The He II radiation appears in the wisps and blobs in the outer shell as well as in the inner parts of the nebula.

IV. NGC 2440

NGC 2440, very similar in some respects to NGC 7027, possesses a number of features of interest. A very complete study of the spectrum in the visual and photographic region was carried out by Wyse (1942). Previous abundance estimates of nitrogen and oxygen by Menzel and the writer (1945) and by Seaton (1954) indicated that nitrogen was more abundant than oxygen. This conclusion came about largely because of the great strength of the [N II] lines in this object (Wyse 1942; White 1952) and the assumptions concerning the distribution of atoms among the various ionization stages. Observations of the Balmer decrement (Aller 1951) indicate that the correction for space absorption must be small.

From the [N II] and [O III] lines Seaton, who used Wyse's eye-estimates of intensity, derived $T_\epsilon = 17000°$ K and $N_\epsilon = 4 \times 10^4$ electrons/cm³. On the other hand, I have adopted the intensities from the Lick and Mount Wilson photometric work as given in the accompanying table. From these data are obtained $T_\epsilon = 13000°$ K and $N_\epsilon = 2.9 \times 10^4$. The lower electron temperature and density arise primarily from the differ-

λ 4363............	[O III]	2.90	λ 5755...........	[N II]	2.8
(N_1+N_2).........	[O III]	300	λ 6548, 84........	[N II]	104

ent values adopted for the line intensities. Table 10 has been computed on the basis of this lower N_ϵ and T_ϵ. Columns 1, 2, and 3 are self-explanatory. Column 4 gives the adopted intensity. The starred values are taken from Wyse's eye-estimates. Unfortunately, the photometric measurements in the visual region were obtained from only one

TABLE 10

CONCENTRATION OF IONS IN NGC 2440

Ion (1)	Trans. (2)	λ (3)	I (4)	C_λ (5)	η (6)	N_i (7)
N I.........	*N*	5200	19.5	625	0.909	1.18
N II........	*N*	6560	1034	9.20	.213	2.35
O I.........	*N*	6322	10.4	3.88	.0212	0.11
O II........	*N*	3727	83	60.1	.783	1.30
Ne III.......	*N*	3895	55	0.0832	.00127	0.66
Ne IV.......	*A*	4712	3*	0.0183	.00032	1.32
Ne V........	*N*	3400	220	0.0366	.00070	2.98
S II.........	*N*	6722	{42, 12*}	7.82	.260	0.020
S II.........	*TA*	4072	{3.0, 8*}	0.0329	.00136	0.021
S III........	*A*	6311	12*	0.0491	.00020	0.60
Cl III.......	*N*	5522	8n*	1.64	.0635	0.016
A IV........	*N*	4720	23	0.441	0.0152	0.073
O III........	....	....	3000	....	....	11.0

plate. Column 5 gives the factor C_λ. NGC 2440 has prominent [N I] and [O I] lines. The cross-sections for the collisional excitation of these lines do not vary as v^{-2}, but the necessary collisional parameters are given by Seaton (1954). Column 6 gives the correction factor for the Bolzmann formula, while column 7 gives the ionic concentrations based on a density of 11 O^{++} ions/cm³. The O^{++} density is derived from the intensity ratio $I(H\beta)/I(N_1 + N_2)$.

The next step is the estimation of the concentration of atoms among the different stages of ionization. It is here that our present estimates differ so substantially from those made earlier. The ionization-distribution graph prepared in 1945 showed a rise to a flat maximum near 40 ev, followed thereafter by a slow decline to the higher energies. Measurements of the [Ne V] lines were not then available; ionization stages above 60 ev were not observed. The recent photometric work shows that the [Ne V] lines in this object are very strong, the concentrations of Ne V:Ne IV:Ne III being in the ratio roughly 8:4:2 with our assumed electron temperature of 13000° K. Hence I have drawn the ionization graph with a maximum near 100 ev and a steep fall-off thereafter. The last part of the curve is purely conjectural, being based on the fact that fairly high stages of ionization of iron, e.g., [Fe VI] and [Fe VII], appear to exist, although the coronal lines of [Fe X] have not been observed. The ionization-curve is assumed to fall to zero at about 180 ev. Thus the ionization-curve differs substantially from that for NGC 7027, where the maximum was placed near 50 or 60 ev. The correction factors for a number of elements—nitrogen, sulfur, chlorine, and argon—are unpleasantly large (see Table 11).

An inspection of the results in the last column of Table 11 shows another difficulty The ratio of hydrogen or helium to the heavier elements is cut down by a factor of about 10 as compared with other nebulae. Possibly the ionization-curve has been drawn too high at the high energies, or possibly there exist at least two types of filaments, one a low-excitation type, in which the radiations of [O I], [N I], [O II], and [N II] are produced, and another a high-excitation type, in which the [O III], [Ne III], [Ne V], etc., emissions appear. The two-filament-type model is thus akin to that suggested for the solar chromosphere where "hot" and "cool" structures are required to account for the observed features of solar-line profiles and spectroheliograms. If the two-element model is adopted, there is no possibility of estimating abundances. Rather, one should assume the abundances and attempt to obtain the filament characteristics. No direct observational evidence exists in NGC 2440 for such filaments, and they may, in fact, be intertwined, but the simultaneous strength of low- and high-excitation features is remarkable. The [O II] λ 3727 and the [Ne V] lines both attain great strength in this nebula, and it seems difficult to reconcile both with a state of affairs resembling those believed to obtain in more typical objects, such as the ring nebulae NGC 7009 and NGC 7662. In the absence of any observational basis for postulating filaments of different excitation types, however, I have carried out the analysis on the assumption that all lines are produced in filaments of essentially the same temperature and density. Such a hypothesis leads to a consistent picture of the nebulae, better agreement being obtained than for NGC 7662, but low relative abundances of hydrogen and helium are found. The spectrum of NGC 2440 cannot be reconciled with a "standard" composition and at the same time a unique electron temperature and density for the production of all its spectral features. Until the hypothetical filaments can actually be verified, further speculation is unjustified. A search for the forbidden lines of highly ionized metals would be profitable for this bright planetary. Further spectrophotometry on the lines is also needed.

TABLE 11

NGC 2440

Element	Obs. No. of Ions	Correction Factor	Relative No. of Atoms	Element	Obs. No. of Ions	Correction Factor	Relative No. of Atoms
H	2.9×10^4	1.0	2.6×10^6	Ne	5.0	1.6	880
He	5.1×10^3	1.0	4.6×10^5	S	0.61	9.8	680
N	3.51	16	6500	Cl	0.016	19	35
O	12.4	7	10000	A	0.073	11	91

A nebula remarkably similar to NGC 2440, although of a somewhat lower surface brightness is Anon $21^h31^m + 39°22'$. The electron density is about $8 \times 10^3/cm^3$ and $T_\epsilon \sim 20000°$ K. The [Ne V] lines are stronger than in any other planetary so far observed, but the [N II] lines are also strong. If allowances are made for the distribution of atoms among various stages of ionization in any plausible way, nitrogen is either as abundant as oxygen or perhaps more abundant. The finally estimated abundances for this object, appended in Table 15, are in good agreement with those found for other planetaries for hydrogen, helium, neon, and argon.

V. NGC 2165

The bright southern planetary NGC 2165 has been studied by Wyse (1942), whose observations were discussed also by Menzel and the writer (1945). The [Ne V] lines are present, although weaker than in NGC 2440, and the whole level of excitation in this object appears to be somewhat below that of NGC 7027, 7662, or 2392.

The first uncertainty in the analysis of this nebula stems from the fact that the electron density cannot be estimated from the forbidden lines (with the now available data) or from the surface brightness, since no photoelectric measurements of the surface brightness have been made. Earlier estimates (Aller and Menzel 1945) gave $N_\epsilon = 2.4 \times 10^3$ electrons/cm^3, and we shall use this value for a preliminary estimate of the electron temperature. This temperature, derived from the ratio of the λ 4363 and the green nebular line, turns out to be 14500° K. We shall adopt $T_\epsilon = 14000°$ K for the following analysis.

The observed values of the intensities of the [S II] lines cannot be reconciled with the N_ϵ and T_ϵ values we have used. The measurements for the red [S II] lines obtained by Minkowski and the writer are in sharp disaccord with those of Wyse; hence, since the former are based on very limited observational material, we are justified in disregarding the [S II] line material until better observations are obtained. In Table 12 we give for each line the adopted intensity (asterisks refer to values taken from the paper of Wyse [1942]), the factor η, and, finally, the ionic concentrations, computed on the basis of $N(O^{++}) = 0.454$. This value is derived from a comparison of the green nebular lines and Hβ, with $N_\epsilon = 2.4 \times 10^3$ and $T_\epsilon = 14000°$ K.

The adopted ionization-curve rises steeply from around 10 ev to a maximum between 40 and 60 ev. Thereafter it falls off gradually to zero at about 140 ev. The correction factor is very large and uncertain for nitrogen, moderate for sulfur and argon, and small for oxygen and neon. The final concentrations are estimated to be as follows: nitrogen, 0.74; oxygen, 1.55; neon, 0.28; sulfur, 0.22; argon, 0.016 atoms/cm^3. If the same temperature is assumed to hold in the regions where the hydrogen, He I, and He II emissions are produced, the hydrogen/helium ratio is found to be about 5.7.

Isophotic contours in the brighter monochromatic radiations have been measured on slitless plates secured at Mount Wilson by the technique described by Wilson. The nebula shows a small disk with some internal structure, but there seem to be no pronounced filamentary concentrations. Unfortunately, the southern declination of this object and its position in the winter skies make it difficult to secure images with critical definition.

VI. NGC 1535

The beautiful double-ring planetary NGC 1535 somewhat resembles NGC 7662 and NGC 3242, except that the level of excitation is lower. The rings are somewhat smoother, and the filaments are less marked. Attempts have been made, using Crossley reflector data (Aller 1951), to establish the density distribution in the two shells. Isophotic contours have been measured from Mount Wilson slitless spectrographs. The present study, which refers to the inner shell, is based on spectrophotometric measurements on slitless plates secured in 1944 at the Lick Observatory (Aller 1951) and plates obtained with the Mount Wilson 60-inch reflector (Minkowski and Aller 1956). Improved values of T_ϵ (based on the Seaton cross-sections) and of N_ϵ have been obtained from the photoelectric measurements of surface brightness (Liller and Aller 1954). The electron density turns out to be 2.8×10^3, and the electron temperature to be 14300° K. There is no means of calculating an independent relation between N_ϵ and T_ϵ with the aid of forbidden lines and our present data. The 3727 [O II] radiation appears to be relatively weak in this nebula.

The ionic concentrations of oxygen, neon, and argon are calculated on the assumption of a density of 3×10^3 hydrogen ions/cm^3. The ionization-curve is relatively well defined. It rises from a small value near 10 ev to a maximum near 50 ev, falls off rapidly at the higher energies, and drops essentially to zero just beyond 100 ev. The total numbers of observed ions of each element and the factors by which these numbers are to be multiplied is shown in the small Table 13. The [Ne V] lines are very weak, He II λ 4686

TABLE 12

IONIC CONCENTRATIONS IN PLANETARIES

ION	$\bar{\lambda}$	Trans.	I	η	$N_i \times 10^4$
			NGC 2165		
N II	6560	N	150*	0.0202	200
O II	3727	N	31	0.293	65
O III	5000	N	1810	0.00214	4540
Ne III	3895	N	113	0.0001	900
Ne IV	4724	A	2*	0.0000142	800
Ne V	3400	N	80	0.000055	760
S II	{6722	N	9.3	0.000024	7.2
	{4072	TA	2.1	0.032	2.4
S III	6312	A	5*, 7.4	1.31×10^{-5}	250
A IV	4720	N	13.0	0.00126	27
			NGC 1535		
O II	3727	N	8.0	0.288	3
O III	5000	N	1550	2.55×10^{-3}	4700
Ne III	3895	N	105	0.125×10^{-3}	960
Ne V	3400	N	<8	0.69×10^{-4}	<85
A IV	4720	N	12	1.58×10^{-3}	29
			NGC 2022		
O II	....	....	15.4	0.140	1.9
O III	....	....	743	1.06×10^{-3}	700
Ne III	....	....	47	5×10^{-5}	131
Ne V	....	....	79	2.8×10^{-5}	247
A IV	....	....	23	0.64×10^{-3}	17.4

is present, but He I is weak. The ratio of hydrogen to helium turns out to be in the neighborhood of 6.5.

VII. NGC 2022

The nebula NGC 2022 was selected for detailed study because of its high excitation. Unfortunately, its low surface brightness precluded our measurement of anything except the strongest lines, so that our analysis of the heavier elements is restricted to oxygen, neon, and argon. The [N II] lines have not been observed.

From a surface brightness in N_1 measured by Liller and the writer at Mount Wilson in 1954, the photographically determined intensity ratio $N_1/H\beta$, Berman's parallax, and an assumed relative shell thickness $d/A = 0.2$, we have derived, as density, $N_\epsilon = 840$ electrons/cm³. The electron temperature obtained simultaneously from the λ 4363/$(N_1 + N_2)$ ratio is 16000° K.

Direct photographs demonstrate a filamentary structure in this nebula. Hence, for the subsequent calculations, I have taken $N_\epsilon = 1.3 \times 10^3/\text{cm}^3$, corresponding to an increase in density by a factor of 1.5, a conservative estimate of the influence of the filamentary structure on the nebular densities.

Notice that Ne V is about twice as abundant as Ne III and that O II is negligible in comparison with O III. Similarly, doubly ionized helium is about three times as abundant

TABLE 13

NGC 1535

Element	Total No. Observed Atoms	Correction Factor for Ionization	Relative No. Atoms
O...............	0.47	2.7	1.3
Ne..............	.10	2.0	0.20
A...............	0.0029	3.0	0.0087

TABLE 14

Ionic Concentrations in NGC 6741 and IC 5217

Ion	Trans.	NGC 6741		IC 5217	
		I	N_i	I	N_i
N I..........	N	8	9	....	....
N II..........	N	640	12	60	0.016
O I..........	N	73	4	12	0.024
O II..........	N	103	5	20	0.013
O III..........	N	1940	40	2000	1.0
Ne III..........	N	132	10	112	0.22
Ne V..........	N	<80	<8	....	....
S II..........	TA	7	6	3.5	0.004
S III..........	A	16	4	6	0.054
Cl III..........	N	18	2.5	....	....
A IV	N	21	4	15	0.006
A V..........	N	12	2	....	....

as singly ionized helium. It is difficult to draw an ionization distribution-curve for this object, since the high-energy portion of such a curve is naturally conjectural.

Table 14 includes some results for NGC 6741 and IC 5217, based on the observations by Wyse (1942) and the writer (Aller 1951). In NGC 6741 it is possible to solve for N_ϵ and T_ϵ simultaneously with the aid of the forbidden-line data of [N II] and [O III]. The resultant electron density, 1×10^5 electrons/cm³, is much higher than that indicated by the surface brightness, $\sim 2 \times 10^3$ electrons/cm³, but I have adopted the higher density in the numerical analysis. Then the electron temperature is 11000° K. The behavior of the nitrogen lines suggests that in this object also there may be filaments of differing excitation. Observations of the ultraviolet [Ne V] lines shows that this nebula has a lower excitation than NGC 2440, NGC 7662, or NGC 2392. In IC 5217 I have adopted $T_\epsilon = 11500° K$ and $N_\epsilon = 3 \times 10^3$ electrons/cm³.

VIII. DISCUSSION OF THE RESULTS

Table 15 summarizes the results we have obtained. For comparison, the composition of NGC 7027 (Minkowski and Aller 1956) is also listed. The concentrations are all given in terms of oxygen as 10000.

The most striking feature of this tabulation is the scatter of the relative abundances. The argon/oxygen ratio ranges over a factor of about 4. Chlorine is observed in only four of these nebulae. It puts in a weak appearance in NGC 7662 and is strong in NGC 7027. Sulfur ranges over a factor of about 9 except for NGC 6741, where it seems to have a very high concentration, possibly because of filamentary effects. It is prominent in those nebulae with strong [S II] lines, such as NGC 2392; the "abundance" is relatively independent of the level of excitation of the nebula.

TABLE 15

COMPOSITIONS OF PLANETARIES

Element	NGC 7027	NGC 7662	NGC 2392	NGC 2440	NGC 2022	NGC 1535	NGC 2165	NGC 6741	IC 5217	Anon 21^h31^m $+39°22'$
H	1.17×10^7	1.4×10^7	3×10^7	0.263×10^7	2.7×10^7	2.3×10^7	1.6×10^7	1×10^7	2×10^7	1.7×10^7
He*	2.1×10^6	2.5×10^6	4.5×10^6	0.46×10^6	4.8×10^6	3.6×10^6	2.7×10^6	2.4×10^6	4.4×10^6	3.8×10^6
N	5450	206†	....	6500	....	....	4800:‡	11000:	870	12000
O	10000	10000	10000	10000	10000	10000	10000	10000	10000	10000
F	5	....	....	....	....	....	....	....	....	....
Ne	1430	1260	1430	880	1900	1545	1840	3800	2400	1900
S	1270	220	1870	675	....	....	1400	3600	1360	740
Cl	120	9	....	35	....	....	....	150	....	....
A	236	91	180	91	310	68	100	81	106	123
K	16	....	....	....	....	....	....	....	....	....
Ca	16	....	....	....	....	....	....	....	....	....

* Based on b values calculated by Mathis (1956).

† Very weak lines.

‡ : = very uncertain.

The neon/oxygen ratio is surprisingly stable, although neon appears enhanced in NGC 6741. Both elements are usually observed in two or more stages of ionization, and the extrapolation procedure is less likely to lead to gross errors than in the case of an element such as nitrogen that is observed in only one stage of ionization, particularly a low stage. The neon/oxygen ratio appears to be definitely lower than in the stars. No object has been found in which neon is as abundant as oxygen, nor can any juggling of the data increase the neon abundance by a factor of the order of 5 or 7. One possibility remains: a gross error in the collision cross-section could account for the discordance, but it seems unlikely that Seaton's cross-sections for neon could be overestimated by so large a factor.

Fluorine, unfortunately, depends on a pair of weak lines observed in one object, NGC 7027, with the 200-inch coudé. Potassium and calcium are also observed only in NGC 7027.

The behavior of the nitrogen [N II] lines is so erratic that it must be a clue of considerable importance to the structure and physical state of these gaseous nebulae. Wyse's investigation showed how strikingly the intensities of the [N II] lines varied from one object to another, without regard to the level of excitation of the nebula. They are strong, for example, in NGC 7027 and NGC 2440 and extremely weak in NGC 7662 and NGC 2392. Some years ago White (1951) undertook a systematic survey of these lines in a number of planetaries. His studies confirm the capricious behavior characteristics of these lines. Minkowski has obtained direct photographs with appropriate plate-filter combinations to isolate the [N II] plus Hα radiations. These photographs show that the filaments and fine structure have a crisp appearance as compared with the [O III] radiations. Baade (1942) called attention to the concentration of the [N II] and [O III] radiations to different filaments in the shell around DQ Herculis. Thus it is suggested that, in objects such as NGC 2440, NGC 6741, or Anon $+21^h31^m$ $+39°$, we cannot tell whether nitrogen is really more abundant than oxygen or whether it is primarily an excitation effect involving filaments of different levels of excitation.

Perhaps the nitrogen abundance can be established most reliably (if it can be ascertained at all) only in low-excitation objects where a concentration in higher ionization stages is unlikely. Probably IC 418 would be the most fruitful object for a study of this sort; I have obtained a tentative oxygen/nitrogen ratio of about 3 for this object from a discussion of its structural features. Although IC 418 has a large-scale structural pattern, there is no direct observational suggestion of small-scale filaments such as appear to exist in other planetaries.

The determination of the ratio of hydrogen to helium is of special interest since both stages of ionization can be observed. In his analysis of the statistical equilibrium of the

TABLE 16

MEAN COMPOSITION OF PLANETARY NEBULAE

Element	Present Investigation	Aller and Menzel (1945)
H.	1.7×10^7	4.0×10^7
He.	3.2×10^6	4×10^6
N.	4000:	8000
O.	10000	10000
F.	4	4
Ne.	1500	400
S.	900	1400
Cl.	34	80
A.	130	60

He I lines in a planetary nebula, Goldberg (1941) treated all the high triplet levels as a single level. Hence the variation of the *b*-factors from one level to another is not taken into account. Furthermore, we have assumed that the He I, He II, and H lines are all formed in strata of the same temperature and density. By selecting special regions of the nebula, it may be possible to do better than by taking the radiation from the whole nebular image.

The utilization of Goldberg's (1941) theory leads to an appreciable variation of the hydrogen/helium ratio from one object to another. Thus NGC 6886 and IC 5217 were found to give ratios of about 12, while NGC 2165, J 900, IC 2003, and NGC 6309 all gave a ratio of about 9. The smallest ratio, between 6.3 and 7 was given by the high-excitation objects NGC 6741, NGC 2440, Anon 21^h31^m, and NGC 2022. Part of the fluctuation may be real, but some of it is probably spurious. The lowest ratios tended to be given by the nebulae showing the strongest 4686 lines. I assumed that Menzel and Baker's case B holds for both hydrogen and helium.

Very recently, Mathis (1956) has computed greatly improved *b*-values for the triplet levels in helium. His *b*-values are smaller than those given by Goldberg, since he allows for the finite lifetime of the 2^3S level. Hence substantially smaller hydrogen/helium ratios tend to be found, but, what is equally interesting, the hydrogen/helium ratio is now more nearly constant. Tentatively, we may accept a mean value of 5.3 as the hydrogen/helium ratio for the planetaries discussed in this paper.

Table 16 gives the suggested mean abundances calculated on the scale (oxygen =

10000). For comparison I also give the numbers found in the earlier survey (Aller and Menzel 1945). Much of the difference is to be attributed to the employment of Seaton's greatly improved cross-sections for collisional excitation and to the availability of improved observational data over a more extended spectral range. As mentioned earlier, the largest difference with respect to the stars seems to lie in the smaller concentration of neon in the nebulae. Sulfur appears as though it were more abundant in the planetaries and in the Orion Nebula (Liller and Aller 1956) than in the stars, while both the stellar and the nebular data for chlorine and argon are uncertain. Forbidden lines of fluorine, calcium, and potassium were observed only in NGC 7027, and I have not repeated the data for calcium and potassium here.

A substantial improvement in these numbers must await a better understanding of the filamentary structure of these nebulae. On the one hand, direct photographs may yield significant clues, while theoretical intensity ratios for all the observed density-sensitive forbidden-line doublets of ions with a p^3 configuration will prove most useful. We would then be able to construct more realistic filamentary models than is possible at present.

What is urgently needed, in addition, is a hydrodynamical theory that is capable of taking the kinematical data, the information on ionization, mean densities, and temperatures and then predicting the properties of the individual filaments, including their sizes and density and temperature fluctuations. The shapes of many nebulae indicate the presence of some order-creating agency, probably a magnetic field, so that it is likely that what we need is a magnetohydrodynamical theory of nebular structure.

This investigation was assisted by a grant from the National Science Foundation. I am deeply grateful to Dr. John Mathis, who made a copy of his thesis available to me in advance of publication.

REFERENCES

Aller, L. H. 1941, *Ap. J.*, **93**, 236.
———. 1951, *ibid.*, **113**, 125.
———. 1953, *Astrophysics* (New York: Ronald Press Co.), **2**, 327, 329.
———. 1954, *Ap. J.*, **120**, 401.
———. 1956, *Gaseous Nebulae* (New York: John Wiley & Sons).
Aller, L. H., and Menzel, D. H. 1945, *Ap. J.*, **102**, 239.
Baade, W. 1942, *Pub. A.S.P.*, **54**, 244.
Berman, L. 1930, *Lick Obs. Bull.*, **15**, 86.
———. 1937, *ibid.*, **18**, 73.
Bowen, I. S., and Wyse, A. B. 1939, *Lick Obs. Bull.*, **19**, 1.
Goldberg, L. 1941, *Ap. J.*, **93**, 244.
Liller, W. 1955, *Ap. J.*, **122**, 240.
Liller, W., and Aller, L. H. 1954, *Ap. J.*, **120**, 48.
———. 1956, *A.J.*, **61**, 183.
MacRae, D., and Stock, J. 1954, *Nature*, **173**, 589.
Mathis, J. 1956, Thesis, California Institute of Technology.
Menzel, D. H. 1951, *Trans. I.A.U.*, **7**, 471.
Menzel, D. H., and Aller, L. H. 1941*a*, *Ap. J.*, **93**, 178.
———. 1941*b*, *ibid.*, **94**, 436.
Menzel, D. H., and Baker, J. G. 1938, *Ap. J.*, **88**, 52.
Minkowski, R., and Aller, L. H. 1956*a*, *Ap. J.*, **124**, 93.
———. 1956*b*, *ibid.*, **124**, 110.
Page, T. L. 1936, *M.N.*, **96**, 604.
———. 1942, *Ap. J.*, **96**, 78.
Seaton, M. J. 1953, *Ann. d'ap.*, **17**, 74.
———. 1954, *M.N.*, **114**, 154.
White, M. L. 1952, *Ap. J.*, **115**, 71.
Whitford, A. E. 1948, *Ap. J.*, **107**, 105.
Wilson, O. C. 1950, *Ap. J.*, **111**, 279.
Wurm, K. 1954, Liége conference: *Les Processus nucléaires dans les astres* (Liége, 1954).
Wyse, A. B. 1942, *Ap. J.*, **95**, 356.

PART 4. THE INTENSITIES OF "FORBIDDEN" SPECTRAL LINES

Reprinted from THE PHYSICAL REVIEW, Vol. 57, No. 3, 225–234, February 1, 1940

The Computation of Quadrupole and Magnetic-Dipole Transition Probabilities

GEORGE H. SHORTLEY
Harvard College Observatory, and Mendenhall Laboratory of Physics, Ohio State University, Columbus, Ohio
(Received November 29, 1939)

The theory of quadrupole radiation is formulated in such a fashion as to make available all the methods of computation and sum rules which have been developed for electric-dipole radiation. Formulas for the strengths are given for the SLM_SM_L scheme as well as the $SLJM$ scheme. The problem of the relative strengths of the multiplets in a transition array is treated, and formulae are given that explicitly solve this problem for all possible transitions between two-electron configurations. Values for $np\,n'p \rightleftarrows np\,n''p$, $np\,n'p \rightleftarrows np^2$, $np\,n'p \rightarrow np\,n'p$ and $np^2 \rightarrow np^2$ are tabulated. The spectroscopic stability method and the eigenfunction method of obtaining relative multiplet strengths are given a convenient formulation; and values are computed for $np^3 \rightarrow np^3$. A root $S^{\frac{1}{2}}$ of the line strength is defined in such a way that it transforms like a Hermitian matrix component. This enables convenient computation of line strengths in intermediate coupling. The J-file sum rule is shown to hold in intermediate coupling with the same degree of generality as for the electric-dipole case. Finally, closed formulas for the magnetic-dipole strengths in LS coupling are given and the method of transformation to intermediate coupling is indicated.

THE following considerations were developed in preparation for a complete tabulation of all quadrupole and magnetic-dipole strengths within the configurations p^2, p^3, and p^4, for all possible cases of intermediate coupling, by a group at the Harvard College Observatory.[1]

I. QUADRUPOLE RADIATION

For the case of quadrupole transitions, the *relative* strengths of the Zeeman components of a line and of the lines of a multiplet were first computed by Rubinowicz[2] and later, from group theory, by Brinkman.[3] The J sum rule was shown to be obeyed within the multiplet. But nothing has been done, from a general point of view, in connection with the relative or absolute strengths of the multiplets in a transition array in LS coupling, and in connection with other coupling schemes of importance, such as intermediate coupling, the SLM_SM_L scheme, and jj coupling. Condon[4] has computed the absolute strengths of most of the lines of p^2, p^3, and p^4, for LS coupling and small

[1] G. H. Shortley, L. H. Aller, J. G. Baker and D. H. Menzel, "Tabulation of Strengths of Forbidden Lines in p^2, p^3, p^4 as a Function of Coupling," to appear soon in the *Astrophysical Journal.*
[2] A. Rubinowicz, Zeits. f. Physik **61**, 338; **65**, 662 (1930).
[3] H. C. Brinkman, *Zur Quantenmechanik der Multipolstrahlung*, Dissertation, Utrecht, 1932.
[4] E. U. Condon, Astrophys. J. **79**, 217 (1934).

departures therefrom; but always by tediously writing explicitly one of the intermediate-coupling eigenfunctions in the *SLJM* scheme for each separate level and computing from these the line strengths. It turns out, however, that all the simplifications which are used in the computation of ordinary dipole radiation[5] are available also in the case of quadrupole radiation; similar methods of computation may be used and similar sum rules are obeyed. In this note we give a formulation of the theory which is very convenient in this connection, and outline, frequently by analogy with the dipole case, the necessary proofs.

Preliminaries

The strength $\mathcal{S}_q(\alpha J, \alpha' J')$ of a quadrupole line, in terms of which the probability of spontaneous transition from the upper level αJ to the lower $\alpha' J'$ has the value

$$A_q(\alpha J, \alpha' J') = \frac{1}{2J+1}\,\frac{32\pi^6\sigma^5}{5h}\,\mathcal{S}_q(\alpha J, \alpha' J'),$$

is defined by

$$\mathcal{S}_q(\alpha J, \alpha' J') = \mathcal{S}_q(\alpha' J', \alpha J) = \sum_{M,M'} |(\alpha J M|\mathfrak{N}'|\alpha' J' M')|^2. \tag{1}$$

Here $\mathfrak{N}'$ is the dyadic

$$\mathfrak{N}' = -e\sum_i (\boldsymbol{r}_i\boldsymbol{r}_i - \tfrac{1}{3}r_i^2\mathfrak{J}), \qquad (\mathfrak{J} = \boldsymbol{ii}+\boldsymbol{jj}+\boldsymbol{kk}) \tag{2}$$

the summation running over the electrons. In (1), the absolute square of a dyadic means the double dot product of the dyadic and its complex conjugate, which is the sum of the absolute squares of the nine elements of the dyadic.

The formulas of Rubinowicz for the elements of the Hermitian matrix occurring in (1), but in a notation patterned after that used by TAS for the dipole components, and which will prove extremely convenient, are

$$\begin{aligned}
(\alpha\, J\, M|\mathfrak{N}'|\alpha'\, J\, M\pm2) &= (\alpha\, J\vdots N\vdots\alpha'\, J)\tfrac{1}{2}[(J\mp M)(J\mp M-1)(J\pm M+1)(J\pm M+2)]^{\frac{1}{2}}\,\mathfrak{K}(\pm2),\\
(\alpha\, J\, M|\mathfrak{N}'|\alpha'\, J\, M\pm1) &= (\alpha\, J\vdots N\vdots\alpha'\, J)\tfrac{1}{2}(2M\pm1)[(J\mp M)(J\pm M+1)]^{\frac{1}{2}}\,\mathfrak{K}(\pm1),\\
(\alpha\, J\, M|\mathfrak{N}'|\alpha'\, J\, M) &= (\alpha\, J\vdots N\vdots\alpha'\, J)6^{-\frac{1}{2}}[3M^2-J(J+1)]\,\mathfrak{K}(0),\\
(\alpha\, J\, M|\mathfrak{N}'|\alpha'\, J-1\, M\pm2) &= (\alpha\, J\vdots N\vdots\alpha'\, J-1)\\
&\qquad(\pm\tfrac{1}{2})[(J\mp M)(J\mp M-1)(J\mp M-2)(J\pm M+1)]^{\frac{1}{2}}\,\mathfrak{K}(\pm2),\\
(\alpha\, J\, M|\mathfrak{N}'|\alpha'\, J-1\, M\pm1) &= (\alpha\, J\vdots N\vdots\alpha'\, J-1)\tfrac{1}{2}(J\pm2M+1)[(J\mp M)(J\mp M-1)]^{\frac{1}{2}}\,\mathfrak{K}(\pm1),\\
(\alpha\, J\, M|\mathfrak{N}'|\alpha'\, J-1\, M) &= (\alpha\, J\vdots N\vdots\alpha'\, J-1)M[\tfrac{3}{2}(J-M)(J+M)]^{\frac{1}{2}}\,\mathfrak{K}(0),\\
(\alpha\, J\, M|\mathfrak{N}'|\alpha'\, J-2\, M\pm2) &= (\alpha\, J\vdots N\vdots\alpha'\, J-2)\tfrac{1}{2}[(J\mp M)(J\mp M-1)(J\mp M-2)(J\mp M-3)]^{\frac{1}{2}}\,\mathfrak{K}(\pm2),\\
(\alpha\, J\, M|\mathfrak{N}'|\alpha'\, J-2\, M\pm1) &= (\alpha\, J\vdots N\vdots\alpha'\, J-2)(\pm)[(J-M)(J+M)(J\mp M-1)(J\mp M-2)]^{\frac{1}{2}}\,\mathfrak{K}(\pm1),\\
(\alpha\, J\, M|\mathfrak{N}'|\alpha'\, J-2\, M) &= (\alpha\, J\vdots N\vdots\alpha'\, J-2)[\tfrac{3}{2}(J-M)(J+M)(J-1-M)(J-1+M)]^{\frac{1}{2}}\,\mathfrak{K}(0).
\end{aligned} \tag{3}$$

Here the $\mathfrak{K}(\Delta M)$ represent the dyadics

$$\mathfrak{K}(\pm2) = \tfrac{1}{2}[\boldsymbol{ii}-\boldsymbol{jj}\pm i(\boldsymbol{ij}+\boldsymbol{ji})], \quad \mathfrak{K}(\pm1) = \tfrac{1}{2}[\boldsymbol{ki}+\boldsymbol{ik}\pm i(\boldsymbol{kj}+\boldsymbol{jk})], \quad \mathfrak{K}(0) = \sqrt{\tfrac{2}{3}}\,[\boldsymbol{kk}-\tfrac{1}{2}\boldsymbol{ii}-\tfrac{1}{2}\boldsymbol{jj}]. \tag{4}$$

The factors $(\alpha J\vdots N\vdots\alpha' J')$ are expressible in terms of the factors $(\alpha J\vdots r_i\vdots\alpha' J')$ employed in the dipole discussion (TAS, pp. 63, 95) as linear combinations of quantities of the type

$$\sum_i\sum_{\alpha''}(\alpha J\vdots r_i\vdots\alpha'' J'')(\alpha'' J''\vdots r_i\vdots\alpha' J') = D(J, J'', J'), \tag{5}$$

namely

$$\begin{aligned}
(\alpha J\vdots N\vdots\alpha' J) &= D(J,J,J) - D(J,J-1,J) - D(J,J+1,J);\\
(\alpha J\vdots N\vdots\alpha' J-1) &= D(J,J,J-1) + D(J,J-1,J-1);\\
(\alpha J\vdots N\vdots\alpha' J-2) &= D(J,J-1,J-2).
\end{aligned} \tag{6}$$

---

[5] See E. U. Condon and G. H. Shortley, *Theory of Atomic Spectra* (Camb. Univ. Press, 1935), to which we shall refer by the letters TAS.

[6] TAS, p. 93 *et seq.*; E. C. Kemble, *Quantum Mechanics* (McGraw-Hill, 1937), p. 469, p. 462 (footnote).

The factors $(\alpha J\vdots N\vdots\alpha' J')$ *form a Hermitian matrix which transforms like the matrix of an observable:*

$$(\beta J\vdots N\vdots\beta' J') = \sum_{\alpha,\alpha'} (\beta J|\alpha J)(\alpha J\vdots N\vdots\alpha' J')(\alpha' J'|\beta' J'). \tag{7}$$

This may be easily demonstrated from the known transformation properties (TAS $9^3 12$, p. 64) of the separate factors of (5).

The SLM_SM_L scheme

Computations in Russell-Saunders coupling are frequently simplified by using states characterized by the quantum numbers M_SM_L in place of JM. For example, one can obtain the dependence of the matrix components of $\mathfrak{N}'$ on M_SM_L much more simply than on JM. They are diagonal in S and M_S, and by exactly the arguments used in the computation of (3), we see that the dependence on M_L is

$$(\gamma SLM_SM_L|\mathfrak{N}'|\gamma' SL'M_SM_L') = (\gamma SL\vdots N\vdots\gamma' SL') \times \text{factor given by (3) with } JM \text{ replaced by } LM_L. \tag{8}$$

The elements $(\gamma SL\vdots N\vdots\gamma' SL')$ are given by (5) and (6) with α and J replaced by γS and L (cf. $11^3 9$, p. 70 TAS). From this Eq. (8) we see at once that the selection rules forbidding transitions $J=0\rightarrow 0$, $\frac{1}{2}\rightarrow\frac{1}{2}$, $0\leftrightarrows 1$, which are embodied in (3), apply also to L; *transitions* $L=0\rightarrow 0$, $0\leftrightarrows 1$, *i.e.*, $S\rightarrow S$ *and* $S\leftrightarrows P$ *are forbidden.*

The $SLJM$ scheme

The matrix components of $\mathfrak{N}'$ in this scheme should be expressible in terms of the same elements as occur on the right of (8). This may be done by using $11^3 8$, p. 69 TAS to express the factors (6) in terms of $(\gamma SL\vdots N\vdots\gamma' SL')$. This rather tedious calculation was carried out by Rubinowicz; we give a typical formula:

$$(\gamma S\,L\,J\vdots N\vdots\gamma' S\,L-1\;J+1) = -(\gamma S\,L\vdots N\vdots\gamma' S\,L-1)$$
$$\times\frac{[(S+L-J)(S+L-J-1)(J+S-L+1)(J+S-L+2)]^{\frac{1}{2}}}{2J(J+1)(J+2)[(2J+1)(2J+3)]^{\frac{1}{2}}}[J(J+1)-S(S+1)+(J+L)(L+1)]. \tag{9}$$

Sum rules, line and multiple strengths

From (3) we easily evaluate the sums corresponding to $13^3 1$, TAS, which we need for the line strength formulas and the proof of the J-file sum rule:

$$\sum_{M'}(\alpha JM|\mathfrak{N}'|\alpha' J'M') : (\alpha' J'M'|\mathfrak{N}'|\alpha'' J''M'') = \delta_{JJ''}\delta_{MM''}(\alpha J\vdots N\vdots\alpha' J')(\alpha' J'\vdots N\vdots\alpha'' J)\mathrm{H}(J,J'), \tag{10}$$

where

$$\begin{aligned} &\mathrm{H}(J,J) = \tfrac{1}{6}J(J+1)(2J-1)(2J+3),\\ &\mathrm{H}(J,J-1) = \tfrac{1}{2}J(J+1)(J-1)(2J-1), \qquad \mathrm{H}(J,J+1) = \tfrac{1}{2}J(J+1)(J+2)(2J+3),\\ &\mathrm{H}(J,J-2) = J(J-1)(2J-1)(2J-3), \qquad \mathrm{H}(J,J+2) = (J+1)(J+2)(2J+3)(2J+5). \end{aligned} \tag{11}$$

These are not symmetric in J and J', but become so when multiplied by $2J+1$ to accomplish the summation over M. This gives the formula for the *line strength:*

$$\mathcal{S}_q(\alpha J,\alpha' J') = \sum_{M,M'} |(\alpha JM|\mathfrak{N}'|\alpha' J'M')|^2 = (2J+1)\,\mathrm{H}(J,J')|(\alpha J\vdots N\vdots\alpha' J')|^2. \tag{12}$$

By exact analogy with the simple transformation argument used to obtain the similar dipole sum on p. 72 TAS we may write the value of the sum over J' in the $SLJM$ scheme:

$$\sum_{J'M'} (\gamma SLJM|\mathfrak{N}'|\gamma' SL'J'M') : (\gamma' SL'J'M'|\mathfrak{N}'|\gamma'' SL''J''M'')$$
$$= \delta_{JJ''}\delta_{MM''}\delta_{LL''}(\gamma SL\vdots N\vdots\gamma' SL')(\gamma' SL'\vdots N\vdots\gamma'' SL)\,\mathrm{H}(L,L'). \tag{13}$$

Since the right member is independent of J and M, dropping the double primes and multiplying by $2J+1$ to accomplish the summation over M gives us the ordinary *J sum rule* which holds within a multiplet. If instead we multiply by $(2S+1)(2L+1)$ to sum over J and M, we get the total *multiplet strength*

$$\mathbb{S}_q(\gamma SL,\gamma' SL') = \sum_{J,J'} \mathbb{S}_q(\gamma SLJ,\gamma' SL'J') = (2S+1)(2L+1)\,\mathrm{H}(L,L')|(\gamma SL\vdots N\vdots\gamma' SL')|^2. \quad (14)$$

The strengths of the individual lines of the multiplet are to be obtained by substituting in (12) the formulas of the type (9) to obtain, e.g.

$$\mathbb{S}_q(\gamma S\,L\,J,\gamma' S\,L-1\,J+1)$$
$$=(-)^2\frac{(S+L-J)(S+L-J-1)(J+S-L+1)(J+S-L+2)[J(J+1)-S(S+1)+(J+L)(L+1)]^2}{8J(J+1)(J+2)}$$
$$\times|(\gamma S\,L\vdots N\vdots\gamma' S\,L-1)|^2, \quad (15a)$$

or in general

$$\mathbb{S}_q(\gamma SLJ,\gamma' SL'J') = (2J+1)\,\mathrm{H}(J,J')|(\gamma SLJ\vdots N\vdots\gamma' SL'J')|^2 = f_q(SLJ,SL'J')|(\gamma SL\vdots N\vdots\gamma' SL')|^2. \quad (15b)$$

In this notation the value of $f_q(SLJ,SL'J')$ is $\frac{1}{16}$ the coefficient of G^2, H^2, or I^2 in the table of p. 253, TAS.[7]

Multiplet strengths in a transition array. Two different configurations. Coupled groups

We now turn to the question of the relative strengths of the different multiplets in a transition array in LS coupling. The transition array must either connect two different configurations, differing in regard to the nl value of one electron, or be entirely within the same configuration. For two different configurations all the arguments of §3[9], p. 244 TAS may be repeated in the quadrupole case. If the configurations are expressible in the forms I+II and I+IV, respectively,[8] we may reduce the strengths to those connecting II and IV. The parent term $\gamma^{\mathrm{I}}S^{\mathrm{I}}L^{\mathrm{I}}$ cannot change. Only the part, $\mathfrak{N}'^{\mathrm{II}}$, of $\mathfrak{N}'$ which refers to the electrons jumping from II to IV enters the consideration. This part commutes with L^{I} and from the known properties of the dipole components we see that the factor[9] $(\gamma^{\mathrm{I}}S^{\mathrm{I}}L^{\mathrm{I}},\gamma^{\mathrm{II}}S^{\mathrm{II}}L^{\mathrm{II}},SL\vdots N^{\mathrm{II}}\vdots\gamma^{\mathrm{I}}S^{\mathrm{I}}L^{\mathrm{I}},\gamma^{\mathrm{IV}}S^{\mathrm{II}}L^{\mathrm{IV}},SL')$ which enters (14) and (15) is independent of S and is the same function of $L^{\mathrm{I}}L^{\mathrm{II}}L$, $L^{\mathrm{I}}L^{\mathrm{IV}}L'$ as $(\gamma SLJ\vdots N\vdots\gamma' SL'J')$ is of SLJ, $SL'J'$, cf. (9). With this knowledge, a comparison of (14) with (15b) gives us the formula

$$\mathbb{S}_q(\gamma^{\mathrm{I}}S^{\mathrm{I}}L^{\mathrm{I}},\gamma^{\mathrm{II}}S^{\mathrm{II}}L^{\mathrm{II}},SL;\gamma^{\mathrm{I}}S^{\mathrm{I}}L^{\mathrm{I}},\gamma^{\mathrm{IV}}S^{\mathrm{II}}L^{\mathrm{IV}},SL')$$
$$=(2S+1)f_q(L^{\mathrm{I}}L^{\mathrm{II}}L,L^{\mathrm{I}}L^{\mathrm{IV}}L')|(\gamma^{\mathrm{II}}S^{\mathrm{II}}L^{\mathrm{II}}\vdots N^{\mathrm{II}}\vdots\gamma^{\mathrm{IV}}S^{\mathrm{II}}L^{\mathrm{IV}})|^2. \quad (16)$$

That the last factor is independent of $\gamma^{\mathrm{I}}S^{\mathrm{I}}L^{\mathrm{I}}$ is shown by the fact that the matrix components

$$(\gamma^{\mathrm{I}}S^{\mathrm{I}}L^{\mathrm{I}}M_S{}^{\mathrm{I}}M_L{}^{\mathrm{I}},\gamma^{\mathrm{II}}S^{\mathrm{II}}L^{\mathrm{II}}M_S{}^{\mathrm{II}}M_L{}^{\mathrm{II}}|\mathfrak{N}'|\gamma^{\mathrm{I}}S^{\mathrm{I}}L^{\mathrm{I}}M_S{}^{\mathrm{I}}M_L{}^{\mathrm{I}},\gamma^{\mathrm{IV}}S^{\mathrm{II}}L^{\mathrm{IV}}M_S{}^{\mathrm{II}}M_L{}^{\mathrm{IV}}),$$

which are given by formulas like (3) in terms of the same factors, are independent of $\gamma^{\mathrm{I}}S^{\mathrm{I}}L^{\mathrm{I}}$ (see the discussion in §1[8] TAS of vector coupling in antisymmetric states). If we sum (16) over all values of L and L' consistent with the given values of L^{I}, L^{II}, and L^{IV}, we obtain, in analogy with (14)

[7] There is one error in this table, in the ninth row, for the entry given explicitly in (15a), the − sign before $J(L+1)$ in the bracket should be changed to +.

[8] For example, I+II might be $p^2\cdot sd$, I+IV, $p^2\cdot d^2$. The invariant part p^2 may be considered as the parent configuration to which the groups sd and d^2 are added. It is necessary that none of the electrons in either II or IV be equivalent to any in I.

[9] The functional dependence follows clearly from the definition of this factor and the discussion of §11[3], p. 67 TAS. The argument of TAS (top of p. 70, bottom of p. 237) for the independence of S is, however, not clear. That in this case any matrix component of $\mathfrak{N}'$ in the SLM_SM_L scheme, and hence (cf. 8) the factor in question, is independent of S may be shown at once by transformation to a scheme employing quantum numbers $S^{\mathrm{I}}S^{\mathrm{II}}M_S{}^{\mathrm{I}}M_S{}^{\mathrm{II}}$, where the components are known to be independent of $M_S{}^{\mathrm{I}}$ and $M_S{}^{\mathrm{II}}$ by a standard theorem (TAS 3[3]8, p. 49).

and (15), for the total strength of all multiplets of spin S built on $S^{\mathrm{I}}L^{\mathrm{I}}$ by the addition of $S^{\mathrm{II}}L^{\mathrm{II}}$ and $S^{\mathrm{II}}L^{\mathrm{IV}}$ the value

$$\mathcal{S}_{\mathrm{q}}(\gamma^{\mathrm{I}}S^{\mathrm{I}}L^{\mathrm{I}},\gamma^{\mathrm{II}}S^{\mathrm{II}}L^{\mathrm{II}},S;\gamma^{\mathrm{I}}S^{\mathrm{I}}L^{\mathrm{I}},\gamma^{\mathrm{IV}}S^{\mathrm{II}}L^{\mathrm{IV}},S)$$
$$=(2S+1)(2L^{\mathrm{I}}+1)(2L^{\mathrm{II}}+1)\ \mathrm{H}(L^{\mathrm{II}}L^{\mathrm{IV}})\,|(\gamma^{\mathrm{II}}S^{\mathrm{II}}L^{\mathrm{II}}\vdots N^{\mathrm{II}}\vdots\gamma^{\mathrm{IV}}S^{\mathrm{II}}L^{\mathrm{IV}})|^2. \quad (17)$$

But the last two factors here are those in terms of which the multiplet strengths in the transition II→IV are expressed by (14). Hence we may set (17)

$$=\frac{(2S+1)(2L^{\mathrm{I}}+1)}{2S^{\mathrm{II}}+1}\mathcal{S}_{\mathrm{q}}(\gamma^{\mathrm{II}}S^{\mathrm{II}}L^{\mathrm{II}},\gamma^{\mathrm{IV}}S^{\mathrm{II}}L^{\mathrm{IV}}). \quad (17')$$

This completes the reduction of the strengths to those of II→IV. In particular it is interesting to note that in the case in which I involves only closed shells, so that $S^{\mathrm{I}}=L^{\mathrm{I}}=0$, $S=S^{\mathrm{II}}$, $L=L^{\mathrm{II}}$, $L'=L^{\mathrm{IV}}$, (17′) reduces to an equation which says that the closed shells do not influence the strengths.

Addition of an electron to an ion

For the case in which the jumping electron is not equivalent to any in the ion in either the initial or the final configuration, the total strength of a supermultiplet is given by (17) and the strengths of the multiplets in a supermultiplet by (16). We let $\gamma^{\mathrm{I}}S^{\mathrm{I}}L^{\mathrm{I}}$ represent the parent term of the ion, $\gamma^{\mathrm{II}}L^{\mathrm{II}}=nl$ and $\gamma^{\mathrm{IV}}L^{\mathrm{IV}}=n'l'$, the nl values of the jumping electron in the initial and final states ($nl\neq n'l'$), and $S^{\mathrm{II}}=\frac{1}{2}$. The multiplet strength is

$$\mathcal{S}_{\mathrm{q}}(\gamma^{\mathrm{I}}S^{\mathrm{I}}L^{\mathrm{I}}\ nl\ SL,\gamma^{\mathrm{I}}S^{\mathrm{I}}L^{\mathrm{I}}\ n'l'\ SL')=(2S+1)f_{\mathrm{q}}(L^{\mathrm{I}}lL,L^{\mathrm{I}}l'L')|(nl\vdots N\vdots n'l')|^2. \quad (18)$$

The relative strengths of the multiplets in a supermultiplet are thus the same functions of $L^{\mathrm{I}}lL$, $L^{\mathrm{I}}l'L'$ as the relative strengths of the lines in a multiplet are of SLJ, $SL'J'$. To obtain absolute strengths we need to evaluate the last factor of (18), which is just the factor which arises in connection with intensities in one-electron spectra. This is easily done by expressing one of the 9 components of the dyadic $-e\boldsymbol{rr}$ in terms of this factor by (8) and then expressing this component as an integral. In this way we find

$$(n\,l\vdots N\vdots n'\,l) \quad =\frac{2e}{(2l-1)(2l+3)}\int_0^\infty r^2\,R(n\,l)\,R(n'\,l)\,dr=s_{\mathrm{q}}(n\,l,n'\,l),$$
$$(n\,l\vdots N\vdots n'\,l-1)=0, \quad (19)$$
$$(n\,l\vdots N\vdots n'\,l-2)=-\frac{}{(2l-1)[(2l-3)(2l+1)]^{\frac{1}{2}}}\int_0^\infty r^2\,R(n\,l)\,R(n'\,l-2)\,dr=s_{\mathrm{q}}(n\,l,n'\,l-2).$$

The short s_{q}-notation for these components is in practice convenient.

Two-electron configurations. Transitions involving an electron jump

If neither initial nor final configuration is composed of equivalent electrons, the multiplet strengths are given at once by (18). If the two electrons of *one* of the configurations are equivalent, Eq. $6^8 17$, p. 232 TAS, shows that the strengths of the allowed multiplets have just *twice* the values given by (18). The numerical values for $np\ n'p \rightleftharpoons np\ n''p$ and $np\ n'p \rightleftharpoons np^2$ are given in Table II.

Summary

For transitions involving an electron jump, the jumping electron not equivalent to any in the ion (but see the previous paragraph), the computation of the line strengths in terms of s_{q} (19) involves successive application of the following three formulas, which are very convenient as soon as tables of $F_{\mathrm{q}}(SLJ,SL'J')$ are available:

Line strength in terms of multiplet strength, from (14) and (15)

$$\mathcal{S}_q(\gamma SLJ,\gamma' SL'J') = \frac{f_q(SLJ,SL'J')}{(2S+1)(2L+1)\mathrm{H}(L,L')}\mathcal{S}_q(\gamma SL,\gamma' SL') = F_q(SLJ,SL'J')\ \mathcal{S}_q(\gamma SL,\gamma' SL'). \quad (20)$$

Multiplet strength in terms of supermultiplet strength, from (17) and (18)

$$\mathcal{S}_q(\gamma^{\mathrm{I}}S^{\mathrm{I}}L^{\mathrm{I}}\ nl\ SL,\gamma^{\mathrm{I}}S^{\mathrm{I}}L^{\mathrm{I}}\ n'l'\ SL') = \frac{f_q(L^{\mathrm{I}}lL,L^{\mathrm{I}}l'L')}{(2L^{\mathrm{I}}+1)(2l+1)\ \mathrm{H}(l,l')}\mathcal{S}_q(\gamma^{\mathrm{I}}S^{\mathrm{I}}L^{\mathrm{I}}\ nl\ S,\gamma^{\mathrm{I}}S^{\mathrm{I}}L^{\mathrm{I}}\ n'l'\ S)$$

$$= F_q(L^{\mathrm{I}}lL,L^{\mathrm{I}}l'L')\ \mathcal{S}_q(\gamma^{\mathrm{I}}S^{\mathrm{I}}L^{\mathrm{I}}\ nl\ S,\gamma^{\mathrm{I}}S^{\mathrm{I}}L^{\mathrm{I}}\ n'l'\ S). \quad (21)$$

Strength of supermultiplet, from (17)

$$\mathcal{S}_q(\gamma^{\mathrm{I}}S^{\mathrm{I}}L^{\mathrm{I}}\ nl\ S,\gamma^{\mathrm{I}}S^{\mathrm{I}}L^{\mathrm{I}}\ n'l'\ S) = (2S+1)(2L^{\mathrm{I}}+1)(2l+1)\ \mathrm{H}(l,l')\ s_q^2(nl,n'l'). \quad (22)$$

Spectroscopic stability method. Any transition array

For any transition array we may obtain the relative strengths of the multiplets by a method, based on the principle of spectroscopic stability, similar to that sketched in §4[9], p. 249 TAS. This method, while quick and convenient, has the disadvantage of giving only the amplitudes and not the phases in the matrix of $\mathfrak{N}'$. Hence it is of no value for departures from LS coupling. For this reason we do not give detailed formulas. Roughly, in this method, one first computes the absolute squares of all matrix components of $\mathfrak{N}'$ in the $nl\ m_s m_l$ orbital scheme in terms of the s_q's, making use of (8) for the one-electron components. Then the quantities $|(\gamma SLM_SM_L|\mathfrak{N}'|\gamma' S'L'M_S'M_L')|^2$ are expressed in terms of $|(\gamma SL\vdots N\vdots\gamma' S'L')|^2$ by (8). Applying the principle of spectroscopic stability to each M_SM_L partition then gives equations to determine $|(\gamma SL\vdots N\vdots\gamma' S'L')|^2$, and hence from (14) the multiplet strengths, in terms of the s_q's.

Eigenfunction method. Any transition array

This method is the only one available in cases not falling in the categories covered by the matrix methods when it is desired to relate the phases of the strengths (see definition later) to those of the eigenfunctions. It requires writing just one eigenfunction in the SLM_SM_L scheme *for each term* of the initial and final configuration. From these eigenfunctions we may calculate one component of $\mathfrak{N}'$ in the SLM_SM_L scheme for each multiplet, employing the standard techniques for reducing the matrix components of an observable of this type to one-electron components and evaluating these from (8) in terms of s_q. This same matrix component is, on the other hand, *directly* expressible by (8) in terms of $(\gamma SL\vdots N\vdots\gamma' SL')$. This enables us to obtain this last quantity in terms of the s_q's with proper phase for the particular choice of eigenfunctions. The multiplet strengths and line strengths are then obtained from this by (14) and (20).

The methods of the previous two paragraphs are applicable in cases of transition between two terms of the same configuration as well as in cases in which an electron jumps. Before considering matrix methods available for such cases, it will be desirable to consider the transformation to intermediate coupling.

Intermediate coupling. Computation of line strengths

In any coupling, the strength of a line αJ, $\alpha' J'$ is given in terms of $(\alpha J\vdots N\vdots\alpha' J')$ by (12). The transformation properties of $(\alpha J\vdots N\vdots\alpha' J')$ are known from (7). If we introduce a quantity $\mathcal{S}_q^{\frac{1}{2}}(\alpha J,\alpha' J')$ *defined as the square root of the line strength taken with the sign of the matrix component* $(\alpha J\vdots N\vdots\alpha' J')$, we obtain the transformation formula

$$\mathcal{S}_q^{\frac{1}{2}}(\beta J,\beta' J') = \sum_{\alpha,\alpha'}(\beta J|\alpha J)\ \mathcal{S}_q^{\frac{1}{2}}(\alpha J,\alpha' J')(\alpha' J'|\beta' J'). \quad (23)$$

This enables us, in particular, to obtain the line strengths in intermediate coupling in terms of those in *LS* coupling if the intermediate-coupling eigenfunctions are known in terms of the *LS*-coupling functions and the values of $S_q^{\frac{1}{2}}$ are known for the *same* Russell-Saunders functions *with the same phases.*

For the matrix computation summarized above, the phases of $S_q^{\frac{1}{2}}(\gamma SLJ, \gamma' SL'J')$ are given, in terms of the standard system of eigenfunction phases employed in TAS, by (20), (21), (22) if in taking the roots we use for the phases of $f_q^{\frac{1}{2}}$ or of $F_q^{\frac{1}{2}}$ those in the table of p. 253 of TAS. These phases are those in the column marked "Sign" multiplied by the sign of the quantity in the bracket which occurs squared in some of the elements.

Intermediate coupling. Sum rules

The *J-group sum rule* of course holds because of the principle of spectroscopic stability. This sum rule says that the sum of the strengths of the lines connecting all the levels of a given J in the initial configuration with all the levels of a given J' in the final configuration is independent of coupling. For the case in which the initial and final configuration are identical, it is clear that in those groups with $J=J'$ we must include the strengths of the "lines" connecting a level with itself and must double the strength of all other lines.

In case the initial and final configuration are different, the *J-file sum rule* obtains. A *J file* is the set of all lines connecting a single given level of one configuration with all levels of the other configuration. The sum rule states that *for any coupling, the strengths of the J files referring to the levels of the initial (final) configuration are proportional to* $2J+1$ *provided that the jumping electron is not equivalent to any other in the final (initial) configuration.* The proof parallels exactly that for the dipole case on pp. 279–281 of TAS since the sum relations (10) and (13) are similar to the corresponding dipole sums. In particular, if the nl electron which jumps is equivalent to $k-1$ other nl electrons in the configuration to which the J files refer, and jumps to an $n'l'$ orbit, the invariant strengths of the J files are

$$k\,(2J+1)\,\mathrm{H}(l,l')\,s_q^2(nl,n'l'). \tag{24}$$

Matrix calculation for transitions between states of the same configuration (no-electron jumps)

For a configuration I+II consisting of inequivalent coupled groups, the strengths may be reduced to those within configuration I and II separately by methods similar to those used for the spin-orbit interaction in §1[11] p. 266 TAS. The procedure will be clear from the discussion of the important two-electron case, to which we proceed at once.

In the case of transition $nl\,n'l' \to nl\,n'l'$ the roots of the multiplet strengths are (cf. 14)

$$S_q^{\frac{1}{2}}(nl\,n'l'\;SL,\,nl\,n'l'\;SL') = [(2S+1)(2L+1)\,\mathrm{H}(L,L')]^{\frac{1}{2}}\,(nl\,n'l'\;L\vdots N\vdots nl\,n'l'\;L'), \tag{25}$$

the last factor being independent of S because $\mathfrak{N}'$ commutes with the spins of both electrons. The arguments of pp. 216, 232 TAS show that the last factor may be evaluated by ordinary vector-coupling methods, writing $N=N_1+N_2$, giving the first electron the quantum numbers nl, the second the quantum numbers $n'l'$, and ignoring considerations of antisymmetry. *This is true whether or not nl and n'l' are equivalent.*

The part $(nl\,n'l'\;L\vdots N_2\vdots nl\,n'l'\;L')$ of this factor is the same function of $ll'L$, $ll'L'$ as $(\gamma SLJ\vdots N\vdots \gamma' SLJ')$ is of SLJ, SLJ'. Hence its contribution to $S_q^{\frac{1}{2}}$ is, cf. (15b), $(2S+1)^{\frac{1}{2}}\,f_q^{\frac{1}{2}}(ll'L,\,ll'L')\,s_q(n'l',n'l')$. The factor $(nl\,n'l'\;L\vdots N_1\vdots nl\,n'l'\;L')$ is the same function of $l'lL$, $l'lL'$ as $(\gamma SLJ\vdots N\vdots \gamma' SLJ')$ is of SLJ, SLJ', *except for phase.* The reversal of the rôles of the added vectors makes a phase change of $(-1)^{l+l'-L+l+l'-L'}=(-1)^{L+L'}$ (cf. 14[3]7, p. 78 TAS). In this way we find for the total root multiplet strength

$$S_q^{\frac{1}{2}}(nl\,n'l'\;SL,nl\,n'l'\;SL') = (2S+1)^{\frac{1}{2}}[f_q^{\frac{1}{2}}(ll'L,ll'L')\,s_q(n'l',n'l') + (-1)^{L+L'}f_q^{\frac{1}{2}}(l'lL,l'lL')\,s_q(nl,nl)], \tag{26}$$

in terms of which the root line strength with the standard phase choice is given by (20).

In this case the J-file sum rule does not hold unless either $s_q(nl,nl)$ or $s_q(n'l',n'l')$ vanishes.

TABLE I. $F_q(SLJ,\ SL'J')$, *which gives the relative strengths of the lines in a quadrupole multiplet, and the relative strengths of the multiplets in a supermultiplet. The sign of* $F_q^{\frac{1}{2}}$ *is indicated.*

	$^2D_{3/2}$	$^2D_{5/2}$	
$^2P_{1/2}$	+3	−2	$\frac{1}{15}$
$^2P_{3/2}$	+3	+7	

	^1Y_Y
^1X_X	+1

	$^3P_0$	$^3P_1$	$^3P_2$	
$^3P_0$	0	0	+4	
$^3P_1$	0	−3	−9	$\frac{1}{36}$
$^3P_2$	+4	−9	+7	

	$^3D_1$	$^3D_2$	$^3D_3$	
$^3S_1$	+3	+5	+7	$\frac{1}{15}$

	$^3D_1$	$^3D_2$	$^3D_3$	
$^3P_0$	0	−20	0	
$^3P_1$	+27	+5	−28	$\frac{1}{180}$
$^3P_2$	−9	+35	+56	

	$^3D_1$	$^3D_2$	$^3D_3$	
$^3D_1$	+21	−35	+4	
$^3D_2$	−35	+25	−40	$\frac{1}{300}$
$^3D_3$	+4	−40	+96	

Transition arrays $np\, n'p \to np\, n'p$ **and** $np^2 \to np^2$.

In these cases the f_q's occurring in (26) are given by the entries of Table I for $^3P \to {}^3P$ and the relation $f_q(^3P_J, {}^3P_{J'}) = 15\ F_q(^3P_J, {}^3P_{J'})$. If we write s_q for $s_q(np,np)$ and s_q' for $s_q(n'p,n'p)$, $(s_q'+s_q)^2=\alpha$, $(s_q'-s_q)^2=\beta$, we find for the multiplet strengths of $np\, n'p \to np\, n'p$:

$$
\begin{aligned}
&\mathbb{S}_q(^1S,{}^1S) = \mathbb{S}_q(^1S,{}^1P) = 0,\\
&\mathbb{S}_q(^1S,{}^1D) = (+)^2\, \tfrac{5}{3}(s_q'+s_q)^2 = (+)^2\, \tfrac{5}{3}\alpha,\\
&\mathbb{S}_q(^1P,{}^1P) = (-)^2\, \tfrac{5}{4}(s_q'+s_q)^2 = (-)^2\, \tfrac{5}{4}\alpha,\\
&\mathbb{S}_q(^1P,{}^1D) = (-)^2\, \tfrac{15}{4}(s_q'-s_q)^2 = (-)^2\, \tfrac{15}{4}\beta,\\
&\mathbb{S}_q(^1D,{}^1D) = (+)^2\, \tfrac{35}{12}(s_q'+s_q)^2 = (+)^2\, \tfrac{35}{12}\alpha,\\
&\mathbb{S}_q(^3S,{}^3S) = \mathbb{S}_q(^3S,{}^3P) = 0,\\
&\mathbb{S}_q(^3S,{}^3D) = (+)^2\, 5(s_q'+s_q)^2 = (+)^2\, 5\alpha,\\
&\mathbb{S}_q(^3P,{}^3P) = (-)^2\, \tfrac{15}{4}(s_q'+s_q)^2 = (-)^2\, \tfrac{15}{4}\alpha,\\
&\mathbb{S}_q(^3P,{}^3D) = (-)^2\, \tfrac{45}{4}(s_q'-s_q)^2 = (-)^2\, \tfrac{45}{4}\beta,\\
&\mathbb{S}_q(^3D,{}^3D) = (+)^2\, \tfrac{35}{4}(s_q'+s_q)^2 = (+)^2\, \tfrac{35}{4}\alpha.
\end{aligned}
\qquad (27)
$$

TABLE II. *Line strengths in quadrupole transition arrays (The sign of* $\mathbb{S}^{\frac{1}{2}}_q$ *is indicated.) The array labeled* $np\, n'p \leftrightarrows np\, n''p$ *is applicable to arrays of the following four types, according to the choice of* α *and* β:

$np\, n'p \leftrightarrows np\, n''p$
  $\alpha=\beta=s_q^2(n'p,\, n''p)$;
$np\, n'p \leftrightarrows np^2$
  $\alpha=\beta=2s_q^2(n'p,\, np)$ *for allowed lines;*
$np\, n'p \to np\, n'p$ [cf (27)]
  $\alpha=(s_q'+s_q)^2$; $\beta=(s_q'-s_q)^2$;
$np^2 \to np^2$
  $\alpha=4s_q^2(np,\, np)$ *for allowed lines.*

$np\, n'p \leftrightarrows np\, n''p$

	$^1S_0$	$^1P_1$	$^1D_2$	
$^1S_0$	0	0	+4α	
$^1P_1$	0	−3α	−9β	$\frac{5}{12}$
$^1D_2$	+4α	−9β	+7α	

	$^3P_0$	$^3S_1$	$^3P_1$	$^3D_1$	$^3P_2$	$^3D_2$	$^3D_3$	
$^3P_0$	0	0	0	0	−100α	+300β	0	
$^3S_1$	0	0	0	+240α	0	+400α	+560α	
$^3P_1$	0	0	+75α	−405β	+225α	−75β	+420β	
$^3D_1$	0	+240α	−405β	+147α	+135β	−245α	+28α	$\frac{1}{240}$
$^3P_2$	−100α	0	+225α	+135β	−175α	−525β	−840β	
$^3D_2$	+300β	+400α	−75β	−245α	−525β	+175α	−280α	
$^3D_3$	0	+560α	+420β	+28α	−840β	−280α	+672α	

$np^2 \to np^2$

	$^1S_0$	$^3P_0$	$^3P_1$	$^3P_2$	$^1D_2$	
$^1S_0$	0	0	0	0	+16	
$^3P_0$	0	0	0	−4	0	
$^3P_1$	0	0	+3	+9	0	$\frac{5}{12}s_q^2$
$^3P_2$	0	−4	+9	−7	0	
$^1D_2$	+16	0	0	0	+28	

$np^3 \to np^3$

	$^4S_{3/2}$	$^2P_{1/2}$	$^2P_{3/2}$	$^2D_{3/2}$	$^2D_{5/2}$	
$^4S_{3/2}$	0	0	0	0	0	
$^2P_{1/2}$	0	0	0	−6	+4	
$^2P_{3/2}$	0	0	0	−6	−14	s_q^2
$^2D_{3/2}$	0	−6	−6	0	0	
$^2D_{5/2}$	0	+4	−14	0	0	

The strengths of the multiplets actually occurring in $np^2 \to np^2$ are given by these same formulas with $s_q = s_q' = s_q(np,np)$; $\alpha = 4\, s_q^2(np,np)$.

From (27), (20), and Table I we immediately obtain the line strengths given in Table II. The strengths of the lines connecting a level to itself are of course of no interest in pure LS coupling, but play an important rôle in Eq. (23) in transformations to intermediate coupling.

Transition array $np^3 \to np^3$

This transition cannot be handled by these matrix methods. The multiplet strengths are found very simply, however, by the eigenfunction method given above, using the eigenfunctions of 4^86j, p. 224 TAS. It turns out that there is only one nonvanishing multiplet,[10] namely ^2P-^2D, for which $(^2P\vdots N\vdots {}^2D) = -3^{-\frac{1}{2}}s_q(np,np)$ and $\mathbb{S}_q(^2P,{}^2D) = (-)^2\ 30\ s_q^2(np,np)$. This, with the $F_q(^2P_J,{}^2D_{J'})$ of Table I, gives the line strengths of Table II.

[10] That the strengths of the "diagonal" multiplets of p^3 vanish is analogous to the vanishing of the Landé intervals, and follows from the almost-closed-shell considerations given below in the following way: When we consider p^3 as starting to fill the shell we denote it by $\mathfrak{L}$; when we consider it as three electrons missing from a closed shell we denote it by $\mathfrak{N}$. The considerations of almost-closed shells shows that $(\mathfrak{L}^2P\vdots N\vdots \mathfrak{L}^2P) = -(\mathfrak{N}^2P\vdots N\vdots \mathfrak{N}^2P)$; $(\mathfrak{L}^2D\vdots N\vdots \mathfrak{L}^2D) = -(\mathfrak{N}^2D\vdots N\vdots \mathfrak{N}^2D)$. Now from the way the correlation of the states of $\mathfrak{L}$ and $\mathfrak{N}$ are made, it is not necessary that $\mathfrak{L}^2P$ and $\mathfrak{N}^2P$, $\mathfrak{L}^2D$ and $\mathfrak{N}^2D$ be identical; but because they really represent the same physical state we must have $\mathfrak{L}^2P = \pm\mathfrak{N}^2P$ (in our case actually −), $\mathfrak{L}^2D = \pm\mathfrak{N}^2D$ (actually +). From this it follows at once that the above *diagonal elements* vanish. Such general arguments give no information about the value of the nondiagonal element $(\mathfrak{L}^2P\vdots N\vdots \mathfrak{L}^2D)$, which equals $-(\mathfrak{N}^2P\vdots N\vdots \mathfrak{N}^2D)$.

Transition arrays involving almost-closed shells. The array $np^4 \to np^4$

When we investigate the relation between transition arrays involving almost-closed shells and the corresponding simpler arrays, we find just as in the dipole case (p. 316 TAS) that if the transition is between two *different* configurations containing the *same* almost-closed shell the strengths are the same as for the simpler array in which the almost-closed shell is replaced by the missing electrons. But for quadrupole radiation we have another case to consider, the case of *no-electron jumps.* Here we find, by an examination of the corresponding matrix components in the $m_s m_l$-orbital scheme that *the strengths are the same for a pure almost-closed shell array (such as $p^4 \to p^4$) and for the complementary array (such as $p^2 \to p^2$).* The matrix components of $\mathfrak{N}'$ and hence *the phases of $\mathcal{S}_q^{\frac{1}{2}}$ are of opposite sign* in the two cases. Hence the strengths are not the same in an array such as $p^4d \to p^4d$ and in the corresponding array $p^2d \to p^2d$. Rather, the one set of strengths will be obtained from the other by reversing the sign of $s_q(p,p)$. This is analogous to the rule which holds for spin-orbit interaction.

jj coupling

It is clear from the similarity of the above formulation to that for the dipole case that the methods used in the latter to obtain strengths in jj coupling (see TAS p. 264) are directly available here.

II. Magnetic-Dipole Radiation

The strength $\mathcal{S}_m(\alpha J, \alpha' J')$ of a magnetic-dipole line, in terms of which the probability of spontaneous transition from the upper level αJ to the lower level $\alpha' J'$ has the value

$$A_m(\alpha J, \alpha' J') = \frac{1}{2J+1} \frac{64\pi^4\sigma^3}{3h} \mathcal{S}_m(\alpha J, \alpha' J'), \tag{28}$$

is defined by

$$\mathcal{S}_m(\alpha J, \alpha' J') = \mathcal{S}_m(\alpha' J', \alpha J) = \sum_{M,M'} |(\alpha J M|\boldsymbol{M}|\alpha' J' M')|^2, \quad \text{where} \quad \boldsymbol{M} = -\frac{e}{2\mu c}(\boldsymbol{L}+2\boldsymbol{S}). \tag{29}$$

In LS coupling, magnetic-dipole transitions occur only between two levels of the same term. It does not seem to have been generally realized[11] that simple closed formulas can be obtained[12] for these strengths in LS coupling, and that these strengths can be used in transformations to intermediate coupling. By the same procedure as in the electric-dipole case (p. 99 TAS) we can write

$$\mathcal{S}_m(\alpha J, \alpha' J') = (2J+1)\, \Xi(J,J') |(\alpha J \vdots M \vdots \alpha' J')|^2. \tag{30}$$

The last factor can be evaluated from $10^3 2$ pp. 64, 66 TAS to give

---

[11] For example, by Condon, reference 4.

[12] These formulas, at least for the nondiagonal elements, are given by Brinkman, reference 3. They are of considerable direct interest in connection with the excitation of the ground term in rarefied atmospheres. For example, Dunham, Nature **139**, 246 (1937), found four interstellar absorption lines from the lowest level $^4F_{3/2}$ of Ti II and none from the next level $^4F_{5/2}$ in spite of its low excitation potential of only 0.012 volt. The reason for the absence of absorption by $^4F_{5/2}$ is its short mean life for magnetic-dipole radiation of 7.6 hours, as determined from (28) and (31), whereas collisions occur much less frequently. This mean life was computed by Houston from Brinkman's formula and reported by Dunham, but with a numerical error of a factor of 10^3, as Houston has kindly verified.

$$\mathcal{S}_{\mathrm{m}}(SL\,J,SL\,J+1)=[+]^2\frac{(J-S+L+1)(J+S-L+1)(J+S+L+2)(S+L-J)}{4(J+1)}\left(-\frac{e\hbar}{2\mu c}\right)^2,$$

$$\mathcal{S}_{\mathrm{m}}(SL\,J,SL\,J)\quad=[g(SLJ)]^2\,J(J+1)(2J+1)\left(-\frac{e\hbar}{2\mu c}\right)^2, \tag{31}$$

where g is the Landé factor for the Zeeman effect. If we define $\mathcal{S}_{\mathrm{m}}^{\frac{1}{2}}(\alpha J,\, \alpha' J')$ as the root of $\mathcal{S}_{\mathrm{m}}$ taken with the sign of $(\alpha J \vdots M \vdots \alpha' J')$ when measured in units $(-e\hbar/2\mu c)$, the transformation of coupling is effected exactly as in (23). In LS coupling, the sign of $\mathcal{S}_{\mathrm{m}}^{\frac{1}{2}}$ is the sign of the quantity in square brackets in (31), which is + except in rare instances.

The magnetic-dipole strengths do not satisfy the ordinary J sum rule within a multiplet. Instead, one obtains the formula

$$\sum_{J'}\mathcal{S}_{\mathrm{m}}(SLJ,\, SLJ')=[2J(J+1)+2S(S+1)-L(L+1)](2J+1). \tag{32}$$

In intermediate coupling, only the J-group sum rule is obeyed.

It is interesting to note that while one easily obtains the simple closed formulas (31) for the magnetic-dipole strengths in LS coupling, the strengths in jj coupling are relatively difficult to obtain; for many configurations the last factor of (30) cannot be evaluated by matrix methods and recourse must be had to the eigenfunctions or to transformation from LS coupling.